H.-J. Rossberg · B. Jesiak · G. Siegel

Analytic Methods of Probability Theory

Mathematische Lehrbücher und Monographien
Herausgegeben von der Akademie der Wissenschaften der DDR
Karl-Weierstraß-Institut für Mathematik

II. Abteilung
Mathematische Monographien
Band 67

Analytic Methods of Probability Theory

von H.-J. Rossberg, B. Jesiak, G. Siegel

# Analytic Methods of Probability Theory

by Hans-Joachim Rossberg, Bernd Jesiak, Gerhard Siegel

Akademie-Verlag Berlin
1985

Verfasser:
Prof. Dr. habil. Hans-Joachim Rossberg
Dr. rer. nat. Bernd Jesiak
Dr. sc. nat. Gerhard Siegel
Karl-Marx-Universität Leipzig
Sektion Mathematik

ISSN 0076-5430

Erschienen im Akademie-Verlag, Berlin, DDR-1086 Berlin, Leipziger Straße 3—4

Lizenznummer: 202 · 100/400/85
Printed in the German Democratic Republic
Gesamtherstellung: VEB Druckerei „Thomas Müntzer“, 5820 Bad Langensalza
Lektor: Dr. Reinhard Höppner
Einband und Schutzumschlag: Dietmar Kunze
LSV 1075
Bestellnummer: 763 420 3 (6868)

04800

# Preface

*To the late Ingrid Rossberg*
*for continuous encouragement*
*during the writing of this book.*

There are many outstanding monographs dealing with analytic methods of probability theory. Venturing to add a further contribution to this collection we pursue the aim to acquaint the interested reader with some topics which have been one subject of research work in recent years. These investigations were stimulated by new results on the normal distribution and the central limit theorem.

The crucial idea of the present book was proposed by V. M. ZOLOTAREV during his visit to Karl Marx University in 1973. It concerns a sequence $(F_n)$ of distribution functions of sums of independent random variables which converges to the normal distribution $\Phi$ on a half line. Loosely speaking, Zolotarev's conjecture implied that this assumed "restricted convergence" continues to the whole line so that we have complete convergence of $(F_n)$ to $\Phi$, i.e., a new version of the central limit theorem. This has turned out to be completely right. It is now formulated as Theorem 10.4.1 and can be considered the beginning of a new branch of the summation theory of random variables, which is included in part III. Moreover, the plan of the corresponding proof, which is explained in Sect. 10.4, necessitated solving two quite different problems: A *compactness problem* and a *continuation problem for the normal distribution in the set of infinitely divisible distributions.* The first led to new investigations on the relative compactness of sets of distribution functions. The second proved later to be a problem of A. N. KOLMOGOROV; it can be considered to be the starting point of the *continuation theory of distribution functions.*

A considerable part of these investigations is fairly elementary. Therefore it seemed, to be natural to combine the form of a monograph with that of a textbook. Thus, we outline foundations which are always wider than what will be used in the sequel. Of course, a book that leads the reader from introductory sections up to unsolved problems demands a concise style in later chapters.

The first part consists of an introduction to handling distribution functions and characteristic functions. We do not claim in each case much originality in methods and presentation of material and express our indebtedness to W. FELLER and E. LUKACS on whose well known works we have relied to a certain extent. But we direct the reader's attention to our explanations about the concentration function, characteristic functions, and relative compactness which contain recent results.

In the second part we deal with important special classes of distribution functions. Much of the material is well known to experts. The main part of what we have added concerns the above mentioned idea that a distribution function $F$ the values of which are given on a certain point set can be uniquely continued provided that $F$ belongs to an appropriate class. It is needless to add that analytic distribution functions play a considerable role in this context; they have not been treated in detailed form so far. In the special case of the normal distribution it turned out that even weak assumptions concerning only the asymptotic behaviour of one or both tails are sufficient to characterize the normal distribution in the set of infinitely divisible distributions and even in larger sets.

Finally, the third part presents a brief introduction to the summation theory for independent random variables. Particular attention is paid to the central limit theorem. Moreover, we present new results assuming restricted convergence of a sequence $(F_n)$ of distribution functions of sums on certain real point sets. We show that, under appropriate assumptions, restricted convergence implies complete convergence of $(F_n)$. It is for theorems of this kind that we need many results of parts I and II. For this reason it is justified to include the theory of sums in a book on analytic probability theory.

The supplements contain more information and hints for further studies. In this way it is possible to lead the reader up to the present state of research.

A few words are in order concerning an analytic tool. We emphasize that theorems belonging to the theory of Phragmén and Lindelöf have been used only sporadically in probability theory so far. In the present book a special attempt has been made to display the possibilities provided by that theory. Appendix C contains a brief introduction and shows typical methods of handling these tools.

We wrote this book for students and for research workers as well and expect the reader only to have attended an introductory course to probability theory. We wished on one hand to stimulate readers to further studies on analytic probability theory and, on the other hand, to promote further developments of the fields under consideration.

We know that there hardly exists any book without mistakes. So we will be grateful to every reader who will kindly direct our attention towards ours.

We have done our best to give credit to authors wherever possible, but doubt that we have been able to render justice to all authors whose results touch the subjects of the present book, and for this we apologize.

We have benefited from comments by D. Bischoff, M. Dewess, G. Laue and M. Riedel, who also assisted us in eliminating inaccuracies and obscurities, and helped to make the manuscript more readable.

We should like to express our gratitude to G. Laue who contributed Sect. 4.9 to the present book.

Our warm thanks go to J. Behtmann, D. Bischoff, M. Dewess, B. Fritzsche, B. Kirstein, G. Laue, M. Riedel, and R. Scharm for their assistance in reading the proof. Mrs. Lehmann deserves our special thanks for her efficient and error-free typing. Finally, we express our indebtedness to Dr. Höppner for his understanding and cooperation.

# Contents

# Abbreviations, notations, and symbols

## Generalities

(i) The numbering of statements and formulae begins anew in each section. Throughout any section, statements will be numbered according to their sequence, and without respect to their kind. References are given with the abbreviations Chap. (Chapter), Cor. (Corollary), Def. (Definition), Ex. (Example), L. (Lemma), Prop. (Proposition), Rem. (Remark), Sect. (Section), and Th. (Theorem).

(ii) Functions will often be used without specifying their arguments. Then $F = G$ means that $F(x) = G(x)$ for all $x$ belonging to the common domain of definition of $F$ and $G$; $F \neq G$ means that there exists an $x$ such that $F(x) \neq G(x)$. The notations $F \geqq G$, $F > G$, etc. have analogous meaning. In all these notations the function $G$ may be a constant.

(iii) For asymptotic relations we preferably use the Landau symbols $O$ and $o$.

Let $f$ and $g$ be (real or complex) functions, and let $x_0$ be a real number or $x_0 = \pm\infty$. Then we write

$$f(x) = O(g(x)), \quad x \to x_0, \quad \text{if} \quad |f(x)/g(x)| \leqq C < \infty \quad \text{as} \quad x \to x_0,$$

$$f(x) = o(g(x)), \quad x \to x_0, \quad \text{if} \quad |f(x)/g(x)| \to 0 \quad \text{as} \quad x \to x_0.$$

## Abbreviations

| | |
|---|---|
| a.c.f. | analytic characteristic function(s) |
| a.d.f. | analytic distribution function(s) |
| b.c.f. | boundary characteristic function(s) |
| c.f. | characteristic function(s) |
| CLT | central limit theorem |
| d.f. | distribution function(s) |
| i. | independent |
| i.d. | identically distributed |
| iff | if and only if |
| inf.div. | infinitely divisible, infinite divisibility |
| LLN | law of large numbers |
| r.v. | random variable(s) |
| r-convergence | restricted convergence |

## Notations

We list a number of general and special notations which are used in larger parts of the book. Most of them occur with affixes; these will be specified in the corresponding context.

| | |
|---|---|
| $a, A$ | location parameters |
| $\mathcal{A}$ | $\sigma$-algebra of a probability space |
| $\arg z$ | argument of the complex number $z$ |
| $b, B$ | scale parameters |
| $c, C$ | general (finite) constants |
| $d, D$ | special (finite) constants |

| | |
|---|---|
| $d_G$ | $= 1 - (G(\infty) - G(-\infty))$; defect of the monotone function $G$, $0 \leqq G \leqq 1$ |
| $D$ | domains of the complex plane |
| $\mathsf{E}$ | expectation |
| $\exp z$ | $= e^z$ |
| $F$ | distribution functions, special affixes: |
| $F_-(x)$ | $= 1 - F(-x + 0)$ |
| $F^s$ | $= F * F_-$ |
| $F'(x)$ | $= F(bx + a), b > 0$ |
| ${}^nF$ | $n$-th "root" of an inf.div.d.f. $F$ |
| $\tilde{F}$ | accompanying laws (see Sect. 10.3) |
| $\boldsymbol{F}$ | general sets (classes) of d.f. |
| $f$ | c.f. of the d.f. $F$ (with the same affixes as $F$, if any) |
| $G, g$ | monotone functions and their Fourier-Stieltjes transforms |
| $H$ | monotone functions |
| $H$ | Lévy-Hinčin spectral functions |
| $\mathscr{H}(a, H)$ | Lévy-Hinčin representation of an inf.div.d.f. |
| $\operatorname{Im} z$ | imaginary part of the complex number $z$ |
| $K(t, x)$ | see Sect. 7.2 |
| $k(z, u)$ | see Sect. 7.3 |
| $k_1(z, u)$ | see Sect. 7.3 |
| $k_\psi$ | $= \lim_{s\to\infty} \frac{\log \lvert\psi(is)\rvert}{s}$ |
| $l_\nu(F)$ | see Sect. 6.1 |
| $\mathscr{L}(a, \gamma^2, M, N)$ | Lévy representation of an inf.div.d.f. |
| lext $(F)$ | left extremity of $F$ |
| $M$ | left Lévy spectral functions |
| $M_\eta$ | absolute moments of order $\eta > 0$ |
| $m$ | median |
| $\max_k$ | $\max_{1\leqq k\leqq k_n}$ (cf. p. 199) |
| $N$ | right Lévy spectral functions |
| $\mathsf{P}$ | probability |
| $p$ | density (with the same affixes as the corresponding d.f., if any) |
| $p_x$ | jump of a d.f. at the atom $x$ |
| $p_{\mu,\sigma}$ | normal density |
| $P_\lambda$ | Poisson d.f. with parameter $\lambda > 0$ |
| $\boldsymbol{P}$ | see Def. 6.2.1 |
| $Q$ | concentration functions |
| $\boldsymbol{R}_1$ | set of all real numbers |
| $\boldsymbol{R}_n$ | $n$-dimensional Euclidean space |
| $\operatorname{Re} z$ | real part of the complex number $z$ |
| rext $(F)$ | right extremity of $F$ |
| $S$ | subsets of $\boldsymbol{R}_1$ |
| $S(s_1, s_2)$ | $= \{z: s_1 < \operatorname{Im} z < s_2\}$ |
| $S(s_1, s_2]$ | $= \{z: s_1 < \operatorname{Im} z \leqq s_2\}$ |
| $S[s_1, s_2)$ | $= \{z: s_1 \leqq \operatorname{Im} z < s_2\}$ |
| $S[s_1, s_2]$ | $= \{z: s_1 \leqq \operatorname{Im} z \leqq s_2\}$ |
| $S(s)$ | $= S(-s, s) = \{z: \lvert\operatorname{Im} z\rvert < s\}$ |
| $\mathscr{S}(\alpha, \beta; a, b)$ | stable d.f. with parameters $\alpha$, $\beta$, $a$, and $b$ |
| $S_n$ | sums of i.r.v. |
| $T_F$ | $= 1 - F(x) + F(-x + 0)$; tail function of the d.f. $F$ |
| $U_a$ | degenerate d.f. with jump 1 at the atom $a$ |
| $\mathsf{Var}$ | variance |
| $w(x)$ | $= \min\{1, x^2\}, x \in \boldsymbol{R}_1$ |
| $X, Y$ | r.v. (with the same affixes as the corresponding d.f.) |
| $X'$ | $= \frac{X - a}{b}$ |
| $X^s$ | symmetrized r.r. cf. Sect. 1.4. |

| | |
|---|---|
| $\alpha$ | characteristic exponent of a stable d.f. |
| $\beta$ | "symmetry" parameter of a stable d.f. |
| $\gamma^2$ | Gaussian parameter of an inf.div.d.f. |
| $\Gamma$ | $\Gamma$-function |
| $\mu_k$ | moment of order $k \geqq 1$ |
| $\mu_F$ | first moment of the d.f. $F$ |
| $\sigma_F^2$ | variance of the d.f. $F$ |
| $\sum_k$ | $\sum_{k=1}^{k_n}$ (cf. p. 199) |
| $\Phi_{\mu,\sigma}$ | normal d.f. with expectation $\mu$ and variance $\sigma^2$ |
| $\Phi$ | $= \Phi_{0,1}$ |
| $\varphi_{\mu,\sigma}$, $\varphi$ | c.f. of $\Phi_{\mu,\sigma}$, $\Phi$ |
| $\chi(F)$ | see Sect. 6.1 |
| $\Omega$ | set of "elementary events" |
| $\bigcup_k$ | $\bigcup_{k=1}^{k_n}$ (cf. p. 199) |

## Symbols

| | |
|---|---|
| $*$ | convolution |
| $F^{*n}$ | $n$-fold convolution of $F$ |
| $\xrightarrow{w}$ | weak convergence |
| $\xrightarrow{c}$ | complete convergence |
| $\xrightarrow[S]{}$ | restricted convergence on the set $S \subseteqq R_1$ |
| $\sim$ | subject to, distributed as |
| $[a]$ | entier |

# I. Foundations

## Chapter 1. Distribution functions

In this chapter we introduce some of the terminology and state several important facts for ease of future reference. It is assumed that the reader is familiar with most of the material contained in Sect. 1.1.—1.3, and therefore only some short proofs are given.

Though the fundamental concepts of probability theory have their roots in measure theory, we do not stress this aspect since we focus our attention on analytic methods.

### 1.1. Definitions and basic facts

Let a probability space $(\Omega, \mathcal{A}, \mathsf{P})$ be given, i.e. a non-empty fundamental set $\Omega$ of *elementary events*, a $\sigma$-algebra $\mathcal{A}$ of subsets of $\Omega$ — called *random events* — and a *probability measure* $\mathsf{P}$ defined on $\mathcal{A}$. Such a space is the mathematical model of the phenomena to be studied in probability theory and underlies every probabilistic investigation.

In applications we incessantly meet random quantities that are determined by an experiment. Using a probability space they can be described by a well-known mathematical concept.

**Definition 1.1.1.** A *random variable* (r.v.) on $(\Omega, \mathcal{A}, \mathsf{P})$ is an $\mathcal{A}$-measurable function $X: \Omega \to \mathbb{R}_1$, i.e., a function $X$ such that $\{\omega \in \Omega: X(\omega) < x\} \in \mathcal{A}$ for all $x \in \mathbb{R}_1$. □

To avoid endless repetitions we make the convention that all our r.v. are defined on a fixed, but otherwise arbitrary probability space.

The definition of a r.v. makes it possible to introduce the notion of a distribution function which will prove to be an essential tool of analytic probability theory.

**Definition 1.1.2.** Let $X$ be a r.v. Then the function

$$F(x) := \mathsf{P}(\{\omega \in \Omega: X(\omega) < x\}), \qquad x \in \mathbb{R}_1,$$

is said to be the *distribution function* (d.f.) of $X$; for convenience we write $F(x) = \mathsf{P}(X < x)$. We will also express this by writing $X \sim F$ and saying that $X$ is subject to $F$. □

For examples see the table at the end of this book.

Throughout this book r.v. are denoted by capital letters $X, Y, \ldots$ and d.f. by $F, G, H, V,$ with or without affixes.

Note that two different r.v. $X$ and $Y$ can have the same d.f.; then we say that $X$ and $Y$ are *identically distributed* (i.d.) and write $X \sim Y$.

There is a one-to-one correspondence between a d.f. $F$ and a certain probability measure $\Pi$ defined on the Borel sets of $\mathbb{R}_1$. If $F$ is given, then we define $\Pi = \Pi_F$ by its restriction $\Pi_F[x_1, x_2) = F(x_2) - F(x_1)$ to the class of half-open finite intervals. In view of the extension theorem this domain of definition can be extended to the class of all Borel sets. — Conversely, if we are given a probability measure $\Pi$ on all Borel sets of $\mathbb{R}_1$, then the corresponding d.f. $F = F_\Pi$ is defined by $F_\Pi(x) = \Pi(-\infty, x)$, $x \in \mathbb{R}_1$. The measure $\Pi_F$ corresponding to $F$ will often also be denoted by $F$.

D.f. are of great methodological importance for the following reason. Measures are set functions and not easy to handle with the tools of classical analysis, for methods of analysis were developed to deal primarily with finite point functions on $\mathbb{R}_1$. It is the one-to-one correspondence described above that establishes a link between the notion of measure and the classical notions.

We agree to say that a d.f. $F$ is *concentrated on the set* $S \subseteq \mathbb{R}_1$ if $\Pi_F(S) = 1$. If there exists a bounded set $S$ on which $F$ is concentrated, then we call $F$ *finite*; $F$ is said to be *degenerate* if $S = \{a\}$ consists of only one point; then we write $F = U_a$.

Starting with Def. 1.1.2 it is not hard to prove that any d.f. possesses the following characteristic properties.

**Proposition 1.1.3.** *A real-valued function $F$ on $\mathbb{R}_1$ is a d.f. iff it has the following properties:*

(i) *$F$ is non-decreasing;*

(ii) *$F$ is left-continuous, i.e.,* $F(x) = F(x - 0) = \lim\limits_{\varepsilon \to 0} F(x - \varepsilon)$;

(iii) $F(\infty) = \lim\limits_{x \to \infty} F(x) = 1\,, \qquad F(-\infty) = \lim\limits_{x \to -\infty} F(x) = 0\,.$ □

The limit $F(x + 0) = \lim\limits_{\varepsilon \to 0} F(x + \varepsilon)$ plays a particular role in the case where $p_x := F(x + 0) - F(x) = \mathsf{P}(X = x) > 0$; then $x$ is called an *atom* of $F$ and $p_x$ the *jump* of $F$ at $x$.

There exist at most 2 jumps $\geqq 1/2$, at most 4 jumps $\geqq 1/4$, and so on. This fact leads to the following statement.

**Proposition 1.1.4.** *Let $F$ be a d.f. Then its set of atoms is countable.* □

A point $x \in \mathbb{R}_1$ is said to be a *point of increase* of $F$ if $F(x + \varepsilon) - F(x - \varepsilon) > 0$ for all $\varepsilon > 0$; $x$ is called a *point of strict increase* of $F$ if $F(x - \varepsilon) < F(x) < F(x + \varepsilon)$ for all $\varepsilon > 0$.

A d.f. $F$ such that $F(x) = 0$ for some $x \in \mathbb{R}_1$ is said to be *bounded to the left* and its *left extremity* is defined by

$$\text{lext}\,(F) := \sup\,\{x\colon F(x) = 0\}\,;$$

otherwise, if $F > 0$ we write lext $(F) = -\infty$. — Analogously, if $F(x) = 1$ for some $x \in \mathbb{R}_1$, we say that $F$ is *bounded to the right* with *right extremity*

$$\text{rext}\,(F) := \inf\,\{x\colon F(x) = 1\}\,;$$

otherwise, if $F < 1$ we write rext $(F) = \infty$. — Both kinds are referred to as *one-sided* d.f. If $-\infty <$ lext $(F) \leqq$ rext $(F) < \infty$, then $F$ is obviously finite.

We will often deal with linear transforms of a r.v. $X$. Putting $X_- = -X$ and $X' = (X - a)/b$, where $a, b \in \mathbb{R}_1$, $b > 0$, we see that

$$F_-(x) := \mathsf{P}(X_- < x) = 1 - F(-x + 0) \tag{1.1.1}$$

and

$$F'(x) := \mathsf{P}(X' < x) = F(bx + a)\,. \tag{1.1.2}$$

We say that $F$ is *symmetric* if $F = F_-$, i.e., if $X \sim X_-$; $F$ is said to be *symmetric with respect to* $a \in \mathbb{R}_1$, if $F(x + a)$ is symmetric. Moreover, we say that $F$ and $F'$ in (1.1.2) *belong to the same type*; $a$ and $b$ are called *location* and *scale parameters*, resp.

Continuation theorems for d.f. will be an important subject of ours. The following initial contribution to this field is a simple consequence of the left-continuity of all d.f.

**Proposition 1.1.5.** *Two d.f. $F_1$, $F_2$ coincide if they coincide on a set $S$ which is everywhere dense on $\mathbb{R}_1$.* □

If $S$ is not everywhere dense, then we can state non-trivial continuation theorems only under additional assumptions which imply that the d.f. under consideration belong to a certain class. For example, we have the following result.

**Proposition 1.1.6.** *Two d.f. $F_1$, $F_2$ belonging to the same type coincide if they coincide on an open interval $I$ which contains two points of increase $x_1 \neq x_2$ of $F_1$.*
Proof. By assumption we have $F_2(x) = F_1(bx + a)$ for all $x \in \mathbb{R}_1$ and certain $a \in \mathbb{R}_1$, $b > 0$. Now we assume, without loss of generality, that

$$F_1(x_1 - \varepsilon) < F_1(x_1) \leqq F_1(x_1 + \varepsilon) \quad \text{for all} \quad \varepsilon > 0\,. \tag{1.1.3}$$

Choosing $\varepsilon$ small enough we have $x_1 \pm \varepsilon \in I$ and therefore, using $F_1(x) = F_2(x) = F_1(bx + a)$, $x \in I$, we get from (1.1.3)

$$F_1(b(x_1 - \varepsilon) + a) < F_1(x_1) \leqq F_1(b(x_1 + \varepsilon) + a)\,.$$

This yields $b(x_1 - \varepsilon) + a < x_1 \leqq b(x_1 + \varepsilon) + a$, and on letting $\varepsilon \to 0$ we get $x_1 = bx_1 + a$. In the same manner we can derive $x_2 = bx_2 + a$, whence $a = 0$, $b = 1$ follows. This proves the assertion. □

We give still a useful modification of this result; for an application see Prop 12.1.5.

**Proposition 1.1.7.** *Two d.f. $F_1$, $F_2$ belonging to the same type coincide provided that they coincide at two points of strict increase $x_1 \neq x_2$ of $F_1$.*
Proof. By assumption the inequalities (1.1.3) are true but this time without the equality sign. Thus, using $F_1(x_1) = F_2(x_1) = F_1(bx_1 + a)$, we can apply a similar reasoning as above. □

As stated above, probability theory is intimately connected with measure theory. But it is not simply a part of it. In particular, it is the notion of *independence* that is peculiar to probability theory, since it is not of interest in measure theory but plays an outstanding role in our field.

We remind the reader of the following definition.

**Definition 1.1.8.** The r.v. $X_1, \ldots, X_n$ are called *independent* if for all $(x_1, \ldots, x_n) \in \mathbb{R}_n$

$$\mathsf{P}(X_1 < x_1, X_2 < x_2, \ldots, X_n < x_n) = \mathsf{P}(X_1 < x_1)\, \mathsf{P}(X_2 < x_2) \ldots \mathsf{P}(X_n < x_n)\,.$$

We say that $(X_k)$ is a *sequence of independent random variables* (i.r.v.) if the r.v. of every finite subset are independent. □

## 1.2. Classification of distribution functions

A d.f. $F$ is said to be *purely discrete* if it is concentrated on a countable set $S$. In particular, if all points of $S$ can be represented as $(a + kb)$, $a \in \mathbb{R}_1$, $b > 0$, $k = 0, \pm 1, \ldots$, then we call $F$ a *lattice* d.f.; $b$ is called a *span* of the lattice. If $S = \{a\}$, then, as mentioned above, $F$ is degenerate and we write $F = U_a$.

We agree to say that a d.f. $F$ is *absolutely continuous* if there exists a Lebesgue integrable function $p$ such that

$$F(x) = \int_{-\infty}^{x} p(u)\,du\,, \qquad x \in R_1\,.$$

Then obviously $p \geqq 0$ almost everywhere and

$$\int_{-\infty}^{\infty} p(u)\,du = 1\,;$$

the function $p$ is called a *density* of $F$. In nearly all applications, the above integrals are Riemann integrals. An absolutely continuous d.f. $F$ is obviously concentrated on an interval $I = (x_1, x_2)$ if $p(x) = 0$ for almost all $x \notin I$. This means that lext $(F) \geqq x_1$, rext $(F) \leqq x_2$.

Finally, we call a d.f. *singular* if it is continuous and concentrated on a Borel set $S \subset R_1$ of Lebesgue measure 0. A well-known example is the Cantor d.f. which is concentrated on $[0, 1]$ in such a way that the intervals of constancy add up to length 1. The derivatives of singular d.f. exist almost everywhere and equal 0.

We remind the reader of the following decomposition theorem.

**Proposition 1.2.1** (Lebesgue). *An arbitrary d.f. $F$ can be uniquely represented as*

$$F = c_1F_1 + c_2F_2 + c_3F_3$$

*where $c_k \geqq 0$, $c_1 + c_2 + c_3 = 1$, and the d.f. $F_1$, $F_2$, $F_3$ are respectively discrete, absolutely continuous, and singular.* □

## 1.3. Moments

The complete description of a r.v. $X$ is given by its d.f. $F$; it indicates not only which values $X$ can assume, but also with what probabilities. However, in many cases we need much less information, namely only certain numerical characteristics of r.v. and their d.f., which we call *moments*. The huge number of propositions in which they occur indicates their importance.

**Definition 1.3.1.** The *mathematical expectation* of a r.v. $X$ with d.f. $F$ is defined by

$$\mathsf{E}X = \int_{-\infty}^{\infty} x\,dF(x)$$

provided that this integral converges absolutely. Otherwise, we say that the expectation of $X$ does not exist. Since $\mathsf{E}X$ depends only on $F$ we also write $\mathsf{E}X = \mu_F$ and call this the *expectation of $F$*. □

The expectation of $F$ has a mechanical analogue. This is that the d.f. $F$ distributes the mass 1 over $R_1$ such that the portion $F(x_2) - F(x_1)$ is attributed to the interval $[x_1, x_2)$; thus $\mathsf{E}X$ is obviously the *centre of gravity* of that mass. The relevance of the expectation for stochastic phenomena appears particularly impressively in the law of large numbers, see Prop. 3.7.1 and Sect. 3.8.7.

More information about a d.f. $F$ is provided by its "higher moments"; this is seen, for instance, from the Markov inequality given in Prop. 1.3.7.

**Definition 1.3.2.** We call

$$\mu_k := \mathsf{E}X^k = \int_{-\infty}^{\infty} x^k \, \mathrm{d}F(x) \,, \qquad k = 1, 2, \ldots,$$

the *moment of order* $k$ of the r.v. $X$ and

$$M_\eta := \mathsf{E}\,|X|^\eta = \int_{-\infty}^{\infty} |x|^\eta \, \mathrm{d}F(x) \,, \qquad \eta > 0 \,,$$

its *absolute moment of order* $\eta$ (provided that these integrals exist). □

A finite d.f. has trivially moments of all orders.

The deviations of $X$ from $\mathsf{E}X$ deserve particular attention; the following moment gives some insight into them.

**Definition 1.3.3.** The *variance* of $X$ is given by

$$\mathsf{Var}\, X = \mathsf{E}(X - \mathsf{E}X)^2 = \int_{-\infty}^{\infty} (x - \mathsf{E}X)^2 \, \mathrm{d}F(x)$$

provided, that this integral exists. We also write $\mathsf{Var}\, X = \sigma_F^2$ and call this the variance of $F$. □

According to this formula $\mathsf{Var}\, X$ is the *moment of inertia with respect to* $\mathsf{E}X$ of the above mentioned mass distribution.

We have the following well-known rules.

**Proposition 1.3.4.** *Let $X$ and $Y$ be two r.v. and $a, b \in R_1$. Then*

$$\mathsf{E}(bX + a) = b\mathsf{E}X + a \,, \qquad \mathsf{Var}\,(bX + a) = b^2 \,\mathsf{Var}\, X$$

$$(\mathsf{E}X + Y) = \mathsf{E}X + \mathsf{E}Y \,,$$

*provided that the expectations on the right exist. — Under the same assumption we have for i.r.v. $X$ and $Y$*

$$\mathsf{E}(XY) = \mathsf{E}X\mathsf{E}Y \,, \qquad \mathsf{Var}\,(X + Y) = \mathsf{Var}\, X + \mathsf{Var}\, Y \,.$$

*If* $\mathsf{Var}\, X$ *exists then* $\mathsf{Var}\, X = \mathsf{E}X^2 - (\mathsf{E}X)^2 \geqq 0$. □

**Proposition 1.3.5.** (i) *Assume that $X \sim F$ and* lext $(F) = a$, $\mathsf{E}X = a$. *Then $F$ is degenerate and concentrated on $\{a\}$.*

(ii) *If $X \sim F$ and $\sigma_F^2 = \mathsf{Var}\, X = 0$, then we also have $F = U_a$, but the number $a$ remains undefined.*

Proof. (i) By assumption we have for all $\varepsilon > 0$

$$0 = \int_{-\infty}^{\infty} (x - a) \, \mathrm{d}F(x) = \int_{a}^{\infty} (x - a) \, \mathrm{d}F(x) \geqq \int_{a+\varepsilon}^{\infty} (x - a) \, \mathrm{d}F(x)$$

$$\geqq \varepsilon\big(1 - F(a + \varepsilon)\big) \geqq 0 \,,$$

so that $F = U_a$ follows.

(ii) Put $Y = (X - \mathsf{E}X)^2 \sim G$, say. Then by assumption $\mathsf{E}Y = \mathsf{Var}\, X = 0$ and $l = \text{lext}\,(G) \geqq 0$. For $l > 0$ we obtain $\mathsf{E}Y > 0$, a contradiction. Hence $' = 0$ so that the first statement applies and gives us $G = U_0$, i.e. $\mathsf{P}(Y = 0) = 1$. This yields the second assertion. □

**Proposition 1.3.6.** *The moment $\mu_k$ exists for a certain $k$ iff the absolute moment $M_k$ exists; in this case, the moments $M_\eta$ exist for all orders $\eta < k$.* □

The existence of certain moments restricts the probability of large values of $X$ as is shown by the following Markov inequality which for $\eta = 2$ reduces to the celebrated Čebyšev inequality.

**Proposition 1.3.7.** *If $M_\eta$ exists, then*

$$\mathsf{P}(|X| \geqq x) \leqq \frac{M_\eta}{|x|^\eta}, \qquad x > 0 .$$

*In particular, if* $\mathsf{Var}\, X = \sigma_F^2$ *exists, then we have*

$$\mathsf{P}(|X - \mathsf{E}X| \geqq x) = \frac{\sigma_F^2}{x^2}, \qquad x > 0 .$$

Proof. The first inequality follows from

$$M\eta \geqq \int_{|y|>x} |y|^\eta \,\mathrm{d}F(y) \geqq |x|^\eta \,\mathsf{P}(|X| \geqq x) .$$

For the second put $\eta = 2$ and replace $X$ by $X' := X - \mathsf{E}X$ so that $M_2 = \mathsf{E}(X'^2) = \mathsf{Var}\, X$. □

We also need the following well-known inequalities.

**Proposition 1.3.8** (Cauchy-Schwarz inequality). *Let $X_1$ and $X_2$ be two r.v. with $\mathsf{E}X_1^2 < \infty$ and $\mathsf{E}X_2^2 < \infty$. Then $\mathsf{E}(X_1X_2)$ exists, and we have*

$$(\mathsf{E}(X_1X_2))^2 \leqq \mathsf{E}X_1^2\mathsf{E}X_2^2 .$$ □

**Proposition 1.3.9** (Jensen inequality). *Let $X$ be a r.v. and $I \subseteqq \mathbb{R}_1$ an open interval such that $\mathsf{P}(X \in I) = 1$. Consider, moreover, a convex function $h$ defined on $I$. If $\mathsf{E}X$ and $\mathsf{E}h(X)$ exist, then*

$$\mathsf{E}h(X) \geqq h(\mathsf{E}X) .$$ □

For the proof see Feller (1971); it holds also if $I$ is closed and $h$ bounded on $I$. In Prop. 1.3.9 it is trivial that $h(X)$ is a r.v. Finally, we state a very general result.

**Proposition 1.3.10.** *Let $X \sim F$ be a r.v. and $h$ a Borel measurable function on $\mathbb{R}_1$. Then $h(X)$ is also a r.v. and*

$$\mathsf{E}h(X) = \int_{-\infty}^{\infty} h(x)\, F(\mathrm{d}x) \tag{1.3.1}$$

*provided that this integral exists. — Let $X_1, \dots, X_n$ be i.r.v., subject to $F_1, \dots, F_n$, resp., and let $h$ be Borel measurable on $\mathbb{R}_n$. Then $h(X_1, \dots, X_n)$ is also a r.v. and we have the analogous formula*

$$\mathsf{E}h(X_1, \dots, X_n) = \int_{-\infty}^{\infty} \cdots \int_{-\infty}^{\infty} h(x_1, \dots, x_n)\, F_1(\mathrm{d}x_1) \dots F_n(\mathrm{d}x_n) .$$ □

It is the point of these formulae that the knowledge of the d.f. $F$ and $F_1, \dots, F_n$, resp., suffices to calculate these general expectations. Formula (1.3.1) comprises all moments introduced above: We obtain $\mu_k$ and $M_\eta$ by setting $h(x) = x^k$ and $h(x) = |x|^\eta$, resp. Putting $h(x) = (x - \mu_1)^2$ we get $\mathsf{E}h(X) = \mathsf{Var}\, X$.

## 1.4. Convolutions and mixtures

In the present section we construct d.f. by means of other d.f. which we consider to be given.

### 1.4.1. Convolutions

Let $F_1$ and $F_2$ be two d.f. We introduce the monotone function

$$F(x) := \int_{-\infty}^{\infty} F_1(x-y)\, F_2(\mathrm{d}y)\,, \qquad x \in \mathbb{R}_1\,. \tag{1.4.1}$$

Here the integral is in general a Lebesgue-Stieltjes integral; but it is a Riemann integral if $F_1$ or $F_2$ is continuous. By dominated convergence (cf. App. A) it can easily be shown that $F$ is a d.f.

**Definition 1.4.1.** The d.f. $F$ given in (1.4.1) is said to be the *convolution* of $F_1$ and $F_2$; we write $F = F_1 * F_2$. □

**Proposition 1.4.2.** *Let $X_1$ and $X_2$ be two i.r.v. subject to $F_1$ and $F_2$, resp. Then $F_1 * F_2$ is the d.f. of the sum $X_1 + X_2$.* □

**Corollary 1.4.3.** *The convolution is commutative and associative, i.e. for arbitrary d.f $F_1, F_2, F_3$ we have*

$$F_1 * F_2 = F_2 * F_1\,, \qquad F_1 * (F_2 * F_3) = (F_1 * F_2) * F_3\,.$$

The proof is immediate from Prop. 1.4.2 since the addition of r.v. is commutative and associative. □

It is remarkable that the first relation also follows from Fubini's theorem, see App. A; put

$$h(x, y) = h(x + y) = \begin{cases} 1, & x + y < t\,, \\ 0 & \text{otherwise}\,. \end{cases}$$

Sums $S_n := X_1 + \ldots + X_n$ of i.r.v. with a common d.f. $F$ will frequently occur. The d.f. of $S_n$ is the $n$-fold convolution of $F$ with itself; it will be denoted by $F^{*n}$. Thus

$$F^{*1} = F\,, \qquad F^{*(n+1)} = F^{*n} * F\,.$$

For the sake of brevity, we put $F^{*0} = U_0$.

**Proposition 1.4.4.** *If $F_1$ or $F_2$ is (absolutely) continuous then so is $F := F_1 * F_2$.*

Proof. It suffices to assume that $F_1$ is (absolutely) continuous. If $F_1$ is continuous then it is uniformly continuous on $\mathbb{R}_1$ whence the continuity of $F$ directly follows If $F_1$ has a density $p_1$, then we introduce the function

$$p(u) = \int_{-\infty}^{\infty} p_1(u-y)\, F_2(\mathrm{d}y) \geqq 0\,. \tag{1.4.2}$$

Next we integrate over $(-\infty, x)$ and conclude from (1.4.1) by Fubini's theorem that

$$F(x) = \int_{-\infty}^{x} p(u)\, \mathrm{d}u\,.$$ □

The following result is now easily seen.

**Corollary 1.4.5.** *If $F_1$ has density $p_1$, then the density $p$ of $F := F_1 * F_2$ is given by (1.4.2). If $F_2$, too, possesses a density $p_2$, then*

$$p(u) = \int_{-\infty}^{\infty} p_1(u-y)\, p_2(y)\, \mathrm{d}y\,.$$ □

Now let $X_1$ and $X_2$ be independent with the common d.f. $F$. Then by Prop. 1.4.2 the difference $X^s := X_1 - X_2$ has the d.f. $F^s$ given by

$$F^s = F * F_- = F_- * F \tag{1.4.3}$$

where $F_-$ is as defined in (1.1.1). Formula (1.4.3) indicates that $X_1 - X_2 \sim X_2 - X_1$. Accordingly, $F^s$ is a symmetric d.f.; we will say that it is obtained from $F$ by *symmetrization*. Though this operation is very simple it serves as an essential tool in many considerations; namely, it often permits us to avoid lengthy arguments.

**Example 1.4.6.** Symmetrization of the exponential d.f. leads to the *bilateral exponential (Laplace) d.f.*; the *triangular d.f.* is obtained in this way from the uniform d.f. on $(-b/2, b/2)$, $b > 0$; compare the table at the end of this book. □

To make use of symmetrization we will need the following proposition in which a *median* of the r.v. $X$ occurs; this is a number $m$ which satisfies

$$\mathsf{P}(X \leqq m) \geqq 1/2\,, \qquad \mathsf{P}(X \geqq m) \geqq 1/2\,.$$

**Proposition 1.4.7.** *Let $h \geqq 0$ be a non-decreasing function on $[0, \infty)$, and let $m$ stand for a median of the i.r.v. $X_1, X_2 \sim F$. Then for $X^s := X_1 - X_2$ we have*

$$\mathsf{E}h(|X_1 - m|) \leqq 2\mathsf{E}(|hX^s|)\,. \tag{1.4.4}$$

*Further, if there exists $c \geqq 1$ such that for all $x_1, x_2 \geqq 0$*

$$h(x_1 + x_2) \leqq c\big(h(x_1) + h(x_2)\big) \tag{1.4.5}$$

*then for arbitrary real $a$*

$$\mathsf{E}h(|X^s|) \leqq 2c\mathsf{E}h(|X_1 - a|)\,. \tag{1.4.6}$$

Proof. By Prop. 1.3.10 we see that

$$\mathsf{E}h(|X^s|) = \mathsf{E}h(|X_1 - X_2|) = \int_{-\infty}^{\infty}\int_{-\infty}^{\infty} h(|x - y|)\, F(\mathrm{d}x)\, F(\mathrm{d}y)\,. \tag{1.4.7}$$

To prove (1.4.4) we may assume $m = 0$ since otherwise we would consider $X_1 - m$ instead of $X_1$. — In view of $h \geqq 0$ we get from (1.4.7)

$$\mathsf{E}h(|X^s|) \geqq \int_{y \leqq 0}\int_{x \geqq 0} + \int_{y \geqq 0}\int_{x < 0} h(|x - y|)\, F(\mathrm{d}x)\, F(\mathrm{d}y)\,.$$

We can see that the integrand in these two integrals satisfies $h(|x - y|) \geqq h(|x|)$ since $h$ is non-decreasing. Hence

$$\mathsf{E}h(|X^s|) \geqq \int_{y \leqq 0}\int_{x \geqq 0} + \int_{y \geqq 0}\int_{x < 0} h(|x|)\, F(\mathrm{d}x)\, F(\mathrm{d}y) \geqq \tfrac{1}{2}\int_{-\infty}^{\infty} h(|x|)\, F(\mathrm{d}x) = \tfrac{1}{2}\,\mathsf{E}h(|X_1|)\,.$$

This yields (1.4.4).

To prove the second assertion we again start with (1.4.7) and obtain

$$\mathsf{E}h(|X^s|) \leqq \int_{-\infty}^{\infty}\int_{-\infty}^{\infty} h(|x| + |y|)\, F(\mathrm{d}x)\, F(\mathrm{d}y);$$

now assumption (1.4.5) yields

$$\leqq c \int_{-\infty}^{\infty}\int_{-\infty}^{\infty} \big(h(|x|) + h(|y|)\big)\, F(\mathrm{d}x)\, F(\mathrm{d}y) = 2c\mathsf{E}h(|X_1|)\,.$$

This consideration can be repeated with $X_1 - a$ instead of $X_1$ so that we obtain (1.4.6). □

**Corollary 1.4.8.** *For the function*

$$w(x) := \min\{1, x^2\}, \qquad x \in \mathbb{R}_1,$$

*and an arbitrary* $a \in \mathbb{R}_1$ *we have the inequalities*

$$\tfrac{1}{2}\,\mathsf{E}w(|X_1 - m|) \leqq \mathsf{E}w(|X^s|) \leqq 4\mathsf{E}w(|X_1 - a|).$$

Proof. The function $w$ is monotone on $[0, \infty)$ and satisfies (1.4.5) for $c = 2$ so that Prop. 1.4.7 yields the assertion. □

Now we obtain the *symmetrization inequalities* they permit us to generalize certain results which we have so far obtained only for $X^s$.

**Corollary 1.4.9.** *For an arbitrary real a*

$$\frac{1}{2}\mathsf{P}(|X - m| \geqq x) \leqq \mathsf{P}(|X^s| \geqq x) \leqq 2\mathsf{P}\left(|X - a| \geqq \frac{x}{2}\right), \qquad x \in \mathbb{R}_1. \tag{1.4.8}$$

Proof. The first inequality is (1.4.4) for $h = U_x$ (degenerate d.f.). — On the other hand, we note that the random event $\{|X^s| \geqq x\}$ cannot occur unless either $\{|X_1 - a| \geqq x/2\}$ or $\{|X_2 - a| \geqq x/2\}$, and hence the second inequality follows. □

Roughly speaking, the inequalities (1.4.8) mean that the tails of the d.f. $F$ and $F^s$ are of comparable magnitude.

### 1.4.2. Mixtures

Suppose that the real numbers $p_0, p_1, \ldots$ (finitely or infinitely many) satisfy the conditions

$$p_k \geqq 0, \qquad \sum_k p_k = 1.$$

Let $G_0, G_1, \ldots$ be d.f. and put

$$F = \sum_k p_k G_k.$$

Clearly, $F$ is a d.f. even in the case that this is an infinite series since the convergence is uniform on $\mathbb{R}_1$. $F$ is called the *discrete mixture* of the d.f. $G_0, G_1, \ldots$ with weights $p_0, p_1, \ldots$

Mixtures are, for example, provided by the formula of total probability. We illustrate this by discussing a particularly interesting case. Consider the sequence $(X_j)$, $j \geqq 1$, of i.i.d.r.v. with d.f. $G$, and let $N$ stand for an integer valued r.v. indepen dent of all $X_j$ such that $\mathsf{P}(N = k) = p_k$, $k \geqq 0$. Then for the sum

$$S_N := \sum_{j=1}^{N} X_j, \qquad S_0 := 0,$$

we get the d.f.

$$F(x) := \mathsf{P}(S_N < x) = \sum_{k=0}^{\infty} \mathsf{P}(S_N < x \mid N = k)\, p_k = \sum_{k=0}^{\infty} p_k G^{*k}(x).$$

In the special case that $p_k = e^{-\lambda}\frac{\lambda^k}{k!}$, $\lambda > 0$, we obtain the compound Poisson d.f.

$$F = e^{-\lambda} \sum_{k=0}^{\infty} \frac{\lambda^k}{k!} G^{*k}.$$

Putting $G = U_1$ we get the Poisson d.f. $P_\lambda$.

$F$ can, for instance, be considered to be the d.f. of the accumulated amount of the random number $N$ of claims $X_j$ within the time interval $(0, x)$ against an insurance institution. We need only assume that $N$ is subject to a Poisson d.f. and that the individual claims have the d.f. $G$, lext $(G) = 0$.

We turn now to a more general case.

Let $\{G(x, y)\}$ be a set of functions which have the following properties:

(i) For each value of $y$ the function $G(x, y)$ is a d.f. in $x$;
(ii) for fixed $x$, $G(x, y)$ is a Borel measurable function of $y$.

With an arbitrary d.f. $V$, we now form the expression

$$F(x) := \int_{-\infty}^{\infty} G(x, y)\, V(dy)\,. \tag{1.4.9}$$

Then it is seen by dominated convergence that $F$ is a d.f.; we call it the (continuous) *mixture of the set* $\{G\}$ *with respect to* $V$. For details see LUKACS (1970).

## 1.5. The concept of uniqueness theorems for distribution functions

From Sect. 4.7 on uniqueness theorems for d.f. will be an important subject of ours. They play, for instance, a particular role in the context of limit theorems under the assumption of restricted convergence which will be introduced in Sect. 10.4. The general problem we have in mind is the following.

**Uniqueness problem.** Let $\boldsymbol{F}_1$ and $\boldsymbol{F}_2$, $\boldsymbol{F}_1 \subseteqq \boldsymbol{F}_2$, be two sets of d.f., and let $F_1 \in \boldsymbol{F}_1$ and $F_2 \in \boldsymbol{F}_2$. Which interrelations between $F_1$ and $F_2$ ensure the coincidence of these d.f.? □

Such interrelations can be of very different kind. Two of the most important conditions occurring in our theorems will be the followings:

(i) Coincidence of $F_1$ and $F_2$ on a set $S \subset \mathbb{R}_1$; cf., for example, Sect. 4.7 and 8.4.
(ii) Asymptotic relations between $F_1$ and $F_2$ as $x \to x_0 \in \mathbb{R}_1$ or $|x| \to \infty$; cf., for example, Sect. 7.6 and 8.3.

If the only interrelation between $F_1$ and $F_2$ has the form (i), then we call the stated uniqueness theorem a *continuation theorem*. Accordingly, it is the point of a continuation theorem that a d.f. $F_1 \in \boldsymbol{F}_1$ can be uniquely continued from the set $S$ to the whole real axis within the class $\boldsymbol{F}_2$.

As a first example for such a continuation theorem we mention Prop. 1.1.5 which can be reformulated in the following way. Let $\boldsymbol{F}_1 = \boldsymbol{F}_2 = \boldsymbol{F}$ be the set of all d.f., and assume $F_1, F_2 \in \boldsymbol{F}$; if $F_1(x) = F_2(x)$, $x \in S$, where $S$ is dense in $\mathbb{R}_1$, then $F_1 = F_2$.

The reader can easily reformulate Prop. 1.1.6 and 1.1.7 in an analogous way.

We will obtain a less trivial result if, for example, $\boldsymbol{F}_1$ consists only of the normal d.f. $\Phi$ and $\boldsymbol{F}_2$ stands for the set of all convolution powers $F^{*n}$, $F \in \boldsymbol{F}$, for some fixed $n \geqq 2$: If $\Phi(x) = F^{*n}(x)$ on $S = (-\infty, x_0)$ then $\Phi = F^{*n}$ (Cor. 4.7.6).

The first problem of this kind was posed by KOLMOGOROV; it is solved in Th. 7.6.1 and, more generally in Prop. 4.7.3.

We will also formulate assertions the point of which is to conclude not the identity $F_1 = F_2$ but merely the equality of certain quantities (e.g. parameters) related to

$F_1$ and $F_2$. Such theorems will also be called uniqueness theorems (in a large sense), cf. Sect. 7.6.

There are only two continuation theorems in the present book which do not quite fit in the above uniqueness problem since neither $\boldsymbol{F}_1 \subseteqq \boldsymbol{F}_2$ nor $\boldsymbol{F}_2 \subseteqq \boldsymbol{F}_1$, see Prop. 6.5.1 and 6.5.2. Here it is the point of the statement that $F_1, F_2 \in \boldsymbol{F}_1 \cap \boldsymbol{F}_2$ and, moreover, $F_1 = F_2$.

## Notes and bibliographical data

The notions and most results of Sect. 1.1 through 1.3 can also be found in elementary textbooks on probability theory; we have added only the simple Prop. 1.1.6 and 1.1.7. Sect. 1.4 contains more special results which we will need in the sequel. Prop. 1.4.7, Cor. 1.4.8 are one-dimensional special cases of a result due to SIEGEL (1981a). Cor. 1.4.8, first inequality, is also contained in GNEDENKO and KOLMOGOROV (1959).

# Chapter 2. Convergence of monotone functions

We start with the notions of weak and complete convergences, which are generally accepted. Further, we introduce the new and more general idea of restricted convergence (r-convergence) which has its origin in Zolotarev's idea concerning the central limit theorem (Th. 10.4.1). This exercises a slight influence even on the formulation of several well-known theorems such as Helly's selection theorem. But a crucial role will be played by r-convergence in part III of the present book.

Typical for our presentation is also the early treatment of the relative compactness and tightness of sets of d.f. This subject is particularly essential in connection with r-convergence for the following reason: An r-converging sequence of d.f. always contains a subsequence which weakly converges to some limit function; then it is of great interest to know conditions guaranteeing that we also have complete convergence to this function.

Finally, we study functions of regular variation which are useful analytic tools in modern probability theory. For instance, they permit us to find straightforward proofs for several results connected with stable d.f.

## 2.1. Weak, complete, and restricted convergence

**Definition 2.1.1.** Let $G_n$, $n \geqq 1$, and $G$ be non-decreasing functions on $\mathbb{R}_1$. The sequence $(G_n)$ is said to *converge weakly to* $G$ if, for all continuity points $x$ of $G$,

$$\lim_{n\to\infty} G_n(x) = G(x); \tag{2.1.1}$$

we express this by writing $G_n \xrightarrow{w} G$.

If $G_n$ and $G$ are also bounded satisfying $G_n \xrightarrow{w} G$ and $\lim\limits_{n\to\infty} G_n(\pm\infty) = G(\pm\infty)$, then we say that $(G_n)$ *converges completely to* $G$ and write $G_n \xrightarrow{c} G$. □

Since we assume here that $G$ is monotone we can apply an argument similar to the one which led us to the proof of Prop. 1.1.4; hence $G$ has — even if it is unbounded — at most a countable set $S$ of discontinuity points where (2.1.1) is possibly violated. Thus, if we additionally assume that the function $G$ in Def. 2.1.1 is left-continuous, then $G$ is uniquely defined by weak convergence; moreover, we may speak of *atoms* $x$ and the corresponding *jumps* $p_x = G(x + 0) - G(x)$.

It goes without saying that in most of our considerations on weak convergence the $G_n$ will be d.f. The following simple example shows that even in this case weak convergence does not imply complete convergence. In other words, if $G_n \xrightarrow{w} G$ for a sequence of d.f. $G_n$, then it can happen that

$$d_G := 1 - \big(G(\infty) - G(-\infty)\big) > 0;$$

in this case $G$ is called a *defective* d.f. with *defect* $d_G$.

**Example 2.1.2.** Let $G_n$ stand for a purely discrete d.f. with atoms $-n$ and $n$ and jumps $p$ and $1 - p$, resp. Then we have

$$G_n \xrightarrow{w} G = p\,.$$

Here the whole probability mass of the $G_n$ "escapes to infinity" in an obvious and intuitive way, and we have $d_G = 1$. □

The following statement gives a condition necessary and sufficient for weak convergence.

**Proposition 2.1.3.** *The sequence $(G_n)$ converges weakly to a non-decreasing function $G$ iff*

$$G(x-0) \leqq \varliminf_{n\to\infty} G_n(x)\,, \qquad \varlimsup_{n\to\infty} G_n(x) \leqq G(x+0)\,, \qquad x \in R_1\,. \tag{2.1.2}$$

Proof. If (2.1.2) holds true for all $x \in R_1$, then we get for an arbitrary continuity point $x$ of $G$,

$$G(x) = G(x-0) \leqq \varliminf G_n(x) \leqq \varlimsup G_n(x) \leqq G(x+0) = G(x)\,;$$

thus we have $G_n \xrightarrow{w} G$ by Def. 2.1.1.

Assume now $G_n \xrightarrow{w} G$. Then (2.1.2) follows immediately for all continuity points $x$ of $G$. Otherwise, let $x$ be an atom of $G$. Then we are able to select a sequence $(x_k)$ of continuity points $x_k$ of $G$ such that $x_k \to x - 0$. So we get, for fixed $k$,

$$\varliminf_{n\to\infty} G_n(x) \geqq \varliminf_{n\to\infty} G_n(x_k) = G(x_k)\,;$$

and on letting $k \to \infty$ we obtain the first of the relations (2.1.2). The second can be shown similarly. □

**Corollary 2.1.4.** *Assume $G_n \xrightarrow{w} G$ where $G$ is left-continuous. Then we have, for any atom $x$ of $G$ with jump $p_x$,*

$$\begin{aligned} p_x &= \lim_{\varepsilon\to 0} \varliminf_{n\to\infty} \big(G_n(x+\varepsilon) - G_n(x-\varepsilon)\big) \\ &= \lim_{\varepsilon\to 0} \varlimsup_{n\to\infty} \big(G_n(x+\varepsilon) - G_n(x-\varepsilon)\big)\,. \end{aligned}$$

Proof. Obviously,

$$\begin{aligned} p_x &= G(x+0) - G(x) = \lim_{\varepsilon\to 0} \big(G(x+\varepsilon) - G(x-\varepsilon+0)\big) \\ &= \lim_{\varepsilon\to 0} \big(G(x+\varepsilon+0) - G(x-\varepsilon)\big)\,. \end{aligned} \tag{2.1.3}$$

We recall that for arbitrary real sequences $(a_n)$ and $(b_n)$

$$\begin{aligned} \varliminf\,(a_n - b_n) &\geqq \varliminf a_n - \varlimsup b_n\,, \\ \varlimsup\,(a_n - b_n) &\leqq \varlimsup a_n - \varliminf b_n\,. \end{aligned}$$

Now we replace in (2.1.2) $x$ by $x \pm \varepsilon$, $\varepsilon > 0$, and note that $G(x-0) = G(x)$ by assumption; thus we get the two chains of inequalities

$$\begin{aligned} G(x+\varepsilon) - G(x-\varepsilon+0) &\leqq \varliminf_{n\to\infty} G_n(x+\varepsilon) - \varlimsup_{n\to\infty} G_n(x-\varepsilon) \\ &\leqq \varliminf_{n\to\infty} \big(G_n(x+\varepsilon) - G_n(x-\varepsilon)\big) \end{aligned}$$

and

$$G(x+\varepsilon+0) - G(x-\varepsilon) \geqq \overline{\lim_{n\to\infty}}\, G_n(x+\varepsilon) - \underline{\lim_{n\to\infty}}\, G_n(x-\varepsilon)$$

$$\geqq \overline{\lim_{n\to\infty}}\, \big(G_n(x+\varepsilon) - G_n(x-\varepsilon)\big)\,.$$

Combining this with (2.1.3) we obtain

$$\lim_{\varepsilon\to 0}\, \overline{\lim_{n\to\infty}}\, \big(G_n(x+\varepsilon) - G_n(x-\varepsilon)\big) \leqq p_x \leqq \lim_{\varepsilon\to 0}\, \underline{\lim_{n\to\infty}}\, \big(G_n(x+\varepsilon) - G_n(x-\varepsilon)\big)$$

whence the assertion follows by an easy argument. □

Next we introduce a further convergence concept for non-decreasing functions, which we call *restricted convergence* or briefly *r-convergence*. It describes a special type of convergence of $(G_n)$ to a function $G$ on a set $S \subseteqq R_1$ which will be crucial for part III. To simplify our definition we use the following expression: For a given set $S \subseteqq R_1$ we say that $x \in S$ is a *two-sided limit point* of $S$ if $(x-\varepsilon, x) \cap S \neq \emptyset$ and $(x, x+\varepsilon) \cap S \neq \emptyset$ for all $\varepsilon > 0$. To give an example, we put $S = \{\pm 1/k: k = 1, 2, \ldots\}$, then 0 is a two-sided limit point; see also Rem. 2.1.6, (iii).

**Definition 2.1.5.** Let $G_n$, $n \geqq 1$, and $G$ be non-decreasing functions defined at least on a non-empty set $S \subseteqq R_1$. We say that the sequence $(G_n)$ *r-converges* to $G$ on $S$ if

(i) $G(x-0) \leqq \underline{\lim}_{n\to\infty} G_n(x)\,,\qquad \overline{\lim}_{n\to\infty} G_n(x) \leqq G(x+0)$

for all two-sided limit points $x$ of $S$,

(ii) $\lim_{n\to\infty} G_n(x) = G(x)$ for all other points $x \in S$;

we express this by writing $G_n \xrightarrow[S]{} G$. □

**Remark 2.1.6.** In most applications we will meet the following special kinds of sets $S$ for which the definition of r-convergence takes a simpler form.

(i) If $S = \{x_k\}$ is a (finite or infinite) countable set containing only isolated points then $G_n \xrightarrow[S]{} G$ is equivalent to

$$\lim_{n\to\infty} G_n(x) = G(x)\,, \qquad x \in S;$$

this is easily seen by Def. 2.1.5.

(ii) If $S = I = (x_1, x_2)$ is a (finite or infinite) open interval, then $G_n \xrightarrow[S]{} G$ holds true iff

$$\lim_{n\to\infty} G_n(x) = G(x)$$

for all continuity points $x \in I$ of $G$. This can be shown in the same way as Prop. 2.1.3. In particular, the limit function $G$ is unique if at is assumed to be left-continous. Moreover, the weak convergence $G_n \xrightarrow{w} G$ is equivalent to $G_n \xrightarrow[R_1]{} G$ and it trivially implies $G_n \xrightarrow[I]{} G$ for any open interval $I$.

(iii) If $S = [x_1, x_2]$ is a closed interval, then $x_1$ and $x_2$ are no two-sided limit points; thus we have $G_n \xrightarrow[S]{} G$ iff

$$G_n \xrightarrow[(x_1, x_2)]{} G \quad \text{and} \quad \lim_{n\to\infty} G_n(x_k) = G(x_k), \qquad k = 1, 2. \qquad \square$$

It is clear that case (i) does not require any special investigation. Therefore, in the following section we will focus our attention on r-convergence on intervals which is merely a simple generalization of weak convergence. More explanations of r-convergence are to be found in Sect. 10.4.

## 2.2. Restricted convergence on intervals

We now investigate continuation and selection problems connected with r-convergence on intervals. Throughout this section we denote by $I = (x_1, x_2) \subseteqq \mathbb{R}_1$ some open interval and by $G_n$, $G$ non-decreasing functions which are defined at least on $I$. The first result shows that r-convergence assumed on an appropriate subset $S \subseteqq I$ continues to r-convergence on $I$.

**Proposition 2.2.1.** *Let $S \subseteqq I$ be dense within $I$. Then $G_n \xrightarrow[S]{} G$ implies $G_n \xrightarrow[I]{} G$.*

Proof. For fixed $x \in I$ we choose $x', x'' \in S$ such that $x' < x < x''$. Then we have by assumption and Def. 2.1.5 (i)

$$G(x' - 0) \leqq \varliminf_{n\to\infty} G_n(x') \leqq \varliminf_{n\to\infty} G_n(x)$$

and also

$$\varlimsup_{n\to\infty} G_n(x) \leqq \varlimsup_{n\to\infty} G_n(x'') \leqq G(x'' + 0).$$

On letting $x', x'' \to x$ (where $x', x'' \in S$) we obtain

$$G(x - 0) \leqq \varliminf_{n\to\infty} G_n(x), \qquad \varlimsup_{n\to\infty} G_n(x) \leqq G(x + 0).$$

Thus we have convergence for all continuity points $x$ of $G$; this means that $G_n \xrightarrow[I]{} G$ by Rem. 2.1.6 (ii). $\square$

In part III we will find deeper results of this type. Then it will turn out that Prop. 2.2.1 has the salient feature that it can be proved without Helly's selection theorem, while thistheorem is essentially needed for limit theorem s involving r-convergence which we treat in Part III.

A particular case of Prop. 2.2.1 is of special interest. Let $I = \mathbb{R}_1$, then it shows that $G_n \xrightarrow[S]{} G$ implies $G_n \xrightarrow{w} G$. Note that in Prop. 2.2.1 the set $S$ may be countable.

Now we are going to prove two results concerning convergence of subsequences of a given sequence $(G_n)$.

**Theorem 2.2.2** (Helly's selection theorem). *Every sequence $(G_n)$ of functions $G_n$ which are non-decreasing and uniformly bounded at least on an interval $I \subseteqq \mathbb{R}_1$ contains a subsequence $(G_{n'})$ such that*

$$G_{n'} \xrightarrow[I]{} G$$

*for a certain function $G$ which is non-decreasing, left continuous, and bounded on $I$.*

**Proof.** First we use Cantor's diagonal method. Let $S = \{x_k\}$ be countable and dense within $I$; for instance, $S$ may be the set of rationals contained in $I$. Then we apply the Bolzano-Weierstrass theorem to the sequence $(G_n(x_1))$ of real numbers which is bounded by assumption; hence, $(G_n)$ contains a subsequence $(G_{1n})$ such that $\lim_{n\to\infty} G_{1n}(x_1) = l(x_1)$, say. Next we consider the sequence $(G_{1n}(x_2))$ and proceed as above so that we obtain a subsequence $(G_{2n}) \subset (G_{1n})$. In this way we can continue inductively and select a double sequence

$$\begin{matrix} G_{11} & G_{12} & G_{13} & \cdots \\ G_{21} & G_{22} & G_{23} & \cdots \\ \vdots & & & \end{matrix}$$

such that the $k$-th line is contained in the foregoing one and

$$\lim_{n\to\infty} G_{k,n}(x_k) = l(x_k)\,,$$

say; this is possible since the $G_n$ are uniformly bounded on $I$. Now consider the diagonal sequence $(G_{n,n})$; it is — but for a finite number of terms — a subsequence of every line of the above double sequence; hence we obtain for all $x_k \in S$

$$\lim_{n\to\infty} G_{n,n}(x_k) = l(x_k)\,.$$

Next we define a function $G$ on $I$ by means of

$$G(x) := \sup_{\substack{x_j < x \\ x_j \in S}} l(x_j)\,, \qquad x \in I\,, \tag{2.2.1}$$

and see that $G$ is non-decreasing and bounded on $I$. Further, from this definition it follows for the restriction of $G$ to $S$ that

$$G(x_k - 0) \leqq l(x_k) \leqq G(x_k + 0)\,, \qquad x_k \in S\,.$$

Now we recall that $S$ is dense, that is, all $x_k \in S$ are two-sided limit points. Thus we have by Def. 2.1.5 (i) $G_{n,n} \xrightarrow[S]{} G$ and Prop. 2.2.1 tells us that $G_{n,n} \xrightarrow[I]{} G$; obviously, $(G_{n,n})$ is a subsequence of $(G_n)$ $r$-converging on $I$.

It remains to show that $G$ is left-continuous. Since the $G_n$ are non-decreasing so is $l$ and, as a consequence, also $G$. Now we choose a number $x \in I$ and a monotone sequence $y_j \to x - 0$; then $G(y_j)$ approaches a limit $c \leqq G(x)$. We carry an indirect proof and assume $c < G(x)$. From (2.2.1) we see that there exists $x_k \in S$, $x_k < x$, such that $c < l(x_k) \leqq G(x)$. For sufficiently large $j$ we have $x_k < y_j < x$ whence $c < l(x_k) \leqq G(x_k + 0) \leqq G(y_j) \to c$ follows. This is a contradiction so that all assertions have now been proved. □

The above theorem will be used time and again to justify the selection of converging subsequences, and it will not always be explicitely mentioned.

**Corollary 2.2.3.** *Let $G_n$, $n \geqq 1$, be non-decreasing functions which are uniformly bounded at least on an interval $I$. If every subsequence of $(G_n)$ $r$-converging on $I$ has the same limit function $G$, then we have $G_n \xrightarrow[I]{} G$.*

**Proof.** Suppose that $G_n \xrightarrow[I]{} G$ is not true. Then by Def. 2.1.5 (i) there exists an $x \in I$ such that

$$G(x - 0) > \varliminf_{n\to\infty} G_n(x) \quad \text{or} \quad \varlimsup_{n\to\infty} G_n(x) > G(x + 0)\,.$$

Thus we can pick a subsequence $(n')$ of natural numbers satisfying

$$G(x-0) > \lim_{n'\to\infty} G_{n'}(x) \quad \text{or} \quad \lim_{n'\to\infty} G_{n'}(x) > G(x+0)\,. \tag{2.2.2}$$

By Helly's selection theorem, $(G_{n'})$ contains a subsequence $(G_{n''})$ which r-converges on $I$ and, by assumption, we have $G_{n''} \xrightarrow[I]{} G$. But this contradicts (2.2.2), whence the assertion follows. □

To close this section we prove several results concerning the convergence of sequences of Stieltjes integrals.

**Theorem 2.2.4** (Helly-Bray theorem). *Let $(G_n)$ be a sequence of non-decreasing and uniformly bounded functions on $S = [x_1, x_2]$ and let $h$ be a (real or complex) continuous function on $S$. Then $G_n \xrightarrow[S]{} G$ implies*

$$\lim_{n\to\infty} \int_S h(x)\,\mathrm{d}G_n(x) = \int_S h(x)\,\mathrm{d}G(x)\,.$$

Proof. First of all, by assumption there exists a constant $C$ such that, for all $x \in S$,

$$|G_n(x)|,\ |G(x)|,\ |h(x)| \leqq C\,. \tag{2.2.3}$$

Now let $\varepsilon > 0$. Since $h$ is continuous it is uniformly continuous on the closed interval $S$; hence, we are able to find a subdivision $x_1 = y_0 < y_1 < \dots < y_N = x_2$ of $S$ such that in every interval $(y_{k-1}, y_k)$ we have the inequality $|h(x) - h(y_k)| < \varepsilon$. Now we introduce a step function $h_\varepsilon$ with atoms $y_k$ putting $h_\varepsilon(x_1) = h(y_1)$ and

$$h_\varepsilon(x) = h(y_k)\,, \qquad y_{k-1} < x \leqq y_k\,, \qquad k = 1, \dots, N;$$

hence

$$|h(x) - h_\varepsilon(x)| < \varepsilon\,, \qquad x \in S\,.$$

Moreover, we may assume that the $y_k$ are continuity points of $G$; thus we have, for all $n \geqq n_0(\varepsilon)$,

$$|G_n(y_k) - G(y_k)| < \frac{\varepsilon}{N}, \qquad 0 \leqq k \leqq N;$$

as to the cases $k = 0$ and $k = N$ see Rem. 2.1.6 (iii). So we obtain the estimates

$$\Delta_1 := \Big|\int_S \big(h(x) - h_\varepsilon(x)\big)\,\mathrm{d}G_n(x)\Big| < \varepsilon 2C\,,$$

$$\Delta_2 := \Big|\int_S \big(h(x) - h_\varepsilon(x)\big)\,\mathrm{d}G(x)\Big| < \varepsilon 2C\,,$$

and for $n \geqq n_0(\varepsilon)$ we have also

$$\Delta_3 := \Big|\int_S h_\varepsilon(x)\,\mathrm{d}\big(G_n(x) - G(x)\big)\Big|$$
$$= \Big|\sum_{k=1}^{N} h_\varepsilon(y_k)\,\big[\big(G_n(y_k) - G(y_k)\big) - \big(G_n(y_{k-1}) - G(y_{k-1})\big)\big]\Big| < C2\varepsilon\,.$$

Now the assertion follows by virtue of

$$\Big|\int_S h(x)\,\mathrm{d}G_n(x) - \int_S h(x)\,\mathrm{d}G(x)\Big| \leqq \Delta_1 + \Delta_2 + \Delta_3 < 6C\varepsilon\,, \qquad n \geqq n_0(\varepsilon)\,. \quad \square$$

**Theorem 2.2.5** (Extended Helly-Bray theorem). *Let $(G_n)$ be a sequence of non-decreasing and uniformly bounded functions on $\mathbb{R}_1$. Assume $G_n \xrightarrow{c} G$. Then we have for any (real or complex) continuous and bounded function $h$*

$$\lim_{n\to\infty} \int_{-\infty}^{\infty} h(x)\,\mathrm{d}G_n(x) = \int_{-\infty}^{\infty} h(x)\,\mathrm{d}G(x)\,.$$

Proof. Our method is a purely technical reduction to Th. 2.2.4. The assumptions ensure that (2.2.3) holds true for all $x \in \mathbb{R}_1$. Now we choose continuity points $x_1$, $x_2$ of $G$ such that, for a given $\varepsilon > 0$,

$$G(x_1) - G(-\infty) < \varepsilon\,, \qquad G(\infty) - G(x_2) < \varepsilon$$

and thus obtain, for sufficiently large $n$,

$$G_n(x_1) - G_n(-\infty) < 2\varepsilon\,, \qquad G_n(\infty) - G_n(x_2) < 2\varepsilon\,.$$

So we get

$$\begin{aligned} \Delta_4 &:= \Big|\int_{-\infty}^{x_1} h(x)\,\mathrm{d}\big(G_n(x) - G(x)\big)\Big| \\ &\leqq C[\big(G_n(x_1) - G_n(-\infty)\big) + \big(G(x_1) - G(-\infty)\big)] < 3C\varepsilon \\ \Delta_5 &:= \Big|\int_{x_2}^{\infty} h(x)\,\mathrm{d}\big(G_n(x) - G(x)\big)\Big| < 3C\varepsilon\,. \end{aligned}$$

Hence, by virtue of

$$\Big|\int_{-\infty}^{\infty} h(x)\,\mathrm{d}G_n(x) - \int_{-\infty}^{\infty} h(x)\,\mathrm{d}G(x)\Big| \leqq \Delta_4 + \Delta_5 + \Big|\int_{x_1}^{x_2} h(x)\,\mathrm{d}\big(G_n(x) - G(x)\big)\Big|\,,$$

the assertion follows from the Helly-Bray theorem. □

With the same notation as before, we also have the following result.

**Corollary 2.2.6.** *Let $G_n$, $n \geqq 1$, be non-decreasing and uniformly bounded on $I = (x_1, x_2) \subseteqq \mathbb{R}_1$ and assume that $h$ is a continuous and bounded function on $I$ satisfying $h(x_1 + 0) = h(x_2 - 0) = 0$. Then*

$$G_n \xrightarrow[I]{} G \quad \textit{implies} \quad \int_I h\,\mathrm{d}G_n \to \int_I h\,\mathrm{d}G.$$

Proof. The method is the same as before. For $\varepsilon > 0$, select a closed subinterval $S \subset I$ such that $\big|\int_{I\setminus S} h\,\mathrm{d}(G_n - G)\big| < \varepsilon$ for all $n$; this is possible by the assumption made on $h$. Now the proof is obvious in view of the Helly-Bray theorem. □

**Remark 2.2.7.** The same assertions hold if $h(x_2 - 0) = 0$ and $G_n(x_1) \to G(x_1)$.

## 2.3. Convergence under change of scale and location parameters

In Chap. 13 we will often meet the necessity of changing scale and location parameters in converging sequences of d.f.; then it will be important to know that such a change does not affect the type of the limit function.

Throughout this section we deal with d.f. $F_n$ and put

$$F_n'(x) = F_n(b_n x + a_n)$$

for given constants $a_n$ and $b_n > 0$. We start with a simple result showing the influence which the convergence of scale and location parameters exercises on the limit function of a weakly converging sequence $(F_n)$.

**Proposition 2.3.1.** *Assume $a_n \to a$, $b_n \to b > 0$, and $F_n \xrightarrow[I]{} G$ for some monotone function $G$ and $I = (x_1, x_2) \subseteqq \mathbb{R}_1$. Then we have $F_n' \xrightarrow[I']{} G'$ where $G'(x) = G(bx + a)$ and* $I' = \left(\dfrac{x_1 - a}{b}, \dfrac{x_2 - a}{b}\right)$.

Proof. We fix a number $x \in I'$ such that $bx + a \in I$ is a continuity point of $G$; this can be done because $b > 0$. Now we choose $\varepsilon > 0$ such that $bx + a \pm \varepsilon \in I$ are also continuity points of $G$. By assumption we obtain, for $n$ sufficiently large,

$$bx + a - \varepsilon < b_n x + a_n < bx + a + \varepsilon$$

and, as a consequence,

$$F_n(bx + a - \varepsilon) \leqq F_n(b_n x + a_n) = F_n'(x) \leqq F_n(bx + a + \varepsilon) .$$

On letting $n \to \infty$ we get from our assumptions

$$G(bx + a - \varepsilon) \leqq \varliminf_{n\to\infty} F_n'(x) \leqq \varlimsup_{n\to\infty} F_n'(x) \leqq G(bx + a + \varepsilon) .$$

Finally, as $\varepsilon \to 0$, we conclude that

$$\lim_{n\to\infty} F_n'(x) = G(bx + a) = G'(x) .$$

This demonstrates the assertion since $x$ is also a continuity point of $G'$. □

In the rest of this section, we will obtain the convergence of $(a_n)$ and $(b_n)$ from certain assumptions on the convergence of $(F_n)$ and $(F_n')$. We first collect together several auxiliary results in the following lemma.

**Lemma 2.3.2.** *Assume $F_n \underset{I}{\to} G$, $F_n' \underset{I}{\to} G'$ for non-decreasing and left-continuous functions $G$, $G'$ defined at least on $I = (x_1, x_2) \subseteqq R_1$.*

(i) *If $G \neq$ const. on $I$ and if there exists $x' \in I$ such that*

$$G(x_1 + 0) < G'(x') < G(x_2) , \tag{2.3.1}$$

*then we have*

$$x_1 < \varliminf_{n\to\infty} (b_n x' + a_n) , \qquad \varlimsup_{n\to\infty} (b_n x' + a_n) < x_2 . \tag{2.3.2}$$

*Moreover, the sequences $(a_n)$, $(b_n)$ are bounded.*

(ii) *If $a_n \to a$, $b_n \to b \geqq 0$, then we have for all $x \in I$*

$$G(bx + a - \varepsilon) \leqq G'(x) \leqq G(bx + a + \varepsilon) , \tag{2.3.3}$$

*provided that $bx + a \pm \varepsilon$ also belong to $I$. If, moreover, $b > 0$, then*

$$G'(x) = G(bx + a) , \qquad x \in I , \qquad bx + a \in I , \tag{2.3.4}$$

*provided that such $x$ exist.*

Proof. (i) First of all, since $G'$ is left-continuous we may suppose that $x'$ is a continuity point of $G'$. — To prove (2.3.2) we first suppose $x_1 > -\infty$. Contrary to the assertion we assume that

$$\varliminf_{n\to\infty} (b_n x' + a_n) \leqq x_1 .$$

Then for given $\varepsilon > 0$ there exists a subsequence $(n')$ such that

$$b_{n'} x' + a_{n'} < x_1 + \varepsilon \quad \text{for all } n' ;$$

without loss of generality we can assume that $x_1 + \varepsilon$ is a continuity point of $G$. Thus we obtain

$$G'(x') = \lim_{n'\to\infty} F_{n'}'(x') = \lim_{n'\to\infty} F_{n'}(b_{n'} x' + a_{n'}) \leqq \lim_{n'\to\infty} F_{n'}(x_1 + \varepsilon) = G(x_1 + \varepsilon)$$

and, on letting $\varepsilon \to 0$, we conclude $G'(x') \leqq G(x_1 + 0)$. But this contradicts (2.3.1); hence, the first one of the relations (2.3.2) has been proved. A slight change of this argument (assuming $b_{n'}x' + a_{n'} \to -\infty$) shows that this remains true even for $x_1 = -\infty$. The second inequality in (2.3.2) can be obtained analogously.

Moreover, by left-continuity there exists a further point $x'' < x'$, $x'' \in I$, such that (2.3.1) holds true both for $x'$ and $x''$. Then (2.3.2) is also true for $x''$ so that the two sequences $(b_n x' + a_n)$, $(b_n x'' + a_n)$ are bounded. Now we see immediately that $(a_n)$ and $(b_n)$ are also bounded.

(ii) We choose some continuity point $x \in I$ of $G'$ and suppose that $bx + a \pm \varepsilon$ belong also to $I$ and are continuity points of $G$. As in the proof of Prop. 2.3.1 we obviously have, for $n$ sufficiently large,

$$bx + a - \varepsilon < b_n x + a_n < bx + a + \varepsilon .$$

This implies

$$F_n(bx + a - \varepsilon) \leqq F_n'(x) \leqq F_n(bx + a + \varepsilon) ,$$

and, on letting $n \to \infty$, we obtain (2.3.3) for all $x$ under consideration. By left-continuity, these inequalities also hold true for all other $x \in I$ with $bx + a \pm \varepsilon \in I$. Assume now $b > 0$ and, moreover, that $bx + a \in I$ for some $x \in I$. Then $x$ can be chosen such that $bx + a$ is a continuity point of $G$; further, we have $bx + a \pm \varepsilon \in I$ for sufficiently small $\varepsilon$, and (2.3.3) implies $G'(x) = G(bx + a)$. Finally, (2.3.4) follows from the left-continuity of $G$ and $G'$. □

Having proved the necessary lemma we now turn to the announced results.

**Proposition 2.3.3.** *Assume $F_n \xrightarrow{c} G$, $F_n' \xrightarrow{c} G'$ for certain non-degenerate d.f. $G, G'$. Then we have the relations*

$$a_n \to a , \qquad b_n \to b > 0 , \qquad G'(x) = G(bx + a) \quad \textit{for} \quad x \in \mathbb{R}_1 .$$

Proof. We apply L. 2.3.2 for $I = \mathbb{R}_1$ so that $x_1 = -\infty$, $x_2 = \infty$. Since $G'$ is non-degenerate (2.3.1) is satisfied for a certain $x'$. Therefore the sequences $(a_n)$, $(b_n)$ are bounded. Thus we are able to select a subsequence $(n')$ such that $a_{n'} \to a$, $b_{n'} \to b$, say, where $b \geqq 0$.

Assume now $b = 0$. Then from (2.3.3) it follows that

$$G(a - \varepsilon) \leqq G'(x) \leqq G(a + \varepsilon) , \qquad x \in \mathbb{R}_1 , \qquad \varepsilon > 0 .$$

But this implies $G(a - \varepsilon) = 0$, $G(a + \varepsilon) = 1$, contrary to the supposition that $G$ is non-degenerate. So we see that $b > 0$; therefore (2.3.4) holds true for all $x \in \mathbb{R}_1$.

Finally we show that $a_n \to a$, $b_n \to b$. If $a_{n''} \to \bar{a}$, $b_{n''} \to \bar{b}$ for a certain subsequence $(n'')$ different from $(n')$, then we obtain as before $G'(x) = G(\bar{b}x + \bar{a})$ or, which is the same by (2.3.4),

$$G(x) = G\left(\frac{\bar{b}}{b} x + \bar{a} - \frac{\bar{b}}{b} a\right).$$

Since $G$ has at least two points of increase in $I = \mathbb{R}_1$, Prop. 1.1.6 tells us that $\bar{a} = a$, $\bar{b} = b$. Thus all assertions have been proved. □

**Proposition 2.3.4.** *Let $G, G'$ be non-decreasing and left-continuous on $I = (-\infty, x_2) \subseteqq \mathbb{R}_1$ such that $G(-\infty) = G'(-\infty) = 0$ and $G > 0$, $G' > 0$. If $F_n \xrightarrow[I]{} G$, $F_n' \xrightarrow[I]{} G'$, then* *we have the relations*

$$a_n \to a\,, \quad b_n \to b > 0\,, \quad G'(x) = G(bx + a) \quad \textit{for} \quad x < \min\left\{x_2, \frac{x_2 - a}{b}\right\}.$$

Proof. The assumptions made on $G$ and $G'$ ensure that (2.3.1) is satisfied for a certain $x' \in I$. Therefore $(a_n)$, $(b_n)$ are bounded, and, assuming $a_{n'} \to a$, $b_{n'} \to b$, we have again (2.3.3). Suppose now $b = 0$. Then (2.3.2) tells us that $a < x_2$. This means that $a - \varepsilon \in I$ for all $\varepsilon > 0$, and (2.3.3) leads to $G(a - \varepsilon) \leqq G'(x)$. Letting $x \to -\infty$ we arrive at a contradiction because $G > 0$ by assumption. So we obtain $b > 0$, and the remaining part of the proof proceeds like that of Prop. 2.3.3. □

**Proposition 2.3.5.** *Let* $G, G'$ *be non-decreasing and left-continuous on* $I = (0, x_2) \subseteqq (0, \infty)$ *such that* $G(+0) = G'(+0) = 0$, *and* $G > 0, G' > 0$ *on* $I$. *If* $F_n \xrightarrow[I]{} G$, $F_n' \xrightarrow[I]{} G'$ *then we have the relations*

$$a_n \to 0\,, \quad b_n \to b > 0\,, \quad G'(x) = G(bx) \quad \textit{for} \quad 0 < x < \min\left\{x_2, \frac{x_2}{b}\right\},$$

Proof. As before we see that $(a_n)$, $(b_n)$ are bounded and select $(n')$ such that $a_{n'} \to a$, $b_{n'} \to b$. If $b = 0$, then (2.3.2) implies $0 < a < x_2$, and from (2.3.3) we obtain $G(a - \varepsilon) \leqq G'(x)$; letting $x \to +0$ we get $G(a - \varepsilon) = 0$, a contradiction. Therefore $b > 0$, and we again get $G'(x) = G(bx + a)$ provided that $x \in I$, $bx + a \in I$. The latter conditions are indeed compatible. Namely, for $x \in I$ sufficiently small (2.3.1) is satisfied because $G > 0$, $G' > 0$ on $I$, and $G'(+0) = 0$; hence (2.3.2) tells us that we also have $bx + a \in I$. Accordingly, on letting $x \to +0$ we see that $0 = G'(+0) = G(a)$, i.e. $a = 0$ so that the proof is complete. □

## 2.4. Compactness and tightness

Let us consider a sequence $(F_n)$ of d.f. As was shown in Ex. 2.1.2, the assumption $F_n \xrightarrow{w} F$ does not imply $F_n \xrightarrow{c} F$ because $F$ might be a defective d.f. Now we are interested in excluding the occurrence of a defect.

We will introduce the notions compact and tight for general sets $\boldsymbol{F}$ of d.f., but it goes without saying that our definitions can be carried over to the sequences $(F_n)$ of d.f.

**Definition 2.4.1.** A non-empty set $\boldsymbol{F}$ of d.f. is called *relatively compact* if any sequence $(F_n)$, $F_n \in \boldsymbol{F}$, contains a subsequence which completely converges to a certain d.f. $F$. □

Note that by Helly's selection theorem this definition can by simplified since the existence of a weakly converging sequence is always granted: The set $\boldsymbol{F}$ is relatively compact if for every weakly converging sequence $(F_n)$, $F_n \in \boldsymbol{F}$, the limit function $F$ has no defect, i.e. if $F_n \xrightarrow{w} F$ implies $F_n \xrightarrow{c} F$.

**Example 2.4.2.** We consider the three sets of normal d.f.

$$\boldsymbol{F}_1 := \{\Phi_{\mu,1} : |\mu| \leqq c\}\,, \qquad \boldsymbol{F}_2 := \{\Phi_{\mu,1} : \mu \in \boldsymbol{R}_1\}\,, \qquad \boldsymbol{F}_3 := \{\Phi_{0,\sigma} : \sigma > 0\}\,.$$

The set $\boldsymbol{F}_1$ is obviously relatively compact since $\Phi_{\mu,1}(x) = \Phi(x - \mu)$ so that $\Phi_{\mu_n,1} \xrightarrow{w} F$ implies the existence of $\lim\limits_{n\to\infty} \mu_n = \mu$, say, whence $F = \Phi_{\mu,1}$ follows. — But the sets $\boldsymbol{F}_2$ and $\boldsymbol{F}_3$ are not relatively compact since

$$\Phi_{\mu_n,1} \xrightarrow{w} 0\,, \qquad \mu_n \to \infty\,, \quad \text{and} \quad \Phi_{0,\sigma_n} \xrightarrow{w} \tfrac{1}{2}\,, \qquad \sigma_n \to \infty\,.$$

□

Now we turn to conditions necessary and sufficient for the relative compactness of a set $\boldsymbol{F}$ of d.f. Let us first introduce the *tail function* $T_F$ of a d.f. $F$ by putting

$$T_F(x) = 1 - F(x) + F(-x+0) = \int_{|y| \geqq x} \mathrm{d}F(y), \qquad x > 0.$$

Clearly, we have $T_F(x) = \mathsf{P}(|X| \geqq x)$ for any r.v. $X \sim F$. The notation $T$ will be used with or without affixes.

**Definition 2.4.3.** A non-empty set $\boldsymbol{F}$ of d.f. $F$ is called *tight* if, to every $\varepsilon > 0$, there corresponds $x_\varepsilon > 0$ such that

$$\sup_{\boldsymbol{F}} T_F(x) < \varepsilon, \qquad x \geqq x_\varepsilon,$$

in other words, if $\sup T_F(x) \to 0$ as $x \to \infty$. □

To prove the tightness of $\boldsymbol{F}$ it obviously suffices to verify the condition $T_F(x_\varepsilon) < \varepsilon$ for all $F \in \boldsymbol{F}$; this is so because $T_F$ is non-increasing.

**Theorem 2.4.4.** *A set $\boldsymbol{F}$ of d.f. is tight iff it is relatively compact.*

Proof. If $\boldsymbol{F}$ is tight, then we consider an arbitrary sequence $(F_n)$, $F_n \in \boldsymbol{F}$, satisfying

$$F_n \xrightarrow{\mathrm{w}} G$$

for some monotone function $G$. By assumption we have, for every $\varepsilon > 0$, $T_{F_n}(x) < \varepsilon$, $x \geqq x_\varepsilon$; hence, it follows that

$$1 - G(x) + G(-x+0) \leqq \varepsilon, \qquad x \geqq x_\varepsilon.$$

This means that $G$ is non-defective, i.e. $\boldsymbol{F}$ is relatively compact.

On the other hand, assume that $\boldsymbol{F}$ is not tight, i.e. $\lim_{n\to\infty} \sup_{\boldsymbol{F}} T_F(n) =: \eta > 0$. Hence, we can select a sequence $(F_n)$ such that $T_{F_n}(n) \geqq \eta/2$ for $n \geqq 1$. By Helly's selection theorem there exists a subsequence $(F_{n'})$ of $(F_n)$ satisfying $F_{n'} \xrightarrow{\mathrm{w}} G$ for a certain function $G$. Now we conclude that, for all continuity points $\pm x$ of $G$,

$$1 - G(x) + G(-x) = \lim_{n'\to\infty} T_{F_{n'}}(x) \geqq \underline{\lim}_{n'\to\infty} T_{F_{n'}}(n') \geqq \eta/2;$$

hence $G$ is defective, i.e. $\boldsymbol{F}$ is not relatively compact. Thus the proof is complete. □

**Definition 2.4.5.** Let $\boldsymbol{F}$ be a non-empty set of d.f. We call $\boldsymbol{F}$ *shift compact* if for every $F \in \boldsymbol{F}$ there exists a constant $a_F$ such that the set $\boldsymbol{F}'$ of d.f. $F'(x) = F(x + a_F)$ is relatively compact (that is also tight). The numbers $a_F$ are then called *shift constants*. □

**Remark 2.4.6.** The relative compactness of a set $\boldsymbol{F}$ trivially implies the shift compactness of $\boldsymbol{F}$, while the converse is not true; this is shown by the set $\boldsymbol{F}_2$ of Ex. 2.4.2, which is obviously shift compact with $a_F = \mu$. Moreover, we now recognize a remarkable difference between the sets $\boldsymbol{F}_2$ and $\boldsymbol{F}_3$: The latter is not even shift compact.

**Proposition 2.4.7.** *Assume that $\boldsymbol{F}$ is shift compact. Then the following assertions are equivalent:*

(i) *$\boldsymbol{F}$ is relatively compact;*
(ii) *Any bounded set $\{a_F : F \in \boldsymbol{F}\}$ is a set of shift constants;*
(iii) *Any set $\{a_F' : F \in \boldsymbol{F}\}$ of shift constants is bounded;*
(iv) *Neither 0 nor 1 are limit functions of weakly converging sequences $(F_n)$, $F_n \in \boldsymbol{F}$.*

Proof. We show the implications (i) ⇒ (iv) ⇒ (iii) ⇒ (ii) ⇒ (i), the first of which is trivial.

(iv) ⇒ (iii) We carry out an indirect proof and assume first that there is a set $\{a'_F\colon F \in \boldsymbol{F}\}$ of shift constants satisfying $\sup_{\boldsymbol{F}} a'_F = \infty$. Then there exists a sequence $a_n := a'_{F_n} \to \infty$. In this way we have also selected a sequence $(F_n)$. Since $\boldsymbol{F}$ is shift compact we have $F_{n'}(x + a_{n'}) \xrightarrow{c} F(x)$ for a certain subsequence $(F_{n'})$. On the other hand, by Helly's selection theorem and assumption (iv) there exists a subsequence $(n'')$ of $(n')$ such that $F_{n''} \xrightarrow{w} G \neq 0$; accordingly, $F_{n''}(x_0) \to G(x_0) > 0$ for some $x_0 \in \boldsymbol{R}_1$. By virtue of $a_{n''} \to \infty$ we trivially have $F_{n''}(x + a_{n''}) \geqq F_{n''}(x_0)$ for all $x$ and sufficiently large $n''$. Hence $F \geqq G(x_0) > 0$. This is a contradiction since $F$ is a d.f.

The case $\inf_{\boldsymbol{F}} a'_F = -\infty$ can be ruled out analogously using the assumption $G \neq 1$. In this way (iii) follows.

(iii) ⇒ (ii) Clearly, the assumption of shift compactness implies the tightness of the set of all d.f. $F(x + a'_F)$, that is, the relation

$$1 - F(x + a'_F) + F(-x + a'_F + 0) < \varepsilon\,, \qquad x \geqq x_\varepsilon\,,$$

for all $F \in \boldsymbol{F}$ and for certain shift constants $a'_F$. By assumption (iii) we have $|a'_F| \leqq C' < \infty$ for all $F \in \boldsymbol{F}$. Let $\{a_F\colon F \in \boldsymbol{F}\}$ be an arbitrary bounded set, i.e. $|a_F| \leqq C$ for all $F \in \boldsymbol{F}$. Then we get, for $x \geqq x_\varepsilon + C + C'$,

$$\begin{aligned} 1 - F(x + a_F) + F(-x - a_F + 0) &= 1 - F\big(x + a'_F + (a_F - a'_F)\big) \\ + F\big(-x - a'_F - (a_F - a'_F) + 0\big) &\leqq 1 - F(x + a'_F - C - C') \\ + F(-x - a'_F + C + C' + 0) &< \varepsilon; \end{aligned}$$

hence, $\{F(x + a_F)\colon F \in \boldsymbol{F}\}$ is tight and therefore relatively compact as asserted.

(ii) ⇒ (i) is trivial; we simply choose $a_F = 0$ for all $F \in \boldsymbol{F}$. □

The case in which the variances of all $F \in \boldsymbol{F}$ exist is of particular interest in connection with a modification of the Lindeberg-Feller theorem, see Sect. 12.3.

**Proposition 2.4.8.** *Let $\boldsymbol{F}$ be a set of d.f. $F$ with variance $\sigma_F^2$ such that*

$$\sup_{\boldsymbol{F}} \sigma_F^2 =: C < \infty; \tag{2.4.1}$$

*then $\boldsymbol{F}$ is shift compact, and the mathematical expectations $\mu_F$ can serve as shift constants.*

Proof. In view of Th. 2.4.4 it is enough to show that the set $\boldsymbol{F}'$ of d.f. $F'(x) = F(x+\mu_F)$ is tight. This follows from Čebyšev inequality; namely, we have

$$T_{F'}(x) = \int\limits_{|y| \geqq x} \mathrm{d}F(y + \mu_F) \leqq \frac{\sigma_F^2}{x^2} \leqq \frac{C}{x^2} < \varepsilon$$

for arbitrary $\varepsilon > 0$, and sufficiently large $x$. □

The intuitive meaning of Prop. 2.4.8 is particularly obvious from the following result.

**Corollary 2.4.9.** *Assume (2.4.1) and consider a weakly converging sequence $(F_n)$ tending to a monotone function $G$, $F_n \xrightarrow{w} G$. Then either $G$ is a d.f. or $G = 0$ or $G = 1$; in other words, no proper part of the mass 1 distributed over $\boldsymbol{R}_1$ according to $F_n$ escapes to infinity.*

Proof. By Prop. 2.4.8 $\boldsymbol{F} := \{F_n\colon n \geqq 1\}$ is shift compact; hence Prop. 2.4.7 applies and, comparing its assertions (i) and (iv) we see that $\boldsymbol{F}$ is either relatively compact or the limit function $G = 0$ or $G = 1$ occur. □

In part III we will need the following easy consequence of Cor. 2.4.9.

**Corollary 2.4.10.** *Assume* (2.4.1) *and, moreover, that*

$$0 < \inf_{\boldsymbol{F}} F(x_1)\,, \qquad \sup_{\boldsymbol{F}} F(x_2) < 1$$

*for certain* $x_1, x_2 \in \boldsymbol{R}_1$. *Then* $\boldsymbol{F}$ *is relatively compact.* □

We next state a condition necessary and sufficient for shift compactness.

**Proposition 2.4.11.** *A set* $\boldsymbol{F}$ *of d.f.* $F$ *is shift compact iff the corresponding set* $\boldsymbol{F}^s$ *of d.f.* $F^s$ *obtained from* $F$ *by symmetrization is relatively compact.*

Proof. For every d.f. $F \in \boldsymbol{F}$ we choose a r.v. $X \sim F$ and denote by $m$ a median of $F$. Then we write the symmetrization inequality given in Cor. 1.4.9 in the form

$$\frac{1}{2}\,\mathsf{P}(|X - m| \geqq x) \leqq \mathsf{P}(|X^s| \geqq x) \leqq 2\mathsf{P}\left(|X - m| \geqq \frac{x}{2}\right)$$

and see that $\boldsymbol{F}^s$ is tight iff the corresponding set $\boldsymbol{F}'$ of d.f. $F'(x) = F(x + m)$ is tight or, equivalently, iff the set $\boldsymbol{F}$ is shift compact. □

The proof of Prop. 2.4.11 shows that the medians can serve as shift constants provided $\boldsymbol{F}$ is shift compact.

It is not difficult to carry all the above notions and statements over to the case of non-decreasing and bounded functions. We will give an important example.

**Proposition 2.4.12.** *A set* $\boldsymbol{G}$ *of uniformly bounded and non-decreasing functions is relatively compact iff it is tight, that is, if to every* $\varepsilon > 0$ *there corresponds* $x_\varepsilon > 0$ *such that*

$$\sup_{\boldsymbol{G}} \big(G(\infty) - G(x) + G(-x + 0) - G(-\infty)\big)$$
$$= \sup_{\boldsymbol{G}} \int_{|y| \geqq x} \mathrm{d}G(y) < \varepsilon, \qquad x \geqq x_\varepsilon\,.$$ □

## 2.5. Regular variation

**Definition 2.5.1.** Let $H > 0$ be a function defined on $(x_0, \infty)$, $x_0 \geqq 0$. $H$ is said to *vary regularly at infinity with exponent* $r \in \boldsymbol{R}_1$ if

$$\lim_{t \to \infty} \frac{H(tx)}{H(t)} = x^r\,, \qquad x > 0\,. \ \square \tag{2.5.1}$$

**Remark 2.5.2.** (i) Equation (2.5.1) is required to be valid for all $x > 0$ although $H$ is defined only on $(x_0, \infty)$. Such a requirement is possible since, for given $x > 0$, $H(tx)$ is defined for $t > x_0/x$.

(ii) In Def. 2.5.1, $H$ is assumed to be an arbitrary positive function on $(x_0, \infty)$. But we will often deal with monotone functions $H$. In this case we have not only (2.5.1) but also

$$\lim_{t \to \infty} \frac{H\big(t(1 + h(t))\,x\big)}{H(t)} = x^r\,, \qquad x > 0\,,$$

for any function $h$ satisfying $\lim_{t \to \infty} h(t) = 0$. In fact, if $H$ is for example increasing, then this follows from the inequalities

$$H\big(t(1 - \varepsilon)\,x\big) \leqq H\big(t(1 + h(t))\,x\big) \leqq H\big(t(1 + \varepsilon)\,x\big)\,,$$
$$0 < \varepsilon < 1\,, \qquad t \geqq t_0(\varepsilon, x)\,.$$

(iii) If $r > 0$ and $H$ is non-decreasing, then it is unbounded. Indeed, for any $x > 0$ and some $t_0 > x_0$ we have

$$x^r = \lim_{t\to\infty} \frac{H(tx)}{H(t)} \leqq \frac{1}{H(t_0)} \lim_{t\to\infty} H(tx) = \frac{H(\infty)}{H(t_0)},$$

and on letting $x \to \infty$ we find $H(\infty) = \infty$. □

It is the point of the notion of regular variation that we can obtain (2.5.1) from surprisingly weak sufficient conditions the first of which is formulated in the following theorem.

**Theorem 2.5.3.** *A function $H > 0$ defined on $(x_0, \infty)$ varies regularly at infinity iff the limit*

$$\psi(x) := \lim_{t\to\infty} \frac{H(tx)}{H(t)}, \qquad x \in I = (x_1, x_2), \qquad 0 < x_1 < x_2 < \infty,$$

*is positive and bounded.*

Proof. The only-if part is trivial by virtue of Def. 2.5.1. To prove the if-part we introduce the functions

$$\underline{\psi}(x) := \varliminf_{t\to\infty} \frac{H(tx)}{H(t)}, \qquad \overline{\psi}(x) := \varlimsup_{t\to\infty} \frac{H(tx)}{H(t)}, \qquad x > 0.$$

Then we have by assumption $\underline{\psi}(x) = \overline{\psi}(x) = \psi(x)$ for $x \in I$. Now we will show step by step that for all $x > 0$

$$\underline{\psi}(x) = \overline{\psi}(x) := \psi(x). \tag{2.5.2}$$

(i) For $x \in I^{-1} := (1/x_2, 1/x_1)$ we have $1/x \in I$; thus we obtain

$$\underline{\psi}(x) = \varliminf_{t\to\infty} \frac{H(tx)}{H(t)} = \left[ \varlimsup_{t\to\infty} \frac{H\left(tx\dfrac{1}{x}\right)}{H(tx)} \right]^{-1} = \frac{1}{\overline{\psi}\left(\dfrac{1}{x}\right)} = \frac{1}{\psi\left(\dfrac{1}{x}\right)} > 0$$

and also $\overline{\psi}(x) = 1/\psi(1/x) < \infty$. Therefore (2.5.2) holds true for $x \in I^{-1}$.

(ii) Every $x \in I_1 := (x_1/x_2, x_2/x_1)$ can be represented as $x = x'x''$ for certain numbers $x' \in I$, $x'' \in I^{-1}$. Thus we get

$$\underline{\psi}(x) = \varliminf_{t\to\infty} \frac{H(tx''x')}{H(tx'')} \frac{H(tx'')}{H(t)} \geqq \underline{\psi}(x')\, \underline{\psi}(x'') = \psi(x')\, \psi(x'') > 0$$

and also $\overline{\psi}(x) \leqq \psi(x')\, \psi(x'') < \infty$ so that (2.5.2) is verified for $x \in I_1$; note that $x_1/x_2 < 1$, $x_2/x_1 > 1$.

(iii) If $x \in I_2 := (x_1^2/x_2^2, x_2^2/x_1^2)$ we have $x = x'x''$ for certain $x', x'' \in I_1$. Using the same argument as in (ii) we see that (2.5.2) is true for $x \in I_2$. Proceeding in this way, we finally obtain the assertion (2.5.2).

Moreover, the function $\psi$ defined in (2.5.2) has the properties $0 < \psi < \infty$ and

$$\psi(x'x'') = \lim_{t\to\infty} \frac{H(tx''x')}{H(tx'')} \frac{H(tx'')}{H(t)} = \psi(x')\, \psi(x''), \qquad x', x'' > 0. \tag{2.5.3}$$

Now we put $\psi_0(x) = \psi(e^x)$, $x \in \mathbb{R}_1$. Since $\psi$ is positive and bounded on $I$ the function $\psi_0$ has the same properties on $(\log x_1, \log x_2)$. Further, from (2.5.3) we conclude that

$$\psi_0(x' + x'') = \psi_0(x')\, \psi_0(x''), \qquad x', x'' \in \mathbb{R}_1.$$

Hence, Prop. A.6 shows that $\psi_0(x) = e^{rx}$ for a certain real $r$, whence $\psi(x) = \psi_0(\log x) = x^r$ follows. □

**Remark 2.5.4.** Theorem 2.5.3 shows that the convergence assumed only on $I$ continues to $(0, \infty)$ such that even the limit function is explicitely known. The same is true under the suppositions of the next corollary, which also contains a sufficient condition for regular variation.

**Corollary 2.5.5.** *Let $H > 0$ be monotone on $(x_0, \infty)$. Assume that there exist sequences $(b_n)$, $(c_n)$ of positive numbers satisfying $b_n \to \infty$ and $c_n/c_{n+1} \to 1$ such that the limit*

$$\psi(x) := \lim_{n\to\infty} c_n H(b_n x)\,, \qquad x \in I = (x_1, x_2)\,, \qquad 0 < x_1 < x_2 < \infty\,, \tag{2.5.4}$$

*is positive and finite. Then $H$ varies regularly at infinity with exponent $r$, say. Further, the limit in (2.5.4) exists for all $x > 0$ and equals $cx^r$ for a certain $c > 0$.*

Proof. We assume that $H$ is increasing; the other case can be treated analogously. Let $x' \in I$ be fixed. Then for given $t > x_0$, there exists a smallest number $n$ such that $b_n x' < t \leqq b_{n+1}x'$. So we get for all $x \in (x_1/x', x_2/x')$

$$\frac{c_{n+1}}{c_n}\,\frac{c_n H(b_n x' x)}{c_{n+1}H(b_{n+1}x')} \leqq \frac{H(tx)}{H(t)} \leqq \frac{c_n}{c_{n+1}}\,\frac{c_{n+1}H(b_{n+1}x' x)}{c_n H(b_n x')}\,.$$

By assumption, the outer terms converge to $\psi(x'x)/\psi(x')$ as $n \to \infty$, which is equivalent to $t \to \infty$. Therefore $H$ satisfies the suppositions of Th. 2.5.3, which tells us that $H$ is of regular variation with a certain $r \in \mathbb{R}_1$.

Finally, from (2.5.1) and (2.5.4) we obtain for every $x > 0$ and an arbitrary fixed $x' \in I$

$$\lim_{n\to\infty} c_n H(b_n x) = \lim_{n\to\infty}\left[c_n H(b_n x')\,\frac{H\left(b_n x'\,\dfrac{x}{x'}\right)}{H(b_n x')}\right] = \psi(x')\left(\frac{x}{x'}\right)^r = cx^r$$

where $c = \psi(x')/x'^r$. The assertions have now been proved. □

**Remark 2.5.6.** With the very same ideas we can introduce regular variation at 0. Then we obtain, for instance, an analogue to Cor. 2.5.5 with $b_n \to 0$ instead of $b_n \to \infty$. □

## 2.6. Supplement

**2.6.1.** Let $(F_n)$ be a sequence of d.f. and $F$ a continuous d.f. If $F_n \xrightarrow{c} F$, then the convergence is uniform on $\mathbb{R}_1$. This result is due to PÓLYA (see PETROV (1975) I. 4); it serves for an easy proof of the following statement: Let $F$ be a continuous d.f.; then $F_n \xrightarrow{c} F$ is equivalent with the convergence $\sup_x |F_n(x) - F(x)| \to 0$, as $n \to \infty$. This is, for example, important in the central limit theorem.

**2.6.2.** The extended Helly-Bray theorem can be converted. The corresponding statement is sometimes used as the definition of complete convergence. It is useful since it makes sense also for probability measures defined on the Borel sets of complete separable metric spaces; see BILLINGSLEY (1968).

**2.6.3.** Prop. 2.4.8 was proved by means of Čebyšev inequality. Making use of Markov inequality instead we obtain the following statement: *Let $F$ be a set of d.f. and assume*

$$\sup_{F} \int_{-\infty}^{\infty} |x|^{\eta} \, dF(x) < \infty \quad \textit{for some} \quad \eta > 0 .$$

*Then $F$ is relatively compact* (Loève (1963), Sect. 11).

**2.6.4.** Theorem 2.4.4 can be carried over to probability measures defined on complete separable metric spaces; this version is often called the *theorem of Prohorov* (see Prohorov (1956); the proof is also contained in, amongst others, Laha-Rohatgi (1979)).

**2.6.5.** A function $H > 0$ varies regularly at infinity with exponent $r$ iff $H(x) = x^r L(x)$ where $L$ is a function satisfying $\frac{L(tx)}{L(x)} \to 1$, as $x \to \infty$, for every $t > 0$ (Feller (1971)).

## Notes and bibliographical data

Prop. 2.3.3 is due to Gnedenko and Hinčin, see Gnedenko and Kolmogorov (1959), Sect. 10; for the counterparts, Prop. 2.3.4 and Prop. 2.3.5 see Rossberg (1979) and (1983), resp.

It was Karamata who in 1930 introduced functions regularly varying at infinity and elaborated a fruitful theory on this. An improved version of it is presented in Feller (1971) with applications to stable d.f., the extreme members of a variational series, and other subjects. A selfcontained up to date representation of the basic theory of regularly varying functions is given in Seneta (1976). For recent applications to renewal theory see Callaert and Cohen (1972); the result of these authors was generalized by Dewess (1980). Regular variation was applied to a Wiener-Hopf factorization in Dewess (1978b).

For Th. 2.5.3 and Cor. 2.5.5 compare Feller (1971).

# Chapter 3. Characteristic functions

Many important theorems of probability theory can be derived by means of characteristic functions. This is the reason why we now aim at giving an introductory presentation of this analytic tool. The main results are the uniqueness, inversion, convolution, and continuity theorems.

It is the first that permits us, in principle, to replace studies on d.f. by an equivalent investigation of the corresponding characteristic functions. In this way the problem under consideration can be tackled by means of a quite different method. Applying inversion formulae we can return to the d.f. under consideration or at least conclude some of their properties. The convolution theorem is often needed; therefore, characteristic functions are mainly used when i.r.v. are involved.

The continuity theorem permits us to study the complete convergence of d.f. Its power will be seen particularly in part III; but first applications are given already in the present chapter. For the case that complete convergence is not granted we present a "weak convergence criterion" for "integral characteristics".

It is noteworthy that the relative compactness of sets of d.f. proves to be equivalent to the equicontinuity of the corresponding characteristic functions. This fact also sheds some new light on the proofs of the continuity theorem and of its corollaries.

## 3.1. Definition and basic facts

**Definition 3.1.1.** Let $F$ be a d.f. Then the function

$$f(t) := \int_{-\infty}^{\infty} e^{itx}\, dF(x) = \int_{-\infty}^{\infty} \cos tx\, dF(x) + i \int_{-\infty}^{\infty} \sin tx\, dF(x) \tag{3.1.1}$$

is called the *characteristic function* (c.f.) of $F$. □

Note that this improper Riemann-Stieltjes integral exists for all d.f. $F$ since $|e^{itx}| = 1$.

Sometimes we will also say that a r.v. $X \sim F$ has the c.f. $f$ although the uniqueness theorem 3.2.4 only justifies the use of the expression as given in Def. 3.1.1. — For examples of c.f. we refer to the table at the end of this book.

Next we state several elementary facts, the proofs of which we omit owing to their simplicity. We use here the horizontal bar atop of $f$ to denote the complex conjugate of $f$.

**Proposition 3.1.2.** (i) *For any c.f. $f$ the following assertions hold true:*

$$f(0) = 1\,, \qquad |f| \leqq 1\,, \qquad f(-t) = \overline{f(t)}\,, \qquad t \in R_1\,,$$

(ii) *the d.f.* $F'(x) := F(bx + a)$, $b > 0$, *has the c.f.*

$$f'(t) = e^{-i\frac{a}{b}t} f\left(\frac{t}{b}\right);$$

*the d.f.* $F_-(x) := 1 - F(-x + 0)$ *has the c.f.* $f_-(t) = f(-t)$. □

**Lemma 3.1.3.** *For all c.f. f we have*

$$|f(t + h) - f(t)|^2 \leqq 2(1 - \operatorname{Re} f(h)) \quad \textit{for} \quad t, h \in R_1,$$

*here* Re $f$ *is the real part of* $f$.

Proof. By Def. 3.1.1 we see that

$$|f(t + h) - f(t)|^2 = |\int_{-\infty}^{\infty} e^{itx} (e^{ihx} - 1) \, dF(x)|^2$$

$$\leqq (\int_{-\infty}^{\infty} |e^{ihx} - 1| \, dF(x))^2 \leqq \int_{-\infty}^{\infty} |e^{ihx} - 1|^2 \, dF(x) \qquad (3.1.2)$$

(by Schwarz inequality). Now the assertion follows from

$$\int_{-\infty}^{\infty} |e^{ihx} - 1|^2 \, dF(x) = 2 \int_{-\infty}^{\infty} (1 - \cos hx) \, dF(x) = 2(1 - \operatorname{Re} f(h)) . \qquad □$$

**Proposition 3.1.4.** *A c.f. f is uniformly continuous on* $R_1$.

Proof. For arbitrary $\varepsilon > 0$ we can satisfy the inequality

$$1 - \operatorname{Re} f(h) = \int_{|x| \leqq x_0} (1 - \cos hx) \, dF(x) + \int_{|x| > x_0} (1 - \cos hx) \, dF(x) < \varepsilon$$

by first choosing $x_0$ sufficiently large (such that $2T_F(x_0) < \varepsilon/2$) and then $h$ sufficiently small (such that $1 - \cos hx < \varepsilon/2$, $|x| \leqq x_0$). Now the assertion follows from L. 3.1.3. □

We note that Prop. 3.1.4 follows also from (3.1.2) by dominated convergence since $|e^{ihx} - 1|^2 \leqq 4$ and this bound is integrable with respect to $F$.

Now we prove two useful inequalities the first of which is often called *truncation inequality*.

**Lemma 3.1.5.** *For a d.f. F with tail function* $T_F$ *and c.f. f we have the two inequalities*

$$T_F\left(\frac{1}{x}\right) \leqq \frac{7}{x} \int_0^x (1 - \operatorname{Re} f(t)) \, dt , \qquad x > 0 ,$$

$$\int_{|y| \leqq 1/x} y^2 \, dF(y) \leqq \frac{3}{x^2} (1 - \operatorname{Re} f(x)) , \qquad x > 0 .$$

Proof. In view of (3.1.1) and Fubini theorem we have for $x > 0$

$$\frac{1}{x} \int_0^x (1 - \operatorname{Re} f(t)) \, dt = \frac{1}{x} \int_0^x \int_{-\infty}^{\infty} (1 - \cos ty) \, dF(y) \, dt =$$

$$= \frac{1}{x} \int_{-\infty}^{\infty} \int_0^x (1 - \cos ty) \, dt dF(y) = \int_{-\infty}^{\infty} \left(1 - \frac{\sin xy}{xy}\right) dF(y)$$

$$\geqq \int_{|xy| \geqq 1} \left(1 - \frac{\sin xy}{xy}\right) dF(y) \geqq (1 - \sin 1) \, T_F\left(\frac{1}{x}\right);$$

thus the first inequality follows since $(1 - \sin 1)^{-1} < 7$. To show the second we write

$$\frac{1}{x^2}(1 - \operatorname{Re} f(x)) = \frac{1}{x^2}\int_{-\infty}^{\infty}(1 - \cos xy)\, dF(y)$$

$$\geq \int_{|xy| \leq 1} \frac{1 - \cos xy}{x^2y^2} y^2\, dF(y) \geq (1 - \cos 1) \int_{|xy| \leq 1} y^2\, dF(y)$$

and note that $(1 - \cos 1)^{-1} < 3$. □

C.f. are uniquely appropriate in the study of sums of i.r.v. in view of the following well-known result.

**Theorem 3.1.6** (Convolution theorem). *If $F = F_1 * F_2$ is the convolution of the d.f. $F_1$ and $F_2$, then $f = f_1 f_2$.* □

Hence the genuine multiplication of the c.f. and the symbolic multiplication of the d.f. correspond to each other. For the proof we refer to Lukacs (1970) or Laha and Rohatgi (1979).

**Example 3.1.7.** Let $S_n := X_1 + \ldots + X_n$, where $X_k$ are i.i.d.r.v. the d.f. of which has atoms 0 and 1 with jumps $q$ and $p$, resp. Then $q + p \exp(it)$ is the c.f. of $X_1$ and $S_n$ is subject to the binomial d.f. with c.f. $(q + p \exp(it))^n$. □

Theorem 3.1.6 implies the following fact.

**Corollary 3.1.8.** *The product of two c.f. is also a c.f. In particular, the d.f. $F^s = F * F_-$ has the c.f. $f^s = ff_- = |f|^2$.* □

**Proposition 3.1.9.** *Let $f$ be the c.f. of a non-degenerate d.f. $F$. Then there exist positive numbers $\delta$, $\eta$ such that*

$$|f(t)| \leq 1 - \eta t^2, \qquad |t| \leq \delta.$$

Proof. We carry an indirect proof and assume that the assertion is not true. This means that for an arbitrary pair $\delta > 0$, $\eta > 0$ there exists a number $t_1$, $|t_1| < \delta$, such that $1 - |f(t_1)| < \eta t_1^2$. Similarly we conclude that there is a $t_2$, $|t_2| < \delta/2$, satisfying $1 - |f(t_2)| < \frac{\eta}{2} t_2^2$. Continuing this reasoning we obtain by induction that there exists a sequence $t_k \to 0$ such that

$$0 \leq 1 - |f(t_k)| < \frac{\eta}{k} t_k^2, \qquad k = 1, 2, \ldots;$$

accordingly, $|f(t_k)| = 1 + o(t_k^2)$, or, which is the same, $f^s(t_k) = |f(t_k)|^2 = 1 + o(t_k^2)$. Now we use the second inequality of L.3.1.5 and write

$$\int_{y \leq 1/t_k} y^2\, dF^s(y) \leq \frac{3}{t_k^2}(1 - \operatorname{Re} f^s(t_k)) = \frac{3}{t_k^2}(1 - f^s(t_k)) = o(1).$$

On letting $k \to \infty$ we see that the second moment $\mu_2^s$ of $F^s$ vanishes. Since the variance $\sigma_{F^s}^2 = \mu_2^s - (\mu_1^s)^2 \geq 0$ it follows that $\sigma_{F^s}^2 = \mu_1^s = 0$; therefore, by Prop. 1.3.5 (ii), $F^s = U_0$ (because $a = \mu_1^s = 0$). From Cor. 1.4.9 we see now that $F$ is also degenerate. This contradiction completes the proof. □

We state now an easy consequence of Prop. 3.1.9.

**Corollary 3.1.10.** *If for an arbitrary c.f. $f$ and a sequence $t_k \to 0$ we have $|f(t_k)| = 1 + o(t_k^2)$, then $|f| = 1$ and $F = U_a$ for a certain $a \in \mathbb{R}_1$.* □

We also prove a sort of counterpart to Prop. 3.1.9.

**Proposition 3.1.11.** *Let $f$ be the c.f. of a non-degenerated d.f. $F$ with finite variance $\sigma^2$. Then*

$$|f(t)|^2 \geqq 1 - \sigma^2 t^2, \qquad 0 \leqq |t| \leqq \sigma^{-2}.$$

*Further, if $f$ is real, then we have even*

$$f(t) \geqq 1 - \frac{\sigma^2}{2} t^2, \qquad 0 \leqq |t| \leqq 2\sigma^{-2}.$$

Proof. (i) We consider the symmetrized d.f. $F^s$ with variance $2\sigma^2$, vanishing expectation, and c.f. $f^s = |f|^2$. Then using $\cos u \geqq 1 - u^2/2$, we have

$$f^s(t) = \int_{-\infty}^{\infty} \cos tx \, dF^s(x) \geqq 1 - \frac{t^2}{2} \int_{-\infty}^{\infty} x^2 \, dF^s(x) = 1 - \sigma^2 t^2, \quad t \in \mathbb{R}_1;$$

This proves the first assertion, and the second follows analogoulsy; here we anticipate from Cor. 3.2.6 that $F$ is symmetric so that $\mu_F = 0$. Hence

$$f(t) = \int_{-\infty}^{\infty} \cos tx \, dF(x) \geqq 1 - \frac{t^2}{2} \int_{-\infty}^{\infty} x^2 \, dF(x) = 1 - \frac{\sigma^2}{2} t^2, \qquad t \in \mathbb{R}_1. \qquad \square$$

Next we consider the *mixture*

$$F = \sum_{k=0}^{\infty} p_k G_k,$$

where $(G_k)$ is a sequence of d.f. and $p_k \geqq 0$, $\sum_{k=0}^{\infty} p_k = 1$.

**Proposition 3.1.12.** *The c.f. $f$ of the d.f. $F$ is given by*

$$f = \sum_{k=0}^{\infty} p_k g_k.$$

Proof. The assertion is trivial if $p_k = 0$ for $k \geqq n_0$. Otherwise, we introduce the non-decreasing functions

$$F_n := \sum_{k=0}^{n} p_k G_k, \qquad n = 0, 1, \ldots$$

By assumption we have $F_n \xrightarrow{c} F$. Therefore, the assertion is an immediate consequence of the extended Helly-Bray theorem; indeed, we have

$$\sum_{k=0}^{\infty} p_k g_k(t) = \lim_{n\to\infty} \left(p_0 g_0(t) + \ldots + p_n g_n(t)\right)$$

$$= \lim_{n\to\infty} \int_{-\infty}^{\infty} e^{itx} \, dF_n(x) = \int_{-\infty}^{\infty} e^{itx} \, dF(x) = f(t). \qquad \square$$

Note that a similar passage to the limit will yield us one part of the continuity theorem 3.4.1.

**Example 3.1.13.** Let $g$ be a certain c.f. Putting $p_k = e^{-\lambda}\lambda^k/k!$, $\lambda > 0$, and $g_k = g^k$, $k = 0, 1, \ldots$, we obtain the c.f. of the compound Poisson d.f. (cf. Sect. 1.4.2)

$$f = e^{-\lambda} \sum_{k=0}^{\infty} \frac{\lambda^k}{k!} g^k = e^{\lambda(g-1)}. \qquad \square$$

**Example 3.1.14.** If $g$ is the c.f. of the d.f. $G$, then $\operatorname{Re} g$ is also a c.f. It corresponds to the d.f. $F := (G + G_-)/2$ where $G_-(x) = 1 - G(-x + 0)$. This is immediate from Prop. 3.1.12 and 3.1.2 (ii) since the c.f. $f$ of $F$ is given by

$$f = \tfrac{1}{2}(g + g_-) = \tfrac{1}{2}(g + \bar{g}) = \operatorname{Re} g. \qquad \square$$

## 3.2. Inversion formulae and the uniqueness theorem

Let $F$ and $G$ be two d.f. with c.f. $f$ and $g$, resp. By Fubini theorem we immediately get the well-known Parseval equation

$$\int_{-\infty}^{\infty} e^{-ixt} f(x)\, dG(x) = \int_{-\infty}^{\infty}\int_{-\infty}^{\infty} e^{ix(y-t)}\, dG(x)\, dF(y) = \int_{-\infty}^{\infty} g(y-t)\, dF(y). \quad (3.2.1)$$

In the sequel it will be important that the right side — but for a positive factor — proves to be a density for the special choice $g(t) = \exp(-t^2\sigma^2/2)$, i.e., for $G = \Phi_{0,\sigma}$. Namely, $(\sigma/\sqrt{2\pi})\, g$ is then the density of $\Phi_{0,1/\sigma}$. Therefore we multiply (3.2.1) by $\sigma/\sqrt{2\pi}$ and obtain the expression

$$\frac{1}{2\pi} \int_{-\infty}^{\infty} e^{-ixu} f(x)\, e^{-\frac{1}{2}\left(\frac{x}{\sigma}\right)^2} dx = \frac{\sigma}{\sqrt{2\pi}} \int_{-\infty}^{\infty} e^{-\frac{\sigma^2}{2}(u-y)^2}\, dF(y) =: p_\sigma(u). \quad (3.2.2)$$

Now we see from formula (1.4.2) that this is the density of the convolution $\Phi_{0,1/\sigma} * F$.

**Lemma 3.2.1.** *For an arbitrary d.f. $F$ we have, as $\sigma \to \infty$,*

$$F * \Phi_{0,1/\sigma} \xrightarrow{c} F.$$

Proof. For arbitrary $\varepsilon > 0$ and $\sigma > 0$ we have

$$F * \Phi_{0,1/\sigma}(x) = \int_{-\infty}^{\infty} F(x-y)\, d\Phi(\sigma y) = \int_{|y|<\varepsilon} \ldots + \int_{|y|\geqq\varepsilon} \ldots \leqq F(x+\varepsilon) + T_\Phi(\sigma\varepsilon)$$

and, as a consequence,

$$\overline{\lim_{\sigma\to\infty}}\, F * \Phi_{0,1/\sigma}(x) \leqq F(x+\varepsilon).$$

Similarly,

$$F * \Phi_{0,1/\sigma}(x) \geqq \int_{-\infty}^{\varepsilon} F(x-y)\, d\Phi(\sigma y) \geqq F(x-\varepsilon)\, \Phi(\sigma\varepsilon)$$

which implies

$$\underline{\lim}_{\sigma\to\infty}\, F * \Phi_{0,1/\sigma}(x) \geqq F(x-\varepsilon).$$

Thus $F * \Phi_{0,1/\sigma}(x) \to F(x)$ for all continuity points of $F$ which was to be proved. $\square$

Now we turn to the first inversion formula.

**Theorem 3.2.2** (Fourier inversion theorem). *Let the c.f. $f$ of the d.f. $F$ be absolutely integrable on $\mathbb{R}_1$. Then $F$ has a bounded and uniformly continuous density $p$ given by*

$$p(u) = \frac{1}{2\pi} \int_{-\infty}^{\infty} \mathrm{e}^{-itu} f(t)\, \mathrm{d}t = \frac{1}{\pi} \int_{0}^{\infty} \mathrm{Re}\left(\mathrm{e}^{-itu} f(t)\right) \mathrm{d}t\,. \tag{3.2.3}$$

Proof. Since $f$ is absolutely integrable the limit on the left side of (3.2.2) exists, as $\sigma \to \infty$, by dominated convergence; moreover, the passage to the limit under the integral sign is permissible. Therefore, we can define a function

$$p(u) := \lim_{\sigma\to\infty} p_\sigma(u)\,, \qquad u \in \mathbb{R}_1\,,$$

and see from (3.2.2) that $p$ has the representation (3.2.3).

We next prove that $p$ is the density of $F$. Since $p_\sigma$ is the density of $F * \Phi_{0,1/\sigma}$ we see from L. 3.2.1 that

$$\int_{-\infty}^{x} p_\sigma(u)\, \mathrm{d}u = F * \Phi_{0,1/\sigma}(x) \xrightarrow{c} F(x)\,, \qquad \sigma \to \infty\,.$$

Therefore, we obtain for arbitrary continuity points $x_1 < x_2$ of $F$, by dominated convergence,

$$F(x_2) - F(x_1) = \lim_{\sigma\to\infty} \int_{x_1}^{x_2} p_\sigma(u)\, \mathrm{d}u = \int_{x_1}^{x_2} p(u)\, \mathrm{d}u$$

so that $p$ is the density of $F$.

Finally, using (3.2.3) we write

$$|p(u+h) - p(u)| \leq \frac{1}{2\pi} \int_{-\infty}^{\infty} |\mathrm{e}^{-ihx} - 1|\, |f(t)|\, \mathrm{d}t\,.$$

This bound does not depend on $u$. Again applying dominated convergence we see now that $p$ is uniformly continuous. □

The usefulness of the proposition just proved is obvious since most of the densities in our table and many more can readily be obtained in this way.

Now we are going to derive a general inversion formula.

**Theorem 3.2.3.** *Let $f$ be the c.f. of the d.f. $F$. Then we have, for all continuity points $x$ of $F$,*

$$F(x) = \frac{1}{2} - \frac{1}{\pi} \int_{0}^{\infty} \mathrm{Im}\left(f(t)\, \mathrm{e}^{-itx}\right) \frac{\mathrm{d}t}{t}\,. \tag{3.2.4}$$

*This implies that the improper integral exists.*

Proof. Let $x$ be an arbitrary continuity point of $F$. Then we introduce the d.f. $F'(y) = F(y + x)$ which is obviously continuous at 0; further, $F'$ has the c.f. $f'(t) = f(t)\, \mathrm{e}^{-itx}$ by Prop. 3.1.2. Therefore we obtain, for any $t_0 > 0$, using Fubini theorem

$$I(t_0) := \frac{2}{\pi} \int_{-\infty}^{\infty} \left[\int_{0}^{t_0} \frac{\sin ty}{t}\, \mathrm{d}t\right] \mathrm{d}F'(y) = \frac{2}{\pi} \int_{0}^{t_0} \int_{-\infty}^{\infty} \sin ty\, \mathrm{d}F'(y) \frac{\mathrm{d}t}{t}$$

$$= \frac{2}{\pi} \int_{0}^{t_0} \mathrm{Im}\, f'(t) \frac{\mathrm{d}t}{t} = \frac{2}{\pi} \int_{0}^{t_0} \mathrm{Im}\left(f(t)\, \mathrm{e}^{-itx}\right) \frac{\mathrm{d}t}{t}\,.$$

Now Prop. A6 tells us that, as $t_0 \to \infty$, the term in brackets tends to $\pm \pi/2$ according as $y > 0$ or $y < 0$. So we obtain by dominated convergence (with respect to the continuity of $F'$ at 0)

$$\lim_{t_0\to\infty} I(t_0) = -\int_{-\infty}^{0} dF'(y) + \int_{0}^{\infty} dF'(y) = 1 - 2F'(0) = 1 - 2F(x). \qquad \square$$

We now immediately get the following basic result to which the adjective "characteristic" is due.

**Corollary 3.2.4** (Uniqueness theorem). *Two d.f. $F_1$, $F_2$ coincide iff their c.f. $f_1$, $f_2$ coincide.*
Proof. Clearly, the only-if part follows from Def. 3.1.1. — On the other hand, if $f_1 = f_2$, then Th. 3.2.3 shows that $F_1(x) = F_2(x)$ for all $x$ which are continuity points both of $F_1$ and $F_2$. Now the assertion follows by left-continuity, see Prop. 1.1.5. $\square$

Assuming $f_1 = f_2$ we see from (3.2.2) that $F_1 * \Phi_{0,1/\sigma} = F_2 * \Phi_{0,1/\sigma}$ so that L. 3.2.1 yields a second proof. — A corollary is given in App. C.2.3; it is particularly natural in the framework of the continuation theory of d.f.

**Remark 3.2.5.** Let $G$ be some monotone, left-continuous, and bounded function on $R_1$ and $g(t) := \int_{-\infty}^{\infty} e^{itx}\, dG(x)$ its Fourier-Stieltjes transform. Then it is easily seen that $g$ determines $G$ up to an additive constant.

Indeed, we have $g(0) = G(\infty) - G(-\infty)$ and if $g(0) = 0$, then $G = $ const. On the other hand, if $g(0) \neq 0$, then we apply the uniqueness theorem to the d.f. $\dfrac{G(x) - G(-\infty)}{G(\infty) - G(-\infty)}$ whose c.f. is $\dfrac{g(t)}{g(0)}$. Thus the desired conclusion follows from an indirect argument. $\square$

Now we return to d.f. and prove two consequences of the previous results.

**Corollary 3.2.6.** *The d.f. $F$ is symmetric with respect to $a \in R_1$ iff $f(t)\, e^{-ita}$ is real.*
Proof. It obviously suffices to show the assertion for $a = 0$; otherwise we consider the d.f. $F(x + a)$ with the c.f. $f(t)\, e^{-ita}$.

If $F$ is symmetric, then $F = F_-$ and therefore $f = f_-$. Now Prop. 3.1.2 implies $f(t) = f(-t) = \overline{f(t)}$, i.e., $f$ is real. Conversely, let $f$ be real. Then $f = \overline{f} = f_-$, and the uniqueness theorem tells us that $F = F_-$ so that $F$ is symmetric. $\square$

**Corollary 3.2.7.** *Let $\pm x$ be continuity points of $F$. Then*

$$F(x) - F(-x) = \frac{2}{\pi} \int_0^{\infty} \sin xt \operatorname{Re} f(t) \frac{dt}{t},$$

$$F(x) + F(-x) = 1 - \frac{2}{\pi} \int_0^{\infty} \cos xt \operatorname{Im} f(t) \frac{dt}{t}.$$

The proof is immediate by virtue of Th. 3.2.3. $\square$

Finally, we wish to derive an expression also for the jumps of a d.f.

**Proposition 3.2.8.** *For all $x \in R_1$ we have the representation*

$$F(x + 0) - F(x) = \lim_{T\to\infty} \frac{1}{2T} \int_{-T}^{T} e^{-itx} f(t)\, dt.$$

Proof. From Fubini theorem we obtain, for fixed $T > 0$,

$$\frac{1}{2T}\int_{-T}^{T} e^{-itx} f(t)\, dt = \frac{1}{2T}\int_{-T}^{T} e^{-itx} \int_{-\infty}^{\infty} e^{ity}\, dF(y)\, dt =$$

$$= \frac{1}{2T}\int_{-\infty}^{\infty}\int_{-T}^{T} e^{it(y-x)}\, dt\, dF(y) = \int_{-\infty}^{\infty} \frac{\sin T(y-x)}{T(y-x)}\, dF(y) =$$

$$= F(x+0) - F(x) + \int_{|y-x|>0} \frac{\sin T(y-x)}{T(y-x)}\, dF(y)\,.$$

Since the integrand on the right converges to 0 as $T \to \infty$, the assertion follows by dominated convergence. □

## 3.3. Characteristic functions and compactness

Let us consider a non-empty set $\boldsymbol{F}$ of d.f. and let $\boldsymbol{f}$ be the corresponding set of c.f. The aim of the present section is to express the relative and the shift compactness of $\boldsymbol{F}$ in terms of $\boldsymbol{f}$; cf. the Def. 2.4.1 and 2.4.5, resp.

We first introduce a new notion which concerns an arbitrary set $\boldsymbol{g}$ of real or complex functions on $\mathbb{R}_1$.

**Definition 3.3.1.** The set $\boldsymbol{g}$ is said to be *equicontinuous at* $t_0$ if, for any $\varepsilon > 0$, there exists a number $\delta = \delta(\varepsilon, t_0)$ such that

$$|g(t_0 + h) - g(t_0)| < \varepsilon \tag{3.3.1}$$

for all $g \in \boldsymbol{g}$ and $|h| \leq \delta$. The set $\boldsymbol{g}$ is said to be *equicontinuous on* $\mathbb{R}_1$, if for any $\varepsilon > 0$, there exists $\delta(\varepsilon)$ such that (3.3.1) is satisfied for all $g \in \boldsymbol{g}$ and all $t_0 \in \mathbb{R}_1$, $|h| \leq \delta$. □

We illustrate this notion by an example.

**Example 3.3.2.** The set $\boldsymbol{f} = \left\{\exp\left(-\frac{\sigma^2 t^2}{2}\right) : \sigma > 0\right\}$ of normal c.f. is equicontinuous for every $t_0 \neq 0$; but for $t_0 = 0$ this property is violated. Note that each function $f \in \boldsymbol{f}$ is uniformly continuous on $\mathbb{R}_1$ (see Prop. 3.1.4). □

The importance of the new concept becomes clear from the next result.

**Proposition 3.3.3.** *The following conditions are equivalent:*

(i) $\boldsymbol{F}$ *is relatively compact.*

(ii) $\boldsymbol{f}$ *is equicontinuous at the origin* 0.

(iii) *The set* $\boldsymbol{u} := \{\operatorname{Re} f : f \in \boldsymbol{f}\}$ *is equicontinuous at the origin* 0.

Proof. (i) ⇒ (ii) If $\boldsymbol{F}$ is relatively compact, then it is also tight by Th. 2.4.4. Accordingly, for any $\varepsilon > 0$, there exists an $x_0(\varepsilon) > 0$ such that $2T_F(x_0) < \varepsilon/2$ for all $F \in \boldsymbol{F}$. Moreover, we have $|e^{iu} - 1| \leq |u|$ for all real $u$ (see App. B.4). Hence we get, for $|h| \leq \varepsilon/2x_0$ and any $f \in \boldsymbol{f}$,

$$|1 - f(h)| \leq \int_{|x|<x_0} |1 - e^{ihx}|\, dF(x) + \int_{|x|\geq x_0} |1 - e^{ihx}|\, dF(x)$$

$$\leq \int_{|x|<x_0} |xh|\, dF(x) + 2T_F(x_0) = |h|\, x_0 \int_{|x|<x_0} dF(x) + \varepsilon/2 < \varepsilon\,.$$

Thus condition (ii) is satisfied.

(ii) ⇒ (iii) is obvious since $0 \leqq 1 - \operatorname{Re} f(t) \leqq |1 - f(t)|$.

(iii) ⇒ (i) If $\boldsymbol{u}$ is equicontinuous at 0, then we use the truncation inequality (L. 3.1.5)

$$T_F\left(\frac{1}{x}\right) \leqq \frac{7}{x} \int_0^x (1 - \operatorname{Re} f(t))\, dt\,, \qquad x > 0\,.$$

By assumption we have $(1 - \operatorname{Re} f(t)) < \varepsilon/7$ for $|t| < \delta(\varepsilon)$ and all $F \in \boldsymbol{F}$ whence $T_F(1/x) < \varepsilon$ follows for $|x| < \delta$. Accordingly, $\boldsymbol{F}$ is tight, that is relatively compact. □

The equivalence of (ii) and (iii) can also be proved by a consideration in terms of the corresponding d.f. provided that the equivalence of (i) and (ii) is guaranteed. Indeed, by Ex. 3.1.14, $u = \operatorname{Re} f$ is the c.f. of $G = \frac{1}{2}(F + F_-)$. Moreover, we see that $T_F = T_{F_-} = T_G$. Hence tightness of $\boldsymbol{F}$ is the same as tightness of $\boldsymbol{G}$. By assumption the first is equivalent to the equicontinuity of $\boldsymbol{f}$ at 0, while the second is equivalent to the same property of $\boldsymbol{u}$.

**Remark 3.3.4.** If $\boldsymbol{f}$ is equicontinuous at the origin, then $\boldsymbol{f}$ is equicontinuous on $\mathbb{R}_1$; the same conclusion is true for $\boldsymbol{u}$. These facts are easily seen from the inequalities

$$[\operatorname{Re}(f(t+h) - f(t))]^2 \leqq |f(t+h) - f(t)|^2 \leqq 2(1 - \operatorname{Re} f(h)) \leqq 2\,|1 - f(h)|$$

which hold for all $t, h \in \mathbb{R}_1$ by L. 3.1.3. □

We also need a criterion which holds for sequences of d.f. $(F_n)$; for the corresponding c.f. we introduce the function

$$\psi(t) := \varliminf_{n\to\infty} \operatorname{Re} f_n(t)\,, \qquad t \in \mathbb{R}_1 \tag{3.3.2}$$

**Proposition 3.3.5.** The continuity of $\psi$ at 0 is equivalent to the relative compactness of $(F_n)$.

Proof. (i) Let $\psi$ be continuous at 0 and $x > 0$. By the truncation inequality (L. 3.1.5) we may write

$$\varlimsup_{n\to\infty} T_{F_n}\left(\frac{1}{x}\right) \leqq 7\left(1 - \frac{1}{x} \varliminf_{n\to\infty} \int_0^x \operatorname{Re} f_n(t)\, dt\right). \tag{3.3.3}$$

Fatou's lemma (Prop. A.1) and the continuity of $\psi$ at 0 yield now estimates

$$\leqq \frac{7}{x} \int_0^x (1 - \psi(t))\, dt < \varepsilon\,, \qquad 0 < x \leqq x_0(\varepsilon)\,.$$

Thus the set $\{n : T_{F_n}(x_0^{-1}) \geqq 2\varepsilon\}$ is empty or at most finite whence it follows that there exists $x_1 \geqq x_0^{-1}$ such that $T_{F_n}(x_1) < 2\varepsilon$ $n \geqq 1$. Accordingly, $(F_n)$ is tight, i.e., also relatively compact.

(ii) Let $(F_n)$ be relatively compact. Then we write similarly as in the proof of Prop. 3.3.3, (i) ⇒ (ii),

$$\psi(0) - \psi(h) = \varlimsup_{n\to\infty} (1 - \operatorname{Re} f_n(t)) \leqq \varlimsup_{n\to\infty} \int_{|x| \leqq x_0} |xh|\, dF_n + \varlimsup_{n\to\infty} 2T_{F_n}(x_0)$$

so that the continuity of $\psi$ at 0 directly follows from the assumption. □

**Proposition 3.3.6.** *The set of d.f. $\boldsymbol{F}$ is shift compact iff the set of real functions $\{|f| : f \in \boldsymbol{f}\}$ is equicontinuous at 0.*

Proof. By Prop. 2.4.11, $\boldsymbol{F}$ is shift compact iff the corresponding set $\boldsymbol{F}^s$ of symmetrizized d.f. $F^s = F * F_-$ is relatively compact. From Prop. 3.3.3 we conclude that $\boldsymbol{F}^s$ is relatively compact iff $\boldsymbol{f}^s$ is equicontinuous at 0. But in view of $f^s = |f|^2$, the latter fact is true iff $\{|f| : f \in \boldsymbol{f}\}$ is equicontinuous at 0. □

**Example 3.3.7.** Let $F$ be a d.f. Then $\boldsymbol{F} = \{F^{*n} : n = 1, 2, \ldots\}$ is shift compact iff $F$ is degenerate. Indeed, if $F$ is degenerate, then the shift compactness of $\boldsymbol{F}$ is trivial. — Conversely, if $F$ is non-degenerate, then Prop. 3.1.9 implies the estimate $|f(t)| \leqq 1 - \eta t^2$, $|t| \leqq \delta$, for some $\delta, \eta > 0$, and, as a consequence,

$$\psi^s(t) := \lim_{n\to\infty} f_n^s(t) = \lim_{n\to\infty} |f(t)|^{2n} = \begin{cases} 0 & \text{for} \quad 0 < |t| \leqq \delta\,, \\ 1 & \text{for} \quad t = 0\,. \end{cases}$$

Therefore, Prop. 3.3.5 tells us that $\boldsymbol{F}^s$ is not relatively compact. Accordingly, we see from Prop. 2.4.11 that $\boldsymbol{F}$ is not shift compact. For sharper asymptotic statements for convolution powers we refer the reader to Sect. 5.4. □

**Remark 3.3.8.** We can now reconsider the subject of Prop. 2.4.8 where we treated a set $\boldsymbol{F}$ of d.f. $F$ with finite variances $\sigma_F^2$. Now we recognize in another way that the condition $\sup_{F\in\boldsymbol{F}} \sigma_F^2 = C$ implies the shift compactness of $\boldsymbol{F}$. Namely, we see from Prop. 3.1.11 that

$$\inf_{F\in\boldsymbol{F}} |f(t)|^2 \geqq 1 - Ct^2\,, \qquad |t| < C^{-1}\,.$$

Thus it follows that we have, for all $f \in \boldsymbol{F}$,

$$1 - |f(h)| \leqq 1 - |f(h)|^2 \leqq Ch^2\,, \qquad |h| < C^{-1};$$

hence the set $\{|f| : f \in \boldsymbol{f}\}$ is equicontinuous at 0 and the desired conclusion follows from Prop. 3.3.7. □

We finish this section with a statement which we need in the proof of Cor. 3.4.4. In this way we avoid disrupting our probabilistic reasoning in Sect. 3.4.

**Lemma 3.3.9.** *Let $\boldsymbol{g}$ be equicontinuous on $\mathbb{R}_1$ and assume that there exists a sequence $(g_n) \subseteqq \boldsymbol{g}$ which satisfies $g_n \to g$ pointwise on $\mathbb{R}_1$. Then $g$ is continuous everywhere, and the convergence is uniform in every finite interval.*

Proof. Let $\varepsilon > 0$ be given. We decompose an arbitrary interval $[x_1, x_2]$ by $x_1 = y_1 < y_2 < \cdots < y_N = x_2$. In view of the assumed equicontinuity it is possible to carry this out such that

$$\max_{1\leqq k\leqq N-1} (y_{k+1} - y_k) \leqq \delta = \delta(\varepsilon)\,,$$

and for all $n$

$$|g_n(x) - g_n(y_k)| < \varepsilon \quad \text{for} \quad |x - y_k| \leqq \delta\,, \qquad k = 1, 2, \ldots, N\,. \tag{3.3.4}$$

By assumption the Cauchy condition holds at $y_1, \ldots, y_N$ so that

$$|g_n(y_k) - g_m(y_k)| < \varepsilon \quad \text{for} \quad n, m \geqq n_0(\varepsilon) \qquad k = 1, 2, \ldots, N\,. \tag{3.3.5}$$

Now let $x \in [x_1, x_2]$ be arbitrary and select $y_j$ such that $|x - y_j| \leqq \delta$. Then we immediately obtain from (3.3.4) and (3.3.5), for $n, m \geqq n_0$,

$$|g_n(x) - g_m(x)| \leqq |g_n(x) - g_n(y_j)| + |g_n(y_j) - g_m(y_j)| + |g_m(y_j) - g_m(x)| < 3\varepsilon\,.$$

Thus we get, as $n \to \infty$, $|g(x) - g_m(x)| \leqq 3\varepsilon$ for $m \geqq n_0$ and all $x \in [x_1, x_2]$. Accordingly, the supposed convergence is uniform on $[x_1, x_2]$. — Finally, using this uniformity we see that $g$ is continuous since the $g_m$ are continuous. □

## 3.4. The continuity theorem

We are now going to prove that the one-to-one correspondence between d.f. and c.f. is continuous in a certain sense. From now on $(F_n)$ stands for a sequence of d.f. with the sequence $(f_n)$ of corresponding c.f.; further, $f_n \to f$ denotes pointwise convergence on $\mathbb{R}_1$.

**Theorem 3.4.1** (Continuity theorem; Lévy). *Let $F$ be a d.f. and $f$ its c.f. Then we have $F_n \xrightarrow{c} F$ iff $f_n \to f$.*

Proof. If $F_n \xrightarrow{c} F$, then the extended Helly-Bray Theorem 2.2.4 can be applied so that we obtain $f_n \to f$.

Conversely, let us assume $f_n \to f$. Since $f$ is continuous at 0, we see that the function $\psi = \lim_{n\to\infty} \operatorname{Re} f_n = \operatorname{Re} f$ is also continuous at 0; hence $(F_n)$ is relatively compact by Prop. 3.3.5.

According to Helly's selection theorem we can find a subsequence $(F_{n'})$ of $(F_n)$ such that $F_{n'} \xrightarrow{w} G$ where $G$ is non-decreasing, left-continuous, and satisfies the inequalities $0 \leqq G \leqq 1$ on $\mathbb{R}_1$. Then the relative compactness of $(F_n)$ implies that $G$ is a d.f., and we have $F_{n'} \xrightarrow{c} G$. By the only-if part proved above, we see that $f_{n'} \to g$ where $g$ is the c.f. of $G$. Since also $f_n \to f$ we obtain $g = f$, and the uniqueness theorem tells us that $G = F$. Therefore, all subsequences $(F_{n'})$ weakly converging on the line tend to $F$ so that $F_n \xrightarrow{c} F$ follows (by Cor. 2.2.3 for $I = \mathbb{R}_1$). □

The assumption of the if-part of Th. 3.4.1 can be considerably weakened; namely, repeating the proof nearly literally we can even obtain the following result.

**Corollary 3.4.2.** *Let $\hat{f}$ be an arbitrary complex function on $\mathbb{R}_1$ which is continuous at 0. If $f_n \to \hat{f}$, then $\hat{f} = f$ is a c.f., and for the corresponding d.f. $F$ we have $F_n \xrightarrow{c} F$.* □

In the following reformulation the compactness condition is expressed separately, but the proof remains essentially the same.

**Corollary 3.4.3.** *Let $(F_n)$ be relatively compact. If $f_n \to \hat{f}$ for an arbitrary function $\hat{f}$ then $\hat{f} = f$ is a c.f. and for the corresponding d.f. $F$ we have $F_n \xrightarrow{c} F$.* □

**Corollary 3.4.4.** *If $f_n \to f$ for a c.f. $f$, then the convergence is uniform in every finite interval.*

Proof. Th. 3.4.1 permits us to conclude from the assumption $f_n \to f$ that $(F_n)$ is relatively compact. Now Prop. 3.3.3 and Rem. 3.3.4 tell us that $(f_n)$ is equicontinuous on $\mathbb{R}_1$. Accordingly, the assertion follows from the general L. 3.3.9. □

In the if-part of the continuity theorem we have assumed the convergence of $(f_n)$ to a c.f. $f$ for all $t \in \mathbb{R}_1$. This had to be done since the restriction $f_I$ of a c.f. $f$ to a finite interval $I = (-t_0, t_0)$ does not define $f$ uniquely in general. Examples are provided by the Pólyatype c.f. Prop. 3.6.5.

On the other hand, in Chap. 4 it will turn out that there do exist large classes of c.f. for which the continuation of $f_I$ to the line is unique in the set of c.f. Therefore, it is worth-while to state the following result the proof of which runs in analogy to that of the if-part of the continuity theorem.

**Proposition 3.4.5.** *Let $f$ be a c.f. which is uniquely defined by its restriction $f_I$ to some interval $I = (-t_0, t_0)$. If $f_n(t) \to f(t)$, $t \in I$, then $F_n \xrightarrow{c} F$ and $f_n \to f$.* □

Thus we have found a further case in which the convergence of a sequence of functions assumed on a finite interval continues to the line; we already met two such cases, see Rem. 2.5.4

The results of the present section obtained so far exclusively concern complete convergence. But later it will sometimes occur that we have weak convergence, which is possibly not complete, so that the limit function may be a defective d.f. This is the reason why we now present a *weak convergence criterion* which permits us to treat this case by means of c.f.

Let $G$ be a possibly defective d.f. and $g(t) = \int_{-\infty}^{\infty} e^{itx}\, dG(x)$ its Fourier-Stieltjes transform. Then we introduce the function

$$g^*(u) := \int_0^u g(t)\, dt = \int_{-\infty}^{\infty} \frac{e^{iux}-1}{ix}\, dG(x) \tag{3.4.1}$$

which we call the *integral characteristic* of $G$. Note that the second equality sign in (3.4.1) is obtained upon replacing $g$ by its defining integral and interchanging the integrations, which is permissible in view of Fubini's theorem.

Clearly, since there is a one-to-one correspondence between $g^*$ and its continuous derivative $g$ it follows that there is a one-to-one correspondence between $g^*$ and $G$ up to an additive constant, c.f. Rem. 3.2.5. With this in mind, we introduce a generalized concept of weak convergence; here and in the following, $(G_n)$ stands for a sequence of non-decreasing, left-continuous, and uniformly bounded functions on $\mathbb{R}_1$.

**Definition 3.4.6.** We say that $(G_n)$ *weakly converges to a monotone function $G$ up to an additive constant* if there is a sequence $(c_n)$ of real numbers such that

$$G_n + c_n \xrightarrow{w} G .$$

We express this by writing $G_n \xrightarrow{w*} G$. □

The limit function $G$ is obviously unique but for an additive constant. Hence, without loss of generality, we may put $G(-\infty) = 0$.

Now the weak convergence criterion runs as follows.

**Proposition 3.4.7.** *Let $(g_n^*)$ be the sequence of integral characteristics corresponding to $(G_n)$. Then we have $G_n \xrightarrow{w*} G$ iff $g_n^* \to h$ for a certain function $h$ on $\mathbb{R}_1$. In either case $h = g^*$ is the integral characteristic of $G$.*

Proof. *Only-if part*: Assume $G_n + c_n \xrightarrow{w} G$ for some monotone function $G$. Since $(e^{iux} - 1)/ix \to 0$, as $|x| \to \infty$, we can apply Cor. 2.2.6 to the extended Helly-Bray theorem and obtain

$$g_n^*(u) = \int_{-\infty}^{\infty} \frac{e^{iux}-1}{ix}\, d\big(G_n(x) + c_n\big) \to g^*(u)\,, \qquad u \in \mathbb{R}_1\,,$$

which was to be proved.

*If-part*: Let $g_n^* \to h$ for a certain function $h$. Choose an arbitrary subsequence $(G_{n'})$ of $(G_n)$ which satisfies $G_{n'} \xrightarrow{w} G$, say. Then the only-if part of the proof implies $g_{n'} \to g^*$ so that by assumption $h = g^*$. Now we select a continuity point $x_0$ of $G$ and define the sequence $(c_n)$ by

$$G_n(x_0) + c_n = G(x_0)\,, \qquad n \geqq 1\,, \tag{3.4.2}$$

and our proof will be terminated by proving $G_n + c_n \xrightarrow{w} G$.

To this end we select an arbitrary subsequence $(G_{n''})$ of $(G_n)$ such that $G_{n''} + c_{n''} \xrightarrow{w} G_0$, say. Using the only-if part once again we see that $g^*_{n''} \to g^*_0$ whence $g^*_0 = h = g^*$ follows. Since $g^*_0$ determines $G_0$ up to an additive constant we get $G_0 = G + c$ for a certain $c$. Therefore $x_0$ is also a continuity point of $G_0$ and, by virtue of (3.4.2), we obtain

$$G(x_0) = \lim_{n'' \to \infty} \left(G_{n''}(x_0) + c_{n''}\right) = G_0(x_0) = G(x_0) + c \,,$$

i.e. $c = 0$. So every convergent subsequence of $(G_n + c_n)$ has the same limit $G$, hence the desired relation follows from Cor. 2.2.3. □

**Example 3.4.8.** We put $G_{2k} = U_0 + U_k$ and $G_{2k+1} = U_0 + U_{-k}$, $k = 1, 2, \ldots$, where $U_j$ stands for the degenerate d.f. with atom $j$. Then we have, as $k \to \infty$, $G_{2k} \xrightarrow{w} U_0$ but also $G_{2k+1} - 1 \xrightarrow{w} U_0$ so that

$$G_n \xrightarrow{w^*} U_0 \,.$$

The Fourier-Stieltjes transforms corresponding to $(G_n)$ are $g_{2k}(t) = 1 + e^{itk}$ and $g_{2k+1}(t) = 1 + e^{-itk}$, resp., and we see that

$$g^*_n(t) = \int_0^u g_n(t)\, dt \to u \,, \qquad u \in \mathbb{R}_1 ;$$

this is indeed the integral characteristic of $U_0$. □

## 3.5. Characteristic functions and moments

There is a close connection between c.f. and the moments of the corresponding d.f. which we have introduced in Sect. 1.3. Note that the moment $\mu_k$ exists iff the absolute moment $M_k$ exists.

**Proposition 3.5.1.** *If the moment $\mu_k$ exists for some $k \geqq 1$, then the k-th derivative $f^{(k)}$ of $f$ exists and is a uniformly continuous function given by*

$$f^{(k)}(t) = i^k \int_{-\infty}^{\infty} e^{itx} x^k \, dF(x) \,. \tag{3.5.1}$$

*Moreover, we have $f^{(k)}(0) = i^k \mu_k$.*

Proof. The difference ratio of $f$ is obviously

$$\frac{f(t+h) - f(t)}{h} = \int_{-\infty}^{\infty} e^{itx} \frac{e^{ihx} - 1}{hx} x \, dF(x) \,.$$

Prop. B.4.1 tells us that the integrand is dominated by $|x|$, hence for $k = 1$ the existence of $f^{(1)}$ and formula (3.5.1) follow by dominated convergence.

The general case can be obtained by induction using

$$\frac{f^{(k)}(t+h) - f^{(k)}(t)}{h} = i^k \int_{-\infty}^{\infty} e^{itx} \frac{e^{ihx} - 1}{hx} x^{k+1} \, dF(x) \,.$$

Finally, we treat the continuity of $f^{(k)}$ assuming the existence of $\mu_k$. We have, in view of (3.5.1),

$$|f^{(k)}(t+h) - f^{(k)}(t)| = \left| \int_{-\infty}^{\infty} e^{itx} (e^{ihx} - 1)\, x^k \, dF(x) \right| \leqq \int_{-\infty}^{\infty} |e^{ihx} - 1|\, |x^k| \, dF(x) .$$

This bound does not depend on $t$ and tends to zero as $h \to 0$ by dominated convergence whence uniform continuity trivially follows.

The very last assertion is obvious from (3.5.1). □

Now we turn to the Taylor expansion of a c.f. $f$ about the origin 0.

**Corollary 3.5.2.** *If $\mu_n$ exists for some $n \geqq 1$ then*

$$f(t) = 1 + \sum_{k=1}^{n} \mu_k \frac{(it)^k}{k!} + o(t^n) , \qquad t \to 0 .$$

Proof. By virtue of Prop. 3.5.1, $f^{(n)}$ exists and is continuous. In a neighbourhood of $t = 0$ we develop $\operatorname{Re} f$ and $\operatorname{Im} f$ into a Taylor expansion so that

$$f(t) = \operatorname{Re} f(t) + i \operatorname{Im} f(t) = 1 + \sum_{k=1}^{n} f^{(k)}(0) \frac{t^k}{k!} + R_n(t) ;$$

using Lagrange's form of the remainder term we obtain for $0 < \theta, \theta' < 1$

$$R_n(t) = \frac{t^n}{n!} \left( [(\operatorname{Re} f)^{(n)} (\theta t) - (\operatorname{Re} f)^n (0)] + i[(\operatorname{Im} f)^{(n)} (\theta' t) - \operatorname{Im} f)^{(n)} (0)] \right) .$$

Since $f^{(n)}$ is continuous we see that $R_n(t) = o(t^n)$ as $t \to 0$. The last assertion of Prop. 3.5.1 yields now the desired formula. □

Unfortunately Prop. 3.5.1 cannot be converted; this is shown by Zygmund's example, see Kawata (1972). It is only for moments of even orders that we give a result with necessary and sufficient conditions. (For odd orders see Suppl. 3.8.13.)

Before formulating this proposition we note that for $\mu_{2k} < \infty$ the expression

$$f_{2k} := (-1)^k f^{(2k)} / \mu_{2k} , \qquad k \geqq 0 , \tag{3.5.2}$$

(where $\mu_0 = 1$) is the c.f. corresponding to the d.f.

$$F_{2k}(x) := \frac{1}{\mu_{2k}} \int_{-\infty}^{x} y^{2k} \, dF(y) .$$

We obviously have $\mu_{2k+2} < \infty$ iff $F_{2k}$ has the moment of order 2.

**Proposition 3.5.3.** *The d.f. $F$ possesses a finite moment $\mu_{2k+2}$, $k \geqq 0$, iff $\mu_{2k}$ exists and*

$$\frac{1 - \operatorname{Re} f_{2k}(t)}{t^2} < C , \qquad t \neq 0 , \tag{3.5.3}$$

*for some constant $C > 0$.*

Proof. Let us first suppose that $\mu_{2k+2} < \infty$ so that the second moment of $F_{2k}$ exists. Accordingly, we can apply Cor. 3.5.2 (for $n = 2$) to the c.f. $f_{2k}$; thus we see that (3.5.3) holds. This proves the only-if part of the assertion.

Conversely, let us assume that $\mu_{2k}$ exists for some $k \geqq 0$ and (3.5.3) is true. Then the d.f. $F_{2k}$ is defined and

$$C > \frac{1 - \operatorname{Re} f_{2k}(t)}{t^2} = \int_{-\infty}^{\infty} \frac{1 - \cos tx}{t^2 x^2} x^2 \, dF_{2k}(x)$$

$$\geqq \int_{-\delta}^{\delta} \frac{1 - \cos tx}{t^2 x^2} x^2 \, dF_{2k}(x) \to \frac{1}{2} \int_{-\delta}^{\delta} x^2 \, dF_{2k}(x)$$

as $t \to 0$. Since $\delta > 0$ is arbitrary we see that $F_{2k}$ has a second moment so that $\mu_{2k+2}$ exists. □

Since $\mu_0 = 1$ always exists we have a noteworthy particular case: the condition (3.5.3), $k = 0$, that is

$$\frac{1 - \operatorname{Re} f(t)}{t^2} < C, \qquad t \neq 0, \tag{3.5.4}$$

is equivalent to $\mu_2 < \infty$.

**Corollary 3.5.4.** *The d.f. $F$ possesses a finite moment $\mu_{2r}$ ($r \geqq 1$) iff the $2r$-th derivative of $\operatorname{Re} f$ exists at the origin.*

Proof. In view of Prop. 3.5.1 we have to prove only the if-part of the statement. For this purpose we consider first the case $r = 1$ and set $u = \operatorname{Re} f$ ($= \operatorname{Re} f_0$). The supposition "$u''(0)$ exists" implies that the first derivative $u'$ exists near the origin; in particular, $u'(0) = 0$ since $u$ is even. Thus by the mean value theorem there exists a $\theta$ such that $0 < \theta < 1$ and

$$0 \leqq \frac{1 - \operatorname{Re} f(t)}{t^2} = \frac{u'(\theta t)}{t} < \frac{u'(\theta t) - u'(0)}{\theta t} \to u''(0)$$

as $t \to 0$. Hence (3.5.4) holds so that $\mu_2 < \infty$ follows.

The general case is obtained by induction assuming that the assertion is true for a certain $r \geqq 1$. Let us suppose that the derivative $u^{(2r+2)}(0)$ exists. Then $u^{(2r)}(0)$ also exists and by assumption it follows that $\mu_{2r} < \infty$. Thus $F_{2r}$ is defined and from (3.5.2) it is seen that $u_{2r} := \operatorname{Re} f_{2r}$ has a second derivative at 0. Thus our above argument shows that $F_{2r}$ has a second moment whence the desired conclusion $\mu_{2r+2} < \infty$ follows. □

For one-sided d.f. we have a better result; see Prop. 4.2.9.

## 3.6. Characteristic functions of lattice and absolutely continuous distribution functions

D.f. occurring in applications are in the overwhelming majority of cases either lattice or absolutely continuous d.f. or mixtures of these two. This is the reason why we now briefly consider properties of the corresponding c.f. $f$.

**Proposition 3.6.1.** *Let $f$ be the c.f. of the d.f. $F$. For $h > 0$ the following statements are equivalent:*

(i) $f(h) = 1$;

(ii) *$f$ is periodic with period $h$, i.e. $f(t) = f(t + h)$ for all $t \in \mathbb{R}_1$;*

(iii) *All points of increase of $F$ are contained in the lattice $\{kb : k = 0, \pm 1, \ldots\}$ with span $b = 2\pi/h$.*

Proof. Assume that (i) is true. Then on one hand we get (ii) by L. 3.1.3. On the other hand, we have

$$0 = 1 - \operatorname{Re} f(h) = \int_{-\infty}^{\infty} (1 - \cos hx)\, dF(x).$$

The integrand vanishes at all points $x_k = 2\pi k/h$, $k = 0, \pm 1, \ldots$, but in every interval $(x_k, x_{k+1})$ it is positive. Thus all these intervals do not contain any point of increase of $F$ whence (iii) follows.

Assume now that (iii) holds so that $F$ attributes mass $p_k$, say, to $kb$. Then

$$f(t) = \sum_k e^{ikbt} p_k$$

so that (ii) follows; this in turn yields (i). □

Next we treat a more general case.

**Proposition 3.6.2.** *We have $|f(h)| = 1$ for a certain $h > 0$ iff there exists $a \in R_1$ such that $F$ is concentrated on the lattice $\{a + kb\colon k = 0, \pm 1, \ldots\}$ where $b = 2\pi/h$.*

Proof. *Only-if part*: By assumption there is a real number a such that $f(h) = e^{iah}$. We also note that the d.f. $F'(x) = F(x + a)$ has the c.f.

$$f'(t) = f(t)\, e^{-iat} \tag{3.6.1}$$

and that it satisfies $f'(h) = 1$. Now we see from Prop. 3.6.1 that $F'$ is concentrated on $\{kb\}$ where $b = 2\pi/h$; this is equivalent to the assertion on $F$.

*If-part*: By assumption the d.f. $F'$ is concentrated on $\{kb\}$. Then Prop. 3.6.1 tells us that $f'(h) = 1$ for $h = 2\pi/b$. Now $|f(h)| = 1$ follows from (3.6.1). □

Let us now have a look at the c.f. corresponding to absolutely continuous d.f. Basic to the treatment of this case is the well-known Riemann-Lebesgue lemma as stated in Prop. A.5.

**Proposition 3.6.3.** *Assume that the d.f. $F$ has density $p$ and c.f. $f$. Then $f(t) \to 0$ as $|t| \to \infty$. If, moreover, $p$ has integrable derivatives $p^{(n)}$, then $f(t) = o(t^{-n})$ as $|t| \to \infty$.*

Proof. By assumption $f(t) = \int_{-\infty}^{\infty} e^{itx} p(x)\, dx$ and in view of Prop. A.5 we have to prove only the second part of the proposition. If $p'$ exists and is integrable, we may integrate by parts and obtain

$$f(t) = \frac{1}{it} \lim_{\substack{x_1 \to -\infty \\ x_2 \to \infty}} \left[ p(x_2)\, e^{itx_2} - p(x_1)\, e^{itx_1} - \int_{x_1}^{x_2} e^{itx} p'(x)\, dx \right].$$

By hypothesis the integral on the right has a limit as $x_2 \to \infty$. Hence $p(x_2)\, e^{itx_2} \to p_\infty$, say, whence $p(x_2) \to |p_\infty|$ follows. These two relations contain a contradiction for $p_\infty \neq 0$. — On letting $x_1 \to -\infty$ in a similar way, we obtain

$$f(t) = -\frac{1}{it} \int_{-\infty}^{\infty} e^{itx} p'(x)\, dx.$$

Now Prop. A.5 yields $tf(t) = o(1)$ as $|t| \to \infty$ whence the assertion follows for $n = 1$. The proof is completed by induction. □

**Corollary 3.6.4.** *Let the c.f. $f$ of the d.f. $F$ be real and non-negative. Then $F$ has a bounded density $p$ iff $f$ is integrable on $R_1$.*

Proof. The if-part is clear by Th. 3.2.2. To prove the only-if part we start with the assumption $p \leqq C$ and put $u = 0$ in (3.2.2) so that

$$\frac{1}{2\pi} \int_{-\infty}^{\infty} f(x)\, e^{-\frac{1}{2}\left(\frac{x}{\sigma}\right)^2} dt = \frac{\sigma}{\sqrt{2\pi}} \int_{-\infty}^{\infty} e^{-\frac{\sigma^2 y^2}{2}} p(y)\, dy \leqq C\,.$$

Now by monotone convergence the desired result follows as $\sigma \to \infty$. □

Next we derive a useful criterion for a function $f$ to be a c.f. We recall that a real function $u$ defined on an open interval $I$ is said to be *convex* if all its cords lie above or on the graph of $u$ so that $(1 - s)\, u(x_1) + su(x_2) \geqq u\big((1 - s)\, x_1 + sx_2\big)$ for $x_1, x_2 \in I$ and $0 \leqq s \leqq 1$.

**Proposition 3.6.5.** *Let $f$ be a real, non-negative, and even function which satisfies the following conditions:*
(i) *$f$ is convex on* $(0, \infty)$;
(ii) *$f$ is continuous at* $0$, $f(0) = 1$;
(iii) *$f$ is bounded.*

*Then $f$ is a c.f.*

Proof. We carry the proof in several steps.

(a) First we show indirectly that $f$ is non-increasing on $(0, \infty)$. Indeed, if $f(t_1) < f(t_2)$ for certain numbers $0 \leqq t_1 < t_2$, then there exists $\eta > 0$ such that $f(t_2) \geqq f(t_1) + \eta$. Consider now the points $t_k = t_2 + k(t_2 - t_1)$, $k \geqq 3$. Then the convexity of $f$ tells us that

$$f(t_1) + \eta \leqq f(t_2) = f\left(\frac{t_k + kt_1}{k+1}\right) \leqq \frac{1}{k+1} f(t_k) + \frac{k}{k+1} f(t_1)$$

so that $f(t_k) \geqq f(t_1) + (k+1)\,\eta \to \infty$ as $k \to \infty$. But this contradicts (iii) whence the assertion follows.

As a matter of fact, the monotonicity of $f$ shows that $f(\infty)$ exists where $0 \leqq f(\infty) \leqq 1$. Further, if $f(\infty) = 1$, then from $f(0) = 1$ it follows that $f = 1$ so that $f$ is the c.f. of the degenerated d.f. $U_0$. Therefore, from now on we suppose $0 \leqq f(\infty) < 1$.

(b) In addition to the conditions of the proposition, we suppose now that there exists $t_0$, $0 < t_0 < \infty$, such that $f > 0$ on $(0, t_0)$ and $f = 0$ on $(t_0, \infty)$.

($b_1$) If $f$ is linear on $(0, t_0)$, then obviously $f(t) = g(t/t_0)$ where

$$g(t) = \begin{cases} 1 - |t|\,, & |t| \leqq 1\,, \\ 0, & |t| > 1\,. \end{cases}$$

This is a c.f., cf. the table at the end of this book, hence $f$ is also a c.f.

($b_2$) If $f$ is linear both on $(0, t_1)$ and $(t_1, t_0)$, $0 < t_1 < t_0$, then we can easily check that there exist numbers $p_0$, $p_1$ such that

$$f(t) = p_0 g\left(\frac{t}{t_0}\right) + p_1 g\left(\frac{t}{t_1}\right) = \begin{cases} p_0\left(1 - \dfrac{t}{t_0}\right) + p_1\left(1 - \dfrac{t}{t_1}\right), & 0 \leqq t < t_1\,, \\ p_0\left(1 - \dfrac{t}{t_0}\right), & t_1 \leqq t < t_0; \end{cases}$$

putting $t = t_1$, we get

$$p_0 = f(t_1)\,(1 - t_1/t_0)^{-1} < 0\,.$$

Moreover, the convexity of $f$ tells us that $f(t) \leqq 1 - t/t_0$, $0 \leqq t < t_0$, whence $p_0 \leqq 1$ follows. Further, for $t = 0$ we obtain $p_0 + p_1 = 1$ which implies that $0 \leqq p_1 < 1$. So it comes out that $f$ is the mixture of two c.f.; accordingly, it is itself a c.f.

($b_3$) Proceeding in this manner we can conclude that every convex function which is piecewise linear on $(0, t_0)$ is a c.f.

($b_4$) If $f$ is arbitrary on $(0, t_0)$, then we can approximate its graph by a sequence of inscribed polygons which are obviously convex. Thus we are able to choose a sequence $(f_k)$ of convex functions $f_k$ which are piecewise linear on $(0, t_0)$ such that $f_k \to f$. By virtue of ($b_3$), the $f_k$ are c.f., and since $f$ is continuous at 0 (by assumption (ii)) the Cor. 3.4.2 to the continuity theorem shows that $f$ is a c.f.

(c) To treat the general case we define functions $f_n$ by means of

$$f_n(t) = \begin{cases} \dfrac{f(t) - f(n)}{1 - f(n)}, & 0 \leqq t < n, \\ 0, & t \geqq n, \end{cases}$$

$$f_n(t) = f_n(-t), \qquad t \leqq 0.$$

It is easily seen that the $f_n$ satisfy all conditions of the proposition and, moreover, the additional supposition made in part (b); therefore they are c.f. Further, for fixed $t$ we have

$$f_n(t) \to \frac{f(t) - f(\infty)}{1 - f(\infty)} = g(t),$$

say. Since $g$ is continuous at 0, it is a c.f. by Cor. 3.4.2 to the continuity theorem. Hence

$$f = \bigl(1 - f(\infty)\bigr)\, g + f(\infty)$$

is also a c.f.; namely, it belongs to the d.f.

$$F = \bigl(1 - f(\infty)\bigr)\, G + f(\infty)\, U_0 \tag{3.6.2}$$

where $G$ corresponds $g$. □

Functions which satisfy the above criterion will be called *Pólyatype* c.f.

Formula (3.6.2) gives an insight into the structure of a d.f. $F$ belonging to a Pólya type c.f. It is easy to see that $g$ is also a Pólyatype c.f. which satisfies the additional condition $g(\infty) = 0$.

The next statement shows that $G$ is always absolutely continuous.

**Proposition 3.6.6** (PÓLYA). *Let $g$ be a Pólyatype c.f. such that $g(\infty) = 0$. Then the corresponding d.f. $G$ is absolutely continuous and its density $p$ is given by*

$$p(u) = \frac{1}{2\pi} \int_{-\infty}^{\infty} e^{-itu}\, g(t)\, dt, \qquad u \neq 0.$$ □

For a general proof we refer the reader to LUKACS (1970), Sect. 4.3, or LAHA and ROHATGI (1979), Sect. 3.6; note that the assertion follows trivially from the Fourier inversion theorem if $g$ is integrable on $R_1$.

**Example 3.6.7.** The functions

$$f_1(t) = \frac{1}{1 + t}, \qquad t \in R_1,$$

$$f_2(t) = e^{-|t|^\alpha}, \qquad 0 < \alpha \leqq 1, \qquad t \in R_1$$

are obviously Pólyatype c.f. which satisfy $f_k(\infty) = 0$, $k = 1, 2$. Accordingly, they are c.f. of absolutely continuous d.f. $F_k$ the densities $p_k$ of which can be expressed by means of the Fourier inversion formula. It is noteworthy that $f_1$ is not integrable on $\mathbb{R}_1$ so that $p_1(u)$ is unbounded as $|u| \to 0$, cf. Cor. 3.6.4. □

## 3.7. Applications to sums of independent random variables

We give three well-known applications of the results of the present chapter. They yield us particularly simple cases of laws belonging to the very core of probability theory.

**Theorem 3.7.1** (Hinčin's weak law of large numbers). *Let $(X_k)$ be a sequence of i.i.d.r.v. and put $S_n = X_1 + \dots + X_n$. If $\mathsf{E}X_1 = \mu$ exists, then*

$$\lim_{n\to\infty} \mathsf{P}\left(\left|\frac{S_n}{n} - \mu\right| \geqq \varepsilon\right) = 0 \quad \textit{for all } \varepsilon > 0 .$$

Remark. First one word is in order on the interpretation of the notion of probability in applications. Here we have the following principle. Let $(\Omega, \mathcal{A}, \mathsf{P})$ be a probability space which describes a certain experiment; assume that the probability $\mathsf{P}(A)$ of a certain random event $A \in \mathcal{A}$ be very small; then one can be practically certain that $A$ would not occur in one realization of that experiment. — This is the reason why probabilities close to zero are of particularly great importance. It is thus clear that one of the principal problems of probability theory should be the establishment of events having small probabilities. The most important results of this type are provided by the law of large numbers which we treat in the present book at several occasions.

It is in particular the version given in Th. 3.7.1 that illustrates the intuitive importance of the mathematical expectation of a r.v. Moreover, if $\mu$ is an unknown constant to be determined, then Th. 3.7.1 teaches us that it is useful to repeat observations of i.i.d.r.v. with $\mathsf{E}X_1 = \mu$ and to use the sample mean $S_n/n$ as an estimate of $\mu$. This is the rule of the arithmetic mean that is constantly employed in praxis.

Proof of Theorem 3.7.1. Let $f$ stand for the common c.f. of the $X_k$. Then the c.f. $f_n$ of $S_n/n$ is given by $f_n(t) = f^n(t/n)$. Further, in our situation the Taylor expansion of $f$ reads $f(\tau) = 1 + \mathrm{i}\mu\tau\big(1 + o(1)\big)$ as $\tau \to 0$. Putting $\tau = t/n$ with $t \in \mathbb{R}_1$ fixed, we obtain

$$f_n(t) = \left(1 + \mathrm{i}\mu \frac{t}{n}\big(1 + o(1)\big)\right)^n \to \mathrm{e}^{\mathrm{i}\mu t}, \quad \text{as} \quad n \to \infty .$$

Accordingly, the continuity theorem yields

$$\lim_{n\to\infty} \mathsf{P}\left(\frac{S_n}{n} < x\right) = U_\mu(x), \qquad x \neq \mu ,$$

which is equivalent to the assertion. □

A similar proof will lead us now to the second law.

**Theorem 3.7.2** (The central limit theorem for i.i.d.r.v. with finite variances; Lévy) *Let $(X_k)$ be a sequence of i.i.d.r.v. with $\mathsf{E}X_1 = 0$, $\mathsf{Var}\, X_1 = 1$. Then we have, uniformly on the whole line,*

$$\lim_{n\to\infty} \mathsf{P}\left(\frac{S_n}{\sqrt{n}} < x\right) = \Phi_{0,1}(x) .$$

Remark. Note that the norming factor is chosen such that $\mathsf{Var}\,(S_n/\sqrt{n}) = 1$ for all $n$. But the important thing to notice in this result is that the limiting d.f. of $S_n/\sqrt{n}$ does not depend on the d.f. of $X_1$. Moreover, it tells us how to calculate approximately the probability of the deviations of $S_n/n$ from $\mathsf{E}(S_n/n) = 0$ which always occur for finite $n$. Namely, we obtain for large $n$

$$\mathsf{P}\left(x_1 < \frac{S_n}{n} < x_2\right) \approx \Phi_{0,1}(x_2\sqrt{n}) - \Phi_{0,1}(x_1\sqrt{n}) .$$

In this way Th. 3.7.2 is an important corollary to Th. 3.7.1.

Proof of Theorem 3.7.2. The c.f. $f_n$ of $S_n/\sqrt{n}$ is given by $f_n(t) = f^n(t/\sqrt{n})$. Further, in our case the Taylor expansion of $f$ has the form $f(\tau) = 1 - \tau^2/2\,\big(1 + o(1)\big)$ as $\tau \to 0$. Then, for fixed $t$, we put $\tau = t/\sqrt{n}$ and see that, as $n \to \infty$,

$$f_n(t) = f^n\left(\frac{t}{\sqrt{n}}\right) = \left(1 - \frac{t^2}{2n}\,(1 + o(1))\right)^n \to \exp\left(-\frac{t^2}{2}\right).$$

Now the required formula follows from the continuity theorem.

As to the asserted uniformity see Suppl. 2.6.1. □

Finally, we introduce a generalization of the above model. We consider a *triangular array*

$$\begin{array}{l} X_{11} \\ X_{21}, X_{22} \\ \dots \\ X_{n1}, X_{n2}, \dots, X_{nn} \\ \dots \end{array}$$

of r.v. which are i.i.d. in each row taking values 0 and 1, where $\mathsf{P}(X_{nk} = 0) = q_n$, $\mathsf{P}(X_{nk} = 1) = p_n$, $q_n + p_n = 1$. Then the sum $S_n := X_{n1} + \dots + X_{nn}$ has c.f.

$$f_n(t) = (q_n + p_n\,\mathrm{e}^{it})^n = \exp\big(n \log\,(1 + p_n(\mathrm{e}^{it} - 1))\big) . \tag{3.7.1}$$

For this model we can state a limit theorem with necessary and sufficient conditions. To this end we remind the reader of the Poisson d.f. $\mathrm{P}_\lambda$ which has c.f. $\exp\big(\lambda(\mathrm{e}^{it} - 1)\big)$.

**Theorem 3.7.3** (Poisson limit theorem). *We have, as $n \to \infty$,*

$$\mathsf{P}(S_n < x) \xrightarrow{c} \mathrm{P}_\lambda(x) ,$$

*or, equivalently,*

$$\lim_{n\to\infty} \mathsf{P}(S_n = k) = \mathrm{e}^{-\lambda}\frac{\lambda^k}{k!}, \qquad k = 0, 1, \dots ,$$

*iff $np_n \to \lambda$.*

Proof. The first two limit relations are equivalent since $\{S_n = k\} = \{k - \varepsilon < S_n < k + \varepsilon\}$ for $0 < \varepsilon < 1$, $k \geqq 0$. Accordingly, by the continuity theorem it suffices to show that $np_n \to \lambda$ iff

$$f_n(t) \to \exp\big(\lambda(\mathrm{e}^{it} - 1)\big) . \tag{3.7.2}$$

For this purpose we first note that $\log\,(1 + x) = x(1 + O(x))$, $x \to 0$. Hence, if $p_n \to 0$, we conclude from (3.7.1) that

$$f_n(t) = \exp\big(np_n(\mathrm{e}^{it} - 1)\,(1 + O(p_n)\big) , \qquad n \to \infty . \tag{3.7.3}$$

Let now $np_n \to \lambda > 0$; then $p_n \to 0$ and (3.7.3) implies (3.7.2).

Conversely, let us assume (3.7.2). Then we first conclude from (3.7.1), by an indirect argument, that $p_n \to 0$ so that (3.7.3) is again true. Thus $np_n \to \lambda$ follows. The proof is complete. □

## 3.8. Supplement

3.8.1. *Let $f$ be the c.f. of the d.f. $F$. Then we have the inversion formula*

$$F(x_2) - F(x_1) = \frac{1}{2\pi} \lim_{T\to\infty} \int_{-T}^{T} \frac{e^{-itx_1} - e^{-itx_2}}{it} f(t)\, dt \tag{3.8.1}$$

*provided that $x_1, x_2$ are continuity points of $F$.*

3.8.2. *Let $(g_n)$ be an equicontinuous sequence of functions, $|g_n| \leqq 1$. Then there exists a subsequence $(g_{n'})$ converging to a continuous limit function* (theorem of Arzela-Ascoli). — This is a selection theorem for the class of equicontinuous and uniformly bounded functions; for the proof see FELLER (1971).

3.8.3. For the special convergence $F_n \xrightarrow{c} \Phi$ we need remarkably weak sufficient conditions on the corresponding c.f. $f_n$: *Assume that $\psi = \lim_{n\to\infty} \operatorname{Re} f_n$ is continuous at the origin 0 and*

$$\lim_{n\to\infty} f_n(t_k) = \exp(-t_k^2/2), \qquad k = 1, 2, \dots, \tag{3.8.2}$$

*where $t_k \to 0$ is a real point set. Then $F_n \xrightarrow{c} \Phi$.* — For, in view of Prop. 3.3.5 the first assumption implies the relative compactness of $(F_n)$. Thus the result follows from the fact that all converging subsequences $(F_{n'})$ of $(F_n)$ converge to $\Phi$; this is a consequence of Suppl. 4.9.3.

3.8.4. The continuity of $\psi$ at 0 required in 3.8.3 cannot be omitted. In fact, consider the sequence

$$f_n(t) = \exp(-t^2/2) \cos(2^n t), \qquad n = 1, 2, \dots,$$

of c.f. for which $\psi$ is not continuous at 0, cf. Ex. 3.3.6. Now put $t_k = \pi/2^{k-1}$. Then

$$f_n(t_k) = \exp(-t_k^2/2), \qquad k = 1, 2, \dots, n,$$

whence (3.8.2) follows. In spite of that we have for the corresponding d.f.

$$F_n(x)\ \tfrac{1}{2}\left(\Phi(x + 2^n) + \Phi(x - 2^n)\right) \xrightarrow{w} \tfrac{1}{2}.$$

3.8.5. *Let* lext $(F) \geqq 0$ *and $n \geqq 0$ be an integer. Further, let $H$ be some function varying regularly at infinity with exponent $-r$, $n < r < n + 1$. Then $1 - F(x) = H(x)$, as $x \to \infty$, iff*

$$f(t) = 1 + \frac{\mu_1(it)}{1!} + \dots + \frac{\mu_n(it)^n}{n!} + AH\left(\frac{1}{t}\right)(1 + o(1)), \quad t \to 0.$$

*In either case $A = -\Gamma(n + r)\, e^{i\pi r/2}$* (FELLER (1971), Sect. XVII.12).

3.8.6. *Let $n$ be a natural number and $2n < r < 2n + 1$. The c.f. $f$ corresponding to $F$ admits the expansion*

$$f(t) = 1 + \sum_{k=1}^{2n} \frac{c_k(it)^k}{k!} + o(t^r) \quad as \quad t \to 0,$$

*where $c_1, \dots, c_{2n}$ are certain real numbers, iff $T_F(x) = o(x^{-r})$ as $x \to \infty$* (WOLFE (1973)).

**3.8.7.** The results of Sect. 3.5 are not satisfactory for $k$ odd. But the existence of the first derivative $f'$ of a c.f. $f$ can also be characterized and has a noteworthy intuitive meaning. Namely, it is closely connected with the law of large numbers for a sequence $(X_n)$ of i.i.d.r.v. with the common d.f. $F$. We quote the following result from FELLER (1971), Chap. XVII; note that $f'(0)$ is necessarily purely imaginary since $\operatorname{Re} f$ is even:

*Each of the following three conditions implies the other two.*

(i) *The derivative* $f'(0) = i\hat{\mu}$ *exists;*

(ii) $xT_F(x) \to 0$, $\int_{-x}^{x} y \, dF(y) \to \hat{\mu}$ *as* $x \to \infty$;

(iii) *The law of large numbers holds true in the form* $\mathsf{P}\left(|(S_n/n) - \hat{\mu}| \geqq \varepsilon\right) \to 0$ *for a certain* $\hat{\mu} \in \mathbb{R}_1$.

For one-sided d.f. this implies that $\mu_F$ exists iff $f'(0)$ exists. From this we easily see: *$\mu_k$ exists iff $f^{(k)}(0)$ exists.*

**3.8.8.** *Let $F$ be the mixture of a set $\{G(x, y)\}$ of d.f. with respect to the d.f. $V$, cf.* (1.4.9). *Then its c.f. $f$ is given by*

$$f(t) = \int_{-\infty}^{\infty} g(t, y) \, dV(y)$$

*where $g$ stands for the c.f. of $G$* (LUKACS (1970)).

**3.8.9.** A c.f. $F$ is said to be *unimodal* if there exists at least one value $a$ such that $F(x)$ is convex for $x < a$ and concave for $x > a$. Such a point $a$ is called a *vertex* of $F$. As examples we mention the normal d.f. and the Cauchy d.f.

The following two results are proved in LUKACS (1970).

(i) *A d.f. $F$ is unimodal with vertex at $a = 0$ iff its c.f. $f$ can be represented as*

$$f(t) = \frac{1}{t} \int_0^t g(y) \, dy \,, \qquad t \neq 0 \,,$$

*where $g$ is a c.f.*

(ii) *A unimodal d.f. $F$ with vertex $a = 0$ is — but for a possible jump in 0 — absolutely continuous with density $p$ given by*

$$p(x) = \int_{-\infty}^{x} \frac{dG(u)}{u}, \quad x < 0 \,, \quad p(x) = \int_{x}^{\infty} \frac{dG(u)}{u}, \quad x > 0 \,,$$

*where $G$ is the d.f. corresponding to the c.f. $g$ according to* (i).

**3.8.10.** *Let $g$ be an arbitrary c.f. and $p \geqq 1$ a real number. The function*

$$f(t) = \frac{p}{t^p} \int_0^t g(u) \, u^{p-1} \, du$$

*is then a c.f.* (LUKACS (1970)).

We quote the following two results from FELLER (1971).

**3.8.11.** *Let $F$ be a d.f. with vanishing expectation and c.f. $f$, and let $G$ be a function of bounded variation such that $G(-\infty) = 0$, $G(\infty) = 1$. Finally, suppose that $G$ has a derivative $G'$ satisfying $|G'| \leqq c$ and has a continuously differentiable Fourier-Stieltjes* *transform g such that $g'(0) = 0$. Then we have for all $T > 0$*

$$\sup_x |F(x) - G(x)| \leqq \frac{1}{\pi} \int_{-T}^{T} \left| \frac{f(t) - g(t)}{t} \right| dt + \frac{24c}{\pi T}$$

(Esseen).

**3.8.12.** *Let the $X_k$, $k \geqq 1$, be i.i.d.r.v. such that $\mathsf{E}X_k = 0$, $\mathsf{Var}\, X_k = \sigma^2 > 0$, $M_3 = \mathsf{E}\,|X_k|^3 < \infty$ and let $F_n$ stand for the d.f. of the normalized sum $S_n/\sigma\sqrt{n}$. Then there exists an absolute constant $C$ such that for all $n$*

$$\sup_x |F_n(x) - \Phi(x)| \leqq CM_3/\sigma^3 \sqrt{n}\,.$$

This is the famous Berry-Esseen bound for the discrepancy $F_n - \Phi$ which can be derived from 3.8.11. It is noteworthy that it depends only on the first three moments. The best upper bound for $C$ known so far is 0.79.

**3.8.13.** *The d.f. $F$ possesses a finite moment*

$$\mu_{2k+1}, \qquad k \geqq 0\,, \quad \textit{iff}$$

(i) *$\mu_{2k}$ exists*

(ii) *$\operatorname{Re} f'_{2k}(t)$ exists in $[0, \varepsilon]$*

(iii) $\displaystyle\int_0^{\varepsilon} \frac{1 - \operatorname{Re} f_{2k}}{t^2}\, dt < C$ for some constant $C > 0$ (Laue (1983b)).

*Characteristic functions of one-sided distribution functions.*

We consider c.f. $f$ of d.f. $F$ satisfying lext $(F) > -\infty$; for the sake of brevity we put lext $(F) = 0$. We anticipate from Sect. 4.2 that then $f$ is continuable to the upper half plane $S(0, \infty) = \{z = t + is : s > 0\}$ and that we have

$$f(is) = \int_{-0}^{\infty} e^{-sx}\, dF(x)\,, \qquad s > 0\,.$$

**3.8.14.** *Truncation inequality. We have*

$$1 - F\left(\frac{1}{x}\right) \leqq \frac{e}{x} \int_0^x (1 - f(is))\, ds\,, \qquad x > 0;$$

the proof resembles that of L. 3.1.5.

In the following two numbers we write $\boldsymbol{F}$ for a set of d.f. satisfying lext $(F) = 0$ and $\boldsymbol{f}$ for the set of corresponding c.f.

**3.8.15.** *Compactness criteria.* (i) *The set $\boldsymbol{F}$ is relatively compact iff $\boldsymbol{f}$ is equicontinuous at $+0$. In either case $\boldsymbol{f}$ is equicontinuous for all $s \geqq 0$.*
(ii) *The set $\boldsymbol{F}$ is relatively compact iff the function*

$$\varphi_+(s) := \inf_{f \in \boldsymbol{f}} f(is)\,, \qquad s \geqq 0\,,$$

*is continuous at $+0$.* The proofs resemble those of Prop. 3.3.3 and Prop. 3.3.5, resp.

**3.8.16.** *Feller's continuity theorem. Let $F$ be a d.f. with c.f. $f$. Then $F_n \xrightarrow{c} F$ iff $f_n(is) \to f(is)$, $s \geqq 0$. Moreover, $F_n \xrightarrow{w} G$, $G(\infty) < 1$, iff $f_n(is) \to g(is) = \int_{-0}^{\infty} e^{-sx}\, dG(x)$; in either case we have $g(+i0) = G(\infty)$* (FELLER (1971)).

The proof resembles that of Th. 3.4.1. Suppl. 3.8.14 (ii), serves as compactness criterion while the uniqueness of the limit function is granted by Prop. B.2.8.

**3.8.17.** *Let there exist $A > 0$ and $\lambda$, $0 < \lambda < 1$, such that $\dfrac{F(x+y) - F(y)}{|x|^\lambda} \leqq A$ for all $y > 0$ and $x \in R_1$, $x \neq 0$. Then*

(i) $$F(y) \leqq \frac{2}{\pi} \left| \int_0^T \sin yu \frac{\operatorname{Re} f(u)}{u} \, du \right| + \frac{C_1}{T^\lambda},$$

(ii) $$1 - F(y) \leqq \frac{2}{\pi} \left| \int_0^T \cos yu \frac{\operatorname{Im} f(u)}{u} \, du \right| + \frac{C_2}{T^\lambda}$$

*for certain positive $C_1$ and $C_2$* depending only on $A$ and $\lambda$ (LAUE (1986)).

## Notes and bibliographical data

For a more comprehensive treatment of the theory of c.f. we refer in particular to LUKACS (1970) and RAMACHANDRAN (1967). As to recent advances in the c.f. theory see LUKACS (1983).

L. 3.1.3 is due to RAIKOV (1940).

In FELLER (1971) Parseval's equation (3.2.1) is the starting point of a method which permits to derive surprisingly many conclusions, e.g. the continuity theorem.

Th. 3.2.3 is contained in KAWATA (1972) where it is ascribed to ZOLOTAREV.

The Fourier inversion formula (3.2.3) is true also in some cases in which $f$ is not integrable, cf. Prop. 3.6.6. A thorough investigation of this occurrence was undertaken by LAUE (1984a).

KAWATA (1972) contains some more material on the equicontinuity of c.f.

A detailed discussion of the continuity theorem was also made in KAWATA (1972), Chap. 9 and 12.

Weak convergence up to an additive constant was introduced in LOÈVE (1963).

# Chapter 4. Characteristic functions analytic in a strip

We aim at acquainting the reader with c.f. which can be continued from the real line at least to a certain strip of the complex plane, i.e., we deal with analytic c.f. (a.c.f.) and boundary c.f. (b.c.f.). These classes include many c.f. which are important in probability theory and mathematical statistics.

Our approach is via the c.f. of one-sided d.f. and makes it possible to treat a.c.f. and b.c.f. in a unified way. In Sect. 4.4 and 4.5 we also present a brief exposition on the question how the tail behaviour of a d.f. is connected with the growth of the corresponding c.f. in a half plane or in the plane. Sect. 4.6 is devoted to the factorization theory of c.f.; the main result of this section is Cramér's factorization theorem for the normal d.f. It follows a study of recent modifications of this theorem which have a two-fold interest for us: They are continuation theorems for d.f., and their proofs require tools from the theory of Phragmén and Lindelöf. In Sect. 4.8 we treat Hinčin's phenomenon from a view-point which was stimulated by continuation theory.

A field of research belonging to the subject of this chapter are the c.f. of non-negative r.v. It deserves more attention than it has received so far. In Sect. 4.9 we expose very recent results on the c.f. $f$ of non-negative r.v. starting with the reciprocity relations between $\operatorname{Re} f$ and $\operatorname{Im} f$.

We encourage the reader to become familiar with the content of App. C since we will repeatedly apply it.

## 4.1. Basic definitions

According to Def. 3.1.1 a c.f. $f$ is a function of the real variable $t$ which has, in general, complex values. Therefore, it is a natural question whether $f$ is continuable from $\mathbb{R}_1$ to a function analytic in some domain $D$. The c.f. of the normal d.f. can serve as a first example where $D$ is the whole plane.

Soon we will see that in our field the case is of particular interest in which such a domain $D$ contains a certain horizontal strip

$$S(s_1, s_2) := \{z = t + is\colon s_1 < s < s_2\}\,, \quad -\infty \leqq s_1 \leqq 0 \leqq s_2 \leqq \infty\,, \quad s_1 < s_2\,.$$

Similarly, we will have occasion to use the strip $S[s_1, s_2) = \{z\colon s_1 \leqq s < s_2\}$, and $S(s_1, s_2]$, $S[s_1, s_2]$ have analogous meanings.

The following definition combines the Def. B.1.3, which is useful in connection with Schwarz' reflection principle, with the usual notion of analytic continuation.

**Definition 4.1.1.** A c.f. $f$ is said to be *continuable to the strip* $S(s_1, s_2)$ if there exists a function $\psi$ of the complex variable $z = t + is$ such that

(i) $\psi$ is analytic in $S(s_1, s_2)$,
(ii) $\psi$ is continuous in $S(s_1, s_2) \cup \mathbb{R}_1$,
(iii) $\psi(t) = f(t)$ for $t \in \mathbb{R}_1$.

The widest strip $S(\underline{s}, \overline{s})$ to which $f$ is continuable is called the *strip of regularity* of $f$. □

Obviously, condition (ii) is only needed if $s_1 = 0$ or $s_2 = 0$.

**Remark 4.1.2.** The function $\psi$ is uniquely defined. This is trivial by the uniqueness theorem for analytic functions if $\underline{s} < 0 < \overline{s}$; and from Schwarz' reflection principle (in particular from Cor. B.1.4) it follows that this is true also if either $\underline{s} = 0$ or $\overline{s} = 0$. Therefore, we denote the continuation of $f$ also by $f$. With the very same reasoning we conclude that a c.f. $f$ continuable to a certain strip is uniquely determined by its values on an arbitrary interval $I \subset \mathbb{R}_1$. □

From these two different arguments one sees that Def. 4.1.1 combines two qualitatively different cases. They are often separately dealt with; our approach makes it possible to treat them in a unified way. We now introduce the corresponding terms.

**Definition 4.1.3.** Let $f$ be a c.f. with strip of regularity $S(\underline{s}, \overline{s})$. Then it is called an *analytic* c.f. (a.c.f.) if $\underline{s} < 0 < \overline{s}$. Otherwise, if either $\underline{s} = 0 < \overline{s}$ or $\underline{s} < 0 = \overline{s}$, then $f$ is said to be a *boundary* c.f. (b.c.f.). □

**Example 4.1.4.** (i) The c.f. $f(t) = e^{-t^2/2}$ of the normal distribution is continuable to the whole complex plane $S(-\infty, \infty)$ so that it is even an integral c.f.

(ii) The c.f. $f(t) = \lambda/(\lambda - it)$ of the exponential distribution with parameter $\lambda > 0$ is continuable to the strip of regularity $S(-\lambda, \infty)$, hence it is an a.c.f.

(iii) The c.f. $f(t) = 1/(1 + t^2)$ of the Laplace distribution is an a.c.f. with strip of regularity $S(-1, 1)$.

(iv) Let $f$ be a mixture of exponential c.f. with parameters $\lambda_k > 0$. Then Prop. 3.1.12 tells us that

$$f(t) = \sum_{k=1}^{\infty} p_k \frac{\lambda_k}{\lambda_k - it}, \qquad p_k > 0, \qquad \sum_{k=1}^{\infty} p_k = 1.$$

This series converges uniformly for $z = t + is \in S[0, \infty)$. Thus we see from Weierstrass convergence theorem that $f$ is continuable to $S(0, \infty)$. But the origin 0 can be a singularity; for example, it is a limit point of poles if $\lambda_k = 1/k$, so that $f$ is a b.c.f.

(v) The c.f. $f(t) = e^{-|t|}$ of the Cauchy distribution is not continuable to any strip. This is seen by continuing $f$ from the half line $t > 0$ to the plane. Accordingly, $f$ is neither an a.c.f. nor a b.c.f. □

In the next sections we will derive, step by step, necessary and sufficient conditions which a d.f. must satisfy in order that its c.f. be either an a.c.f. or a b.c.f.

## 4.2. Characteristic functions of one-sided distribution functions

We start our investigation with c.f. of d.f. $F$ such that $l :=$ lext $(F)$ or $r :=$ rext $(F)$ or both are finite. As to the latter case, we have the following result.

**Proposition 4.2.1.** *The c.f. $f$ of a finite d.f. $F$ can be continued to an integral function with the representation*

$$f(z) = \int_{l-0}^{r+0} e^{izx}\, dF(x), \qquad z \in S(-\infty, \infty).$$

Proof. The integral trivially exists, and the Taylor expansion of the integrand converges uniformly in every bounded domain. Thus we obtain for all $z$ the convergent series

$$f(z) = \sum_{k=0}^{\infty} \frac{(iz)^k}{k!} \int_{l-0}^{r+0} x^k \, dF(x)$$

which represents an integral function. It obviously coincides with the given c.f. $f$ of $F$ for $z = t$. This demonstrates the assertion. □

Now we will deal with the general case of one-sided d.f. Clearly, if $F$ is bounded to the right, then $F_-(x) = 1 - F(-x + 0)$ is bounded to the left, and for the corresponding c.f. we have $f_-(t) = f(-t)$. Thus it suffices to consider the case $l > -\infty$; the reformulation of our results to the case $r < \infty$ will be left to the reader. As before we write $z = t + is$.

**Proposition 4.2.2.** (i) *The c.f. $f$ of a d.f. $F$ bounded to the left is continuable to the strip $S(0, \infty)$ and admits here the representation*

$$f(z) = \int_{l-0}^{\infty} e^{izx} \, dF(x) \,. \tag{4.2.1}$$

(ii) *If for a certain $s_1 < 0$ we have additionally*

$$\int_{l-0}^{\infty} e^{-s_1 x} \, dF(x) < \infty \,, \tag{4.2.2}$$

*then $f$ is continuable at least to $S(s_1, \infty)$ and has the form* (4.2.1).

Proof. We first prove part (ii). Since the expression (4.2.1) coincides with $f$ for $z = t$ we have to show that — under the hypothesis (4.2.2) — it represents a function analytic in $S(s_1, \infty)$, $s_1 < 0$, so that the assertion follows by analytic continuation.

We introduce, for arbitrary complex $z$, the functions

$$h_k(z) = \int_{k-0}^{k+1-0} e^{izx} \, dF(x) \,, \qquad k = 0, 1, \ldots \,.$$

Combining (4.2.1) and (4.2.2) we see that

$$|f(z)| \leqq \int_{l-0}^{\infty} e^{-s_1 x} \, dF(x) < \infty \,, \qquad z \in S(s_1, \infty);$$

this means that the integral (4.2.1) converges in $S(s_1, \infty)$. Thus we may write

$$f(z) = \int_{l-0}^{-0} e^{izx} \, dF(x) + \sum_{k=0}^{\infty} h_k(z) \,, \qquad z \in S(s_1, \infty) \,.$$

We see from Prop. 4.2.1 that all these summands are integral. An arbitrary remainder term of this series can be estimated as

$$\left| \sum_{k=K}^{\infty} h_k(z) \right| = \left| \int_{K-0}^{\infty} e^{izx} \, dF(x) \right| \leqq \int_{K-0}^{\infty} e^{-s_1 x} \, dF(x) \,, \qquad z \in S[s_1, \infty) \,.$$

This bound is independent of $z \in S[s_1, \infty)$, and it can be made arbitrarily small by choosing $K$ sufficiently large. Thus the above series converges uniformly in $S[s_1, \infty)$ so that the Weierstrass convergence theorem (App. B.6) yields the desired conclusion.

To prove part (i) of Prop. 4.2.2 we first note that (4.2.2) is generally satisfied for $s_1 = 0$. Thus we can repeat the above argument and see that the function on the right of (4.2.1) is analytic in $S(0, \infty)$. The continuity of $f(z)$ in $S[0, \infty)$ follows (for $\tau \in R_1$) from

$$|f(\tau) - f(z)| = \left|\int_{l-0}^{\infty} e^{i\tau x}\,[1 - e^{i(z-\tau)x}]\,dF(x)\right| \leqq \int_{l-0}^{\infty} |1 - e^{i(z-\tau)x}|\,dF(x) \to 0\,,$$
$$z \to \tau\,,$$

by dominated convergence. Thus $f$ is continuable to $S(0, \infty)$ in view of Def. 4.1.1, as asserted. □

We turn now to a very applicable criterion for the occurrence of case (ii) in Prop. 4.2.2. We emphasize that in the following proof it is the monotonicity of $F$ that plays an essential role.

**Proposition 4.2.3.** *Let the c.f. $f$ correspond to the d.f. $F$ which is bounded to the left, and assume that $f$ is analytic in a certain disk $|z| < r$, $r > 0$. Then $f$ is continuable at least to the half plane $S(-r, \infty)$, and the integral representation* (4.2.1) *holds in this strip.*

Proof. First of all, the assumption ensures that $f$ has derivatives of all orders at $z = 0$. Hence so has Re $f$, and Cor. 3.5.4 implies that all moments of $F$ exist. Also,

$$f^{(k)}(0) = i^k \int_{l-0}^{\infty} x^k\,dF(x)$$

by Prop. 3.5.1. Thus we obtain for the Taylor expansion of $f$ about the origin

$$f(z) = \sum_{k=0}^{\infty} \frac{f^{(k)}(0)}{k!} z^k = \sum_{k=0}^{\infty} \frac{(iz)^k}{k!} \int_{l-0}^{\infty} x^k\,dF(x)\,, \qquad |z| < r\,.$$

Now let $-r < s_1 < 0$. Then we get as in the proof of Prop. 4.2.1 for any positive $x_0 > l$,

$$\int_{l-0}^{x_0} e^{-s_1 x}\,dF(x) = \sum_{k=0}^{\infty} \frac{(-s_1)^k}{k!} \int_{l-0}^{x_0} x^k\,dF\,(x)\,.$$

Using the facts that $F$ is monotone, $x_0 > 0$, and $-s_1 > 0$ we obtain the estimate

$$\int_{l-0}^{x_0} e^{-s_1 x}\,dF(x) \leqq \sum_{k=0}^{\infty} \frac{(-s_1)^k}{k!} \int_{l-0}^{\infty} x^k\,dF(x) = f(is_1)\,;$$

the latter equality follows from the above expansion. Since $x_0 > l$ is arbitrary the condition (4.2.2) is satisfied for all $s_1 > -r$ whence the assertions follow from Prop. 4.2.2 (ii). □

We recall that the symbol $\underline{s}$ determines the boundary of the strip $S(\underline{s}, \infty)$ of regularity; Ex. 4.1.4 shows that all cases $-\infty \leqq \underline{s} \leqq 0$ occur.

**Proposition 4.2.4.** *The c.f. $f$ of a d.f. $F$ bounded to the left is either a b.c.f. or an a.c.f.; it has the strip of regularity $S(\underline{s}, \infty)$ where*

$$\underline{s} = \inf\,\{s\colon \int_{l}^{\infty} e^{-sx}\,dF(x) < \infty\} \leqq 0\,. \tag{4.2.3}$$

*In this strip the integral representation* (4.2.1) *holds true. Further, the point* $z = i\underline{s}$ *is a singularity of* $f$ *no matter if* $\underline{s} < 0$ *or* $\underline{s} = 0$.

Proof. We know from Prop. 4.2.2 (i) that $f$ is always continuable to $S(0, \infty)$. Now the behaviour of $f$ in 0 is essential, and there are only two possibilities: Either $f$ is analytic in a certain circle about 0 so that $f$ is an a.c.f in view of Prop. 4.2.3; or 0 is a singular point so that $\underline{s} = 0$, i.e. $f$ is a b.c.f. This shows the first assertion.

To prove (4.2.3) we see from Prop. 4.2.2 that all strips $S(s_1, \infty)$ for which (4.2.2) is true — and only they — belong to $S(\underline{s}, \infty)$. From this fact the formula (4.2.3) follows. Now it is also easy to conclude from Prop. 4.2.2 that the representation (4.2.1) holds in $S(\underline{s}, \infty)$.

Finally, we show that the point $i\underline{s}$ is singular. We carry an indirect proof and assume that $i\underline{s}$ is not singular. Then we fix $s > 0$ and consider the circle about $is$ with radius $s - \underline{s}$; by assumption it has not any singularity of $f$ on its circumference. Therefore, the radius of convergence of $f$ about $is$ is greater than $s - \underline{s}$. Thus, no matter if $\underline{s} < 0$ or $\underline{s} = 0$, $f$ is regular in some circle $\{z\colon |z| < r\}$ with a certain $r > |\underline{s}|$. Hence, Prop. 4.2.3 implies that the strip of regularity is wider than $S(\underline{s}, \infty)$, a contradiction. □

**Corollary 4.2.5.** *Under the assumptions of Prop. 4.2.4 the derivatives* $f^{(k)}$, $k \geqq 1$, *of* $f$ *can be represented as*

$$f^{(k)}(z) = i^k \int_{l-0}^{\infty} e^{izx} x^k \, dF(x) , \qquad z \in S(\underline{s}, \infty) .$$

Remark. Note that this formula holds in the interior of the strip of regularity. But in the case that $\underline{s} = 0$ the d.f. $F$ need not have any moment so that the above formula is not generally true for $z = t$.

Proof. We consider the case $k = 1$. Since it is clear that the derivative exists at $z = t + is \in S(\underline{s}, \infty)$ it suffices to form the difference ratio

$$\frac{f(z + h) - f(z)}{h} = \int_{l-0}^{\infty} e^{izx} \frac{e^{ihx} - 1}{hx} x \, dF(x) , \qquad z \in S(\underline{s}, \infty) ,$$

for $h > 0$. Then App. B.4 tells us that the ratio in the integrand is not greater than 1 in absolute value. Hence the integral is dominated by

$$\int_{l-0}^{\infty} e^{-sx} |x| \, dF(x) , \qquad s > \underline{s} .$$

Thus the assertion follows for $k = 1$ by dominated convergence. The general statement is obtained by induction. □

Next we deal with the relation between the properties of an a.c.f. or a b.c.f. and the corresponding d.f. We recall from Prop. 4.2.2 that $f$ is trivially continuable to $S(0, \infty)$. Hence, it is the point of the following statement to give conditions on $F$ under which $f$ is also continuable to a certain strip contained in $S(-\infty, 0)$.

**Proposition 4.2.6.** *The c.f.* $f$ *of a d.f.* $F$ *bounded to the left is continuable at least to the strip* $S(-r, \infty)$, $r > 0$, *iff*

$$1 - F(x) = O(e^{sx}) \quad \text{as} \quad x \to \infty \quad \text{for all} \quad s \text{ with } -r < s < 0 . \tag{4.2.4}$$

Proof. *If-part*: Let $-r < s' < s_1 < 0$. Then we find after integration by parts, for any positive $x_0 > l$,

$$\int_{l-0}^{x_0} e^{-s_1 x} dF(x) = -\int_{l-0}^{x_0} e^{-s_1 x} d(1 - F(x)) \leqq e^{-s_1 l} + |s_1| \int_{l}^{x_0} e^{-s_1 x} (1 - F(x)) dx .$$

Now we apply the assumption and obtain the estimate

$$\leqq e^{-s_1 l} + C |s_1| \int_{l}^{\infty} e^{-(s_1 - s')x} dx$$

for a certain $C > 0$.

Accordingly, the integral $\int_{l}^{\infty} e^{-s_1 x} dF(x)$ converges for all $s_1 > -r$ so that $f$ is continuable to $S(-r, \infty)$ by virtue of Prop. 4.2.2, (ii).

*Only-if part*: Let $f$ be continuable to $S(-r, \infty)$, $r > 0$, so that the integral (4.2.2) converges for $s_1 = s > -r$ by Prop. 4.2.4. Then we have

$$1 - F(x) = \int_{x}^{\infty} dF(y) \leqq e^{sx} \int_{x}^{\infty} e^{-sy} dF(y) \leqq e^{sx} f(is) = O(e^{sx}) , \qquad x \to \infty .$$

This yields (4.2.4). □

**Corollary 4.2.7.** *Let $F$ be a d.f. bounded to the left and $f$ its c.f. Then in the strip of regularity of $f$ we have the formula*

$$f(z) = e^{ilz} + iz \int_{l}^{\infty} e^{izx} (1 - F(x)) dx ,$$

*no matter if $f$ is an a.c.f. or a b.c.f.*

Proof. By Prop. 4.2.2 the formula (4.2.1) holds in the considered strip where $z = t + is$. We write it in the form $f(z) = \int_{l-0}^{\infty} e^{izx} d(F(x) - 1)$ and integrate by parts. This is trivially done for $s \geqq 0$; otherwise, if $s < 0$, we have to make use of Prop. 4.2.6. □

Now we prove a result which gives a connection between the behaviour of $f$ in the upper half plane and the left extremity $l$ of $F$.

**Proposition 4.2.8.** *Let $F$ be bounded to the left. Then we have*

$$l = -\lim_{s \to \infty} \frac{\log f(is)}{s} , \qquad F(l + 0) = \lim_{s \to \infty} f(is) e^{sl} .$$

Proof. Let $\varepsilon > 0$ so that $F(l + \varepsilon) > 0$. Then we obtain from (4.2.1) for any $s > 0$

$$f(is) = \int_{l-0}^{\infty} e^{-sx} dF(x) \leqq e^{-sl}$$

but also

$$f(is) \geqq \int_{l-0}^{l+\varepsilon} e^{-sx} dF(x) \geqq e^{-s(l+\varepsilon)} F(l + \varepsilon) .$$

Therefore,

$$-l - \varepsilon \leqq \varliminf_{s \to \infty} \frac{\log f(is)}{s} \leqq \varlimsup_{s \to \infty} \frac{\log f(is)}{s} \leqq -l ,$$

and on letting $\varepsilon \to 0$ we obtain the first assertion. Further, we have by dominated convergence

$$e^{sl} f(\mathrm{i}s) - F(l+0) = \int_{l+0}^{\infty} e^{-s(x-l)}\, dF(x) \to 0 ,$$

as $s \to \infty$, so that the second assertion follows. □

Finally, we turn to a peculiar result on moments which requires a remarkable tool from the theory of Phragmén and Lindelöf.

**Proposition 4.2.9.** *Let the d.f. $F$ be bounded to the left and* lext $(F) = 0$. *Then it possesses a finite moment $\mu_k$, $k \geqq 1$, iff $\mu_{k-1}$ exists ($\mu_0 := 1$) and the difference ratio*

$$\frac{f^{(k-1)}(0) - f^{(k-1)}(t)}{t}, \qquad t \neq 0 , \tag{4.2.5}$$

*is bounded ($f^{(0)} := f$).*

Proof. In view of Prop. 3.5.1 we have to prove only the if-part of the statement. Put $k = 1$.

With some $c > 0$ we introduce the function

$$w(z) := \mathrm{i}c\, \frac{1 - f(z)}{z}, \qquad z \in S[0, \infty) \setminus \{0\} ;$$

it is obviously continuous in the whole domain of definition. In view of the assumption we can choose $c$ such that $|w(t)| \leqq 1$ for all real $t \neq 0$; moreover, we may even suppose that

$$|zw(z)| = c\,|1 - f(z)| \leqq 1 , \qquad |z| \leqq 1 , \qquad s = \mathrm{Im}\, z > 0 .$$

Putting $h(z) = z$ we see that all assumptions of Th. C.1.1 are satisfied, and applying it we obtain $|w(\mathrm{i}s)| \leqq 1$, $0 < s < 1$. Accordingly, we have for any $x_0 > 0$

$$1 \geqq w(\mathrm{i}s) = c \int_0^{\infty} \frac{1 - e^{-sx}}{sx}\, x\, dF(x) \geqq c\, \frac{1 - e^{-sx_0}}{sx_0} \int_0^{x_0} x\, dF(x)$$

$$\to c \int_0^{x_0} x\, dF(x) , \quad \text{as} \quad s \to 0 .$$

Hence $\mu_1$ exists.

The general assertion follows similarly. Assuming that $\mu_{k-1}$, $k > 1$, exists we see from Prop. 3.5.1 that (4.2.5) makes sense. Then the same reasoning as above can be applied since Cor. 4.2.5 tells us that $(-\mathrm{i})^{k-1} f^{(k-1)}/\mu_{k-1}$ is the c.f. of the d.f. $F^{(k-1)}(x)$ $= \mu_{k-1}^{-1} \int_0^x u^{k-1}\, dF(u)$. □

By virtue of Prop. 3.5.1, it is clear that the ratio (4.2.5) — if it is bounded — tends to $\mathrm{i}^k \mu_k$, as $t \to 0$. Further, the case $k = 1$ is noteworthy; then the boundedness of $(1 - f(t))/t$, $t \neq 0$, implies the existence of $\mu_1$.

## 4.3. The general case

Now we turn to the general case by dropping the assumption $l > -\infty$ (or $r < \infty$) made in Sect. 4.2. The method we will use for dealing with this case is very simple but powerful. Namely, we write the c.f. $f$ of an arbitrary d.f. $F$ in the form

$$f(t) = \int_{-\infty}^{-0} e^{itx}\, dF(x) + \int_{-0}^{\infty} e^{itx}\, dF(x) = f^-(t) + f^+(t)\,, \tag{4.3.1}$$

say. (One of the functions $f^\pm$ may, of course, vanish; then we formally consider it as an a.c.f. Otherwise they are, but for positive factors, c.f. of certain one-sided d.f.; for the sake of brevity they will also be called c.f.). Accordingly, the question of whether $f$ is continuable to a strip may be answered in terms of $f^\pm$, to which the results of Sect. 4.2. can be applied.

To begin with, we recall from Prop. 4.2.2 and 4.2.4 that $f^+$ is always continuable to $S(\underline{s}, \infty)$ in the form (4.2.1) where $\underline{s}$ is defined in (4.2.3). Reformulating these propositions to $f^-$ we see that

$$f^-(z) = \int_{-\infty}^{-0} e^{izx}\, dF(x)\,, \qquad z \in S(-\infty, \bar{s})\,,$$

where

$$\bar{s} = \sup\{s\colon \int_{-\infty}^{-0} e^{-sx}\, dF(x) < \infty\} \geqq 0\,.$$

If $f^+$ or $f^-$ vanishes then we put $\underline{s} = -\infty$ or $\bar{s} = \infty$, resp. Now we observe that we are faced with exactly one of the following four cases.

(i) Both $f^+$ and $f^-$ are a.c.f. with strips of regularity $S(-\infty, \bar{s})$ and $S(\underline{s}, \infty)$, resp. Then we see from (4.3.1) by analytical continuation that $f$ is also an a.c.f.; its strip of regularity is

$$S(\underline{s}, \bar{s}) = S(-\infty, \bar{s}) \cap S(\underline{s}, \infty)\,, \qquad -\infty \leqq \underline{s} < 0 < \bar{s} \leqq \infty\,.$$

Moreover, applying Prop. 4.2.4 to $f_+$ and its obvious counterpart to $f_-$ we see that the integral representation (4.3.3) given below holds in this strip and that $i\underline{s}$ and $i\bar{s}$ are singular points of $f$.

(ii) The function $f^+$ is an a.c.f. while $f^-$ is a b.c.f. Then the right side of (4.3.1) is continuable to $S(\underline{s}, 0)$ but $i\underline{s}$ is singular for $f^+$ and 0 is singular for $f^-$. Thus $f$ is a b.c.f. with strip of regularity $S(\underline{s}, 0)$ where (4.3.3) holds, see below.

(iii) The roles of $f^+$ and $f^-$ in (ii) can be interchanged; then $f$ is a b.c.f. with strip of regularity $S(0, \bar{s})$ where (4.3.3) holds.

(iv) Both $f^+$ and $f^-$ are b.c.f. Then the following two lemmas show that in this case $f$ is neither an a.c.f. nor a b.c.f.

**Lemma 4.3.1.** *Let $f$ be an a.c.f. with strip of regularity $S(\underline{s}, \bar{s})$, $-\infty \leqq \underline{s} < 0 < \bar{s} \leqq \infty$ Then $f^\pm$ are a.c.f. with the strips of regularity $S(\underline{s}, \infty)$ and $S(-\infty, \bar{s})$, resp.*

Proof. First of all, $f^-$ is always continuable to $S(-\infty, 0)$. Accordingly, by assumption. the function $f - f^-$ is continuable to $S(\underline{s}, 0)$. By (4.3.1) it coincides with $f^+$ on $\mathbb{R}_1$. Since $f^+$ is always continuable to $S(0, \infty)$ we can apply Schwarz' reflection principle to conclude that $f^+$ can be continued to $S(\underline{s}, 0)$ in the form

$$f^+(z) = f(z) - f_-(z). \tag{4.3.2}$$

Thus $f^+$ is an a.c.f. We know from Prop. 4.2.4 that the singularity of $f^+$ nearest to the real axis is located on the negative imaginary axis. Now (4.3.2) shows that this is true also for $f$ and that $i\underline{s}$ is a common singularity of $f$ and $f^+$. Thus $f^+$ has the asserted strip of regularity.

In the same manner the assertion on $f^-$ can be proved. □

**Lemma 4.3.2.** *Let $f$ be a b.c.f. Then one of the c.f. $f^+$ and $f^-$ is an a.c.f. while the other is a b.c.f. The origin 0 is a singular point of $f$.*

Proof. For the sake of definiteness we assume that $f$ is continuable to $S(\underline{s}, 0)$. Then we conclude as before that $f^+$ is an a.c.f. with strip of regularity $S(\underline{s}, \infty)$ and that (4.3.2) holds here. In analogy to our above reasoning we can now conclude that 0 is a singular point both of $f$ and $f^-$; use the counterpart to Prop. 4.2.4. Thus $f^-$ proves to be a b.c.f.

The other case in which $f$ is continuable to $S(0, \bar{s})$ can be treated analogously. □

From these lemmas we see that the above statements (i)—(iii) can be converted. Therefore, we are able to summarize our most essential results in the following way.

**Theorem 4.3.3.** *A c.f. $f$ which is continuable to a strip has a strip of regularity $S(\underline{s}, \bar{s})$ where*

$$\underline{s} = \inf \{s: \int_0^\infty e^{-sx}\, dF(x) < \infty\}, \qquad \bar{s} = \sup \{s: \int_{-\infty}^0 e^{-sx}\, dF(x) < \infty\},$$

*and the points $i\underline{s}$ and $i\bar{s}$ are singularities of $f$. Further, $f$ has the integral representation*

$$f(z) = \int_{-\infty}^{\infty} e^{izx}\, dF(x), \qquad z \in S(\underline{s}, \bar{s}). \quad \square \tag{4.3.3}$$

Also, we have the following criteria to decide whether $f$ is an a.c.f. or a b.c.f. (cf. Prop. 4.2.4); here we write

$$D_r = \{z: |z| < r, r > 0\}.$$

**Proposition 4.3.4.** *The c.f. $f$ is a b.c.f. if it is continuable to a certain half disk $D_r \cap S(0, \infty) = \{z: |z| < r, \operatorname{Im} z > 0\}$ or $D_r \cap S(-\infty, 0) = \{z: |z| < r, \operatorname{Im} z < 0\}$ (for some $r > 0$) while the origin is a singular point of $f$; the strip of regularity contains this half disk.*

*The c.f. $f$ is an a.c.f. if it is continuable to a certain disk $D_r$; its strip of regularity contains this disk.*

Proof. To prove the first assertion it obviously suffices to consider the case that $f$ is continuable to $D_r \cap S(-\infty, 0)$ since the other can be treated analogously. Then the assumption implies that $f - f^-$ is continuable to the half disk $D_r \cap S(-\infty, 0)$. Since $f^+$ is continuable to $D_r \cap S(0, \infty)$ we can again apply Schwarz' reflection principle to conclude that $f^+$ is continuable to $D_r$ and that (4.3.2) holds in $D_r \cap S(-\infty, 0)$. Now we can apply Prop. 4.2.3 to $f^+$ to conclude that $f^+$ is continuable to $S(-r, \infty)$. Now we see by analytical continuation that (4.3.2) is true in $S(-r, 0)$. This means that the c.f. $f$ is also continuable to $S(-r, 0)$ so that its strip of regularity obviously contains the half disk $D_r \cap S(-\infty, 0)$.

Finally, we note that by assumption the origin 0 is singular for the c.f. $f$; accordingly, $f^+$ is a b.c.f. by Def. 4.1.3.

To prove the last statement we repeat the above reasoning to conclude that $f$ is continuable to $S(-r, 0)$. But in this case we can also apply an analogous argument which shows that $f$ is also continuable to $S(0, r)$. Thus $f$ is analytic in a strip $S(\underline{s}, \bar{s})$ with $\underline{s} \leqq -r < 0 < r \leqq \bar{s}$ so that it is an a.c.f. by Def. 4.1.3. Its strip of regularity obviously contains the disk $D_r$. □

The integral representation (4.3.3) is basic for many investigations on a.c.f. and b.c.f. We give a number of immediate consequences.

**Corollary 4.3.5.** *Let $f$ be an a.c.f. or a b.c.f. with strip of regularity $S(\underline{s}, \bar{s})$. Then it possesses both the ridge property*

$$|f(t + is)| \leqq f(is), \qquad t + is \in S(\underline{s}, \bar{s})$$

*and the reflection property*

$$f(t+\mathrm{i}s) = \overline{f(-t+\mathrm{i}s)}\,, \qquad t+\mathrm{i}s \in S(\underline{s}, \bar{s})\,. \tag{4.3.4}$$

Proof. In the strip under consideration we have the representation (4.3.3) which implies

$$|f(t+\mathrm{i}s)| = \Big|\int_{-\infty}^{\infty} \mathrm{e}^{(\mathrm{i}t-s)x}\,\mathrm{d}F(x)\Big| \leqq \int_{-\infty}^{\infty} \mathrm{e}^{-sx}\,\mathrm{d}F(x) = f(\mathrm{i}s)\,;$$

this is the ridge property. The second assertion is immediate from (4.3.3). □

By the way, the equality (4.3.4) follows also from Schwarz reflection principle since $f(\mathrm{i}s)$ is real.

Cor. 4.3.5 implies the following.

**Corollary 4.3.6.** *An a.c.f. or a b.c.f. has no zeros on the segment of the imaginary axis inside its strip of regularity. Further, its zeros and singularities are located symmetrically with respect to the imaginary axis in the whole domain of regularity.*

Proof. The first assertion is clear from the ridge property by an indirect argument. The second is obvious in $S(\underline{s}, \bar{s})$ in view of (4.3.4); that it is true even in the domain of regularity follows from the reflection principle. □

**Corollary 4.3.7.** *The derivatives $f^{(k)}$, $k \geqq 1$, of an a.c.f. or a b.c.f. can be represented as*

$$f^{(k)}(z) = \mathrm{i}^k \int_{-\infty}^{\infty} x^k\,\mathrm{e}^{\mathrm{i}zx}\,\mathrm{d}F(x)\,, \qquad z \in S(\underline{s}, \bar{s})\,. \tag{4.3.5}$$

Proof. We analytically continue the formula (4.3.1) in the form

$$f(z) = \int_{-\infty}^{-0} \mathrm{e}^{\mathrm{i}zx}\,\mathrm{d}F(x) + \int_{-0}^{\infty} \mathrm{e}^{\mathrm{i}zx}\,\mathrm{d}F(x) = f^-(z) + f^+(z)\,, \qquad z \in S(\underline{s}, \bar{s})\,,$$

and have to show that differentiation under the integral sign is permissible. This follows from Cor. 4.2.5 for the second term. The counterpart of Cor. 4.2.5, which holds for $f^-$, permits us to complete the proof. □

Note that a b.c.f. need not have any moment of positive integral order so that (4.3.5) is not generally true for $z = t$.

Next, we carry over Prop. 4.2.6 to the general case.

**Proposition 4.3.8.** *The c.f. $f$ of the d.f. $F$ is continuable at least to the strip $S(-r, 0)$, $r > 0$, iff, as $x \to \infty$,*

$$1 - F(x) = O(\mathrm{e}^{sx}) \quad \textit{for all} \quad s \text{ with } \quad -r < s < 0\,; \tag{4.3.6}$$

*it is continuable at least to the strip $S(0, r)$, $r > 0$, iff, as $x \to \infty$,*

$$F(-x) = O(\mathrm{e}^{-sx}) \quad \textit{for all} \quad s \text{ with } \quad 0 < s < r \tag{4.3.7}$$

Proof. The method of analytic continuation which we used at the beginning of the present section and in the proofs of L.4.3.1 and L.4.3.2 shows that the c.f. $f = f^- + f^+$ is continuable to $S(-r, 0)$ iff $f^+$ is continuable to this strip. Thus the first assertion follows from Prop. 4.2.6.

To obtain the second we proceed analogously after reformulating Prop. 4.2.6 for c.f. continuable to $S(-\infty, r)$. □

If $\underline{s} < 0 < \bar{s}$, we can combine the two conditions of Prop. 4.3.8 to the following statement.

**Corollary 4.3.9.** *The c.f. $f$ with d.f. $F$ is an a.c.f. whose strip of regularity contains the strip $S(-r, r)$, $r > 0$ iff, as $x \to \infty$,*

$$T_F(x) = O(e^{-sx}) \quad \textit{for all} \quad s \textit{ with } \quad 0 < s < r, \tag{4.3.8}$$

*where $T_F$ is the tail function of $F$.*

Proof. It is very easy to realize that the two relations (4.3.6) and (4.3.7) are equivalent to (4.3.8). Thus we see from Prop. 4.3.8 that (4.3.8) holds iff $f$ is analyticat least in $S(-r, r)$. □

Finally, we compare the a.c.f. and the b.c.f. having mathematical expectation $\mu_1$ with the c.f. $\exp(it\mu_1)$ of the degenerated d.f. $U_{\mu_1}$. We will see that the latter has an extremal property. Note that $\mu_1$ need not exist for b.c.f.

**Proposition 4.3.10.** *Let $f$ be an a.c.f. or a b.c.f. with finite first moment $\mu_1$; then*

$$f(is) \geqq e^{-\mu_1 s}$$

*in the strip of regularity $S(\underline{s}, \overline{s})$.*

Proof. Let the d.f. $F$ corresponding to $f$ be concentrated on the interval $I \subseteq R_1$ and write $X$ for a r.v. subject to $F$, so that $\mathsf{E}X = \mu_1$. Then we choose a real number $s$, $\underline{s} < s < \overline{s}$, and introduce the convex function $h(x) = e^{-sx}$ for $x \in I$. Now we apply the Jensen inequality (Prop. 1.3.9); thus

$$f(is) = \int_{-\infty}^{\infty} e^{-sx}\, dF(x) = \mathsf{E}h(X) \geqq h(\mu_1) = e^{-s\mu_1}.$$

## 4.4. Characteristic functions analytic in a half plane

Now we are going to study c.f. which are continuable at least to the upper half plane It is not hard to see that all our results can be carried over to c.f. continuable to the lower half plane.

Putting $r = \infty$ in Prop. 4.3.8 we see that a condition necessary and sufficient for $f$ to be continuable to $S(0, \infty)$ is

$$F(x) = O(e^{sx}), \qquad \text{as } x \to -\infty, \text{ for all } s > 0.$$

In this case the integral $\int_{-\infty}^{\infty} e^{-sx}\, dF(x)$ converges for all $s > 0$, see Th. 4.3.3.

It is the aim of the present section to investigate more closely the connection between the asymptotic behaviour of $f(is)$, as $s \to \infty$, and that of the left tail of the corresponding d.f. $F$.

**Proposition 4.4.1.** *If $f$ is continuable to the half plane $S(0, \infty)$ and satisfies*

$$f(is) \leqq e^{cs}$$

*for a certain $c \in R$ and sufficiently large $s > 0$, then $F$ is bounded to the left and $l = \operatorname{lext}(F) \geqq -c$. The converse is also true.*

Proof. We carry an indirect proof and assume that $-\infty \leqq \operatorname{lext}(F) < -c$. Accordingly, $F(x_0) > 0$ for some $x_0 < -c$; thus we have

$$f(is) \geqq \int_{-\infty}^{x_0} e^{-sx}\, dF(x) \geqq e^{-sx_0} F(x_0) > e^{sc} \tag{4.4.1}$$

provided that $s\,|x_0 + c| > |\log F(x_0)|$. This contradicts the assumption. The converse statement is obvious in view of Prop. 4.2.2 since

$$f(is) = \int_{l-0}^{\infty} e^{-sx}\,dF(x) \leqq e^{-ls}, \qquad s > 0. \qquad \square$$

**Corollary 4.4.2.** *If $f$ is continuable to $S(0, \infty)$, then we have*

$$\operatorname{lext}(F) = -\lim_{s\to\infty} \frac{\log f(is)}{s} \geqq -\infty.$$

Proof. In view of Prop. 4.2.8 it suffices to consider the case $\operatorname{lext}(F) = -\infty$. Then we have $F(x_0) > 0$ for all $x_0 \in R_1$. Now (4.4.1) is again true, and it implies $\underline{\lim}_{s\to\infty} \frac{\log f(is)}{s} \geqq -x_0$ whence the assertion follows since $x_0$ is arbitrary. $\square$

Now we turn to the case in which $f(is)$ may grow more quickly than $e^{cs}$ for all $c > 0$. The following result will be applied in Sect. 4.5 and 7.7.

**Proposition 4.4.3.** *Let $h > 0$ be a function on some interval $[s_0, \infty)$ which is continuous and strictly increasing to infinity. Further, let $h^{-1}$ be the inverse function of $h$. If $f$ is continuable to $S(0, \infty)$ and satisfies*

$$f(is) \leqq e^{sh(s)}, \qquad s > s_0,$$

*then for every $y$, $0 < y < 1$, we have*

$$F(x) \leqq e^{(1-y)xh^{-1}(-yx)}, \qquad x < x_0(y).$$

Proof. For fixed $y$, $0 < y < 1$, we set $x_0 = -h(s_0)/y$. Now we consider an arbitrary $x < x_0$ and see that $-xy > h(s_0)$. Thus putting $s = h^{-1}(-yx) > s_0$ we get

$$F(x) \leqq e^{sx} \int_{-\infty}^{x} e^{-sy}\,dF(y) \leqq e^{sx} f(is) \leqq e^{s(x+h(s))},$$

and our choice of $s$ implies the assertion. $\square$

We add also something in the nature of a converse to this result.

**Proposition 4.4.4.** *Let $h$, $h^{-1}$ be as in Prop. 4.4.3 and assume*

$$F(x) \leqq e^{xh^{-1}(-x)}, \qquad x < x_0,$$

*for a certain $x_0 \leqq 0$. Then $f$ is continuable to $S(0, \infty)$, and for every $y > 0$ we have*

$$f(is) \leqq e^{(s+y)\,h(s+y)}, \qquad s > s_0(y).$$

Proof. By assumption and in view of $h^{-1}(\infty) = \infty$ we easily see that $F(x) = O(e^{cx})$, $x \to -\infty$, for all $c > 0$, i.e., $f$ is continuable to $S(0, \infty)$, see Prop. 4.3.8 where we have to put $r = \infty$.. Hence, integrating by parts in (4.3.3) we get, for $s > 0$ and $x_1 < 0$,

$$f(is) = s\int_{-\infty}^{\infty} e^{-sx} F(x)\,dx = s\int_{-\infty}^{x_1} e^{-sx} F(x)\,dx + s\int_{x_1}^{\infty} e^{-sx} F(x)\,dx = I_1 + I_2,$$

say. For the latter integral we find

$$I_2 \leqq s\int_{x_1}^{\infty} e^{-sx}\,dx = e^{-sx_1}.$$

To estimate the first we choose $x_1 = -h(y + s)$. Then we get

$$-sx + xh^{-1}(-x) \leqq yx \quad \text{for} \quad x \leqq x_1 < 0 .$$

Since $x_1 < x_0$ for $s$ sufficiently large it follows from our assumption that

$$I_1 \leqq s \int_{-\infty}^{x_1} e^{-sx+xh^{-1}(-x)} dx \leqq s \int_{-\infty}^{x_1} e^{yx} \, dx = \frac{s}{y} e^{yx_1} = e^{yx_1 + o(s)} \leqq e^{o(s)},$$

as $s \to \infty$. Combining these estimates we finally obtain

$$f(is) \leqq e^{o(s)} + e^{sh(y+s)} \leqq 2e^{sh(y+s)} \leqq e^{(s+y)h(y+s)} .$$

□

To close this section, we will prove a special result which establishes a relation between the one-sided asymptotic behaviour of two d.f. and the asymptotic behaviour of the corresponding c.f. To this end, we introduce the notation

$$\underline{I}(F, G) := \varliminf_{x \to -\infty} \frac{G(x)}{F(x)} , \qquad \overline{I}(F, G) := \varlimsup_{x \to -\infty} \frac{G(x)}{F(x)} ,$$

where $F$ and $G$ are d.f. such that lext $(F) = -\infty$.

**Proposition 4.4.5.** *Let $F$ and $G$ be d.f. with c.f. $f$ and $g$, resp. Assume that* lext $(F) = -\infty$ *and $f$ is continuable to the half plane $S(0, \infty)$. If $\overline{I}(F, G) < \infty$, then $g$ is also continuable to $S(0, \infty)$ and we have*

$$\underline{I} \leqq \varliminf_{s \to \infty} \frac{g(is)}{f(is)} , \qquad \varlimsup_{s \to \infty} \frac{g(is)}{f(is)} \leqq \overline{I} .$$

Proof. We are going to apply Prop. 4.3.8 (second part) in either direction. Namely, by assumption we have, as $x \to -\infty$, $G(x) = O(F(x))$ and $F(x) = O(e^{sx})$ for all $s$. Hence $G(x) = O(e^{sx})$ so that $g$ is also continuable to $S(0, \infty)$. Thus the integral representation (4.3.3) is valid in $S(0, \infty)$ for both $f$ and $g$, and integration by parts yields

$$f(is) = s \int_{-\infty}^{\infty} e^{-sx} F(x) \, dx , \qquad g(is) = s \int_{-\infty}^{\infty} e^{-sx} G(x) \, dx , \qquad s > 0 .$$

Let $\varepsilon > 0$. Then $G(x) \leqq (\overline{I} + \varepsilon) F(x)$ for $x \leqq x_0(\varepsilon) < 0$ while we make use of $G(x) \leqq 1$ for $x > x_0$. So we obtain

$$\frac{g(is)}{f(is)} \leqq \frac{(\overline{I} + \varepsilon) \int_{-\infty}^{x_0} e^{-sx} F(x) \, dx + \int_{x_0}^{\infty} e^{-sx} \, dx}{\int_{-\infty}^{x_0} e^{-sx} F(x) \, dx} \leqq \overline{I} + \varepsilon + \frac{\frac{1}{s} e^{-sx_0}}{e^{-sx_0} \int_{\infty}^{x_0} F(x) \, dx}$$

$$= \overline{I} + \varepsilon + O\left(\frac{1}{s}\right), \qquad \text{as } s \to \infty .$$

Thus $\varlimsup_{s \to \infty} \frac{g(is)}{f(is)} \leqq \overline{I} + \varepsilon$ whence the second estimate follows. The first can be shown analogously.

□

## 4.5. Integral characteristic functions

We recall that a c.f. is said to be integral if it is continuable to the whole complex plane $S(-\infty, \infty)$. It follows from Cor. 4.3.9 that this is the case iff the tail function $T_F(x) = 1 - F(x) + F(-x + 0)$ satisfies

$$T_F(x) = O(\mathrm{e}^{-sx}) \quad \text{as} \quad x \to \infty\,, \quad \text{for all} \quad s > 0\,. \tag{4.5.1}$$

Th. 4.3.3 tells us then that

$$f(\mathrm{i}s) = \int_{-\infty}^{\infty} \mathrm{e}^{-sx}\, \mathrm{d}F(x)\,, \qquad s \in R_1\,. \tag{4.5.2}$$

In the present section we wish to study the growth of integral c.f. in terms of the tail behaviour of the corresponding d.f. To this end, we introduce the maximum modulus

$$M(f, r) := \max_{|z| \leqq r} |f(z)|\,, \qquad r > 0\,,$$

of the integral c.f. $f$ on the disk $|z| \leqq r$.

We start by proving several useful properties of $M$.

**Lemma 4.5.1.** *Let $r > 0$. Then we have*

(i) $\quad M(f, r) = \max \{f(-\mathrm{i}r), f(\mathrm{i}r)\}$;

(ii) $\quad M(f, r) \geqq \frac{1}{2}\, \mathrm{e}^{rx}\, T_F(x)\,, \quad x > 0$;

(iii) $\quad M(f, r) \leqq 1 + r \int_0^{\infty} \mathrm{e}^{rx}\, T_F(x)\, \mathrm{d}x \leqq 1 + r \sup_{x>0} \mathrm{e}^{(r+1)x}\, T_F(x) < \infty\,.$

Proof. (i) The maximum modulus principle tells us that the value $M(f, r)$ is attained on the boundary of the disk $|z| \leqq r$. Since $f$ possesses the ridge property (see Cor. 4.3.5) the assertion follows.

(ii) For an arbitrary $x > 0$ we have by virtue of (i)

$$2M(f, r) \geqq f(-\mathrm{i}r) + f(\mathrm{i}r) \geqq \int_{|y| \geqq x} (\mathrm{e}^{yr} + \mathrm{e}^{-yr})\, \mathrm{d}F(y) \geqq \mathrm{e}^{rx}\, T_F(x)\,.$$

(iii) In the integral representation (4.5.2) we integrate by parts — using Prop. 4.3.8 — and obtain, for $r > 0$,

$$f(\mathrm{i}r) = r \int_{-\infty}^{\infty} \mathrm{e}^{-rx}\, F(x)\, \mathrm{d}x \leqq 1 + r \int_{-\infty}^{0} \mathrm{e}^{-rx}\, F(x)\, \mathrm{d}x$$

so that we also have

$$f(\mathrm{i}r) \leqq 1 + r \int_0^{\infty} \mathrm{e}^{rx}\, T_F(x)\, \mathrm{d}x\,.$$

Concluding analogously we see that the same estimate holds true for $f(-\mathrm{i}r)$; hence the first inequality follows by virtue of (i). The second is true in view of

$$M(f, r) \leqq 1 + r \int_0^{\infty} \mathrm{e}^{-x}\, \mathrm{e}^{(r+1)x}\, T_F(x)\, \mathrm{d}x \leqq 1 + r \sup_{x>0} \mathrm{e}^{(r+1)x}\, T_F(x)\,.$$

Note that this supremum is finite in view of (4.5.1). □

Now we are going to apply the notions order and type to integral c.f.; we encourage the reader to get familiar with App. B.3 where the definitions of these notions are given.

**Proposition 4.5.2.** *An integral c.f. $f \neq 1$ possesses an order $\varrho$ such that $1 \leqq \varrho \leqq \infty$. If $\varrho = 1$, then its type $\tau$ satisfies $0 < \tau \leqq \infty$.*

Proof. By assumption there exists $x > 0$ such that $T_F(x) > 0$. Hence all assertions immediately follow from L. 4.5.1 (ii); for instance, we see that

$$\varrho = \overline{\lim_{r\to\infty}} \frac{\log\log M(f, r)}{\log r} \geqq 1. \qquad \square$$

We emphasize that another proof of Prop. 4.5.2 can be given by applying the Phragmén-Lindelöf theorem Cor. C.1.8 to both $S(-\infty, 0]$ and $S[0, \infty)$.

**Proposition 4.5.3.** *The c.f. $f \neq 1$ is integral of order $\varrho = 1$ with a finite type $\tau > 0$, iff the corresponding d.f. $F \neq U_0$ is finite. In either case we have*

$$\tau = \tau_0 := \max\{|\text{lext}(F)|, |\text{rext}(F)|\}.$$

Proof. Let $f$ be integral with order 1 and finite type, i.e., we assume $\tau = \overline{\lim_{r\to\infty}} \frac{\log M(f, r)}{r} < \infty$. Then the assumptions of Prop. 4.4.1 are satisfied (in view of L. 4.5.1 (i)) and applying it we get $\text{lext}(F) > -\infty$; reasoning analogously we also see that $\text{rext}(F) < \infty$. Hence $F$ is finite.

Conversely, if $F \neq U_0$ is finite, then $0 < \tau_0 < \infty$; further, we have $T_F(\tau_0 - \varepsilon) > 0$ and $T_F(\tau_0 + \varepsilon) = 0$ for all $\varepsilon > 0$. So we obtain from L. 4.5.1 (ii), (iii)

$$M(f, r) \geqq \tfrac{1}{2} e^{r(\tau_0 - \varepsilon)} T_F(\tau_0 - \varepsilon) > 0,$$

$$M(f, r) \leqq 1 + r \int_0^{\tau_0} e^{rx} T_F(x) \, dx \leqq 1 + r \int_0^{\tau_0} e^{rx} \, dx = e^{r\tau_0}.$$

Now it can easily be verified that $\varrho = 1$, $\tau = \tau_0$. $\square$

Next we wish to investigate the connection between the growth of an integral c.f. and the tail behaviour of the corresponding d.f. for $T_F > 0$. To this end, we prove the following general result; for the sake of brevity we put $1/0 = \infty$, $1/\infty = 0$, $\log 0 = -\infty$

**Theorem 4.5.4.** *Let $F$ be a d.f. with integral c.f. $f$ and tail function $T_F > 0$. Further, let $V > 0$ and $W > 0$ be functions of regular variation with exponents of regularity $v \geqq 0$ and $w \geqq 0$, resp., which are continuous and strictly increasing to infinity. Then we have*

$$\lim_{x\to\infty} \frac{V\left(-\dfrac{\log T_F(x)}{x}\right)}{W(x)} = l, \qquad 0 \leqq l \leqq \infty,$$

*iff*

$$\overline{\lim_{r\to\infty}} \frac{W\left(\dfrac{\log M(f, r)}{r}\right)}{V(r)} = \frac{d_{v,w}}{l}, \qquad 0 \leqq l \leqq \infty,$$

*where $d_{v,w} := v^v w^w/(v + w)^{v+w}$, if $vw > 0$, and $d_{v,w} := 1$, if $vw = 0$.*

Before proving this theorem we formulate two consequences of it which express the order $\varrho$ and the type $\tau$ of an integral c.f. in terms of the tail behaviour of the corresponding d.f.

**Corollary 4.5.5.** *Let $F$ be a d.f. with integral c.f. and $T_F > 0$; put $\xi = \lim_{x\to\infty} \dfrac{\log(-\log T_F(x))}{\log x}$. Then the order $\varrho$ of its c.f. is given by $\varrho = 1 + \dfrac{1}{\xi - 1}$.*

Proof. We apply Th. 4.5.4 with $V(x) = W(x) = \log x$ so that $v = w = 0$. Thus we obtain

$$l = \varliminf_{x\to\infty} \frac{\log(-\log T_F(x)) - \log x}{\log x} = \xi - 1$$

and, in view of the definition of the order $\varrho$,

$$\frac{1}{l} = \varlimsup_{r\to\infty} \frac{\log\log M(f, r) - \log r}{\log r} = \varrho - 1 .$$

Now the assertion follows from these equalities. □

**Corollary 4.5.6.** *Let $T_F > 0$, $f$ be an integral c.f. of finite order $\varrho$ and $\xi$ be defined as in Cor. 4.5.5.*

(i) *If $\varrho = 1$, then* $\tau = \tau_0 = \max\{|\text{lext}(F)|, |\text{rext}(F)|\} = \infty$.

(ii) *If $\varrho > 1$, then $\tau = (d_{1,\xi-1}/l_\xi)^{1/(\xi-1)}$ where $\xi = 1 + \dfrac{1}{\varrho - 1}$ and $l_\xi := \varliminf\limits_{x\to\infty} \dfrac{-\log T_F(x)}{x^\xi}$.*

Proof. (i) is an immediate consequence of Prop. 4.5.3 even if $\tau_0 = \infty$. (ii) We choose $V(x) = x$ and $W(x) = x^{\xi-1}$ so that $v = 1$, $w = \xi - 1$. Then Th. 4.5.4 tells us that

$$l = \varliminf_{x\to\infty} \frac{-x^{-1}\log T_F(x)}{x^{\xi-1}} = l_\xi$$

while, in view of the definition of type $\tau$

$$\frac{d_{1,\xi-1}}{l} = \varlimsup_{r\to\infty} \frac{(r^{-1}\log M(f, r))^{\xi-1}}{r} = \varlimsup_{r\to\infty} \left(\frac{\log M(f, r)}{r^\varrho}\right)^{\xi-1} = \tau^{\xi-1} .$$

These relations imply the assertion. □

Proof of Theorem 4.5.4. It is obviously enough to prove that

$$\varliminf_{x\to\infty} V\left(-\frac{\log T_F(x)}{x}\right)\Big/ W(x) \geqq l \tag{4.5.3}$$

iff

$$\varlimsup_{r\to\infty} W\left(\frac{\log M(f, r)}{r}\right)\Big/ V(r) \leqq d_{v,w}/l . \tag{4.5.4}$$

This equivalence is trivial for $l = 0$ so that we assume $0 < l \leqq \infty$.

*If-part*: We select some $l_0 > d_{v,w}/l \geqq 0$. Then (4.5.4) tells us that, for large $r > 0$,

$$W\left(\frac{\log M(f, r)}{r}\right) \leqq l_0 V(r) .$$

Accordingly, using also L. 4.5.1 (i), we get

$$\log f(ir) \leqq \log M(f, r) \leqq r W^{-1}(l_0 V(r))$$

where $W^{-1}$ stands for the inverse function of $W$. The assumed properties of $V$ and $W$ ensure that the function $h(r) := W^{-1}(l_0 V(r))$ strictly increases and tends to infinity as $r \to \infty$; further, we have, $h^{-1}(x) = V^{-1}(W(x)/l_0)$. Now we apply Prop. 4.4.3 and verify that

$$\log F(-x) \leqq -x(1 - \eta)\, V^{-1}\left(\frac{W(\eta x)}{l_0}\right)$$

where $0 < \eta < 1$ is arbitrary and $x > 0$ sufficiently large. The same estimate holds true for the right tail $1 - F(x)$. Hence, by virtue of $T_F(x) \leqq 2 \max \{F(-x), 1 - F(x)\}$, we obtain for large $x$

$$\log T_F(x) \leqq -x(1-\eta)\big(1+o(1)\big)\, V^{-1}\left(\frac{W(\eta x)}{l_0}\right).$$

Consequently,

$$\frac{V\left(-\dfrac{\log T_F(x)}{x}\right)}{W(x)} \geqq \frac{V\left((1-\eta)\big(1+o(1)\big)\, V^{-1}\left(\dfrac{W(\eta x)}{l_0}\right)\right)}{V\left(V^{-1}\left(\dfrac{W(\eta x)}{l_0}\right)\right)} \frac{\dfrac{W(\eta x)}{l_0}}{W(x)}.$$

Finally, we make use of the regular variation of $V$ and $W$ and derive, by virtue of Rem. 2.5.2 (ii),

$$\varliminf_{x\to\infty} V\left(-\frac{\log T_F(x)}{x}\right)\Big/ W(x) \geqq \frac{(1-\eta)^v \eta^w}{l_0}, \qquad 0<\eta<1.$$

On letting $l_0 \to d_{v,w}/l$ and maximizing the right side with respect to $\eta$ we obtain (4.5.3); this is so also if $vw = 0$ since in this case $d_{v,w} = 1$.

*Only-if part*: By (4.5.1) the function $g_r(x) := e^{(r+1)x}\, T_F(x)$, $x > 0$, tends to 0 as $x \to \infty$. We show, reasoning indirectly, that it attains its supremum $\bar{g}$. Let us assume $g_r < \bar{g}$. Then we consider a sequence $(x_{rk})$ such that $g_r(x_{rk}) \to \bar{g}$ as $k \to \infty$. We have a finite limit point $x_r$ since $(x_{rk})$ is trivially bounded. Now we choose some subsequence $(x_{rk'})$ such that $x_{rk'} \to x_r$. Now we get a contradiction if $x_r$ is a continuity point of $g_r$ since $g_r(x_{rk'}) \to g_r(x_r)$ follows as $k' \to \infty$ so that we have $\bar{g} = g(x_r)$; otherwise we see from left continuity that $x_{rk'} \to x_r - 0$, hence we obtain the desired contradiction by an analogous reasoning.

Thus we see from L. 4.5.1 (iii), that $M(f, r) \leqq 1 + r g_r(x_r)$. Using also L. 4.5.1 (ii), we get, for any fixed $x > 0$ and sufficiently large $r$,

$$\tfrac{1}{2} e^{rx}\, T_F(x) \leqq M(f, r) \leqq 2r g_r(x_r) = 2r\, e^{(r+1)x_r}\, T_F(x_r).$$

Accordingly, $x_r \to \infty$ as $r \to \infty$ since we assume $T_F > 0$. Moreover, we obtain

$$\log M(f, r) \leqq r x_r\big(1 + o(1)\big) + \log T_F(x_r), \qquad r \to \infty.$$

Now we choose a number $l_0$, $0 < l_0 < l$. Then we see from (4.5.3) that

$$\log T_F(x_r) \leqq -x_r V^{-1}\big(l_0 W(x_r)\big)$$

for sufficiently large $r$. Combining the last inequalities we get

$$0 < \frac{\log M(f, r)}{r} \leqq x_r\left(1 + o(1) - \frac{s_r}{r}\right) \tag{4.5.5}$$

where

$$s_r = V^{-1}\big(l_0(W(x_r)\big);$$

now wee see that $0 \leqq s_r/r \leqq 1 + o(1)$.

Next we choose a sequence $r' \to \infty$ for which the upper limit in (4.5.4) will be attained. This sequence contains a subsequence $r''$ such that $s_{r''}/r'' = \eta + o(1)$, $0 \leqq \eta \leqq 1$, converges; for the sake of brevity we suppress the primes.

Then for arbitrary $\varepsilon > 0$ and all sufficiently large $r$

$$\eta - \varepsilon < \frac{s_r}{r} < \eta + \varepsilon .$$

Now we make use of $l_0 W(x_r) = V(s_r) \leqq V((\eta + \varepsilon) r)$ and get from (4.5.5)

$$\frac{W(r^{-1} \log M(f, r))}{V(r)} \leqq \frac{W(x_r[1 - \eta + \varepsilon + o(1)])}{l_0 W(x_r)} \frac{V((\eta + \varepsilon) r)}{V(r)}$$

On letting first $r \to \infty$ and then $l_0 \to l$ we obtain

$$\overline{\lim_{r \to \infty}} \frac{W(r^{-1} \log M(f, r))}{V(r)} = \frac{1}{l} (1 - \eta + \varepsilon)^w (\eta + \varepsilon)^v .$$

Now, we let $\varepsilon \to 0$, and maximize then the right hand ride with respect to $\eta$; thus we obtain the desired upper bound $l^{-1} d_{v,w}$. □

Finally we wish to point out that the only-if part of Th. 4.5.4 could also be formulated in a somewhat sharper way.

**Corollary 4.5.7.** *Let $F$ be an arbitrary d.f. If* (4.5.3) *is satisfied for a certain $l > 0$, then $F$ has an integral c.f.*

Proof. From (4.5.3) and $V(\infty) = W(\infty) = \infty$ it is seen that $-1/x \log T(x) \to \infty$, i.e., $T(x) = O(e^{-rx})$ for all fixed $r > 0$; thus the assertion follows from Cor. 4.3.9. □

## 4.6. Factorization of characteristic functions

In view of the convolution theorem (Th. 3.1.6) certain c.f. can be written as the product of two or more c.f. In a trivial way this is even possible for all c.f. $f$; namely, we have

$$f(t) = e^{iat} (f(t) e^{-iat})$$

with $a \in R_1$. In order to exclude such cases from our consideration we introduce a new concept.

**Definition 4.6.1.** A c.f. $f$ is said to be *decomposable* if it can be represented as $f = f_1 f_2$ where $f_1$ and $f_2$ are both c.f. of non-degenerated d.f.; then $f_1$ and $f_2$ are called *factors* of $f$. □

Today the factorization theory of c.f. is a rather vast field, but we confine ourselves to acquainting the reader only with one special problem. Let us consider a decomposable c.f. $f$, then the natural question arises: Which properties does $f$ share with its factors? In particular, we aim at proving Cramér's factorization theorem.

We start with a useful result which holds true both for a.c.f. and b.c.f.

**Proposition 4.6.2** (Rajkov). *Let the c.f. $f$ be decomposable. If it is continuable to a certain strip, then any factor of $f$ is continuable at least to the same strip.*

Proof. Assume $f = f_1 f_2$ and let $F$, $F_k$ stand for the d.f. corresponding to $f$, $f_k$. Then there exist i.r.v. $X_k \sim F_k$, $k = 1, 2$, such that $X_1 + X_2 \sim F$. Clearly,

$$\mathsf{P}(X_1 \geqq x)\, \mathsf{P}(X_2 \geqq y) \leqq \mathsf{P}(X_1 + X_2 \geqq x + y) = 1 - F(x + y) \qquad (4.6.1)$$

and we can choose $y$ such that $\mathsf{P}(X_2 \geqq y) > 0$.

Let us now suppose that $f$ is continuable to $S(\underline{s}, 0)$, $\underline{s} < 0$; then we see from the only-if part of Prop. 4.3.8 that

$$1 - F(x) = O(e^{sx}), \qquad x \to \infty, \quad \text{for all} \quad s \text{ with } \underline{s} < s < 0.$$

By (4.6.1) the d.f. $F_1$ has the same property. Now we apply the if-part of Prop. 4.3.8 and see that $f_1$ is continuable to $S(\underline{s}, 0)$. The other case, in which $f$ is continuable to $S(0, \overline{s})$, is now obvious. □

The reader will easily find examples showing that a factor can have a strip of regularity which is wider than the strip of regularity of $f$. Form, for example, products of c.f.

**Corollary 4.6.3.** *If the c.f. $f \neq 1$ is integral, then any factor is also integral and its order cannot exceed the order of $f$.*

Proof. In view of Prop. 4.6.2 it is enough to prove the second assertion and to assume that the order $\varrho$ of $f$ is finite. Then we note first that by analytic continuation the assumed factorization $f = f_1 f_2$ holds in the whole plane. Accordingly, we have

$$\begin{aligned} f(is) &= f_1(is)\, f_2(is) = f_1(is) \int_{-\infty}^{\infty} e^{-sx}\, dF_2(x) \\ &\geqq f_1(is)\, e^{-|s| x_0} \left(F_2(x_0) - F_2(-x_0)\right) > 0 \end{aligned}$$

for all $s \in R_1$ and a sufficiently large number $x_0 > 0$. This is true also after replacing $s$ by $-s$. Now it follows from L. 4.5.1 (i), that, for $|s| = r$,

$$\left(F_2(x_0) - F_2(-x_0)\right) M(f_1, r) \leqq e^{rx_0} M(f, r).$$

By assumption, the c.f. $f \neq 1$ has order $\varrho$ so that we obtain the estimate

$$\left(F_2(x_0) - F_2(-x_0)\right) M(f_1, r) \leqq \exp\left(rx_0 + r^{\varrho+\varepsilon}\right)$$

where $r$ is sufficiently large. Since $\varrho \geqq 1$ the assertion immediately follows. □

Now we are in a position to turn to Cramér's factorization theorem. Let us consider two normal c.f.

$$\varphi_k(t) = \exp\left(i\mu_k t - \frac{\sigma^2 k}{2} t^2\right), \qquad k = 1, 2.$$

Then the product $\varphi = \varphi_1 \varphi_2$ is again a normal c.f.; hence it is seen from the convolution theorem that the set of normal d.f. is closed under convolution. The converse of this fact is by far deeper.

**Theorem 4.6.4** (Cramér). *The c.f. $\varphi_{\mu,\sigma}(t) = \exp\left(i\mu t - \sigma^2/2\, t^2\right)$ of the normal d.f. $\Phi_{\mu,\sigma}$ has only normal factors.*

Proof. Clearly, $\varphi_{\mu,\sigma}$ is an integral function of order 2 without zeros. Accordingly, Cor. 4.6.3 tells us that any factor $f_1$ of $\varphi_{\mu,\sigma}$ is also integral without zeros and with order not exceeding 2. Hence, it is an easy consequence of the Carathéodory inequality stated as Prop. B.3.6 that implies $f_1 = \exp(P_m)$ where $P_m$ stands for a polynomial of degree $m$. From the definition of the order we see that $m$ is the order of $f_1$ whence $m \leqq 2$ follows. Thus

$$\log f_1(t) = c_0 + c_1 t + c_2 t^2;$$

here $c_2 = 0$ is ruled out by the definition of a factor. In view of $f_1(0) = 1$ we have $c_0 = 0$, while $\overline{f_1(t)} = f_1(-t)$ implies that $c_1$ is purely imaginary and $c_2$ is real. Further, since $f_1$ is bounded we see that $c_2 < 0$. □

## 4.7. Cramér's theorem within the framework of the continuation theory of distribution functions

Th. 4.6.4 can easily be reformulated in terms of d.f. Namely, if we have two non-degenerated d.f. $F_1$ and $F_2$ satisfying

$$F_1 * F_2 = \Phi_{\mu,\sigma} ,$$

then it follows that $F_1$ and $F_2$ are also normal.

In the present section we will deal with the problem of whether an analogous statement can be concluded from the weaker assumption

$$F_1 * F_2(x) = \Phi_{\mu,\sigma}(x) , \qquad x \leqq x_0 . \tag{4.7.1}$$

The following example shows that such a simple generalization of Cramér's theorem does not exist and so we will need an additional supposition.

**Example 4.7.1.** We write $\Phi = \Phi_{0,1}$ and put

$$F_1(x) = \begin{cases} 2 \sum_{n=0}^{\infty} (-1)^n \Phi(x-n) = \sqrt{\frac{2}{\pi}} \sum_{n=0}^{\infty} \int_{x-(2n+1)}^{x-2n} e^{-u^2/2} \, du , & x \leqq 0 , \\ v(x) , & x > 0 , \end{cases}$$

where $v$ is an arbitrary monotone function on $(0, \infty)$ such that $v(+0) = F_1(0)$ and $v(\infty) = 1$. Then it is seen that $F_1$ is a d.f. Further, let $F_2$ denote the d.f. attributing mass $1/2$ to the atoms 0 and 1. Then we have, by virtue of Cramér's theorem, $F_1 * F_2 \neq \Phi$ but, on the other hand,

$$F_1 * F_2(x) = \tfrac{1}{2} \big(F_1(x) + F_1(x-1)\big) = \Phi(x) , \qquad x \leqq 0 . \qquad \square$$

We first present a general continuation theorem for d.f.

**Proposition 4.7.2.** *Let $G_1 > 0$ be a d.f. such that its c.f. $g_1$ is continuable to the half plane $S(0, \infty)$, has no zeros, and satisfies*

$$g_1(\mathrm{i}s) \leqq C \, e^{cs^2} , \qquad s \geqq 0 , \tag{4.7.2}$$

*for certain positive numbers $C$ and $c$. If the d.f. $G_2$ satisfies*

$$G_2(x) = G_1(x) , \qquad x \leqq x_0 , \tag{4.7.3}$$

*for some real $x_0$, and if its c.f. $g_2$ also has no zeros in $S(0, \infty)$, then $G_1 = G_2$.*

Proof. Without loss of generality we set $x_0 = 0$ since the d.f. $G_i(x) := G_i(x + x_0)$, $i = 1, 2$, also satisfy the assumptions of Prop. 4.7.2. We introduce the function

$$d(t) := \tfrac{1}{2} \big(g_1(t) - g_2(t)\big) = \tfrac{1}{2} \int_0^{\infty} e^{\mathrm{i}tx} \, d\big(G_1(x) - G_2(x)\big)$$

where the second equality is obvious by (4.7.3). Its analytic continuation

$$d(z) = \tfrac{1}{2} \big(g_1(z) - g_2(z)\big) = \tfrac{1}{2} \int_0^{\infty} e^{\mathrm{i}zx} \, d\big(G_1(x) - G_2(x)\big) , \qquad z \in S[0, \infty) ,$$

is bounded in absolute value by 1. Thus we have by the ridge property of c.f.

$$|g_2(z)| \leqq 2 \, |d(z)| + |g_1(z)| \leqq 2 + g_1(\mathrm{i}s) , \qquad z \in S[0, \infty) .$$

Hence, from (4.7.2) it is seen that there exists $C' > 0$ such that

$$|g_2(z)| \leqq C' \, e^{cs^2} , \qquad s > 0 . \tag{4.7.4}$$

It is the point of the present proof that we can write

$$d(z) = \Delta(z) \left(\sqrt{g_1(z)} + \sqrt{g_2(z)}\right)$$

where

$$\Delta(z) := \tfrac{1}{2} \left(\sqrt{g_1(z)} - \sqrt{g_2(z)}\right) \tag{4.7.5}$$

and $\sqrt{g_1(0)} = \sqrt{g_2(0)} := 1$. We aim at proving that $\Delta = 0$.

By assumption the above roots represent analytic functions in $S(0, \infty)$. Moreover, we have the estimate

$$\begin{aligned} 1 \geqq |d(is)| &= |\Delta(is)| \left(\sqrt{g_1(is)} + \sqrt{g_2(is)}\right) \\ &\geqq |\Delta(is)| \sqrt{g_1(is)}\,. \end{aligned} \tag{4.7.6}$$

Combining (4.7.2), (4.7.4), and (4.7.5) we obtain by means of the ridge property

$$|\Delta(z)| \leqq (\sqrt{C} + \sqrt{C'})\, e^{cs^2}, \qquad s \geqq 0\,.$$

On the other hand, the assumption $G_1 > 0$ implies

$$\lim_{s \to \infty} s^{-1} \log g_1(is) = \infty \tag{4.7.7}$$

by Cor. 4.4.2. Thus it is clear from (4.7.6) that $|\Delta(is)|$ is bounded for $s \geqq 0$.

Thus $\Delta$ satisfies all assumptions of the Phragmén-Lindelöf theorem Prop. C.1.13, and applying it we see that $\Delta$ is bounded in $S(0, \infty)$.

Suppose now that $\Delta \neq 0$; then we combine (4.7.6) and (4.7.7) to obtain

$$k_\Delta := \overline{\lim_{s \to \infty}}\, s^{-1} \log \Delta(is) = -\infty\,.$$

But this is impossible by Cor. C.1.12. Accordingly, $\Delta = 0$, that is, $d = 0$ follows, which it was our intention to prove. □

We recommend the reader to carry out the final indirect argument by means of (4.7.7) and the Phragmén-Lindelöf theorem, Prop. C.1.14., although Prop. 4.7.2 then looses its elementary character.

The following special case deserves particular attention since it is a more general and at the same time simpler version of the very first continuation theorem which is stated as Th. 7.6.1. Note that the suppositions imply that the c.f. $g$ of $G$ is continuable to $S(0, \infty)$, see Prop. 4.3.8.

**Proposition 4.7.3.** *Let $G$ be a d.f. such that*

$$G(x) = \Phi(x)\,, \qquad x \leqq x_0\,.$$

*If its c.f. $g$ has no zeros in the half plane $S(0, \infty)$, then $G = \Phi$.* □

We turn now to the previously mentioned generalization of Cramér's theorem. Before formulating it, we recall that (4.7.1) implies $F_1(x)\, F_2(y) \leqq \Phi_{\mu,\sigma}(x + y)$ for $x + y \leqq x_0$ so that, by virtue of Prop. 4.3.8, the corresponding c.f. $f_1$ and $f_2$ are continuable to $S(0, \infty)$.

**Proposition 4.7.4.** *If (4.7.1) holds and the c.f. $f_k$ of $F_k$ have no zeros in the half plane $S(0, \infty)$, then $F_k$, $k = 1, 2$, are normal d.f.*

Proof. Putting $G_1 = \Phi_{\mu,\sigma}$ and $G_2 = F_1 * F_2$ we can apply Prop. 4.7.2 to obtain $F_1 * F_2 = \Phi_{\mu,\sigma}$ whence the assertion follows by virtue of Cramér's theorem. □

The proof of Prop. 4.7.2 shows that zeros of $g_1$ and $g_2$ of an even order may be admitted. In fact, it suffices that the roots $\sqrt{g_1}$ and $\sqrt{g_2}$ are analytic functions in $S(0, \infty)$. Further, it is even enough to assume that there exists a natural number $n \geqq 2$ such that $g_k^{1/n}$ $(k = 1, 2)$ is analytic in $S(0, \infty)$; namely, in this case we put $\Delta = g_1^{1/n} - g_2^{1/n}$ and obtain

$$2d = g_1 - g_2 = \Delta \sum_{l=0}^{n-1} g_1^{l/n} g_2^{(n-1-l)/n}$$

and $2\,|d(is)| \geqq |\Delta(is)|\, g_2(is)^{(n-1)/n}$ so that the method applies and yields $d = 0$.

These assumptions which are somewhat more general than those of Prop. 4.7.2 are certainly satisfied if $G_1 = F^{*n}$ and $G_2 = G^{*n}$. Therefore we obtain the following result in the formulation of which zeros are not mentioned.

**Proposition 4.7.5.** *Assume that for two d.f. $F$ and $G$, $G > 0$, and a natural number $n \geqq 2$ we have*

$$F^{*n}(x) = G^{*n}(x)\,, \qquad x \leqq x_0\,. \tag{4.7.8}$$

*If the c.f. $g$ of $G$ is continuable to $S(0, \infty)$ and satisfies (4.7.2) then $F = G$, i.e., relation (4.7.8) continues to the line.* □

The most interesting particular case concerns the normal d.f. and is related to Cramér's theorem. We obtain it by putting $G^{*n} = \Phi_{\mu,\sigma}$.

**Corollary 4.7.6.** *If for a certain natural number $n \geqq 2$*

$$F^{*n}(x) = \Phi_{\mu,\sigma}(x)\,, \qquad x \leqq x_0\,, \tag{4.7.9}$$

*then $F$ is a normal d.f.* □

In preparation for an application in Sect. 7.6, we state now a counterpart to Prop. 4.7.5, for the proof of which we again apply the Phragmén-Lindelöf theorem C.1.14.

**Proposition 4.7.7.** *Assume that for two d.f. $F$ and $G$,* $\operatorname{lext}(F) = \operatorname{lext}(G) = 0$, *and a natural number $n \geqq 2$ we have*

$$F^{*n}(x) = G^{*n}(x)\,, \qquad 0 \leqq x \leqq x_0; \tag{4.7.10}$$

*then $F(x) = G(x)$, $x \leqq x_0$.*

Proof. Writing $f$, $g$ for the c.f. of $F$, $G$ we introduce for $z \in S[0, \infty)$ the functions

$$d_n(z) := \big(f^n(z) - g^n(z)\big)\, e^{-izx_0} = \int_{-0}^{\infty} e^{iz(x-x_0)}\, d\big(F^{*n}(x) - G^{*n}(x)\big)\,,$$

$$h(z) := \sum_{l=0}^{n-1} f^l(z)\, g^{n-l-1}(z)$$

so that $d_n = d_1 h$. Since $h$ is (but for a positive factor) a c.f. whose d.f. satisfies $\operatorname{lext}(H) = 0$ we have by Prop. 4.2.9

$$\lim_{s\to\infty} s^{-1} \log h(is) = 0\,.$$

Clearly, $h$ is bounded in $S[0, \infty)$, and by assumption (4.7.10) so is $d_n$. Hence, all assumptions of Th. C.1.14 are satisfied, and applying it we see that $d_1$ is bounded in $S[0, \infty)$. Therefore,

$$\int_{-0}^{x_0} e^{iz(x-x_0)}\, d\big(F(x) - G(x)\big) = \int_{-\infty}^{0} e^{izu}\, d\big(F(x_0+u) - G(x_0+u)\big)$$

is a bounded integral function, i.e. a constant. Accordingly, an easy application of Prop. B.2.8 (to left-sided Laplace-Stieltjes transforms) yields the desired assertion. □

## 4.8. Hinčin's phenomenon within the framework of the continuation theory of distribution functions

The restriction $f_I$ of a c.f. $f$ to the interval $I = (-T, T)$ does not always admit of a unique continuation to $\mathbb{R}_1$ in the set of all c.f.,, obvious examples are provided by Prop. 3.6.5. This observation leads to the fact which is sometimes called *Hincin's phenomenon*. Namely, if we have the relation

$$F := G * F_1 = G * F_2 \tag{4.8.1}$$

for the d.f. $G$, $F_1$, $F_2$, then we cannot conclude $F_1 = F_2$ in general. This is easily seen since, for the corresponding c.f., we have

$$f = gf_1 = gf_2 \tag{4.8.2}$$

and $f_1 = f_2$ follows trivially provided that $g$ has only isolated zeros. If, however, $g$ is for instance the Pólya type c.f. $g(t) = 1 - |t|$, $|t| \leqq 1$, which vanishes for $|t| > 1$, then (4.8.2) entails only $f_1(t) = f_2(t)$ for $|t| \leqq 1$. Therefore, we are faced with the natural question of under which sufficient conditions the relation (4.8.1) implies $F_1 = F_2$.

A simple answer to this problem is possible since there always exists an interval $(-T, T)$ in which $g$ has no zero, so that $f_1(t) = f_2(t)$, $|t| \leqq T$, follows from (4.8.2). Let now $f_1$ and $f_2$ be either a.c.f. or b.c.f.; then they are continuable to a certain strip, and from Rem. 4.1.2 one immediately obtains $f_1 = f_2$, i.e., $F_1 = F_2$. Obviously, we can state the following even more general result.

**Proposition 4.8.1.** *If the c.f. $f_1$ is uniquely defined by its restriction to an arbitrary interval $(-T, T)$, then* (4.8.1) *implies $F_1 = F_2$.* □

Now we modify the problem and make the weaker assumption that

$$G * F_1(x) = G * F_2(x)\,, \qquad x \leqq x_0\,, \tag{4.8.3}$$

for some $x_0 \in \mathbb{R}_1$. As will be shown, also in this setting of the problem, one-sided asymptotic assumptions permit farreaching conclusions on the components $F_1$ and $F_2$. It is needless to add that the following proposition also sheds new light on the original Hinčin problem.

**Lemma 4.8.2.** *Let $D$ be a function of bounded variation satisfying*

$$|D(x)| = O(e^{sx})\,, \qquad x \to -\infty\,,$$

*for any $s > 0$. Then its Fourier-Stieltjes transform*

$$d(t) = \int_{-\infty}^{\infty} e^{itx}\, dD(x)$$

*is continuable to the half plane $S(0, \infty)$.*

Proof. We write

$$d(t) = \int_{-\infty}^{-0} e^{itx}\, dD(x) + \int_{-0}^{\infty} e^{itx}\, dD(x) = d_-(t) + d_+(t)\,,$$

say, and see by partial integration that $d_-$ is continuable to the half plane $S(-\infty, 0)$ in the form

$$d_-(z) = \int_{-\infty}^{-0} e^{izx}\, dD(x) = D(0) - iz \int_{-\infty}^{0} e^{izx}\, D(x)\, dx\,. \tag{4.8.4}$$

By assumption the right integral converges uniformly even in every half plane $S(-\infty, s)$ so that $d_-$ represents an integral function by the Weierstrass convergence theorem. This proves the assertion because $d_+$ can trivially be continued to $S(0, \infty)$. □

**Proposition 4.8.3.** *Assume* (4.8.3) *where* lext $(G) = -\infty$ *and*

$$G(x) = O(e^{sx}), \qquad x \to -\infty, \quad \text{for all} \quad s > 0. \tag{4.8.5}$$

*If, further, for the function* $D := F_1 - F_2$ *we have*

$$|D(x)| = O(\exp(-cx^2)), \qquad x \to -\infty \tag{4.8.6}$$

*for some* $c > 0$, *then* $F_1 = F_2$.

Proof. Without loss of generality we may suppose $x_0 = 0$ since otherwise we consider $G(x + x_0)$ instead of $G(x)$. We introduce the function

$$H := G * D$$

so that $H(x) = 0$, $x < 0$, by assumption. This means that its Fourier-Stieltjes transform $h$ is continuable to the half plane $S(0, \infty)$ as a bounded analytic function. Writing $g$ for the c.f. of $G$ and $d$ for the transform of $D$ we obtain from the convolution theorem (Prop. B.2.5)

$$h(t) = g(t)\, d(t), \qquad t \in \mathbb{R}_1. \tag{4.8.7}$$

The assumptions (4.8.5) and (4.8.6) imply in view of L. 4.8.2 that $g$ and $d$ are also continuable to $S(0, \infty)$. Schwarz's reflection principle indicates that the continuation of $h$ from $\mathbb{R}_1$ to $S(0, \infty)$ is unique so that

$$h(z) = g(z)\, d(z), \qquad z \in S[0, \infty),$$

follows from (4.8.7).

Since $h$ is bounded in $S(0, \infty)$ we have

$$C \geqq |g(\mathrm{i}s)\, d(\mathrm{i}s)|, \qquad s > 0, \tag{4.8.8}$$

for some $C > 0$; accordingly, $d(\mathrm{i}s)$ is also bounded because lext $(G) = -\infty$ implies $\lim_{s\to\infty} g(\mathrm{i}s) = \infty$ by Cor. 4.4.2.

Further, in view of (4.8.6) we have the estimate

$$\left| \int_{-\infty}^{0} e^{\mathrm{i}zx} D(x)\, dx \right| \leqq \int_{-\infty}^{0} e^{-yx} |D(x)|\, dx$$

$$\leqq C_1 \int_{-\infty}^{0} \exp(-yx - cx^2)\, dx \leqq C_2 \exp(y^2/4c), \quad z \in S(0, \infty),$$

for certain positive constants $C_1$ and $C_2$. Hence, again using (4.8.4), we see that

$$\left| \frac{d(z)}{z + \mathrm{i}} \right| \leqq C_3 \exp\left(\frac{s^2}{4c}\right)$$

for some $C_3 > 0$.

Now we are in a position to apply the theory of Phragmén and Lindelöf. First we see that Prop. C.1.13 applies to the function $d(z)/(z + \mathrm{i})$ whence $|d(z)| \leqq C_4(|z| + 1)$ follows for some $C_4 > 0$. Thus $d$ is bounded in $S[0, \infty)$ by Th. C.1.3. Accordingly, we can repeat the final step of the proof of Prop. 4.7.1 to obtain $d = 0$; here the role of (4.7.6) is taken over by (4.8.8), and (4.7.7) is obviously true also for $g$. □

## 4.9. More about characteristic functions of one-sided distribution functions

We return now to the subject of Sect. 4.2. and, for the sake of simplicity, we set lext $(F) = 0$. In this case Cor. 3.2.7. yields two different inversion formulae, which we state as follows.

**Proposition 4.9.1.** *If $F(0) = 0$, then*

$$F(x) = \frac{2}{\pi} \int_0^\infty \sin xu \frac{\operatorname{Re} f(u)}{u} \, du$$

*and*

$$F(x) = 1 - \frac{2}{\pi} \int_0^\infty \cos xu \frac{\operatorname{Im} f(u)}{u} \, du$$

*for all continuity points $x \geqq 0$ of $F$.* □

From this we immediately conclude a peculiar uniqueness theorem.

**Proposition 4.9.2.** *A d.f. $F$ satisfying $F(0) = 0$ is uniquely defined if either* Re $f$ *or* Im $f$ *is given.*

The two inversion formulae of Prop. 4.9.1. imply also that there exists a one-to-one correspondence between Re $f$ and Im $f$. The following result expresses this fact in explicite formulae which are sometimes called *Hilbert transforms.*

**Proposition 4.9.3.** *Let* lext $(F) = 0$; *then*

$$\operatorname{Im} f(t) = \frac{-1}{\pi} \int_0^\infty \frac{\operatorname{Re} f(t+u) - \operatorname{Re} f(t-u)}{u} \, du \tag{4.9.1}$$

*and, if $F(+0) = 0$,*

$$\operatorname{Re} f(t) = \frac{1}{\pi} \int_0^\infty \frac{\operatorname{Im} f(t+u) - \operatorname{Im} f(t-u)}{u} \, du \, . \tag{4.9.2}$$

Proof. We confine ourselves to proving (4.9.2) since (4.9.1) can be derived in a similar way. By the Fubini theorem we have trivially

$$\int_a^b \frac{\operatorname{Im} f(t+u) - \operatorname{Im} f(t-u)}{u} \, du = 2 \int_0^\infty \cos tx \int_a^b \frac{\sin ux}{u} \, du \, dF(x) \, , \qquad t \in R_1 \, , \tag{4.9.3}$$

for $0 < a < b$. By Prop. A.6 the inner integral tends to $\pi/2$ as $a \to 0$ and $b \to \infty$. Accordingly, dominated convergence permits us to carry out these passages to the limit under the integral sign, so that we obtain (4.9.2). Note that the assumptions of Prop. A.2 are satisfied only in the case that $F(+0) = 0$. □

Returning now to moments, we note that necessary and sufficient conditions guaranteeing the existence of moments of non-negative r.v. are often expressed in terms of the real part of the corresponding c.f. on the one hand and of the imaginary part on the other hand. Examples concerning the first moments are given in the following proposition, the conditions (ii) and (iii) of which are to be found in LAUE (1973) and RÉNYI (1962), resp.

**Proposition 4.9.4.** *Let* lext $(F) = 0$. *Then the following three statements are equivalent:*

(i) *The first moment* $\mu_1$ *exists;*

(ii) $|\mathrm{Im}\, f(t)/t| < C$ *for all* $t \neq 0$ *and some* $C > 0$;

(iii) $\int_0^\varepsilon (1 - \mathrm{Re}\, f(u))/u^2\, du$ *converges for any* $\varepsilon > 0$.

*Corresponding to the statements* (ii) *and* (iii) *we have the two representations*

$$\mu_1 = \lim_{t \to 0} \mathrm{Im}\, f(t)/t \tag{4.9.4}$$

$$= \frac{2}{\pi} \int_0^\infty \big(1 - \mathrm{Re}\, f(u)\big)/u^2\, du\,. \quad \square \tag{4.9.5}$$

The known proofs of these formulas are very different, and it was J. FOCKE who demanded to find the inner connection between (4.9.4) and (4.9.5). As a result it turned out that it is Prop. 4.9.3. that permits us to derive each of these formulae directly from the other.

**Proposition 4.9.5.** *Let* lext $(F) = 0$. *If* $\mu_1$ *exists, then* (4.9.4) *directly follows from* (4.9.5) *and vice versa.*

Proof. By assumption the derivative $f'$ exists and, accordingly, partial integration permits us to restate (4.9.5); namely, from Cor. 3.5.2. we get $\big(1 - \mathrm{Re}\, f(t)\big)/t = o(1)$ as $t \to 0$ so that (4.9.5) takes the form

$$\mu_1 = -\frac{2}{\pi} \int_0^\infty \mathrm{Re}\, f'(u)\, du\,. \tag{4.9.5'}$$

Hence, we have to show the equivalence of (4.9.4) and (4.9.5′). We set

$$\Delta_t(u) = \frac{\mathrm{Re}\, f(u+t) - \mathrm{Re}\, f(u-t)}{2ut}$$

Then we find

$$\Delta_t(u) - \frac{\mathrm{Re}\, f'(u)}{u} = -\int_0^\infty \left(\frac{\sin tx}{tx} - 1\right) \frac{\sin ux}{u}\, x\, dF(x)\,.$$

Now from the Fubini theorem

$$\int_0^U \left(\Delta_t(u) - \frac{\mathrm{Re}\, f'(u)}{u}\right) du = \int_0^\infty \left(1 - \frac{\sin tx}{tx}\right) \int_0^{Ux} \frac{\sin v}{v}\, dv\, x\, dF(x)\,.$$

By Prop. A.6 we have dominated convergence as $U \to \infty$, so that

$$\int_0^\infty \left(\Delta_t(u) - \frac{\mathrm{Re}\, f'(u)}{u}\right) du = \frac{\pi}{2} \int_0^\infty \left(1 - \frac{\sin tx}{tx}\right) x\, dF(x)\,.$$

Applying dominated convergence as $t \to 0$ on the right we see that

$$\lim_{t\to 0} \int_0^\infty \left( \Delta_t(u) - \frac{\operatorname{Re} f'(u)}{u} \right) du = 0 .$$

Now the desired conclusions are obvious by virtue of (4.9.1): For instance, if (4.9.5′) holds, then

$$\mu_1 = -\frac{2}{\pi} \lim_{t\to 0} \int_0^\infty \Delta_t(u)\, du = \lim_{t\to 0} \frac{\operatorname{Im} f(t)}{t}$$

follows. This demonstrates the assertion. □

The one-to-one correspondence between Re $f$ and Im $f$ is one of the pecularities of the c.f. of non-negative r.v. Another one follows from the fact that such c.f. are analytic in $S(0, \infty)$ which was shown in Sect. 4.2. Therefore, we have the following representation formulae for $f$ which we quote from Girlich and Rossberg (1971) and Laue (1973), resp.; note that (4.9.6) is a special case of the Poisson formula for $S(0, \infty)$.

**Proposition 4.9.6.** *Let* lext $(F) = 0$. *Then we have*

$$f(z) = \frac{2z}{i\pi} \int_0^\infty \frac{\operatorname{Re} f(u)}{u^2 - z^2}\, du , \qquad z \in S(0, \infty) , \tag{4.9.6}$$

*and, if* $F(+0) = 0$,

$$f(z) = \frac{2}{\pi} \int_0^\infty \frac{u \operatorname{Im} f(u)}{u^2 - z^2}\, du , \qquad z \in S(0, \infty) . \;\square \tag{4.9.7}$$

We use this result to derive further representation formulae.

**Lemma 4.9.7.** *Let* lext $(F) = 0$. *Then there exist the integrals*

$$\int_0^\infty \frac{\operatorname{Im} f(v)}{v \pm z}\, dv , \qquad z \in S(0, \infty) .$$

Proof. We write

$$I(R) := \int_0^R \frac{\operatorname{Im} f(v)}{v + is}\, dv = \int_0^R \frac{v \operatorname{Im} f(v)}{v^2 + s^2}\, dv - is \int_0^R \frac{\operatorname{Im} f(v)}{v^2 + s^2}\, dv , \qquad s > 0 ,$$

and see from (4.9.7) that even the first integral on the right has a finite limit as $R \to \infty$; hence $I(\infty)$ exists. Next we find by partial integration

$$\int_0^R \frac{\operatorname{Im} f(u)}{u + z}\, du = \int_0^R \frac{u + is}{u + z}\, dI(u) = \frac{R + is}{R + z} I(R) + (is - z) \int_0^R \frac{I(u)}{(u + z)^2}\, du$$

whence it follows that the limit, as $R \to \infty$, on the right exists. This shows the assertion since the same reasoning is possible on replacing $z$ by $-z$. □

**Proposition 4.9.8.** *Let* lext $(F) = 0$. *Then we have for* $z \in S(0, \infty)$

$$f(z) = \frac{1}{\pi} \int_{-\infty}^{\infty} \frac{\operatorname{Im} f(u)}{u - z}\, du \tag{4.9.8}$$

*and*

$$\operatorname{Re} f(z) = \frac{1}{\pi} \int_{0}^{\infty} \frac{u}{u^2 + s^2} \left(\operatorname{Im} f(t + u) - \operatorname{Im} f(t - u)\right) du\,. \tag{4.9.9}$$

Proof. In view of (4.9.2) it suffices to assume $s > 0$. Then (4.9.7) yields

$$f(z) = \frac{1}{\pi} \int_{0}^{\infty} \left(\frac{\operatorname{Im} f(u)}{u + z} + \frac{\operatorname{Im} f(u)}{u - z}\right) du\,.$$

Using $\operatorname{Im} f(u) = -\operatorname{Im} f(-u)$ we see now from L. 4.9.7. that (4.9.8) is true. Accordingly,

$$\operatorname{Re} f(z) = \frac{1}{\pi} \int_{-\infty}^{\infty} \frac{u - t}{(u - t)^2 + s^2} \operatorname{Im} f(u)\, du$$

$$= \frac{1}{\pi} \int_{-\infty}^{\infty} \frac{v}{v^2 + s^2} \operatorname{Im} f(t + v)\, dv$$

whence (4.9.9) follows. □

**Remark 4.9.9.** (i) Splitting (4.9.8) into $\operatorname{Re} f$ and $\operatorname{Im} f$ we get a formula for $\operatorname{Im} f$; it proves to be the well known Poisson formula.

(ii) Formula (4.9.9) reduces to (4.9.2) for $s = 0$ whence it follows that the passage to the limit as $s \to 0$ in (4.9.9) can be carried out under the integral sign. Further, the formulae (4.9.9) and (4.9.7) coincide for $z = is$.

(iii) For a represention of $\operatorname{Im} f(z)$ analogous to (4.9.9) see Suppl. 4.10.28.

By means of the formulae (4.9.6) and (4.9.7), we will see that there is a close connection between the behaviour of $F(x)$, as $x \to +0$, and that of $\operatorname{Re} f(t)$ and $\operatorname{Im} f(t)$, as $t \to \infty$. The following proposition represents a first result in this direction, here $\Gamma$ stands for the Gamma function defined in App. B.5.

**Proposition 4.9.10.** *Let* lext $(F) = 0$.
(i) *If for some* $\gamma$, $0 \leqq \gamma < 1$, *we have*

$$\lim_{t \to \infty} t^{\gamma} \operatorname{Re} f(t) = A \geqq 0 \tag{4.9.10}$$

*then*

$$\lim_{x \to +0} x^{-\gamma} F(x) = \frac{A}{\cos \frac{\gamma \pi}{2} \Gamma(\gamma + 1)}\,.$$

(ii) *If for some* $\gamma$, $0 < \gamma < 2$, *we have*

$$\lim_{t \to \infty} t^{\gamma} \operatorname{Im} f(t) = B \geqq 0 \tag{4.9.11}$$

*and if* $F(+0) = 0$, *then*

$$\lim_{x \to +0} x^{-\gamma} F(x) = \frac{B}{\sin \frac{\gamma\pi}{2} \Gamma(\gamma + 1)} \cdot$$

For the proof we need an essential tool. Namely, an analogous result concerning the connection of $F(x)$, as $x \to +0$, and $f(iy)$, as $y \to \infty$ is well known as an Abelian theorem for the Laplace transform. We quote if from WIDDER (1946).

**Proposition 4.9.11.** *Let* lext $(F) = 0$. *If for some* $\gamma > 0$ *we have*

$$\lim_{y \to \infty} s^{\gamma} f(is) = C > 0 , \tag{4.9.12}$$

*then*

$$\lim_{x \to +0} x^{-\gamma} F(x) = C/\Gamma(\gamma + 1) . \qquad \square$$

Proof of proposition 4.9.10. By virtue of Prop. 4.9.11. it is sufficient to show:
(i) If (4.9.10) for $0 \leqq \gamma < 1$, then (4.9.12) with $C = A/\cos(\gamma\pi/2)$.
(ii) If (4.9.11) for $0 < \gamma < 2$, then (4.9.12) with $C = B/\sin(\gamma\pi/2)$.
To prove (i), we use (4.9.6) from which we conclude

$$s^{\gamma} f(is) = \frac{2}{\pi} s^{\gamma+1} \int_0^R \frac{\operatorname{Re} f(u)}{u^2 + s^2} \mathrm{d}u + \frac{2}{\pi} s^{\gamma+1} \int_R^\infty \frac{\operatorname{Re} f(u)}{u^2 + s^2} \mathrm{d}u .$$

where $0 < R < \infty$ will be determined in (4.9.13). The first term on the right satisfies

$$\frac{2}{\pi} s^{\gamma+1} \left| \int_0^R \frac{\operatorname{Re} f(u)}{u^2 + s^2} \mathrm{d}u \right| \leqq \frac{2}{\pi} s^{\gamma-1} \int_0^R |\operatorname{Re} f(u)| \, \mathrm{d}u \to 0 ,$$

as $s \to \infty$. The second term can be rewritten as

$$\frac{2}{\pi} s^{\gamma+1} \int_R^\infty \frac{u^{\gamma} \operatorname{Re} f(u) - A}{u^{\gamma}(u^2 + s^2)} \mathrm{d}u + \frac{2}{\pi} A \int_{R/s}^\infty \frac{\mathrm{d}v}{v^{\gamma}(1 + v^2)} = I_1(s) + I_2(s) .$$

Let $\varepsilon > 0$. Then by assumption we may choose $R = R(\varepsilon)$ such that

$$|u^{\gamma} \operatorname{Re} f(u) - A| \leqq \varepsilon \quad \text{for all} \quad u \geqq R . \tag{4.9.13}$$

Now we obtain by a well-known formula (cf. RYŽIK-GRADSTEJN (1957)) $\lim_{s \to \infty} I_2(s) = A/\cos(\gamma\pi/2)$, and the same formula yields $|I_1(s)| \leqq \varepsilon/\cos(\gamma\pi/2)$, $s > 0$. Thus we obtain (i), and (ii) follows in an analogous manner using (4.9.7) instead of (4.9.6). $\square$

We pursue our subject and describe now the behaviour of $F$ at 0 by showing under which sufficient conditions certain moments

$$\mu_{F,-\lambda} := \int_0^\infty x^{-\lambda} \, \mathrm{d}F(x) , \qquad \lambda > 0 ,$$

of negative order exist. This question has hardly been treated with c.f. yet.

**Example.** (i) We need such moments when we consider the mixture

$$G(x) = \int_0^\infty (1 - \mathrm{e}^{-xu}) \, \mathrm{d}F(u) , \qquad x \geqq 0 , \tag{4.9.14}$$

where $F(+0) = 0$. Namely, we have obviously

$$\mu_{G,\lambda} = \lambda \int_0^\infty x^{\lambda-1}\big(1 - G(x)\big)\,\mathrm{d}x = \lambda \int_0^\infty x^{\lambda-1} \int_0^\infty \mathrm{e}^{-xu}\,\mathrm{d}F(u)\,\mathrm{d}x$$

$$= \lambda \int_0^\infty \int_0^\infty x^{\lambda-1}\,\mathrm{e}^{-xu}\,\mathrm{d}x\,\mathrm{d}F(u) = \Gamma(\lambda+1)\,\mu_{F,-\lambda}\,.$$

These equalities can also be read from right to left. This means that $\mu_{G,\lambda}$ exists iff $\mu_{F,-\lambda}$ exists.

(ii) Moments of negative orders occur also in connection with unimodal d.f. $F$ satisfying lext $(F) = 0$, $F(+0) = 0$. Namely, according to Suppl. 3.8.9, $F$ has a bounded density, iff for the corresponding d.f. $G$ the moment $\mu_{G,-1}$ exists. □

We first state a preliminary result to study the moment $\mu_{F,-1}$.

**Proposition 4.9.12.** *Let* lext $(F) = 0$. *If the integral* $\int_0^\infty \operatorname{Im} f(u)\,du$ *exists, then so does* $\mu_{F,-1}$ *and we have*

$$\mu_{F,-1} = \int_0^\infty \operatorname{Im} f(u)\,\mathrm{d}u\,. \tag{4.9.15}$$

Proof. From Ryžik-Gradstejn (1957) we take the formula

$$\int_0^\infty \mathrm{e}^{-tv}\sin uv\,\mathrm{d}v = \frac{u}{u^2+t^2}\,. \tag{4.9.16}$$

Let the integral on the right of (4.9.15) exist. Then we consider the mixture (4.9.14) whose c.f. (cf. Suppl. 3.8.11) is given by

$$g(t) = \int_0^\infty \frac{u}{u-it}\,\mathrm{d}F(u)\,.$$

Now (4.9.16) yields

$$\frac{\operatorname{Im} g(t)}{t} = \int_0^\infty \frac{u}{u^2+t^2}\,\mathrm{d}F(u) = \int_0^\infty \int_0^\infty \mathrm{e}^{-tv}\sin uv\,\mathrm{d}v\,\mathrm{d}F(u) > 0\,. \tag{4.9.17}$$

Using Fubini theorem it becomes

$$\frac{\operatorname{Im} g(t)}{t} = \int_0^\infty \mathrm{e}^{-tv}\operatorname{Im} f(v)\,\mathrm{d}v = \int_0^\infty \mathrm{e}^{-tv}\,\mathrm{d}\Big(\int_0^v \operatorname{Im} f(u)\,\mathrm{d}u\Big) =$$

$$= \int_0^\infty \mathrm{e}^{-w}\Big(\int_0^{w/t} \operatorname{Im} f(u)\,\mathrm{d}u\Big)\,\mathrm{d}w$$

Now dominated convergence yields by assumption

$$\lim_{t\to+0} \frac{\operatorname{Im} g(t)}{t} = \int_0^\infty \operatorname{Im} f(v)\,\mathrm{d}v\,.$$

Since we now know that this limit exists we can calculate it also by (4.9.17) using momonotone convergence; this shows that $\mu_{F,-1}$ exists and (4.9.15) is true. □

Putting $f(t) = \mathrm{e}^{it}$ we see that the existence of $\mu_{F,-1}$ does not imply the existence of the integral (4.9.15). But we are now able to give necessary and sufficient conditions for $\mu_{F,-1}$ to exist. With this aim in mind we introduce the function

$$I(t) := \int_0^t \left(1 - \frac{u}{t}\right) \operatorname{Im} f(u) \, du . \tag{4.9.18}$$

**Proposition 4.9.13.** *Let* lext $(F) = F(+0) = 0$. *Then* $\mu_{F,-1}$ *exists iff* $\lim_{t\to\infty} I(t)$ *exists. In either case we have*

$$\lim_{t\to\infty} I(t) = \mu_{F,-1} . \tag{4.9.19}$$

Proof. Let us assume that $\mu_{F,-1} < \infty$. Then it follows from the Fubini theorem that

$$I(t) = \int_0^\infty \int_0^t \left(1 - \frac{u}{t}\right) \sin xu \, du \, dF(x) = \int_0^\infty \left(1 - \frac{\sin tx}{tx}\right) \frac{dF(x)}{x}$$

$$= \mu_{F,-1}\left(1 - \int_0^\infty \frac{\sin tx}{tx} \, dH(x)\right)$$

where $H(x) := \dfrac{1}{\mu_{F,-1}} \displaystyle\int_0^x \frac{dF(u)}{u}$. Since $H(+0) = 0$ we immediately obtain (4.9.19) by dominated convergence. This concludes the first part of the proof.

For the converse we assume that $I(\infty)$ exists. We recall that by Suppl. 3.8.10, the function

$$f_2(t) := \frac{2}{t^2} \int_0^t u \, f(u) \, du$$

is a c.f. Integrating by parts in (4.9.18) we get

$$I(t) = \int_0^t \frac{t-u}{ut} \, d\left(\int_0^u v \operatorname{Im} f(v) \, dv\right) = \frac{1}{2} \int_0^t \operatorname{Im} f_2(u) \, du .$$

Thus our supposition means that $\int_0^\infty \operatorname{Im} f_2(u) \, du$ exists, and according to Prop. 4.9.12. the moment $\mu_{F_2,-1}$ exists. It remains to show that $\mu_{F,-1}$ also exists. We write, using Fubini theorem,

$$\operatorname{Re} f_2(t) = \frac{2}{t^2} \int_0^t u \int_0^\infty \cos ux \, dF(x) \, du = \frac{2}{t^2} \int_0^\infty \int_0^t u \cos xu \, du \, dF(x)$$

$$= 2 \int_0^\infty \int_0^1 v \cos(txv) \, dv \, dF(x) = \int_0^\infty \operatorname{Re} k(tx) \, dF(x) .$$

Here $k$ stands for the c.f. of the d.f. $K$ defined by $K(x) = x^2$, $0 \leq x \leq 1$. From the uniqueness theorem Prop. 4.9.2. we now conclude that $f_2(t) = \int_0^\infty k(tx) \, dF(x)$; note that this integral represents a c.f. by Suppl. 3.8.8. Hence, by Suppl. 3.8.8, we obtain for the d.f. $F_2$ corresponding to $f_2$

$$F_2(x) = \int_0^\infty K\left(\frac{x}{v}\right) \mathrm{d}F(v) = F(x) + \int_x^\infty \left(\frac{x}{v}\right)^2 \mathrm{d}F(v) .$$

Accordingly,

$$\mathrm{d}F_2(x) = 2x\,\mathrm{d}x \int_x^\infty \frac{\mathrm{d}F(v)}{v^2}$$

and

$$\mu_{F_2,-1} = \int_0^\infty \frac{\mathrm{d}F_2(x)}{x} = 2\int_0^\infty \int_x^\infty \frac{\mathrm{d}F(v)}{v^2}\,\mathrm{d}x .$$

Now partial integration yields $\mu_{F_2,-1} = 2\mu_{F,-1}$ which implies the desired result. □

## 4.10. Supplement

**4.10.1.** Let $i\bar{s}$ and $i\underline{s}$ be the singular points of an a.c.f. "nearest to the real axis". Then there do sometimes occur more singular points on the lines $(z=t+is: s=\underline{s})$ and $(z: s = \bar{s})$. Examples with finite and also with infinite sets of poles on these lines are given in ROSSBERG (1968); they provide counterexamples in characterization theory and in queuing theory (for the latter see ROSSBERG (1983a)). These lines can also be natural borders: this subject is treated in a generat way in CUPPENS and LUKACS (1970).

**4.10.2.** The numbers $\underline{s}$ and $\bar{s}$ defined in Sect. 4.2 and 4.3 can be expressed in terms of the asymptotic behaviour of $F$, namely

$$\underline{s} = -\varlimsup_{x\to\infty} \frac{\log(1 - F(x))}{x}, \qquad \bar{s} = -\varlimsup_{x\to\infty} \frac{\log F(-x)}{x}$$

(WIDDER (1946)).

**4.10.3.** From formula (4.3.3) it is seen that the function $\psi(s) := f(is)$ satisfies $\psi''(s) > 0$ for $\underline{s} < s < \bar{s}$ so that $f(is)$ is a convex function of $s$. But there is also the following less trivial convexity property of the functions $f^{(2k)}$. *Let $f$ be a c.f. analytic in the strip $S(\underline{s}, \bar{s})$ and let $k \geq 0$ be an integer. Then $\log f^{(2k)}(is)$ is convex on the interval $(\underline{s}, \bar{s})$* (LUKACS (1970)). This result also applies to b.c.f.

**4.10.4.** *Ridge functions.* The ridge property of a.c.f. suggested that consideration should be given to functions $f$ analytic in a strip $S(s_1, s_2)$ which satisfy

$$|f(t + is)| \leq f(is), \qquad s_1 < s < s_2, \qquad t \in \mathbb{R}_1 .$$

They are called ridge functions. A.c.f. and b.c.f. form proper subclasses of this class. These investigations show, for instance, which properties of a.c.f. are consequences of the ridge property. We briefly discuss only one important aspect of this theory.

By Prop. C.1.7 it is clear that integral ridge functions have at least order 1. Further, it can be shown that all orders $\varrho \geq 1$ occur. But the situation changes radically if we make assumptions on the zeros. We illustrate this fact with a celebrated theorem.

*Let $P_n$ stand for a polynomial of degree n. If* $\exp(P_n)$ *is a ridge function, then* $n = 2$ (MARCINKIEWICZ). — (For the case that $\exp(P_n)$ is a c.f. a simple proof is contained in RAMACHANDRAN (1967) and LAHA and ROHATGI (1979)). This theorem found remarkable applications and has had a stimulating effect; important results in this direction are surveyed in LUKACS (1970), ROSSBERG and JESIAK (1983) and eleborated in detail in KAGAN, LINNIK, and RAO (1972), LINNIK and OSTROVSKIJ (1972). Two recent generalizations of the theorem of MARCINKIEWICZ, assuming special locations of the zeros, are given in OSTROVSKIJ (1982b).

**4.10.5.** *Let f be a red c.f. and $t_k \to 0$ a real point set. Assume, moreover, that g is a function analytic in the circle $\{z: |z| < r\}$ satisfying $g(-t) = \overline{g(t)}$. If $f(t_k) = g(t_k)$, $k = 1, 2, \ldots$, then $f = g$ so that f is an a.c.f.* (Simplified version of Th. 7.2.1 given in LINNIK and OSTROVSKIJ (1972)).

**4.10.6.** In Th. 4.5.4 we can also put $V(x) = x^{\varrho(x)}$ where $\varrho(x) \to 1/\varrho$ is a certain proximate order ($v = 1/\varrho$) and $W(x) = x$ ($w = 1$). Then the results of ROSSBERG (1967) and DEWESS and RIEDEL (1977) can be obtained, where the asymptotic behaviour of $M$ and $T$ is described with proximate orders.

**4.10.7.** We introduce the function

$$\Psi(x) = \log M(f; e^x)$$

and write $\Psi^{-1}$ for the corresponding inverse function. *Let F be a d.f. with an integral c.f. of infinite order. Then we have*

$$\varliminf_{x\to\infty} \frac{\log\left(-x^{-1}\log T(x)\right)}{\Psi^{-1}(x)} = 1$$

(VINNICKIJ (1975)). This result has also been derived from Th. 4.5.4 using properties of convex functions (DEWESS (1977b).

**4.10.8.** A counterpart to Th. 4.5.4 runs as follows. *Let F be a d.f.* log *T being strictly concave, which has an integral c.f. Then, using comparison functions V and W as in Th.* 4.5.4. *we have*

$$\varlimsup \frac{V\left(-x^{-1}\log T(x)\right)}{W(x)} = l$$

*iff*

$$\varliminf \frac{W\left(r^{-1}\log M(f, r)\right)}{V(r)} = \frac{d_{v,w}}{l}$$

(DEWESS (1977b)).

**4.10.9.** Prop. 4.7.4 implies the following elegant particular case. *Let $F_1$, $F_2$ be two inf. div. d.f. satisfying*

$$F_1 * F_2(x) = \Phi(x), \qquad x \leqq x_0, \tag{4.10.1}$$

*then they are normal d.f.*

**Problem** (KRUGLOV). Ex. 4.7.1 shows that the assumptions on the zeros made in Prop. 4.7.4 cannot be dropped. Let now $F$ be a d.f. and put $F_k(x) = F(b_k x)$ ($k = 1, 2$; $0 < b_1 < b_2$). Then the following question arises: Does it follow from (4.10.1) that $F$ is normal? Prop. 4.7.4 grants this only in the special case that the c.f. $f$ of $F$ has no zeros in $S(0, \infty)$.

**4.10.10.** Let $\boldsymbol{H}$ be the class of all functions $H$ of bounded variation defined on $\mathbb{R}_1$ having the following properties:

(i) $H(-\infty) = 0\,, \quad H(x-0) = H(x)\,, \quad x \in \mathbb{R}_1\,,$

(ii) the Fourier-Stieltjes transform

$$h(t) = \int_{-\infty}^{\infty} e^{itx}\, dH(x)$$

is continuable to $S(0, \infty)$ and has no zeros,

(iii) $\operatorname{lext}(H) = -\infty$ (cf. App. B.2).

*Let $H_1$ and $H_2$ be finite linear combinations of functions belonging to $\boldsymbol{H}$. If* $\operatorname{lext}(H_1) = -\infty$ *and $H_1(x) = H_2(x)$, $x < 0$, then $H_1 = H_2$* (OSTROVSKIJ (1979)). — A modification in which $h$ is continuable only to $S(0, \bar{s})$, $0 < \bar{s} < \infty$, is contained in BLANK (1982).

We turn now to three uniqueness theorems which are generalizations of continuation theorems.

**4.10.11.** *Let $C > 0$ and $d = (n+1)/(n-1)$, $n \geqq 2$. If*

$$|F^{*n}(x) - \Phi_{\mu,\sigma}(x)| \leqq C \exp(-dx^2/2)\,, \qquad x < 0\,,$$

*then $F$ is normal* (TITOV (1981)).

**4.10.12.** Prop. 4.7.4 can be extended such that it entails also the analogous theorem of LINNIK on the factorization of

$$\Psi_{a,\sigma,\lambda} := \Phi_{a,\sigma} * P_\lambda\,, \qquad a \in \mathbb{R}_1\,, \qquad \sigma > 0\,, \qquad \lambda \geqq 0$$

(ROSSBERG (1982)). The following result is even still more general. *Let $F_1$ and $F_2$ be d.f. such that, for fixed numbers $a \in \mathbb{R}_1$, $\sigma > 0$, $\lambda \geqq 0$, and some $c > \sigma^{-2}$*

$$|F_1 * F_2(x) - \Psi_{a,\sigma,\lambda}(x)| = O(e^{-cx^2})\,, \quad \text{as} \quad x \to -\infty;$$

*If the c.f. $f_1$ and $f_2$ have no zeros in the upper half plane $S(0, \infty)$, then $F_k = \Phi_{a_k,\sigma_k} * P_{\lambda_k}$ $(k = 1, 2)$ where $a_1 + a_2 = a$, $\sigma_1^2 + \sigma_2^2 = \sigma^2$; $\lambda_1 + \lambda_2 = \lambda$* (SCHARM (1983)).

**4.10.13.** *Let $H \in \boldsymbol{H}$* (notation as in Suppl. 4.10.10) *and assume that*

$$\left|\frac{H(x)}{\Phi(x)} - 1\right| \leqq e^{-|x|\,q(x)}\,, \qquad x < 0\,,$$

*where $q$ is a function on $(-\infty, 0)$ such that* $\lim q(x) = \infty$. *Then $H = \Phi$* (BLANK and OSTROVSKIJ (1981)). — This a modification of RIEDEL's (1976) result. BLANK (1981) gave a more general theorem in which $\Phi$ is also replaced by a function of bounded variation belonging to a certain set.

**4.10.14.** Thorough investigations on the original Hinčin phenomenon were carried out by KAWATA; they are reviewed in LUKACS (1970). They are particularly remarkable in our context since the results are also formulated in terms of the one-sided asymptotic behaviour of the d.f. $F$ occurring in (4.8.1). We quote a result related to Prop. 4.8.3. *Assume* (4.8.3) *and* $\operatorname{lext}(G) = 0$. *If the c.f. $g$ of $G$ has no zeros in $S[0, \infty)$ then $F_1(x) = F_2(x)$, $x \leqq x_0$* (ROSSBERG and CHU DUC (1981)). Suppl. 4.10.10 shows that we may now conclude $F_1 = F_2$ provided that the c.f. $f_1$ and $f_2$ are continuable to $S(0, \infty)$ and have no zeros.

**4.10.15. Problem.** Several modifications of Prop. 4.8.3 probably exist. Namely, if we impose more restrictions on the asymptotic behaviour of $G$ then we can replace assumption (4.8.6) by weaker ones. For instance, let $g$ have order 1 and type $\infty$ and completely regular growth (the theory of such c.f. was initiated in ROSSBERG (1967)). Then we expect that (4.8.6) can be replaced by (4.8.5) which implies that $d_-$ is continuable to $S(0, \infty)$. For the case that $g$ has a higher order, the regular growth of c.f. has to be further investigated.

**4.10.16.** *Let $f$ be the c.f. of a d.f. $F$ with* lext $(F) = 0$. *Then $f$ permits the representation*

$$f = \tilde{f}\chi \tag{4.10.2}$$

*in $S(0, \infty)$ where $\chi = 1$ if $f$ has no zeros in $S(0, \infty)$; otherwise, we have*

$$\chi(z) = \prod_k \frac{z - a_k}{z - \bar{a}_k} \frac{z + \bar{a}_k}{z + a_k}$$

*where $a_k$ are the zeros of $f$ in $S(0, \infty)$ with due count of their multiplicity. Further,*

$$\log |\tilde{f}(z)| = \frac{s}{\pi} \int_{-\infty}^{\infty} \frac{\log |f(u)|}{(u - t)^2 + s^2} du, \qquad z \in S(0, \infty),$$

(ROSSBERG (1968)). —

If $\chi \neq 1$, then $\chi$ is never a c.f. since there is no pole on the imaginary axis, but in all known examples the factor $\tilde{f}$ is a c.f. (see ROSSBERG (1968) and LAUE (1976)).

**4.10.17. Problem.** Is the factor $\tilde{f}$ in (4.10.2) always a c.f.? If an affirmative answer is possible, then the influence of the zeros of $f$ on $F$ becomes clearer since for given $f$ there always exists a "simpler" c.f. $\tilde{f}$ which uniquely corresponds to $f$.

**4.10.18. Problem.** Which properties do the d.f. $F$ and $\tilde{F}$, corresponding to $f$ and $\tilde{f}$, resp., share with each other (provided that $\tilde{f}$ is a c.f.)?

A trivial example is provided by Suppl. 3.8.7; there we stated that a d.f. $F$ with lext $(F) = 0$ possesses the moment of order $k$ iff $f^{(k)}(0)$ exists; thus it follows that $F$ and $\tilde{F}$ have moments of the same orders. This is due to the fact that $\chi$ is analytic in a circle about 0. On the other hand, we have $|\chi(t)| = 1$, $t \in \mathbb{R}_1$; this means that all estimates of the concentration function which are expressed in terms of $|f|$ (see Sect. 5.3) hold for $F$ and $\tilde{F}$ as well.

**4.10.19.** Formula (4.10.2) and the above mentioned fact that $|\chi(t)| = 1$ imply the following result for which Th. C.2.5.3 represents a generalization. *Let $F_1$, $F_2$ be d.f.,* lext $(F_1) =$ lext $(F_2) = 0$. *If the zeros of the corresponding c.f. $f_1$, $f_2$ coincide* in $S(0, \infty)$ *and if $|f_1(t)| = |f_2(t)|$, $t \in \mathbb{R}_1$, then $F_1 = F_2$* (ROSSBERG (1968)).

**4.10.20.** Let $\Psi$ be a function analytic in $S(0, \infty)$ and continuous in $S[0, \infty)$. Assume that $|\Psi| \leq 1$ in $S[0, \infty)$, $\Psi(0) = 1$, and that $\Psi$ is hermitian on $\mathbb{R}_1$. LAUE (1983b) treats the question of under which additional assumptions $\Psi$ proves to be the c.f. of a non-negative r.v.

**4.10.21.** According to Prop. 4.9.3 certain d.f. with c.f. $f$ are already defined if either Re $f$ or Im $f$ is given. Then the following questions arises. a) What connections exist between the special properties of Re $f$ and the corresponding properties of Im $f$? b) Which properties of $F$ are implied by the special properties of Re $f$ or the corresponding properties Im $f$? Some results in this direction are given in LAUE (1985).

**4.10.22.** *Let* lext $(F) = 0$, $F(+0) = 0$.
(i) *If* $\operatorname{Re} f$ *is absolutely integrable over* $R_1$, *then* $F$ *is absolutely continuous; its density* $p$ *is continuous and representable as*

$$p(x) = \frac{2}{\pi} \int_0^\infty \cos xu \operatorname{Re} f(u) \, du \,, \qquad x > 0 \,.$$

(ii) *If* $\operatorname{Im} f$ *is absolutely integrable over* $R_1$, *then* $F$ *is absolutely continuous; its density* $p$ *is continuous and representable as*

$$p(x) = \frac{2}{\pi} \int_0^\infty \sin xu \operatorname{Im} f(u) \, du \,, \qquad x > 0 \,,$$

(Laue 1983a)).

**4.10.23.** *Under the assumptions of Suppl. 4.10.22* (i), (ii) *we have*

$$\int_0^\infty (\operatorname{Re} f(u))^2 \, du = \frac{\pi}{2} \int_0^\infty p^2(x) \, dx$$

*and*

$$\int_0^\infty (\operatorname{Im} f(u))^2 \, du = \frac{\pi}{2} \int_0^\infty p^2(x) \, dx \,,$$

*resp.* (Laue (1983b)).

*Moment conditions*

**4.10.24.** Conditions for the existence of moments can be obtained using the auxiliary functions

$$g_k(t) := \begin{cases} f(t) \,, & k = 1 \,, \\ \dfrac{i}{t} \left( g_{k-1}(t) + (-1)^{k-1} \dfrac{\mu_{k-2}}{(k-2)!} \right), & k = 2, 3, \ldots \,, \end{cases}$$

see Girlich and Rossberg (1971). For non-negative r.v. they yield two sequences of statements implying conditions for the existence of moments; these have the form

a) $\mu_k$ *exists iff* $\mu_{k-1}$ *exists and* $\int_0^1 u^{-1} |\operatorname{Im} g_{k-1}(u)| \, du < \infty$,

b) $\mu_k$ *exists iff* $\mu_{k-1}$ *exists and* $t^{-1} |\operatorname{Im} g_k(t)| < C$ *for all* $t \neq 0$, *where* $k = 1, 2, \ldots$

The first condition can be expressed by $\operatorname{Re} f$ only while the second contains only $\operatorname{Im} f$. For example b) implies for $k = 1$ that $\mu_1$ *exists iff* $|t^{-1} \operatorname{Im} f(t)| < C$ *for all* $t \neq 0$ (Laue (1973)). The formulas of Prop. 4.9.3 make it possible to derive each of these sequences from the other (Laue (1983b)).

**4.10.25.** Moments of fractional positive orders have many applications. Various conditions for the existence of such moments have been given; they are in terms of the corresponding c.f., the Laplace transform, or certain auxiliary functions. Using fractional derivatives many of these conditions can be written in a unified way (Laue (1980), for details see also Lukacs (1983)).

**4.10.26.** Results on non-negative r.v. are often simpler than those on general r.v. This fact can be applied, for example, to stochastic models. Considering a random walk $S_0 := 0$, $S_k := X_1 + \dots + X_k$ with i.i.d.r.v. (subject to the d.f. $F$) we have the factorization

$$U_0 - F = (1 - \zeta)(U_0 - H^+) * (U_0 - H^-)$$

where $0 < \zeta < 1$ (see FELLER (1970)). Decomposing $F$ into the form

$$F = F(+0)\,F^- + \big(1 - F(+0)\big)\,F^+$$

we find an intimate connection between the functions $F^\pm$ and $H^\pm$, resp. For instance: *If the $\mu$-th moment of $H^\pm$ exists, then so does the $\mu$-th moment of $F^\pm$* (for every $\mu > 0$). *In the case of the drift to $\pm\infty$ the converse is also valid* (LAUE (1977)). The connection between $F^\pm$ and $H^\pm$ has also been studied by MÜLLER (1977) using the Phragmén-Lindelöf theorem C.1.1.

**4.10.27.** The asymptotic behaviour of a d.f. $F(x)$, as $|x| \to \infty$, is intimately connected with the behaviour of $f(t)$, as $t \to 0$. If lext $(F) = 0$ the behaviour of $F(x)$, as $x \to +0$, is connected with the behaviour of $f(t)$, as $t \to \infty$; it can be expressed by moments of negative order, cf. Sect. 4.9. Further results on such moments are contained is LAUE (1984b).

**4.10.28.** Prop. 4.9.8 gives a harmonic continuation of (4.9.2) into the half plane $S(0, \infty)$. An analogous continuation of (4.9.1) seems to be possible only under an additional assumption. *If $\mu_{F,-\lambda}$, $0 < \lambda < 1$, exists then*

$$\operatorname{Im} f(z) = -\frac{1}{\pi}\int_0^\infty \frac{u^2}{u^2 + s^2}\,\frac{\operatorname{Re} f(t+u) - \operatorname{Re} f(t-u)}{u}\,du\,, \qquad z \in S(0, \infty)\,,$$

(Oral communication by LAUE).

## Notes and bibliographical data

Prop. 4.2.10 has been obtained for $k = 1$ by LAUE (1973). Prop. 4.4.3 and 4.4.4 are modifications of results given in KRUGLOV and ANTONOV (1982). Prop. 4.4.5 is a slightly generalized version of an earlier result of RIEDEL (1976).

The subject of Sect. 4.5 has been thoroughly elaborated by RAMACHANDRAN (1967) and LUKACS (1970); the Cor. 4.5.5 and 4.5.6 are due to RAMACHANDRAN. His results have been generalized by means of proximate orders in ROSSBERG (1966) and (1967b), DEWESS and RIEDEL (1977), cf. Suppl. 4.10.6. Finally, DEWESS (1977b) and (1978a) has succeeded in deriving the general Th. 4.5.4; it has been applied to a single server model of queuing theory in DEWESS (1968b). Related results have been obtained by JAKOVLEVA (1972); they are of particular interest for c.f. of infinite order; see also JAKOVLEVA (1976), DAVIES (1976), DEWESS (1982). The multiuariate case has been treated in DEWESS (1977a).

Detailed expositions of factorization theory are contained in LUKACS (1970), LINNIK-OSTROVSKIJ (1972), and in the survey article OSTROVSKIJ (1977). Ex. 4.7.1 was orally communicated to one of the authors by A. N. TITOV. The elementary proof of Prop. 4.7.2 and the modification of Cramér's theorem, Prop. 4.7.4 are contained in ROSSBERG (1982). Prop. 4.7.5 and its Cor. 4.7.6 are due to TITOV (1981). The first has been greatly generalized by BLANK (1983), who recognized that condition (4.7.2) can be omitted. Prop. 4.7.7 is to be found in ROSSBERG, JESIAK, and SIEGEL (1981a).

The subject of Sect. 4.8 was broached in ROSSBERG and CHU DUC (1981); this paper contains Prop. 4.8.3, amongst others, cf. Suppl. 4.10.14.

The whole Sect. 4.9 was contributed by LAUE. More material on c.f. of non-negative r.v. is contained in RAMACHANDRAN (1967).

# Chapter 5. Concentration functions

Concentration functions were first introduced by Lévy in 1937. They are an appropriate tool for treating convergence problems; important applications are to be found in the summation theory of i.r.v. Since this theory represents an essential part of the present book it is a matter of course that we give a selection of basic properties of concentration functions, in particular we study estimates.

## 5.1. Preliminary results

**Definition 5.1.1.** Let $X$ be a r.v. with d.f. $F$. Then the real function

$$Q(X, h) = \begin{cases} \sup\limits_{x \in R_1} \big(F(x+h) - F(x-h)\big), & h > 0, \\ 0, & h \leqq 0, \end{cases}$$

is called the *concentration function* of the r.v. $X$. □

Obviously, $Q(X, h)$ depends only on the d.f. $F$ of $X$; it gives the maximal mass that is attributed by $F$ to an interval of length $2h$. Note that the definition of $Q$ implies

$$Q(X, h) - \varepsilon \leqq F(x_0 + h) - F(x_0 - h) = \mathsf{P}\big(X \in [x_0 - h, x_0 + h)\big) \qquad (5.1.1)$$

for all $\varepsilon > 0$, $h > 0$, and some $x_0 \in R_1$ depending on $\varepsilon$, $h$.

**Example 5.1.2.** (i) Let $X$ be a non-negative r.v. with lattice d.f. defined by $p_k = \mathsf{P}(X = k)$, $k \geqq 0$. If $p_k \geqq p_{k+1}$ for all $k \geqq 0$, then

$$Q(X, +0) = p_0, \qquad Q(X, h) = \sum_{j=0}^{k-1} p_j, \qquad k - 1 < 2h \leqq k, \qquad k \geqq 1,$$

that is, $Q$ coincides with the d.f. $F'(x) = F2(x)$ of the r.v. $X' = X/2$. The Poisson d.f. $P_\lambda$ with parameter $\lambda \in (0, 1]$ is an example of such a d.f.

(ii) Suppose that the d.f. $F$ of the non-negative r.v. $X$ is absolutely continuous with density $p$. If $p$ is non-increasing on $(0, \infty)$, then $Q(X, h) = F(2h)$ for every $h > 0$. □

**Proposition 5.1.3.** (i) *$Q(X, h)$ is a non-decreasing and left-continuous function satisfying $Q(X, h) > 0$ for all $h > 0$ and $\lim\limits_{h \to \infty} Q(X, h) = 1$, in other words, $Q$ is a d.f. of a non-negative r.v. with* lext $(Q) = 0$.

(ii) $$Q(X, +0) = \sup_{x \in R_1} \mathsf{P}(X = x).$$

(iii) *For all real numbers $a$, $h$ we have*

$$Q(X + a, h) = Q(X, h), \qquad Q(-X, h) = Q(X, h).$$

(iv) *For all* $h_1, h_2 \geqq 0$ *we have*

$$Q(X, h_1 + h_2) \leqq Q(X, h_1) + Q(X, h_2) .$$

(v) *Let* $X_1$, $X_2$ *be i.r.v. and* $h \in R_1$, *then*

$$Q(X_1, \eta_1 h)\, Q(X_2, \eta_2 h) \leqq Q(X_1 + X_2, h) \leqq \min \{Q(X_1, h), Q(X_2, h)\} ,$$

*where* $\eta_1, \eta_2 > 0$, $\eta_1 + \eta_2 = 1$.

Proof. (i) First we have

$$Q(X, h) \geqq F(x + h) - F(x - h) > 0$$

for all $h > 0$ and a certain $x \in R_1$. Clearly, $Q(X, h)$ is non-decreasing in $h$. Hence $Q(X, h + 0)$, $Q(X, h - 0)$ exist. Moreover, for any $\varepsilon > 0$ there is $h_0 = h_0(\varepsilon)$ such that

$$1 - \varepsilon < F(h_0) - F(-h_0) \leqq Q(X, h_0) \leqq 1$$

whence $Q(X, h) \to 1$, $h \to \infty$, follows.

It remains to show that $Q$ is left-continuous. Indeed, for $\varepsilon > 0$ there exist numbers $x_0 = x_0(\varepsilon, h) \in R_1$, $\delta = \delta(\varepsilon, x_0) > 0$ such that, in view of (5.1.1)

$$\begin{aligned} Q(X, h - 0) &\leqq Q(X, h) \leqq \mathsf{P}\big(X \in [x_0 - h, x_0 + h)\big) + \varepsilon \leqq \\ &\leqq \mathsf{P}(X \in [x_0 - h, x_0 + h - \delta)) + 2\varepsilon \leqq Q(X, h - \delta/2) + 2\varepsilon \\ &\leqq Q(X, h - 0) + 2\varepsilon . \end{aligned}$$

On letting $\varepsilon \to +0$ we arrive at the desired result.

(ii) and (iii) are trivial.

(iv) The assertion follows from Def. 5.1.1 and from

$$\begin{aligned} &F(x + h_1 + h_2) - F(x - h_1 - h_2) \\ &= F\big((x + h_2) + h_1\big) - F\big((x + h_2) - h_1\big) + F\big((x - h_1) + h_2\big) \\ &\quad - F\big((x - h_1) - h_2\big) \leqq Q(X, h_1) + Q(X, h_2) . \end{aligned}$$

(v) It suffices to assume $h > 0$. For $x_1, x_2 \in R_1$ we consider the inclusion

$$\begin{aligned} &\{X_1 \in [x_1 - \eta_1 h, x_1 + \eta_1 h)\} \cap \{X_2 \in [x_2 - \eta_2 h, x_2 + \eta_2 h)\} \\ &\subseteqq \{X_1 + X_2 \in [x_1 + x_2 - h, x_1 + x_2 + h)\} . \end{aligned}$$

Denoting the d.f. of $X_k$ by $F_k$ and setting $F = F_1 * F_2$, we obtain by independence

$$\begin{aligned} \prod_{k=1}^{2} \big(F_k(x_k + \eta_k h) - F(x_k - \eta_k h)\big) &\leqq F(x_1 + x_2 + h) - F(x_1 + x_2 - h) \\ &\leqq Q(X_1 + X_2, h) \end{aligned}$$

Forming now the suprma with respect to $x_1$ and $x_2$ on the left we get the first of the desired inequalities.

Next we write

$$F(x + h) - F(x - h) = \int_{-\infty}^{\infty} \big(F_1(x - y + h) - F_1(x - y - h)\big)\, F_2(dy)$$

so that $Q(X_1 + X_2, h) \leqq Q(X_1, h)$ follows. In just the same way we show $Q(X_1 + X_2, h) \leqq Q(X_2, h)$ whence the second inequality follows. □

Now we give some consequences of Prop. 5.1.3. As usual, we denote by $[\eta]$ the integer part of $\eta$ where $[\eta] \leqq \eta$.

**Proposition 5.1.4.** (i) *$F$ is continuous iff $Q(X, +0) = 0$.*

(ii) *We have $Q(X, \eta h) \leqq (1 + [\eta])\, Q(X, h)$ for every $\eta, h > 0$.*

(iii) *For every $h > 0$*

$$\frac{1}{2} Q(X, h) \leqq \frac{1}{h} \int_0^h Q(X, y)\, dy \leqq Q(X, h) .$$

(iv) *A d.f. $G$ with* lext $(G) = 0$ *represents a concentration function iff*

$$G(h_1 + h_2) \leqq G(h_1) + G(h_2) \quad \textit{for all} \quad h_1, h_2 \geqq 0 .$$

Proof. (i) is obvious from Prop. 5.1.3 (ii). The relation (ii) follows from Prop. 5.1.3 (iv), by induction with respect to $[\eta]$ using the monotonicity of $Q$.

(iii) Prop. 5.1.3 (iv), implies

$$Q(X, h) \leqq Q(X, y) + Q(X, h - y) , \qquad 0 \leqq y \leqq h .$$

Integrating this inequality with respect to $y$ we obtain the first relation, the second is now trivially seen.

(iv) The necessity follows from Prop. 5.1.3 (iv). To prove the sufficiency, let $X'$ be a r.v. subject to $F'(x) = G(x/2)$. Then we rewrite the supposition in the shape

$$F'(2x + 2h) - F'(2x) = G(x + h) - G(x) \leqq G(h) = F'(2h) .$$

Forming the supremum with respect to $x \geqq 0$ on the left, we have

$$Q(X', h) \leqq G(h) = F'(2h) \leqq Q(X', h)$$

whence $Q(X', h) = G(h)$ follows for all $h > 0$. □

To close this section we deal with the right-sided differentiability of concentration functions at $h = 0$. Note that the following result implies that $Q(X, h)$ cannot tend to zero more quickly than $h$.

**Proposition 5.1.5.** *Suppose $Q(X, +0) = 0$. Then*

$$0 < \lim_{h \to +0} \frac{Q(X, h)}{h} = \sup_{h > 0} \frac{Q(X, h)}{h} \leqq \infty$$

*holds in the sense that if one expression is infinite then so is the other.*

Proof. We trivially see that

$$\underline{C} := \varliminf_{h \to +0} \frac{Q(X, h)}{h} \leqq \overline{C} := \varlimsup_{h \to +0} \frac{Q(X, h)}{h} \leqq C_0 := \sup_{h > 0} \frac{Q(X, h)}{h} \leqq \infty .$$

Hence it is enough to show that $\underline{C} \geqq C_0$.

First we assume that $C_0 < \infty$. Let $\varepsilon > 0$. Then there is a positive number $h_0$, depending on $\varepsilon$, such that

$$C_0 - \varepsilon \leqq \frac{1}{h_0} Q(X, h_0) .$$

Now we choose a sequence $(h_k)$ which satisfies $0 < h_k \leqq h_0$, $h_k \to 0$, and

$$\underline{C} = \lim_{k \to \infty} \frac{1}{h_k} Q(X, h_k) .$$

Putting $\eta_k = h_0/h_k$, $k \geqq 1$, and using Prop. 5.1.4 (ii), we obtain

$$C_0 - \varepsilon \leqq \frac{1}{h_0} Q(X, h_0) \leqq \frac{1 + [\eta_k]}{h_0} Q(X, h_k) \leqq \frac{h_k + h_0}{h_0} \frac{Q(X, h_k)}{h_k} \leqq \underline{C} + \varepsilon$$

for sufficiently large $k$. This completes the proof for $C_0 < \infty$.

If $C_0 = \infty$ then we prove our statement by similar considerations. □

## 5.2. Convergence of concentration functions

Consider a sequence $(X_n)$ of r.v. and let $(F_n)$ be the sequence of corresponding d.f.

**Proposition 5.2.1.** *The sequence $(F_n)$ is shift compact iff for any $\varepsilon > 0$, there is a number $h > 0$ such that*

$$\inf_n Q(X_n, h) \geqq 1 - \varepsilon\,, \tag{5.2.1}$$

*in other words, iff the sequence $\big(Q(X_n, h)\big)$ is tight.*

Proof. If $(F_n)$ is shift compact, then $\big(F_n(x + a_n)\big)$ is relatively compact for a certain sequence $(a_n)$. Th. 2.4.4 tells us that $\big(F_n(x + a_n)\big)$ is also tight, so that

$$1 - \varepsilon < \mathsf{P}(|X_n - a_n| < h) \leqq Q(X_n, h)$$

for all $n$ and sufficiently large $h$ whence (5.2.1) follows.

Conversely, from (5.2.1) and (5.1.1) it follows that

$$1 - 2\varepsilon < Q(X_n, h) - \varepsilon \leqq \mathsf{P}(|X_n - a_n| \leqq h)\,, \qquad n \geqq 1\,, \tag{5.2.2}$$

for certain constants $a_n$. Hence $\big(F_n(x + a_n)\big)$ is tight, i.e., relatively compact whence the assertion follows. □

Before formulating a limit theorem for concentration functions, we prove the following statement.

**Lemma 5.2.2.** *Assume that the sequence $(F_n)$ is shift compact and let $(a_n')$ be a sequence of shift constants. Further, let $(a_n'')$ be some sequence of real numbers such that*

$$\inf_n \big(F_n(x_0 + a_n'') - F_n(-x_0 + a_n'')\big) > 0 \tag{5.2.3}$$

*for some $x_0 > 0$. Then $(a_n' - a_n'')$ is bounded and $(a_n'')$ is also a sequence of shift constants.*

Proof. Assumption (5.2.3) guarantees that any limit function $F''$ of $\big(F_n(x + a_n'')\big)$ is not identically constant; in particular, $F'' \neq 0$ and $F'' \neq 1$. By virtue of Prop. 2.4.7 we see that $\big(F_n(x + a_n'')\big)$ is relatively compact. On the other hand, the sequence $\big(F_n(x + a_n')\big)$ is relatively compact by assumption. Hence Prop. 2.4.7 applies to conclude that $(a_n' - a_n'')$ is bounded. □

**Proposition 5.2.3.** *Let $X$ be a r.v. subject to $F$ and assume $F_n \xrightarrow{c} F$; then $Q(X_n, h) \xrightarrow{c} Q(X, h)$.*

Proof. By assumption we see that $(F_n)$ is relatively compact. Hence, Prop. 5.2.1 implies $Q(X_n, h_0) \geqq 1/2$ for all $n$ and some $h_0 > 0$. Now it follows from Prop. 5.1.4 (ii), where $h\eta = h_0$, that

$$0 < \delta := \frac{1}{2}\left(1 + \left[\frac{h_0}{h}\right]\right)^{-1} \leqq Q(X_n, h)\,, \quad \text{for all} \quad h > 0\,.$$

By definition of $Q$ there are numbers $a_n \in \mathbb{R}_1$ such that

$$Q(X_n, h) \leqq \mathsf{P}(|X_n - a_n| \leqq h) + \frac{\delta}{2}.$$

Thus (5.2.3) is satisfied for $x_0 = h$. Applying L. 5.2.2 ($a_n' := 0$, $a_n'' = a_n$) there exists a sequence $(a_{n'})$ such that $a_{n'} \to a \in \mathbb{R}_1$. For all fixed $\varepsilon, \eta > 0$, we get by (5.1.1)

$$\begin{aligned} Q(X_{n'}, h) &\leqq \mathsf{P}(|X_{n'} - a_{n'}| \leqq h) + \varepsilon \\ &\leqq \mathsf{P}(|X_{n'} - a| < h + \eta) + \varepsilon \leqq \mathsf{P}(|X - a| < h + \eta) + 2\varepsilon \end{aligned}$$

provided that $n'$ is sufficiently large and $\pm (h + \eta)$ are continuity points of $F(x + a)$. If $h$ is also a continuity point of $Q(X, h)$, we further obtain

$$\mathsf{P}(|X - a| < h + \eta) \leqq Q(X, h + \eta) \leqq Q(X, h) + \varepsilon$$

for sufficiently small $\eta$. Since the right side is obviously independent of the special subsequence $(a_{n'})$ we have

$$\overline{\lim_{n\to\infty}}\, Q(X_n, h) \leqq Q(X, h).$$

In a similar way, we obtain for any $\varepsilon > 0$ and $\eta(\varepsilon) > 0$

$$\begin{aligned} Q(X, h) &\leqq Q(X, h - \eta) + \varepsilon \leqq \mathsf{P}(|X - a| \leqq h - \eta) + 2\varepsilon \\ &\leqq \mathsf{P}(|X_n - a| \leqq h - \eta) + 3\varepsilon \leqq Q(X_n, h) + 3\varepsilon \end{aligned}$$

provided that $n$ is sufficiently large and $\pm (h - \eta)$ are continuity points of $F(x + a)$ Hence

$$Q(X, h) \leqq \underline{\lim}_{n\to\infty}\, Q(X_n, h). \qquad \square$$

The converse of Prop. 5.2.3 is false without additional conditions. Next we turn to an interesting case in which this conclusion is true.

**Proposition 5.2.4.** *The convergence*

$$\lim_{n\to\infty} \mathsf{P}(|X_n - a_n| \leqq h) = 1, \qquad h > 0, \tag{5.2.4}$$

*holds for an appropriate sequence* $(a_n)$ *iff*

$$\lim_{n\to\infty} Q(X_n, h) = 1, \qquad h > 0. \tag{5.2.5}$$

*The medians* $m_n$ *of* $X_n$ *can serve as such a sequence* $(a_n)$.

Proof. The only-if part follows from Prop. 5.2.3 applied to $X_n - a_n$, where $F = U_0$. Conversely, if we assume (5.2.5), then we can choose $a_n \in \mathbb{R}_1$ such that (5.2.2) holds for sufficiently large $n$ and every $h > 0$ whence (5.2.4) follows. The last assertion follows from the proof of Prop. 2.4.11. $\square$

## 5.3. Lower and upper bounds

First we wish to derive estimates of $Q$ below. Let $X$ be a r.v. subject to $F$ and let $p$ be some density. Then we introduce the functional

$$\psi_p(X, h) := \int_{-\infty}^{\infty} \big(F(x + h) - F(x - h)\big)\, p(x)\, \mathrm{d}x.$$

It enables us to obtain lower bounds for $Q$ since

$$\psi_p(X, h) \leqq Q(X, h) \int_{-\infty}^{\infty} p(x)\, dx = Q(X, h); \tag{5.3.1}$$

hence the problem amounts to choosing the density $p$ appropriately. In the present section we make use of the function

$$p_h(x) := \frac{1}{2h} \left(F(x + h) - F(x - h)\right)$$

for fixed $h > 0$. To see that it is a density we write

$$\begin{aligned} 2h \int_{-\infty}^{\infty} p_h(x)\, dx &= \lim_{T\to\infty} \int_{-T}^{T} \left(F(x + h) - F(x - h)\right) dx \\ &= \lim_{T\to\infty} \left[\int_{-T+h}^{T+h} F(y)\, dy - \int_{-T-h}^{T-h} F(y)\, dy\right] \\ &= \lim_{T\to\infty} \left[\int_{T-h}^{T+h} F(y)\, dy - \int_{-T-h}^{-T+h} F(y)\, dy\right] = 2h\,. \end{aligned}$$

For this special choice of $p_h$ we obtain

$$\psi_{p_h}(X, h) = 2h \int_{-\infty}^{\infty} (p_h(x))^2\, dx =: C(X, h)$$

which is called the *mean concentration* function.

**Lemma 5.3.1.** *Let $F$ be a d.f. with c.f. $f$, and let $F^s$ be the corresponding symmetrization. Then we have, for every $h > 0$,*

$$C(X, h) = \frac{1}{\pi} \int_{-\infty}^{\infty} \frac{\sin^2 ht}{ht^2} |f(t)|^2\, dt = \frac{1}{2h} \int_{0}^{2h} \left(F^s(x) - F^s(-x)\right) dx\,.$$

Proof. First we determine the c.f. $f_h$ of $p_h$. Integrating by parts we obtain

$$f_h(t) = \int_{-\infty}^{\infty} e^{itx}\, p_h(x)\, dx = \frac{1}{2hit} \int_{-\infty}^{\infty} \left(e^{it(x+h)} - e^{it(x-h)}\right) dF(x) = \frac{\sin ht}{ht} f(t)\,.$$

The density $p_h^s$ corresponding to the c.f. $f_h^s = |f_h|^2$ can be obtained from a convolution (cf. Cor. 1.4.5). Hence we get (since $p_k$ is even by Cor. 3.2.6)

$$C(X, h) = 2h \int_{-\infty}^{\infty} p_h^2(x)\, dx = 2h\, p_h^s(0)\,.$$

On the other hand, we have $|f_h(t)|^2 \leqq 1/(h^2t^2)$, $t \neq 0$, i.e., $f_h^s$ is integrable on $R_1$. Thus the Fourier inversion formula implies

$$2h\, p_h^s(0) = 2h \frac{1}{2\pi} \int_{-\infty}^{\infty} f_h^s(t)\, dt = 2h \frac{1}{2\pi} \int_{-\infty}^{\infty} |f_h(t)|^2\, dt$$

whence the first of the asserted relations follows. — To prove the second we note that $|f|^2$ is the c.f. of $F^s$ and that $1/\pi\, (\sin^2 ht)/ht^2$ is a density with c.f. (cf. the table)

$$g(x) = \begin{cases} 1 - \dfrac{|x|}{2h}, & |x| \leqq 2h\,, \\ 0, & |x| > 2h\,. \end{cases}$$

Applying the Parseval relation (3.2.1) to $F^s$ we obtain

$$C(X, h) = \frac{1}{\pi} \int_{-\infty}^{\infty} \frac{\sin^2 ht}{ht^2} |f(t)|^2 \, dt = \int_{-2h}^{2h} \left(1 - \frac{|x|}{2h}\right) dF^s(x),$$

and integration by parts yields the assertion. □

Using the last equation in L. 5.3.1 and representing the jump $F^s(+0) - F^s(0)$ by means of Prop. 3.2.8, we also get the following result.

**Corollary 5.3.2.** *For an arbitrary d.f. $F$ we have*

$$C(X, +0) = F^s(+0) - F^s(0) = \sum_{x_k} p_{x_k}^2 = \lim_{T\to\infty} \frac{1}{2T} \int_{|t| \leq T} |f(t)|^2 \, dt,$$

*where the sum runs over all atoms $x_k$ of $F$.*

Note that the second equality sign follows from the definition of $F^s$.

**Proposition 5.3.3.** *For every $h > 0$ we have*

$$Q(X, h) \geqq \tfrac{1}{5} h \int_{|t| \leqq 1/h} |f(t)|^2 \, dt.$$

Proof. The inequality (5.3.1) and L. 5.3.1 imply

$$Q(X, h) \geqq C(X, h) \geqq \frac{h}{\pi} \int_{|t| \leqq 1/h} \frac{\sin^2 ht}{h^2 t^2} |f(t)|^2 \, dt \geqq \frac{\sin^2 1}{\pi} h \int_{|t| \leqq 1/h} |f(t)|^2 \, dt.$$

Now the assertion follows from $\sin^2 1/\pi > 1/5$. □

Next we turn to upper bounds of $Q$. We consider a certain d.f. $G$ with c.f. $g$ and define the functional

$$\Lambda_G(X) = \int_{-\infty}^{\infty} |f(t)| \, dG(t).$$

Some properties of $\Lambda_G$ are summarized in the following way.

**Proposition 5.3.4.** (i) *If $X_1, X_2$ are i.r.v., then*

$$\Lambda_G(X_1 + X_2) \leqq \min \{\Lambda_G(X_1), \Lambda_G(X_2)\}.$$

(ii) *For all $a \in R_1$ we have $\Lambda_G(X) = \Lambda_G(X + a)$.*

(iii) *Assume $g \geqq 0$, let $c \in (0, 1)$, and put $h_c := \sup \{t > 0 : g(u) \geqq c, \ 0 \leqq u \leqq t\}$. Then*

$$Q(X, h_c) \leqq \frac{1}{c} \Lambda_G(X).$$

Proof. The relations (i) and (ii) are easily seen by the definition of $\Lambda_G$. To prove (iii) we write by the Parseval relation

$$c\big(F(x + h_c) - F(x - h_c)\big) = c \int_{|y-x| \leqq h_c} dF(y) \leqq \int_{|y-x| \leqq h_c} g(y - x) \, dF(y)$$

$$\leqq \int_{-\infty}^{\infty} g(y - x) \, dF(y) = \int_{-\infty}^{\infty} e^{-itx} f(t) \, dG(t) \leqq \Lambda_G(X).$$ □

Now we present two examples for upper bounds of $Q$ by specifying the d.f. $G$ in Prop. 5.3.4 (iii).

**Corollary 5.3.5.** *For every $h > 0$, we have the inequalities*

$$Q(X, h) \leqq \frac{12}{11} h \int_{|t| \leqq 1/h} |f(t)| \, dt ,$$

$$Q(X, h) \leqq \sqrt{\frac{e}{2\pi}} \, h \int_{-\infty}^{\infty} \exp\left(-\frac{1}{2} t^2 h^2\right) |f(t)| \, dt .$$

Proof. First we define $G$ by the density

$$p(x) = \begin{cases} h(1 - h|x|) , & |x| \leqq 1/h , \\ 0, & |x| > 1/h , \end{cases}$$

with the corresponding c.f. $g(t) = \left(\frac{\sin t/2h}{t/2h}\right)^2$ (see Table). Then

$$\frac{11}{12} < c := 4 \sin^2 (1/2) \leqq g(t) \quad \text{for} \quad |t| \leqq h := h_c .$$

Thus we see from Prop. 5.3.4 (iii) that

$$Q(X, h) \leqq \frac{1}{c} \int_{|t| \leqq 1/h} |f(t)| \, p(t) \, dt \leqq \frac{12}{11} h \int_{|t| \leqq 1/h} |f(t)| \, dt$$

which proves the first inequality.

To prove the second we set $G = \Phi_{0, 1/h}$ with the corresponding density

$$\varphi(x) = \frac{h}{\sqrt{2\pi}} \exp\left(-\frac{1}{2} x^2 h^2\right).$$

Further, we see that

$$c := \exp\left(-\frac{1}{2}\right) \leqq g(t) = \exp\left(-\frac{1}{2} \frac{t^2}{h^2}\right), \qquad |t| \leqq h =: h_c ,$$

so that the assertion easily follows. □

Finally, we deal with bounds for $Q(X, h)$ in a right-sided neighbourhood of the origin $h = 0$ in the case that $Q(X, +0) = 0$ and, moreover,

$$C_0 := \lim_{h \to +0} \frac{1}{h} Q(X, h) < \infty .$$

Prop. 5.1.5 tells us that this limit exists and yields the inequality

$$Q(X, h) \leqq h C_0 \quad \text{for all} \quad h > 0 .$$

Since $C_0$ is usually unknown we wish to derive upper bounds for $C_0$.

**Proposition 5.3.6.** *Let $F$ be absolutely continuous with bounded density $p \leqq C$. Then $C_0 \leqq 2C$.*

*Moreover, if $f$ is integrable on $R_1$, then*

$$C_0 \leqq \frac{1}{\pi} \int_{-\infty}^{\infty} |f(t)| \, dt .$$

Proof. We have, for all $x \in R_1$

$$F(x+h) - F(x-h) = \int_{x-h}^{x+h} p(t)\,dt \leqq 2hC$$

so that $Q(X, h) \leqq 2hC$. Now the first statement follows from the definition of $C_0$

The second follow from the Fourier inversion formula (3.2.3) which implies

$$p(x) = \frac{1}{2\pi} \int_{-\infty}^{\infty} e^{-itx} f(t)\,dt \leqq \frac{1}{2\pi} \int_{-\infty}^{\infty} |f(t)|\,dt\,. \qquad \square$$

## 5.4. Upper bounds of the concentration function for sums

Let $(X_k)$ be a sequence of i.r.v., $X_k \sim F_k$. As usual, $X_k^s$ denote the corresponding symmetrized r.v., $X_k^s \sim F_k^s$. In the present section, we give upper bounds of the concentration function for the sum $S_n = X_1 + X_2 + \dots + X_n$. For this purpose we consider some r.v. $X$ subject to $F$ and employ the functional

$$D(X, h) := \frac{1}{h^2} \int_{|x|<h} x^2\,dF(x) + \int_{|x|\geqq h} dF(x) = \int_{-\infty}^{\infty} w\left(\frac{x}{h}\right) dF(x)\,,$$

where $w(x) = \min\{1, x^2\}$.

**Theorem 5.4.1.** *Suppose $0 < h_k \leqq h$, $k = 1, 2, \dots, n$; then there exists an absolut constant $d_0$ such that*

$$Q(S_n, h) \leqq d_0 h \left[\sum_{k=1}^{n} h_k^2 D(X_k^s, h_k)\right]^{-1/2}$$
$$\leqq d_0 h \left[\sum_{k=1}^{n} h_k^2 (1 - Q(X_k, h_k))\right]^{-1/2}. \qquad \square$$

The estimates in Th. 5.4.1 are useful in part III of the present book; they are often called inequalities of the type Doeblin-Kolmogorov-Rogozin-Esseen. A better variant of this result is given in Suppl. 5.5.6.

We omit the proof of the first relation which can be found in PETROV (1975) or HENGARTHNER and THEODORESCU (1973). The second inequality in Th. 5.4.1 follows from the first in view of (cf. Prop. 5.1.3 (v))

$$1 - Q(X, h) \leqq 1 - Q(X^s, h) \leqq \mathsf{P}(|X^s| \geqq h) \leqq D(X^s, h)\,.$$

On letting $h \to +0$, Th. 5.4.1 immediately leads to the following.

**Corollary 5.4.2.** *We have*

$$Q(S_n, +0) \leqq d_0 \left[\sum_{k=1}^{n} (1 - Q(X_k, +0))\right]^{-1/2}.$$

It is noteworthy that Th. 5.4.1 obtains a simple and applicable shape in the special case of i.i.d.r.v. $X_k$ with the common non-degenerate d.f. $F$.

**Proposition 5.4.3.** *Let $X_k$ be i.i.d.r.v. Then we have, for fixed $h > 0$,*

$$Q(S_n, h) \leqq \frac{d_0}{\sqrt{n}} [D(X_1^s, h)]^{-1/2} \leqq \frac{d_0}{\sqrt{n}} [1 - Q(X_1, h)]^{-1/2}. \qquad \square$$

These bounds are useful only for constant $h > 0$. Next we derive a bound depending on $h$ in a simple way.

**Proposition 5.4.4.** *Under the assumption of Prop.* 5.4.3 *we have*

$$Q(S_n, h) \leqq \frac{d_0}{\sqrt{n}} \frac{(1+h)}{\sqrt{D(X_1^s, 1)}}, \qquad h > 0 .$$

Proof. It is easy to see that

$$w\left(\frac{x}{h}\right) \geqq w(x)\,(1+h)^{-2}, \qquad h > 0, \qquad x \in R_1 .$$

Hence

$$D(X_1^s, h) = \int_{-\infty}^{\infty} w\left(\frac{x}{h}\right) \mathrm{d}F^s(x) \geqq \frac{1}{(1+h)^2} \int_{-\infty}^{\infty} w(x)\,\mathrm{d}F^s(x) = \frac{D(X_1^s, 1)}{(1+h)^2}.$$

Now the assertion follows from Prop. 5.4.3. □

**Corollary 5.4.5.** *Under the assumption of Prop.* 5.4.3 *we have* $Q(S_n, h) = o(1)$ *provided that* $h = o(\sqrt{n})$, as $n \to \infty$.

## 5.5. Supplement

**5.5.1.** *Let $X$ be a* r.v. *with* c.f. $f$ *and suppose* $Q(X, +0) = 0$.

(i) *We have*

$$0 < \frac{1}{\pi} \int_{-\infty}^{\infty} |f(t)|^2\,\mathrm{d}t = \lim_{h \to +0} \frac{C(X, h)}{h} = \sup_{h > 0} \frac{C(X, h)}{h} \leqq \infty .$$

This relation can be proved in a similar way to Prop. 5.1.5, using L. 5.3.1 so that

$$\infty \geqq \frac{1}{\pi} \int_{-\infty}^{\infty} |f(t)|^2\,\mathrm{d}t \geqq \frac{C(X, h)}{h} = \frac{1}{\pi} \int_{-\infty}^{\infty} \left(\frac{\sin ht}{ht}\right)^2 |f(t)|^2\,\mathrm{d}t$$

$$\geqq \frac{1-\varepsilon}{\pi} \int_{-T}^{T} |f(t)|^2\,\mathrm{d}t$$

for every $\varepsilon > 0$, $T < \infty$, and $0 < h < h_0(\varepsilon, T)$.

(ii) *Assume* $0 \leqq \eta \leqq 1$; *then the following assertions are equivalent*

$$\int_0^T |f(t)|^2\,\mathrm{d}t = O(T^\eta), \quad \textit{as} \quad T^\eta \to \infty; \tag{5.5.1}$$

$$C(X, h) = O(h^{-\eta+1}), \quad \textit{as} \quad h \to +0 \tag{5.5.2}$$

(for $0 < \eta \leqq 1$ cf. Kawata (1972), for $\eta = 0$ cf. (i)).

**5.5.2.** *Let $X_1$, $X_2$ be arbitrary r.v. with d.f. $F_1$, $F_2$, resp.; then*

$$\sup_{h>0} |Q(X_1, h) - Q(X_2, h)| \leqq 2 \sup_{x \in R_1} |F_1(x) - F_2(x)|$$

(Hengartner and Theodorescu (1980)).

**5.5.3.** *Let $(X_k)$ be a sequence of i.i.d.r.v. with common d.f. $F$ and suppose $\sigma_F^2 < \infty$. Further, we set $S_n = \sum_{k=1}^{n} X_k$. Then we have*

$$\lim_{n\to\infty} \sqrt{n}\, Q(S_n, h) = \frac{h}{\sigma_F}\sqrt{\frac{2}{\pi}}$$

*for every $h > 0$, if $F$ is not lattice; and for the maximal span $h$ of $F$, if $F$ is lattice* (Lotov (1974)).

**5.5.4.** *Let $(X_k)$, $S_n$, $F$ be defined as in 5.5.3, and suppose that $F$ has an absolute moment $M_\eta$, $0 < \eta \leqq 2$. Then*

$$Q(S_n, h) \geqq c(\eta)\, h[h^2 + (nM_\eta(a)^{1/\eta}]^{-1} \geqq c(\eta)\, h[h^2 + M_\eta(a)^{1/\eta}]^{-1}\, n^{-1/\eta},$$

*where $M_\eta(a) = \mathsf{E}\,|X_1 - a|^\eta$ and $a \in \mathbb{R}_1$ is arbitrary. Here*

$$c(\eta) = \begin{cases} \frac{1}{2}(\eta+1)^{-(1+1/\eta)}, & 0 < \eta < 2, \\ \frac{2}{3}\sqrt{3}, & \eta = 2, \end{cases}$$

(Esseen (1968)).

**5.5.5.** *Let $(X_k)$ be a sequence of i.r.v., $X_k \sim F_k$; $0 < h_k \leqq h$, $k = 1, 2, \ldots, n$. Further, let $D$ be defined as in Sect. 5.4. Then we have* (Mirošnikov and Rogozin (1980))

$$Q(S_n, h) \leqq ch\left[\sum_{k=1}^{n} \frac{D(X_k^s, h_k)}{Q(X_k, h_k)^2}\, h_k^2\right]^{-1/2}$$

*for a certain absolute constant $c$. If the r.v. $X_k$ are i.d., then the above estimate implies* (Kesten (1969))

$$Q(S_n, h) \leqq \frac{c}{\sqrt{n}} \frac{Q(X_1, h)}{\sqrt{D(X_1^s, h)}}.$$

**5.5.6.** *Let $(X_k)$ be a sequence of independent random elements in a real separable Hilbert space $H$ with norm $\|.\|$ and set $S_n = \sum_{k=1}^{n} X_k$.*

(i) *If the random elements $X_k$ are i.d. the common distribution $F$ of which is not concentrated on a finite-dimensional space, then*

$$Q(S_n, h) := \sup_{x\in \mathrm{H}} \mathsf{P}(\|S_n - x\| \leqq h) = o(n^{-\eta}), \qquad n \to \infty,$$

*for every $\eta > 0$* (Siegel (1981 b)).

(ii) *Let $X_k$ be subject to $F_k$ and let $F_k^s$ be the corresponding symmetrized distributions, then for all $0 < h_k \leqq h$, $k = 1, 2, \ldots, n$, we have*

$$Q(S_n, h) \leqq d_0 h\,[\sum_{k=1}^{n} (\int_{\|x\|\leqq h_k} \|x\|^2\, F_k^s(\mathrm{d}x) + h_k^2 \int_{\|x\|>h_k} F_k^s(\mathrm{d}x))]^{-1/2},$$

*where $d_0$ is the same constant as in Th. 5.4.1 and Cor. 5.4.2* (Siegel (1981 b)).

## Notes and bibliographical data

A detailed investigation of concentration functions was given in HENGARTNER and THEODORESCU (1973) and PETROV (1975).

Prop. 5.1.5. is due to HILLE and PHILLIPS (1957).

Prop. 5.2.3 can be found, for instance, in HENGARTNER and THEODORESCU (1973); note that this assertion is usually proved in terms of the Lévy metric.

L.5.3.1 and Cor. 5.3.2 were shown in KAWATA (1972). Prop. 5.3.3 and the first inequality in Cor. 5.3.5 are slight reformulations of results due to ESSEEN (1968). In the first inequality of Cor. 5.3.5 the constant 12/11 is greater than the constant 96/95 given in PETROV (1975); but contrary to this paper, the present estimate concerns the concentration function which is defined by the maximal probability on an interval of length $2h$. Prop. 5.3.4 can be found in HENGARTNER and THEODORESCU (1973).

# II. Special classes of distributions

# Chapter 6. Analytic distribution functions

This chapter is intended to acquaint the reader with a subject which has been investigated only sporadically in the past. Here we present a first survey of the theory of analytic d.f. It is far from being complete, and we wish to stimulate the reader to contribute further research work to this field.

A sufficient condition for a d.f. to be analytic will be formulated in Sect. 6.1. It plays a particular role in the theory of sums where appropriate assumptions on the summands ensure that the limit function is analytic. The results on the special class $\boldsymbol{P}$ of symmetric d.f. introduced in Sect. 6.2 serve as tools for later purposes; but this class also deserves interest for other reasons. So in Sect. 6.5 we will see that $\boldsymbol{P}$ has a certain bearing on the continuation theory of d.f. In Sect. 6.3 we deal with strips of boundedness. They prove useful in our context since most of the basic results can be sharpened, provided that such a strip exists.

Many examples illustrate our general investigation.

## 6.1. Definition and basic results

**Definition 6.1.1.** A d.f. $F$ is said to be an *analytic d.f.* (*a.d.f.*) if there exists a function $\Psi$ of the complex variable $z = x + iy$ such that

(i) $\Psi$ is analytic in a simply connected domain $D_+ \subseteqq S(0, \infty)$ the boundary of which contains $\mathbb{R}_1$,

(ii) $\Psi$ is continuous in $D_+ \cup \mathbb{R}_1$,

(iii) $\Psi(x) = F(x)$ for $x \in \mathbb{R}_1$. □

If $F$ is an a.d.f., then the function $\Psi$ continuously approaches real values on $\mathbb{R}_1$. Therefore we immediately obtain the following result by virtue of Schwarz' reflexion principle, see App. B.1.

**Corollary 6.1.2.** *Let $F$ be an a.d.f. Then the function $\Psi$ is analytic in the domain $D = D_+ \cup D_- \cup \mathbb{R}_1$ where $D_- = \{z : \bar{z} \in D_+\}$. For $z \in D$ we have $\Psi(\bar{z}) = \overline{\Psi(z)}$ so that the zeros and singularities of $\Psi$ are located symmetrically with respect to $\mathbb{R}_1$.* □

So it results that Def. 6.1.1 implies what we expect of an a.d.f., namely, that it permits an analytic continuation to a domain $D$ the interior of which contains $\mathbb{R}_1$. From now on, we will denote the analytic continuation of $F$ also by $F$.

By the way, we note that an a.d.f. $F$ is always absolutely continuous and has domain of regularity $D$ iff its density $p$ is continuable to $D$.

Next we give several examples for a.d.f.

**Example 6.1.3.** (i) The normal d.f. $\Phi_{\mu,\sigma}$ is an a.d.f. Indeed, its density is continuable to the whole complex plane so that $\Phi_{\mu,\sigma}$ is even an integral d.f.

(ii) The double exponential distribution $F(x) = \mathrm{e}^{-\mathrm{e}^{-x}}$ is an integral d.f.

(iii) The density $p(x) = \dfrac{1}{\pi}\left(\dfrac{\sin x}{x}\right)^2$ is an integral density. □

**Example 6.1.4.** (i) The Cauchy distribution $F(x) = \dfrac{1}{2} + \dfrac{1}{\pi}\arctan x$ is an a.d.f. whose domain of regularity contains the strip $S(1)$. This is obvious since $p(x) = \dfrac{1}{\pi}\dfrac{1}{1+x^2}$.

(ii) The Fisher distribution $F_{m,n}$ with $(m, n)$ degrees of freedom has the density $p(x) = c\,\mathrm{e}^{mx}\,(n + m\,\mathrm{e}^{2x})^{-(m+n)/2}$. Considered as function of the complex variable $z=x+\mathrm{i}y$, the denominator vanishes only for $z = \frac{1}{2}\log n/m + \mathrm{i}\pi\,(k + \frac{1}{2})$, $k = 0, \pm 1, \ldots$ Therefore $F_{m,n}$ is analytic in $S\,(\pi/2)$. We see that all singularities have the real part $\frac{1}{2}\log n/m$, and for $m \neq n$ they are not located on the imaginary axis. □

**Example 6.1.5.** Consider some function $h$ real-valued on $R_1$ and meromorphic in the whole plane with poles at $n \pm (\mathrm{i}/n)$, $n = 1, 2, \ldots$; such a function can be constructed by the Mittag-Leffler theorem. Then the function $p(x) = c\,\mathrm{e}^{-x^2}\sin^2\big(h(x)\big)$ is an example for an analytic density whose domain of regularity does not contain any horizontal strip. In the same way it is possible to construct an a.d.f. $F$ with the property that $S(s)$ is its widest strip of regularity although no singularity of $F$ is located on the line $|\mathrm{Im}\, z| = s$. □

Now we are going to state two general criteria for the analyticity of a d.f. $F$; one of them is necessary while the other is sufficient.

First of all, we note that the analytic behaviour of $F$ is closely connected with the tail behaviour of the corresponding c.f. $f$. Indeed, by the inversion formula (3.8.1) we have for arbitrary $t_0 > 0$

$$2\pi\mathrm{i}\big(F(x) - F(0)\big) = \int\limits_{|t|<t_0} + \int\limits_{|t|\geqq t_0} (1 - \mathrm{e}^{-\mathrm{i}tx})\, f(t)\,\frac{\mathrm{d}t}{t},$$

provided that $F$ is continuous at $x = 0$. Since the exponential series is uniformly convergent in any bounded domain, the first integral represents an integral function. Accordingly, the analytic properties of $F$ depend only on the second integral, i.e. on the tail behaviour of $f(t)$ as $|t| \to \infty$.

Let us introduce two quantities useful for our purposes. If $F$ is an arbitrary d.f. with c.f. $f$, then for any real number $\nu$ we put

$$l_\nu(F) = \varliminf_{t\to\infty} \frac{-\log|f(t)|}{t^\nu}, \qquad \chi(F) = \sup\{\nu : l_\nu(F) > 0\};$$

here and in the following we take $\log 0 = -\infty$ and $\log\infty = \infty$. Since $|f(t)| \leqq 1$ for all $t$, we have always $0 \leqq l_\nu(F) \leqq \infty$. More precisely, it is easily seen that

$$l_\nu(F) = \begin{cases} \infty & \text{if} \quad \nu < \chi(F), \\ 0 & \text{if} \quad \nu > \chi(F). \end{cases} \tag{6.1.1}$$

We emphasize that $\chi(F)$ is not necessarily non-negative. So for any lattice d.f. $F$ we have $\chi(F) = -\infty$ since $|f(t_n)| = 1$ for a certain sequence $t_n \to \infty$ and therefore $l_\nu(F) = 0$ for all $\nu$.

Next we prove a formula to determine the quantity $\chi(F)$.

**Proposition 6.1.6.** *For a d.f. $F$ with c.f. $f$ we have*

$$\chi(F) = \varliminf_{t\to\infty} \frac{\log(-\log|f(t)|)}{\log t}.$$

Proof. Assume $|\chi| < \infty$. Then we have, by virtue of (6.1.1), for fixed $\varepsilon > 0$ and sufficiently large $t$

$$\frac{-\log|f(t)|}{t^{\chi-\varepsilon}} \geqq 1,$$

so that

$$\frac{\log(-\log|f(t)|)}{\log t} \geqq \chi - \varepsilon.$$

Further, for a certain sequence $t_n \to \infty$ we get

$$\frac{-\log|f(t_n)|}{t_n^{\chi+\varepsilon}} \leqq 1,$$

and therefore

$$\frac{\log(-\log|f(t_n)|)}{\log t_n} \leqq \chi + \varepsilon.$$

Since $\varepsilon$ may be arbitrarily small, the assertion is shown. The cases $|\chi| = \infty$ can be treated in a similar way. □

Now we are able to compute the quantities $\chi(F)$ and $l_\chi(F)$ for the d.f. considered in the previous examples; as to the corresponding c.f. see the table at the end of this book.

**Example 6.1.7.** (i) For the normal d.f. we obtain $\chi(\Phi_{\mu,\sigma}) = 2$ and $l_2(\Phi_{\mu,\sigma}) = \sigma^2/2$.

(ii) We apply Stirling formula (see App. B.5) to the c.f. $f(t) = \Gamma(1 - it)$ of the double exponential distribution and obtain, as $t \to \infty$,

$$\begin{aligned}\log|f(t)| &= \operatorname{Re}\log f(t) = \operatorname{Re}\log\Gamma(1 - it)\\ &= \operatorname{Re}\left[\left(\tfrac{1}{2} - it\right)\log(1 - it)\right] + O(1)\\ &= t\left[\frac{1}{2t}\log|1 - it| + \arg(1 - it) + o(1)\right]\\ &= t\left[-\frac{\pi}{2} + o(1)\right] = -\frac{\pi}{2}t[1 + o(1)].\end{aligned}$$

So we get $\chi(F) = 1$ and $l_1(F) = \pi/2$.

(iii) For the c.f. $f$ of the density $p(x) = \dfrac{1}{\pi}\left(\dfrac{\sin x}{x}\right)^2$ we have $f(t) = 0$, $|t| > 2$, whence $\chi(F) = \infty$ follows.

(iv) For the Cauchy distribution we find $\chi(F) = 1$ and $l_1(F) = 1$.

(v) The method used in (ii) can also be applied to the Fisher distribution $F_{m,n}$, and we obtain $\chi(F_{m,n}) = 1$ and $l_1(F_{m,n}) = \pi/2$.

(vi) For the Laplace distribution, which is no a.d.f., we have $\chi(F) = 0$ and $l_0(F) = \infty$. □

Before formulating a condition necessary that a d.f. $F$ should be analytic, we prove the following simple lemma.

**Lemma 6.1.8.** *If* $\overline{\lim\limits_{t\to\infty}}\, |f(t)| < 1$, *then* $\chi(F) \geqq 0$.

Proof. Since $|f(t)| \leqq 1 - \eta$ for a certain $\eta > 0$ and all sufficiently large $t$, we get

$$\frac{\log(-\log|f(t)|)}{\log t} \geqq \frac{\log(-\log(1-\eta))}{\log t} \to 0, \quad \text{as} \quad t \to \infty. \qquad \square$$

**Proposition 6.1.9.** *If F is an a.d.f., then* $\chi(F) \geqq 0$.

The proof is quite trivial. In fact, since $F$ is absolutely continuous, we have $\lim\limits_{t\to\infty} |f(t)| = 0$ by the Riemann-Lebesgue lemma (see App. A.5), and L. 6.1.8 shows the assertion. □

In the proof just given we made use only of the fact that $\overline{\lim\limits_{t\to\infty}}\, |f(t)| < 1$. Therefore it is not surprising that the condition $\chi(F) \geqq 0$ is not sufficient for the analyticity of $F$. Indeed, there exist always non-analytic d.f. with $\chi(F) = c$ for any $0 \leqq c \leqq 1$, see Ex. 6.1.10, while the condition $\chi(F) > 1$ is sufficient for $F$ to be an a.d.f., see Cor. 6.1.12. Nevertheless, Prop. 6.1.9 cannot be improved, i.e. there do exist a.d.f. $F$ with $\chi(F) = c$ for every $0 \leqq c < 1$, see Suppl. 6.6. 1, (i).

**Example 6.1.10.** We check that there exist non-analytic d.f. with $\chi(F) = c$ for every $0 \leqq c \leqq 1$.

For the Laplace distribution we have $\chi(F) = 0$, cf. Ex. 6.1.7, (vi).

For the Pólya type c.f. $f(t) = \mathrm{e}^{-|t|^\alpha}$, $0 < \alpha < 1$, we have $\chi(F) = \alpha$. In Sect. 8.3 we will see that the corresponding d.f. $F$ are not analytic.

To show that there exists a non-analytic d.f. $F$ with $\chi(F) = 1$, we consider the function $f_0(t) = \exp\left(-\dfrac{t}{\log t}\right)$, $t \geqq t_0 > 0$. It can easily be seen that $f_0(t)$ is convex for $t \geqq \mathrm{e}^2$, and taking $t_0$ sufficiently large it is possible to continue $f_0$ to a Pólya type c.f. In Sect. 6.2 we will be able to show that the corresponding d.f. $F$ cannot be an a.d.f. (see Ex. 6.2.11, (i)) although $\chi(F) = 1$. □

Next we prove a sufficient condition for a d.f. $F$ to be analytic.

**Theorem 6.1.11.** *Assume* $0 < l_1(F) \leqq \infty$. *Then F and its density p are analytic in the strip* $S(l_1)$. *For* $z \in S(l_1)$, *F and p have respectively the integral representation*

$$\left.\begin{aligned} F(z) &= F(0) + \frac{1}{2\pi\mathrm{i}} \int_{-\infty}^{\infty} (1 - \mathrm{e}^{-\mathrm{i}tz})\, f(t)\, \frac{\mathrm{d}t}{t}, \\ p(z) &= \frac{1}{2\pi} \int_{-\infty}^{\infty} \mathrm{e}^{-\mathrm{i}tz} f(t)\, \mathrm{d}t. \end{aligned}\right\} \tag{6.1.2}$$

Proof. Choose some $l'$ such that $0 < l' < l_1$. Then $-(1/t)\log|f(t)| \geqq l'$ for large $t$. Hence we have, for an appropriate $C > 0$,

$$|f(t)| \leqq C\,\mathrm{e}^{-tl'}, \qquad t > 0. \tag{6.1.3}$$

Consider now the integral functions

$$\Psi_n(z) := F(0) + \frac{1}{2\pi\mathrm{i}} \int_{-n}^{n} (1 - \mathrm{e}^{-\mathrm{i}tz})\, f(t)\, \frac{\mathrm{d}t}{t}, \qquad n = 1, 2, \ldots$$

Assuming $|y| \leqq l'' < l'$, $m \geqq n$, and using (6.1.3), we see that

$$|\Psi_m(z) - \Psi_n(z)| = \frac{1}{2\pi}\left|\int_{-m}^{-n} + \int_{n}^{m} (1 - e^{-itz}) f(t) \frac{dt}{t}\right| \leqq \frac{1}{\pi}\int_{n}^{m} (1 + e^{t|y|}) |f(t)| \frac{dt}{t}$$

$$\leqq \frac{C}{n\pi}\int_{n}^{m} (e^{-tl'} + e^{-t(l'-l'')}) dt .$$

For fixed $l'$ and $l''$, this expression can be made arbitrarily small by choosing $m$ and $n$ sufficiently large. This means that the sequence $(\Psi_n)$ converges uniformly in $|y| \leqq l''$; the limit function $\Psi$ is analytic in $|y| < l''$ by the Weierstrass convergence theorem B.6.1. Further, in view of the inversion formula (3.8.1), $\Psi$ agrees with $F$ on the real axis. Since $l''$ may be arbitrarily close to $l_1$, we have shown that $F$ is an a.d.f. whose domain of regularity contains the strip $S(l_1)$ and that the first formula (6.1.2) is valid. The second follows immediately since differentiation under the integral sign is permissible by virtue of (6.1.3). □

**Corollary 6.1.12.** *If* $\chi(F) > 1$, *then $F$ is an integral d.f. and the representations* (6.1.2) *hold true for all $z$.* □

For, by assumption and (6.1.1) we have $l_1(F) = \infty$ so that Th. 6.1.11 is applicable.

To close this section we give several examples which also shed light on the strength of the condition $l_1(F) > 0$.

**Example 6.1.13.** (i) In Ex. 6.1.7 for the normal d.f. $\Phi_{\mu,\sigma}$, the Cauchy d.f. $F$, and Fisher d.f. $F_{m,n}$, we have found $\chi(\Phi_{\mu,\sigma}) = 2$, $l_1(F) = 1$, and $l_1(F_{m,n}) = \pi/2$, resp. This agreess with the facts that $\Phi_{\mu,\sigma}$ is an integral d.f. whereas $F$ and $F_{m,n}$ are analytic in $S(1)$ and $S(\pi/2)$, resp., but not in any wider strip.

(ii) For the double exponential d.f. we have $l_1(F) = \pi/2$ although it is an integral d.f. Thus we see that $F$ may be analytic in a strip wider than $S(l_1)$.

(iii) We consider the a.d.f. constructed in Ex. 6.1.5. Without knowing its c.f., we can recognize that $l_1(F) = 0$. Indeed, since the domain of regularity of $F$ does not contain any horizontal strip, the case $l_1(F) > 0$ is not possible by Th. 6.1.11. This example also shows that the condition $l_1(F) > 0$ is not necessary for the analyticity of $F$.

(iv) We again consider the Pólya type c.f. $f(t) = e^{-|t|^\alpha}$, $0 < \alpha < 1$, which correspond to non-analytic d.f., cf. Ex. 6.1.10. Then we have $l_\alpha(F) = 1$. This shows that Th. 6.1.11 cannot be improved in the sense that we only demand $l_\nu(F) > 0$ for some $\nu < 1$ instead of $l_1(F) < 0$. Moreover, in the definition of $l_1(F)$ the denominator $t$ cannot be replaced by certain other functions converging to infinity more slowly than $t$, see Suppl. 6.1.1 (ii). □

## 6.2. A special class of symmetric distribution functions

Clearly, it is not possible to conclude that $F$ is an a.d.f. from suppositions made only on the analytic behaviour of $F$ on a bounded set. But if we combine certain local suppositions with appropriate additional ones, such a conclusion can possibly be drawn. In the present section we pursue this idea investigating a special class of symmetric d.f. which will also prove useful for later purposes.

**Definition 6.2.1.** A d.f. $F$ with c.f. $f$ belongs to the class $\boldsymbol{P}$ if $f(t)\,\mathrm{e}^{-ita}$ is real and non-negative for some $a \in \mathbb{R}_1$ and all $t \in \mathbb{R}_1$.

Obviously, a d.f. $F \in \boldsymbol{P}$ is symmetric with respect to $a$; in the following we assume without loss of generality that $a = 0$.

First we give some examples.

**Example 6.2.2.** Among the d.f. considered in Sect. 6.1 there are many which belong to $\boldsymbol{P}$; we mention

(i) the normal distribution,

(ii) the d.f. with density $p(x) = \dfrac{1}{\pi}\left(\dfrac{\sin x}{x}\right)^2$,

(iii) the Cauchy distribution,

(iv) the Laplace distribution,

(v) the d.f. constructed in Ex. 6.1.10,

(vi) the Fisher distribution $F_{m,n}$ for $m = n$. In fact, by virtue of $\Gamma(\bar{z}) = \overline{\Gamma(z)}$ (this follows from Schwarz' reflexion principle) we have $f(t) = c\left|\Gamma\left(\dfrac{n - it}{2}\right)\right|^2 \geqq 0$. □

Beside these special d.f. we are able to describe two broad classes of d.f. contained in $\boldsymbol{P}$.

**Example 6.2.3.** (i) All d.f. corresponding to Pólya type c.f. belong to $\boldsymbol{P}$. Indeed, these c.f. are required to be real-valued and non-negative, see Prop. 3.6.5.

(ii) Let $F$ be an arbitrary d.f. and $F^s$ its symmetrization. In view of $f^s = |f|^2 \geqq 0$, we have $F^s \in \boldsymbol{P}$. As an example we mention the Fisher d.f. $F_{m,n}$ which is the symmetrization of the double exponential d.f. (but for a scale factor).

Note that there do exist symmetric d.f. $F \notin \boldsymbol{P}$; for example the uniform distribution on $(-b, b)$ having c.f. $f(t) = \dfrac{\sin bt}{bt}$. □

We start our general investigation on the class $\boldsymbol{P}$ with the following important result.

**Proposition 6.2.4.** *A d.f. $F \in \boldsymbol{P}$ satisfies the Lipschitz condition*

$$F(x) - F(-x) \leqq Cx\,, \qquad x > 0\,, \tag{6.2.1}$$

*for some $C > 0$ iff*

$$\int_{-\infty}^{\infty} f(t)\,\mathrm{d}t < \infty\,. \tag{6.2.2}$$

Proof. *If-part*: From (6.2.2) and Th. 3.2.2 it follows that $F$ has a continuous density; this implies (6.2.1).

*Only-if part*: We consider the density $p_n(x) = \dfrac{1}{\pi n}\left(\dfrac{\sin x/n}{x/n}\right)^2$ the c.f. $g_n(y)$ of which vanishes for $|y| > 2/n$. Now we apply the Parseval equation (3.2.1) with $t = 0$ and thus obtain

$$I_n := \int_{-\infty}^{\infty} f(x)\,p_n(x)\,\mathrm{d}x = \int_{\infty}^{\infty} g_n(y)\,\mathrm{d}F(y) \leqq F\left(\frac{2}{n}\right) - F\left(-\frac{2}{n}\right) \leqq \frac{2C}{n}\,,$$

where the last inequality is due to (6.2.1.). Finally, we make use of the Fatou theorem A.1 wich tells us that

$$2C \geqq \varliminf_{n\to\infty} (nI_n) \geqq \int_{-x_1}^{x_1} f(x) \varliminf_{n\to\infty} \left(np_n(x)\right) dx = \frac{1}{\pi} \int_{-x_1}^{x_1} f(x)\, dx . \qquad \square$$

The result just proved should be illustrated by some examples.

**Example 6.2.5.** (i) The function $f(t) = 1/(1 + |t|)$ is a Pólya type c.f. of an absolutely continuous d.f. $F$, and Ex. 6.2.3, (i), shows that $F \in \boldsymbol{P}$. However, $f$ is not integrable on $\boldsymbol{R}_1$ so that (6.2.1) cannot be satisfied. This means that the corresponding density $p(u)$ tends to infinity as $|u| \to 0$, see also Ex. 3.6.7. Accordingly, $F$ is no a.d.f. but has a singularity at $x = 0$. Note that $\chi(F) = 0$ and $l_0(F) = \infty$.

(ii) There exist lattice d.f. belonging to $\boldsymbol{P}$; for example, the symmetrization of any lattice d.f. Then Prop. 6.2.4 tells us by an indirect argument that such a d.f. must have a jump at $x = 0$. □

Now we are going to prove a statement basic for all that follows.

**Proposition 6.2.6.** *Let $F \in \boldsymbol{P}$ possess a c.f. $f$ integrable on $\boldsymbol{R}_1$ and let $p$ be its density. Then here exists a d.f. $\hat{F} \in \boldsymbol{P}$ with c.f. $\hat{f}$ integrable on $\boldsymbol{R}_1$ and density $\hat{p}$ such that, for a certain $c > 0$ and all $t \in \boldsymbol{R}_1$,*

$$f(t) = c\hat{p}(t)\,, \qquad p(t) = \frac{c}{2\pi}\hat{f}(t)\,. \tag{6.2.3}$$

Proof. We put $c = \int_{-\infty}^{\infty} f(t)\, dt$ so that by assumption $0 < c < \infty$. Now we define the absolutely continuous d.f.

$$\hat{F}(x) := \frac{1}{c} \int_{-\infty}^{x} f(t)\, dt\,.$$

Its density $\hat{p}$ is given by $\hat{p}(t) = f(t)/c$ which implies that $\hat{F}$ is symmetric. Further, for the c.f. $\hat{f}$ of $\hat{F}$ we obtain

$$\hat{f}(t) = \int_{-\infty}^{\infty} e^{itx}\, \hat{p}(x)\, dx = \frac{1}{c} \int_{-\infty}^{\infty} e^{-i(-t)x} f(x)\, dx = \frac{2\pi}{c} p(-t) = \frac{2\pi}{c} p(t)\,;$$

the last two equalities follow from the Fourier inversion formula (3.2.3) and the fact that $p$ is symmetric, resp. Since $\hat{f}(t) = \frac{2\pi}{c} p(t) \geqq 0$ we have $\hat{F} \in \boldsymbol{P}$ and $\hat{f}$ is integrable on $\boldsymbol{R}_1$. □

**Remark 6.2.7.** Obviously, $\hat{F}$ is uniquely defined by $F$ where $\hat{\hat{F}} = F$. Therefore, in $\boldsymbol{P}$ we have pairs $(F, \hat{F})$ of d.f. corresponding each to the other in the sense that the density of $F$ coincides with the c.f. of $\hat{F}$ and conversely (but for positive factors). □

**Example 6.2.8.** (i) For the normal d.f. we have $\hat{\Phi}_{0,\sigma} = \Phi_{0,1/\sigma}$.

(ii) The Cauchy d.f. and the Laplace d.f. form such a pair.

(iii) The symmetrization of the uniform distribution $F$ on $(-1, 1)$ has the c.f. $f(t) = \left(\frac{\sin t}{t}\right)^2$; the corresponding d.f. $\hat{F}$ has been mentioned in Ex. 6.2.2, (ii).

The correspondence between a density $p$ of a d.f. $F \in \boldsymbol{P}$ and a certain c.f. $f$, stated in Prop. 6.2.6, makes it possible to use the theory of a.c.f. for investigating analytic properties of $F$. In this way we obtain our main result about the class $\boldsymbol{P}$.

**Theorem 6.2.9.** *Let $F \in \boldsymbol{P}$. Then the following assertions are equivalent:*

(i) *$F$ is continuable to a disk $|z| < s$,*

(ii) *$F$ is analytic in the strip $S(s)$,*

(iii) $l_1(F) \geqq s$,

(iv) *for all $s' < s$ we have* $\int\limits_{-\infty}^{\infty} e^{s'|t|} f(t) \, dt < \infty$.

Proof. We show the chain of implications (iv) $\Rightarrow$ (iii) $\Rightarrow$ (ii) $\Rightarrow$ (i) $\Rightarrow$ (iv) which proves our theorem.

(iv) $\Rightarrow$ (iii) Since the integrand is continuous we have for any $s' < s$, all $t \geqq 0$, and some constant $C > 0$

$$e^{s't} f(t) \leqq C$$

so that

$$\frac{-\log f(t)}{t} \geqq -\frac{\log C}{t} + s' .$$

This yields $l_1(F) \geqq s'$, and since $s'$ may be arbitrarily close to $s$ we get $l_1(F) \geqq s$.

(iii) $\Rightarrow$ (ii) follows from Th. 6.1.11 while (ii) $\Rightarrow$ (i) is obvious.

(i) $\Rightarrow$ (iv) If (i) is satisfied, then $F$ has a bounded derivative at $x = 0$. This means that the Lipschitz condition (6.2.1) holds true and, by Prop. 6.2.4, that $f$ is integrable on $\boldsymbol{R}_1$. Now we make use of Prop. 6.2.6 which tells us that the density $p$ of $F$ coincides with a certain c.f. $\hat{f}$ but for a positive factor. Since $F$ is continuable to the disk $|z| < s$ so is $p$ and therefore also $\hat{f}$. By virtue of Th. 4.3.3 this is tantamount to

$$\int\limits_{-\infty}^{\infty} e^{s'|t|} \, d\hat{F}(t) < \infty$$

for all $s' < s$. Using the first one of the relations (6.2.3), the assertion follows immediately. □

**Corollary 6.2.10.** *A d.f. $F \in \boldsymbol{P}$ is an a.d.f. iff $l_1(F) > 0$. In either case the strip $S(l_1)$ is the widest strip of regularity of $F$ and, if $l_1 < \infty$, the points $z_\pm = \pm i \cdot l_1$ are singularities of $F$. Further, in $S(l_1)$ the integral representations* (6.1.2) *hold true and the density $p$ of $F$ has the ridge property $|p(x + iy)| \leqq p(iy)$.*

Proof. If $F \in \boldsymbol{P}$ is an a.d.f., then it is continuable to a disk $|z| < s$, $s > 0$, and by Th. 6.2.9 we have $l_1(F) \geqq s > 0$.

Assume now $l_1(F) > 0$. Then in view of Th. 6.1.11 the d.f. $F$ is analytic in $S(l_1)$ and the formulas (6.1.2) are valid in this strip. Further, if there were a wider strip of regularity $S(l_1 + \varepsilon)$, we would obtain the contradiction $l_1(F) \geqq l_1 + \varepsilon$ by Th. 6.2.9. So we still have to prove that $z_\pm$ are singularities of $F$ (or, which is the same, of $p$) and that $p$ has the ridge property. But these assertions are easy consequences of the fact that $p = (c/2\pi) \hat{f}$ where $\hat{f}$ is a c.f., see (6.2.3). Hence, Th. 4.3.3 and Cor. 4.3.5 permit us to terminate the proof. □

**Example 6.2.11.** (i) The d.f. $F$ constructed in Ex. 6.1.10 belongs to $\boldsymbol{P}$ and we have $l_1(F) = 0$. Thus $F$ cannot be an a.d.f. by Cor. 6.2.10.

(ii) For the a.d.f. $F$ constructed in Ex. 6.1.5 we have $l_1(F) = 0$, see Ex. 6.1.13, (iii). Accordingly, $F$ cannot belong to $\boldsymbol{P}$ even it is symmetric. This follows again from Cor. 6.2.10. □

## 6.3. The strip of boundedness. Convolutions

From Ex. 6.1.13, (ii), we see that the quantity $l_1(F)$ does not determine in each case the widest strip of regularity of an a.d.f. The following notion permits a better insight into the meaning of $l_1(F)$; it will also prove useful for later purposes.

**Definition 6.3.1.** The open strip $S(s) = \{z\colon |\mathrm{Im}\, z| < s\}$ contained in the domain of regularity of an a.d.f. $F$ with density $p$ is said to be a *strip of boundedness* of $F$ (resp. of $p$), if for every $s'$, $0 < s' < s$, there exists a finite constant $C_F(s')$ (resp. $C_p(s')$) such that, for $|\mathrm{Im}\, z| \leqq s'$,

$$|F(z)| \leqq C_F(s') \qquad (\text{resp. } |p(z)| \leqq C_p(s')) . \qquad \square$$

Before giving some examples, we prove two basic results.

**Proposition 6.3.2.** *If $S(s)$ is a strip of boundedness of an analytic density $p$, then it is also a strip of boundedness for the corresponding d.f. $F$.*

Proof. Assuming $|y| \leqq s' < s$, we get by assumption

$$|F(x + iy)| = \left|F(x) + \int_x^{x+iy} p(z)\, dz\right| \leqq 1 + |y|\, C_p(s')$$
$$\leqq 1 + s' C_p(s') =: C_F(s') . \qquad \square$$

**Proposition 6.3.3.** *If $0 < l_1(F) \leqq \infty$, then $S(l_1)$ is a strip of boundedness of $p$ and therefore also of $F$.*

Proof. Let $0 < s' < l_1(F)$ and $s' < l' < l_1(F)$. Since under the assumption $l_1(F) > 0$ the formulas (6.1.2) and (6.1.3) hold true we obtain, for $|\mathrm{Im}\, z| \leqq s'$,

$$|p(z)| \leqq \frac{1}{\pi} \int_0^\infty e^{ts'}\, C\, e^{-tl'}\, dt = \frac{C}{\pi} \int_0^\infty e^{-t(l'-s')}\, dt =: C_p(s') .$$

Therefore, $S(l_1)$ is a strip of boundedness of $p$ and, by virtue of Prop. 6.3.2, also of $F$. $\square$

Partial conversions of Prop. 6.3.2 and 6.3.3 are also valid; they will be proved in Prop. 6.3.10.

**Example 6.3.4.** (i) For the normal d.f. we have $l_1(\Phi_{\mu,\sigma}) = \infty$ so that the whole complex plane is its strip of boundedness. Obviously, this means that $|\Phi_{\mu,\sigma}(z)|$ is bounded in every closed strip $|\mathrm{Im}\, z| \leqq s'$ but, of course, not in the whole plane.

(ii) For the Cauchy d.f. $l_1(F) = 1$ so that $S(1)$ is a strip of boundedness of $F$. Further, since this strip is the widest strip of regularity of $F$, it is also its widest strip of boundedness.

(iii) For the Fisher d.f. we have $l_1(F_{m,n}) = \pi/2$, and by the same reasoning as in (ii) we see that $S(\pi/2)$ is the widest strip of boundedness of $F_{m,n}$. $\square$

The examples which follow point out that the widest strip of boundedness need not coincide with the widest strip of regularity.

**Example 6.3.5.** (i) The double exponential d.f. is an integral d.f. with $l_1(F) = \pi/2$ so that $S(\pi/2)$ is a strip of boundedness of $F$. Further, there is no wider strip of boundtedness since

$$|F(x + iy)| = e^{-e^{-x} \cdot \cos y} \to \infty , \quad \text{as} \quad x \to -\infty , \ \frac{\pi}{2} < y < \frac{3\pi}{2} .$$

(ii) For the integral d.f. $F(z) = \exp(1 - e^{e^{-z}})$ we obtain

$$|F(x + iy)| = \exp\{1 - \cos(e^{-x} \sin y)\, e^{e^{-x} \cos y}\} .$$

For fixed $y$, $0 < y < \pi\ 2$, we get $\overline{\lim}_{x \to -\infty} |F(x + iy)| = \infty$, i.e. $F$ is an integral d.f. *without* any strip of boundedness. Accordingly, Prop. 6.3.3 tells us that $l_1(F) = 0$ and therefore $\chi(F) \leq 1$ by virtue of (6.1.1). □

We deal now with another subject; namely, the analytic character of convolutions of d.f. We start with several examples which show how both the strips of regularity and those of boundedness can change under convolution.

**Example 6.3.6.** (i) The Cauchy d.f. $F$ is analytic with strip of boundedness $S(1)$. For its symmetrization we have $F^s(x) = F(x/2)$ so that $F^s$ is analytic with strip of boundedness $S(2)$.

(ii) Let $F_1$ be the double exponential d.f. which is an integral d.f. with widest strip of boundedness $S(\pi/2)$. Further, let $F_2(x) = e^x$, $x \leq 0$, which is no a.d.f. Then we get

$$F_1 * F_2(x) = e^x (1 - e^{-e^{-x}}) , \qquad x \in R_1 ;$$

this is an integral d.f. with widest strip of boundedness $S(\pi/2)$.

(iii) The symmetrization of the double exponential d.f. is, but for the scale factor $\frac{1}{2}$, the Fisher d.f. $F_{2,2}$; therefore it is analytic with widest strip of boundedness $S(\pi)$. Hence, the convolution of two integral d.f. need not be integral.

(iv) Without knowing the explicite form of the symmetrization $F^s$ of the integral d.f. $F$ considered in Ex. 6.3.5, (ii), we are able to show that $F^s$ is no a.d.f., see Rem. 6.3.11, (iii). □

So we also see the surprising fact that the known heuristic rule "a convolution is 'not worse' than the 'best' of its components" can fail with regard to analyticity. Accordingly, we want to find out which properties of (at least) one component imply the analyticity of the convolution. To this end we start with the following simple result the proof of which is left to the reader.

**Proposition 6.3.7.** *Let $F_1$ and $F_2$ be arbitrary d.f. Then we have*

$$l_\nu(F_1 * F_2) \geq l_\nu(F_1) + l_\nu(F_2) , \qquad \nu \in R_1 ,$$

$$\chi(F_1 * F_2) \geq \max\{\chi(F_1), \chi(F_2)\} .$$

*In the case of symmetrization it holds*

$$l_\nu(F^s) = 2 l_\nu(F) , \qquad \nu \in R_1 , \qquad \chi(F^s) = \chi(F) .$$ □

Now Prop. 6.3.3 immediately yields the following.

**Corollary 6.3.8.** (i) *If $l_1(F_1) > 0$, then for any d.f. $F_2$ the convolution $F_1 * F_2$ is an a.d.f. which has $S(l_1(F_1))$ as a strip of boundedness.*

(ii) *If $l_1(F) > 0$, then $F^s$ is an a.d.f. which has $S(2 l_1(F))$ as a strip of boundedness.* □

But it is even possible to replace the supposition "$l_1(F_1) > 0$" by "$F_1$ has a strip of boundedness", which follows from $l_1(F_1) > 0$ by Prop. 6.3.3.

**Proposition 6.3.9.** *Assume that $S(s)$ is a strip of boundedness of $F_1$. Then for any d.f. $F_2$ the convolution $F = F_1 * F_2$ is an a.d.f. which has $S(s)$ as a strip of boundedness and the integral representation*

$$F(z) = \int_{-\infty}^{\infty} F_1(z - u)\, dF_2(u) , \qquad z \in S(s) . \tag{6.3.1}$$

Proof. We consider the functions $\Psi_n(z) := \int_{-n}^{n} F_1(z-u)\,\mathrm{d}F_2(u)$, $n \geqq 1$, which are obviously analytic in $S(s)$. For $|\mathrm{Im}\, z| \leqq s' < s$ we have $|\mathrm{Im}\,(z-u)| \leqq s'$, and since $S(s)$ is a strip of boundedness of $F_1$ we get

$$|\Psi_n(z) - \Psi_m(z)| \leqq C_{F_1}(s') \left| \int_{-m}^{-n} + \int_{n}^{m} \mathrm{d}F_2(u) \right| .$$

This bound can be made arbitrarily small by choosing $m$ and $n$ sufficiently large. So the sequence $(\Psi_n)$ converges uniformly in $|\mathrm{Im}\, z| \leqq s'$, and by Weierstrass' convergence theorem the limit $\Psi$ is analytic in $|\mathrm{Im}\, z| < s'$. Further, $\Psi$ coincides with the convolution $F_1 * F_2$ on the real axis. Therefore, $F = F_1 * F_2$ is an a.d.f. with the representation (6.3.1). Finally, for $|\mathrm{Im}\, z| \leqq s' < s$ we get $|F(z)| \leqq C_{F_1}(s') \int_{-\infty}^{\infty} \mathrm{d}F_2(u) = C_{F_1}(s')$; thus $S(s)$ is a strip of boundedness of $F$. □

Now we are able to prove partial conversions of statements given above; in parentheses we refer to the corresponding statements.

**Proposition 6.3.10.** (i) (cf. Cor. 6.3.8, (ii)) *If $S(s)$ is the widest strip of regularity of $F^s$, then $l_1(F) = s/2$.*

(ii) (cf. Prop. 6.3.3) *If $S(s)$ is the widest strip of boundedness of $F$, then $s/2 \leqq l_1(F) \leqq s$.*

(iii) (cf. Prop. 6.3.2) *If $S(s)$ is a strip of boundedness of $F$, then $S(s/2)$ is a strip of boundedness for the corresponding density $p$.*

Proof. (i) Assume that $S(s)$ is the widest strip of regularity of $F^s$. Since $F^s \in \boldsymbol{P}$ we can apply Cor. 6.2.10 and obtain $l_1(F^s) = s$. In view of $l_1(F^s) = 2l_1(F)$ the assertion is shown.

(ii) If $S(s)$ is the widest strip of boundedness of $F$, then Prop. 6.3.9 tells us that $F^s$ is analytic at least in $S(s)$ and assertion (i) yields $l_1(F) \geqq s/2$. Further, the hypothesis $l_1(F) > s$ contradicts Prop. 6.3.3.

(iii) follows immediately from (ii) and Prop. 6.3.3. □

**Remark 6.3.11.** (i) The assertions of Prop. 6.3.10 formally remain true for $s = 0$ and $s = \infty$.

(ii) Part (ii) of Prop. 6.3.10 and Cor. 6.2.10 show that for any a.d.f. $F \in \boldsymbol{P}$ the widest strip of boundedness coincides with the widest strip of regularity.

(iii) Part (i) of Prop. 6.3.10 makes it possible to complete Ex. 6.3.6, (iv). Namely, if $F^s$ were an a.d.f., then $F$ would have a certain strip of boundedness. Since this is not the case, $F^s$ is no a.d.f. □

The results of this section have been derived in several steps. Finally, we summarize our results in the following way.

**Theorem 6.3.12.** *Let $F$ be a d.f. with density $p$.*

(i) *The following assertions are equivalent:*

($\mathrm{i}_1$) *$F$ is an a.d.f. with some strip of boundedness;*
($\mathrm{i}_2$) *$p$ is an analytic density with some strip of boundedness;*
($\mathrm{i}_3$) *$F^s$ is an a.d.f.;*
($\mathrm{i}_4$) *$l_1(F) > 0$.*

(ii) *If* $l_1(F) > 0$, *then*

(ii$_1$) *the widest strips of boundedness* $S_F$ *and* $S_p$ *of* $F$ *and* $p$, *resp., satisfy*

$$S(l_1(F)) \subseteq S_p \subseteq S_F \subseteq S(2l_1(F))\,;$$

(ii$_2$) $S(2l_1(F))$ *is the widest strip of boundedness and the widest strip of regularity of* $F^s$. □

## 6.4. Integral distribution functions

By Th. 6.1.11, a condition sufficient for $F$ to be an integral d.f. is $l_1(F) = \infty$, and in this case the integral representations (6.1.2) hold true for all $z$. Moreover, according to Prop. 6.3.3, the whole complex plane is the strip of boundedness of $F$. But we again emphasize that the condition $l_1(F) = \infty$ is not necessary for $F$ to be an integral d.f., cf. Ex. 6.1.13, (ii), and 6.3.5, (ii).

It is the aim of the present section to describe the growth of an integral d.f. We start with the following lemma.

**Lemma 6.4.1.** *Let* $F$ *be an integral d.f. of order less than* 2 *or of order* 2 *and type* 0. *Then*

$$k_F := \overline{\lim_{y\to\infty}} \frac{\log |F(iy)|}{y} > 0\,.$$

Proof. We make use of a result of the theory of Phragmén and Lindelöf which is outlined in App. C.

If we suppose $k_F \leqq 0$, then Cor. C.1.11 tells us that $|F(z)| \leqq 1$ for $\operatorname{Im} z \geqq 0$. Since $F(\bar z) = \overline{F(z)}$ for all $z$ we have $|F| \leqq 1$, and Liouville's theorem yields $F =$ const., a contradiction. □

**Proposition 6.4.2.** *An integral d.f.* $F$ *possesses order* $\varrho \geqq 1$. *If* $\varrho = 1$, *then its type* $\tau$ *satisfies* $0 < \tau \leqq \infty$.

Proof. Assume $\varrho < 1$ or $\varrho = 1$, $\tau = 0$. Then we have $|F(z)| \leqq e^{\varepsilon|z|}$ for any $\varepsilon < 0$ and sufficiently large $|z|$. This yields $k_F \leqq \varepsilon$ and therefore $k_F \leqq 0$ which is not possible by L. 6.4.1. □

Now we wish to give estimates for the growth of an integral d.f. We will need the following auxiliary result.

**Lemma 6.4.3.** *Let* $h \geqq 0$ *be a real function on* $(0, \infty)$. *If*

$$h(x) \leqq C\,e^{-cx^\eta}\,, \qquad x > 0\,,$$

*for certain constants* $C > 0$, $c > 0$, *and* $\eta > 1$, *then*

$$I_r := \int_0^\infty e^{rx}\, h(x)\,dx \leqq C \exp\{(d_\eta/c)^{1/(\eta-1)}\,(r+1)^{\eta/(\eta-1)}\}\,, \qquad r > 0\,,$$

*where* $d_\eta = (\eta - 1)^{\eta-1}/\eta^\eta$.

Proof. We have

$$I_r = \int_0^\infty e^{-x}\, e^{(r+1)x}\, h(x)\,dx \leqq \sup_{x>0} e^{(r+1)x}\, h(x) \leqq C \sup_{x>0} e^{(r+1)x - cx^\eta}\,,$$

and the assertion can be shown by using the elementary differential calculus. □

From now on, for the sake of brevity, we put $1/0 = \infty$ and $1/\infty = 0$.

**Proposition 6.4.4.** *The order $\varrho$ of an integral d.f. $F$ satisfies* $1 \leqq \varrho \leqq 1 + \frac{1}{\chi(F) - 1}$.

Proof. The assertion is trivial if $\chi(F) = 1$. Assume $1 < \chi(F) \leqq \infty$. Then the integral representations (6.1.2) hold true for all $z$, see Cor. 6.1.12, and we obtain

$$M(p, r) := \max_{|z|=r} |p(z)| \leqq \frac{1}{\pi} \int_0^\infty e^{rt} |f(t)| \, dt , \qquad r > 0 . \tag{6.4.1}$$

Let now $1 < \chi' < \chi(F)$. Then, by virtue of Prop. 6.1.6, we get

$$|f(t)| \leqq C\, e^{-t^{\chi'}}, \qquad t > 0 ,$$

for a certain $C > 0$. Now we make use of L. 6.4.3 for $h = |f|$ so that, for a constant $d > 0$,

$$M(p, r) \leqq \frac{C}{\pi} \exp \{d(r + 1)^{\chi'/(\chi'-1)}\} , \qquad r > 0 .$$

This implies that the order of $p$ does not exceed $\frac{\chi'}{\chi' - 1}$, and this bound tends to $1 + \frac{1}{\chi(F) - 1}$ as $\chi' \to \chi(F)$. Since the order $\varrho$ of $F$ equals that of $p$ the proof is complete. □

**Proposition 6.4.5.** *Let $F$ be an integral d.f. of finite order $\varrho$.*

(i) *If $\varrho = 1$, then* $0 < \tau \leqq \tau_0 := \sup \{t : |f(t)| \neq 0\} \leqq \infty$.

(ii) *If $\varrho > 1$, then we have for $\nu = 1 + \frac{1}{\varrho - 1}$ and $d_\nu = (\nu - 1)^{\nu-1}/\nu^\nu$*

$$0 \leqq \tau \leqq (d_\nu / l_\nu(F))^{1/(\nu-1)} .$$

Proof. (i) The estimate $\tau > 0$ follows from Cor. 6.4.2. – The inequality $\tau \leqq \tau_0$ becomes trivial if $\tau_0 = \infty$. Otherwise, assuming $\tau_0 < \infty$ we get $l_1(F) = \infty$, and the second one of the integral representations (6.1.2) tells us that

$$M(p, r) \leqq \frac{1}{\pi} \int_0^{\tau_0} e^{rt} |f(t)| \, dt \leqq \frac{\tau_0}{\pi} e^{r\tau_0}$$

whence the assertion immediately follows.

(ii) According to Prop. 6.4.4 we have $\varrho \leqq 1 + \frac{1}{\chi(F) - 1}$. If $\varrho < 1 + \frac{1}{\chi(F) - 1}$, then $\nu = 1 + \frac{1}{\varrho - l} > \chi(F)$. Thus (6.1.1) yields $l_\nu(F) = 0$, and the assertion becomes trivial. Now we suppose that $1 < \varrho = 1 + \frac{1}{\chi(F) - 1} < \infty$. Then $\nu = \chi(F)$ and we have to show that

$$\tau \leqq (d_\chi / l_\chi(F))^{1/(\chi-1)}$$

which is trivial if $l_\chi(F) = 0$. Let $0 < l_\chi(F) \leqq \infty$ and $0 < l' < l_\chi(F)$. Then we have, for a certain $C > 0$,

$$|f(t)| \leqq C\, e^{-l' t^\chi}, \qquad t > 0 ,$$

and the assertion can be derived from (6.4.1) and L. 6.4.3. □

**Remark 6.4.6.** The question if $\tau < (d_\chi / l_\chi(F))^{1/(\chi-1)}$ is possible (in the case that $\varrho = 1 + \dfrac{1}{\chi(F) - 1}$) must be answered in the affirmative; see Rem. 8.3.2 (i). □

## 6.5. Continuation theorems

In this section we will deal with continuation theorems concerning both a.d.f. and d.f. of the class $\boldsymbol{P}$.

Clearly, according to the uniqueness theorem for analytic functions, two a.d.f. $F_1$ and $F_2$ coincide if they agree on a real point set with a finite limit point. But apart from this trivial statement we are able to formulate a deeper result in the suppositions of which only one of the d.f. $F_1$ and $F_2$ under consideration is assumed to be analytic; the analyticity of the other one can be concluded from other conditions.

We first give an illustration for this.

**Proposition 6.5.1.** *Let $F_1 \in \boldsymbol{P}$ and let $F_2$ be an a.d.f. If*

$$F_1(x) = F_2(x), \qquad |x| < x_0, \tag{6.5.1}$$

*for a certain $x_0 > 0$, then $F_1 = F_2$.*

Proof. Since $F_2$ is an a.d.f. it is continuable to a certain disk $|z| < s$ and we may assume that $s \leqq x_0$. Then, by virtue of (6.5.1), so is $F_1$. Now Th. 6.2.9 tells us that $F_1$ is analytic in the strip $S(s)$, and the coincidence of the two analytic functions $F_1$ and $F_2$ on the interval $|x| < x_0$ implies the assertion. □

Now we improve Prop. 6.5.1 using a deep result concerning c.f. which has been formulated in Suppl. 4.10.5.

**Proposition 6.5.2.** *Let $F_1 \in \boldsymbol{P}$ and assume $\int_{-\infty}^{\infty} f_1(t)\, dt < \infty$. If $F_2$ is an a.d.f. such that, for a certain sequence $(x_k)$*

$$F_1(x_k) = F_2(x_k), \qquad F_1(-x_k) = F_2(-x_k), \qquad x_k \to 0, \tag{6.5.2}$$

*then $F_1 = F_2$.*

Proof. We define a d.f. $G$ by means of

$$G(x) = \tfrac{1}{2}\left(F_2(x) + 1 - F_2(-x)\right)$$

which is obviously analytic and symmetric with respect to the origin. Further, by virtue of (6.5.2) we have $G(x_k) = F_1(x_k)$. Obviously, $F_1$ and $G$ are absolutely continuous with densities $p_1$ and $q$, say. Therefore, by Rolle's theorem there exists a sequence $(x_k')$, $x_k' \to 0$, such that $q(x_k') = p_1(x_k')$. Since $G$ is symmetric, $q(-x) = q(x)$ holds true for all $x$. Moreover, Prop. 6.2.6 tells us that the density $p_1$ is a c.f. (but for a positive factor) and from Suppl. 4.10.5, we obtain $q = p_1$. Thus $p_1$ is also an analytic density, i.e., $F_1$ is an a.d.f. and $F_1 = F_2$ follows from (6.5.2). □

Since in both propositions $F_1$ is assumed to be symmetric with respect to the origin 0, supposition (6.5.1) means that $F_2$ is symmetric in a certain interval about 0. Therefore, (6.5.2) is weaker than (6.5.1). On the other hand, the conditions on $F_1$ made in Prop. 6.5.2 seem to be sharper than those in Prop. 6.5.1. Nevertheless, the additional property $\int_{-\infty}^{\infty} f_1(t)\, dt < \infty$ follows from the suppositions of Prop. 6.5.1. Indeed, since

$F_2$ is analytic, (6.5.1) means that $F_1$ satisfies a Lipschitz condition at the origin and since $F_1 \in \boldsymbol{P}$ this is equivalent to $\int_{-\infty}^{\infty} f_1(t)\, dt < \infty$ by Prop. 6.2.4. Accordingly, Prop. 6.5.1 is just a consequence of Prop. 6.5.2.

## 6.6. Supplement

**6.6.1.** *Let $h > 0$ be a function on $(0, \infty)$ such that, as $t \to \infty$, $h(t) \downarrow 0$ and $th(t) \uparrow \infty$; and let $S$, $\emptyset \subset S \subseteqq \mathbb{R}_1$, be some closed set. Then there exists a d.f. $F$ which has $S$ as its set of singularities and whose c.f. $f$ satisfies*

$$\overline{\lim_{t\to\infty}}\, |f(t)|\, e^{2th(t)} > 0\,,$$

$$\overline{\lim_{t\to\infty}}\, |f(t)|\, e^{(1-\varepsilon)th(t)} < \infty\,, \qquad 0 < \varepsilon < 1\,,$$

(Boas and Smithies (1938)).

This result implies the following.

(i) *For every $c$, $0 \leqq c < 1$, there exists an a.d.f. $F$ with $\chi(F) = c$.* Namely, choose $S = \emptyset$ and $h(t) = t^{c-1}$ for $c > 0$ and $h(t) = (\log t)/t$ for $c = 0$.

(ii) There exist non-analytic d.f. $F$ such that

$$\underline{\lim_{t\to\infty}}\, \frac{-\log |f(t)|}{th(t)} > 0$$

where $h$ has the properties specified at the beginning. To verify this, choose $S \neq \emptyset$ and make use of the second inequality.

**6.6.2.** *Let $F \in \boldsymbol{P}$, $\hat{F}$ the d.f. corresponding to $F$ as defined in Prop. 6.2.6, and let $\hat{T}$ be the tail function of $\hat{F}$. Then*

$$\underline{\lim_{t\to\infty}}\, \frac{-\log |f(t)|}{t^{\nu}} \leqq \underline{\lim_{x\to\infty}}\, \frac{-\log \hat{T}(x)}{x^{\nu}}\,, \qquad \nu \in \mathbb{R}_1\,,$$

*and*

$$\underline{\lim_{t\to\infty}}\, \frac{\log(-\log |f(t)|)}{\log t} \leqq \underline{\lim_{x\to\infty}}\, \frac{\log(-\log \hat{T}(x))}{\log x}\,,$$

*and the converse inequalities hold true for the upper limits* (Jesiak (1979a)).

**6.6.3.** *An integral d.f. $F \in \boldsymbol{P}$ has order $\varrho = 1 + \dfrac{1}{\chi(F) - 1}$, provided that the limit $\chi(F) = \lim\limits_{t\to\infty} \dfrac{\log(-\log |f(t)|)}{\log t} \leqq \infty$ exists* (Jesiak (1979a)).

**6.6.4.** *An integral d.f. $F \in \boldsymbol{P}$ of finite order $\varrho > 1$ has type $\tau = (d_\nu/l_\nu(F))^{1/(\nu-1)}$ where $\nu = 1 + \dfrac{1}{\varrho - 1}$ and $d_\nu = (\nu - 1)^{\nu-1}/\nu^{\nu}$, provided that the limit $l_\nu(F) = \lim\limits_{t\to\infty} \dfrac{-\log |f(t)|}{t^{\nu}} \leqq \infty$ exists* (Jesiak (1979a)).

**6.6.5.** *If $F_1 \in \boldsymbol{P}$ is integral with order $\varrho_1$, then for any d.f. $F_2$ the convolution $F_1 * F_2$ is integral with order $\varrho \leqq \varrho_1$* (Jesiak (1978)).

Note that the results formulated in 6.6.3—6.6.5 hold true also for a class of d.f. which is broader than $\boldsymbol{P}$, see JESIAK (1978).

**6.6.6.** *Let $F_1$ be such that the limit* $\lim\limits_{t\to\infty} \dfrac{-\log|f_1(t)|}{t}$ *exists. Then the convolution $F_1 * F_2$ is an a.d.f. with a certain strip of boundedness iff at least one of its components is such a d.f.* This follows immediately after an obvious modification of Prop. 6.3.7.

## Notes and bibliographical data

The theory of a.d.f. has been systematically developed by JESIAK (1978), (1979a), (1979b)

The topic of Th. 6.2.9 has been treated by RAJKOV (1939), to whom are also due some methods applied in Sect. 6.2 and 6.3; in his paper we found Ex. 6.2.5, (ii), which he attri buted to GEL'FOND.

The uniqueness theorems of Sect. 6.5 are due to SIEGEL (1979b) who proved them as consequences of Prop. 6.2.4.

The analytic character of two special classes of d.f. has been intensely investigated; they are the class $\boldsymbol{L}$ (ZOLOTAREV (1963)) and the class of stable d.f. (cf. Sect. 8.3).

# Chapter 7. Infinitely divisible distribution functions

We investigate infinitely divisible (inf.div.) d.f. which play a significant role in many fields of probability theory. In particular, they are basic for the study of limit theorems for sums of i.r.v., and at the very beginning they were treated because of the needs of this theory.

C.f. are the classical tool in this field, and their canonical representation formulas play a crucial role. After some initial elementary studies we derive them in Sect. 7.2. They permit us to deepen our investigation, and all that follows in the present chapter basicly rests on them. In Sect. 7.3 we treat the case in which inf.div.c.f. can be continued to a strip of the complex plane.

In modern research interest has shifted from inf.div.c.f. to the corresponding inf.div.d.f. This is natural since it is the d.f. that have an intuitive probabilistic meaning. Therefore, in Sect. 7.4 we return to the study of d.f. and investigate the interrelation between weak and complete convergence of inf.div.d.f. on one hand and the behaviour of the corresponding spectral functions on the other hand; these results serve us as precious tools in the theory of sums. In Sect. 7.5 we give criterias under which an inf.div.d.f. is analytic; they are of independent interest and play a particular role in the continuation theory of d.f.

Sect. 7.6 is devoted to the unique determination of inf.div.d.f. This theory was initiated by the solution of Kolmogorov's problem, and our outline contained in Sect. 7.6 is meant to give a survey of its present state. The following Sect. 7.7 is related to this subject; it contains conclusions on the left spectral function which can be drawn from appropriate asymptotic assumptions on the left tail of the corresponding d.f. A number of unsolved problems concerning uniqueness is formulated in the supplement.

Finally, in Sect. 7.8 we consider the class of power mixtures which is wider than that of inf.div.d.f.; among others, we obtain representation formulas for inf.div. power mixtures.

## 7.1. Definition and fundamental properties

**Definition 7.1.1.** A d.f. $F$ is said to be *infinitely divisible* (*inf.div.*) if there exists a sequence $({}^nF)$ of d.f. such that

$$F = ({}^nF)^{*n}, \quad \text{for all} \quad n \geqq 2 .$$

The corresponding c.f. $f$ is then also called inf.div. □

We see that $f$ is inf.div. iff there exists a sequence $({}^nf)$ of c.f. so that

$$f = ({}^nf)^n, \quad \text{for all} \quad n \geqq 2 . \tag{7.1.1}$$

Throughout this chapter we denote by ${}^nF$ and ${}^nf$ the above functions.

Starting from (7.1.1), one easily sees that if $F(x)$ is inf.div. then so are $F(bx + a)$, $b > 0$; and $F_-(x) = 1 - F(-x + 0)$. Further, if $F_1$ and $F_2$ are inf.div., then we have for certain d.f. ${}^nF_1$ and ${}^nF_2$

$$F_1 * F_2 = ({}^nF_1)^{*n} * ({}^nF_2)^{*n} = ({}^nF_1 * {}^nF_2)^{*n} .$$

Since ${}^nF_1 * {}^nF_2$ is a d.f. we see that $F_1 * F_2$ is also inf.div.

Next we give several examples for inf.div.c.f.

**Example 7.1.2.** (i) The normal and the degenerate d.f. are inf.div. In fact, for $\sigma^2 \geqq 0$ we have

$$f(t) = e^{iat - \frac{\sigma^2}{2}t^2} = \left[e^{i\frac{a}{n}t - \frac{\sigma^2}{2n}t^2}\right]^n = ({}^nf)^n$$

where ${}^nf$ is again a normal or a degenerate c.f.

(ii) The Cauchy d.f. is inf.div. because of

$$f(t) = e^{iat - b|t|} = \left[e^{i\frac{a}{n}t - \frac{b}{n}|t|}\right]^n = ({}^nf)^n ,$$

and ${}^nf$ is also a Cauchy c.f.

(iii) For the c.f. $f$ of the gamma d.f. we have

$$f(t) = \left(1 - \frac{it}{b}\right)^{-\lambda} = \left[\left(1 - \frac{it}{b}\right)^{-\lambda/n}\right]^n = ({}^nf)^n .$$

Clearly, ${}^nf$ belongs to a gamma d.f.; note that it is not of the same type as $f$ — contrary to (i) and (ii). □

**Example 7.1.3.** To construct a considerably wide class of inf.div.c.f. we consider an arbitrary c.f. $g$. Then, for any $\lambda > 0$, the c.f. $f = e^{\lambda(g-1)}$ belongs to a compound Poisson d.f., see Ex. 3.1.13. Since

$$f = \left[e^{\frac{\lambda}{n}(g-1)}\right]^n$$

we see that $f$ is inf.div. In particular, the Poisson d.f. $P_\lambda$ is inf.div. (put $g(t) = e^{it}$).

More generally, let $G$ be a non-decreasing, left-continuous, and bounded function on $\mathbb{R}_1$ such that $G(-\infty) = 0$, and let $g$ be its Fourier-Stieltjes transform. Then, for every $\lambda > 0$, the function $f = e^{\lambda(g - G(\infty))}$ is an inf.div.c.f. belonging to the d.f.

$$F = e^{-\lambda G(\infty)} \sum_{k=0}^{\infty} \frac{\lambda^k}{k!} G^{*k} .$$

This is obvious if $G(\infty) = 0$ since $G^{*0} := U_0$; otherwise consider the d.f. $G/G(\infty)$. □

It is a natural question if for a given inf.div.c.f. $f$ the "roots" ${}^nf$ in (7.1.1) are uniquely determined. Clearly, in a certain neighbourhood of the origin we have ${}^nf = f^{1/n}$ $= e^{(1/n)\log f}$ where we choose the principal branch so that ${}^nf(0) = 1$. But if there were zeros of $f$, then the continuation of ${}^nf$ to the whole real axis would be ambiguous. Fortunately, we have the following basic fact which implies that the answer to the above question is affirmative.

**Proposition 7.1.4.** *An inf.div.c.f. $f$ has no zeros.*

Proof. Since $f$ is inf.div. there exist c.f. ${}^nf$ such that $f = ({}^nf)^n$; thus ${}^nf\,\overline{{}^nf} = |{}^nf|^2 = |f|^{2/n}$ $\geqq 0$. Now we introduce the function $g(t) := \lim_{n\to\infty} |f(t)|^{2/n}$ and see that

$$g(t) = \begin{cases} 0 & \text{if } |f(t)| = 0 , \\ 1 & \text{if } |f(t)| > 0 . \end{cases}$$

The c.f. $f$ is continuous and $f(0) = 1$. Thus $|f(t)| > 0$ in a neighbourhood of the origin and we conclude that $g(t) = 1$ in the same neighbourhood. Hence $g$ is continuous at 0 and, as a limit of c.f., it is also a c.f. (see Cor. 3.4.2 to the continuity theorem). Accordingly, $g$ is continuous everywhere which implies that $g = 1$; hence $f$ has no zeros. □

**Corollary 7.1.5.** *Let $f = ({}^n f)^n$ be inf.div. Then*

$$n({}^n f - 1) \to \log f\,, \qquad n \to \infty\,.$$

Proof. This follows from

$${}^n f = f^{1/n} = e^{1/n \log f} = 1 + \frac{1}{n} \log f + o\left(\frac{1}{n}\right), \qquad n \to \infty\,.$$ □

## 7.2. Canonical representations

It is the aim of the present section to derive representation formulae for inf.div.c.f. which will prove crucial for the investigation of such c.f. and of the corresponding d.f.

Throughout this section we denote by $H$ (with or without indices) non-decreasing, left-continuous, bounded functions on $\mathbb{R}_1$ which satisfy $H(-\infty) = 0$. Further, for $a \in \mathbb{R}_1$ we write

$$f(a, H; t) = e^{iat + \int_{-\infty}^{\infty} K(t,x) dH(x)}\,, \qquad t \in \mathbb{R}_1\,. \tag{7.2.1}$$

where

$$K(t, x) := \begin{cases} \left(e^{itx} - 1 - \dfrac{itx}{1 + x^2}\right) \dfrac{1 + x^2}{x^2}\,, & x \neq 0\,, \\ -\dfrac{t^2}{2}\,, & x = 0\,. \end{cases}$$

We see that $K$ is bounded and continuous in $x$ for every fixed $t$ so that the integral in (7.2.1) exists. Further, $K$ is continuous in $t$ for every fixed $x$. By dominated convergence, the function $f(a, H; t)$ is continuous in $t$.

We will need the following three lemmas.

**Lemma 7.2.1.** *Every function $f(a, H; t)$ is a c.f.*

Proof. Let $a$, $H$ be given. We put, for any $n \geqq 1$,

$$H_n(x) = \begin{cases} H(x)\,, & |x| > 1/n \\ H\left(-\dfrac{1}{n}\right), & |x| \leqq 1/n\,, \end{cases}$$

and see that $H_n \xrightarrow{c} H$. Therefore, from the extended Helly-Bray theorem we get

$$f(a, H_n; t) = \exp\left\{iat + \int_{-\infty}^{\infty} K(t, x)\, dH_n(x)\right\} \to f(a, H; t)\,.$$

Since $f(a, H; t)$ is continuous our proof amounts to showing that the $f(a, H_n; t)$ are c.f., cf. Cor. 3.4.2 to the continuity theorem. To prove this we put

$$G_n(x) = \int_{-\infty}^{x} \frac{1 + y^2}{y^2}\, dH_n(y)\,, \quad g_n(t) = \int_{-\infty}^{\infty} e^{itx}\, dG_n(x)\,,$$

$$a_n = a - \int_{-\infty}^{\infty} \frac{1}{x}\, dH_n(x)\,.$$

Now we can verify that

$$f(a, H_n; t) = e^{ia_n t} e^{g_n(t) - G_n(\infty)} .$$

The first factor on the right is a degenerate c.f. while the second belongs to a d.f. considered in Ex. 7.1.3. Thus $f(a, H_n; t)$ is a product of c.f. and therefore also a c.f. □

**Lemma 7.2.2.** *Let $f$ be a c.f. with the representation $f(t) = f(a, H; t)$. Then $a$ and $H$ are uniquely determined by $f$.*

Proof. We introduce the function

$$\psi(t) = \log f(t) - \tfrac{1}{2} \int_{t-1}^{t+1} \log f(\tau)\, d\tau , \qquad t \in R_1 .$$

Using $f(t) = f(a, H; t)$ and applying the Fubini theorem, we get

$$\psi(t) = iat + \int_{-\infty}^{\infty} K(t, x)\, dH(x) - \frac{1}{2} \int_{t-1}^{t+1} \left[ ia\tau + \int_{-\infty}^{\infty} K(\tau, x)\, dH(x) \right] d\tau$$

$$= \frac{1}{2} \int_{-\infty}^{\infty} \int_{t-1}^{t+1} [K(t, x) - K(\tau, x)]\, d\tau\, dH(x)$$

$$= \int_{-\infty}^{\infty} e^{itx} \left(1 - \frac{\sin x}{x}\right) \frac{1 + x^2}{x^2}\, dH(x) = \int_{-\infty}^{\infty} e^{itx}\, d\Psi(x)$$

where

$$\Psi(x) = \int_{-\infty}^{x} \left(1 - \frac{\sin y}{y}\right) \frac{1 + y^2}{y^2}\, dH(y) .$$

Clearly, $\Psi$ is non-decreasing, left-continuous, and bounded on $R_1$ and satisfies $\Psi(-\infty) = 0$. Further, we verify

$$H(x) = \int_{-\infty}^{x} \left(1 - \frac{\sin y}{y}\right)^{-1} \frac{y^2}{1 + y^2}\, d\Psi(y) .$$

Now we are faced with the following situation. For given $f$, the function $\psi$ is uniquely determined. Since it is a Fourier-Stieltjes transform, the corresponding function $\Psi$ is uniquely determined by $\psi$, cf. Rem. 3.2.5; note that $\Psi(-\infty) = 0$. Finally, $H$ is uniquely determined by $\Psi$ and therefore also by $f$. Now it comes out that the constant $a$, too, is uniquely determined by $f$. □

**Lemma 7.2.3.** *Assume that for certain sequences $(a_n)$ and $(H_n)$*

$$f(a_n, H_n; t) \to f(t) , \qquad n \to \infty ,$$

*and suppose that $f$ is continuous at the origin. Then there exists a constant $a \in R_1$ and a function $H$ such that*

$$a_n \to a , \qquad H_n \xrightarrow{c} H , \qquad \text{as } n \to \infty ,$$

*and we have $f(t) = f(a, H; t)$.*

Proof. First we conclude from Cor. 3.4.2 that $f$ is a c.f.; and Cor. 3.4.4 tells us that the assumed convergence is uniform in every bounded interval. Accordingly, for given $\varepsilon > 0$ and sufficiently large $n$, we have (with $f_n(t) := f(a_n, H_n, t)$)

$$\log |f_n(t)| \geqq \log |f(t)| - \varepsilon \geqq -2\varepsilon, \qquad |t| \leqq t_0(\varepsilon).$$

Hence

$$-\frac{1}{t_0}\int_0^{t_0} \log |f_n(t)|\, dt \leqq 2\varepsilon, \qquad n \geqq n_0(\varepsilon).$$

Next we show that the sequence $(H_n(\infty))$ is bounded. Indeed, we have

$$2\varepsilon \geqq -\frac{1}{t_0}\int_0^{t_0} \log |f_n(t)|\, dt = \frac{1}{t_0}\int_0^{t_0}\int_{-\infty}^{\infty} (1 - \cos tx)\frac{1+x^2}{x^2}\, dH_n(x)\, dt$$

$$= \int_{-\infty}^{\infty} \left(1 - \frac{\sin t_0 x}{t_0 x}\right)\frac{1+x^2}{x^2}\, dH_n(x).$$

Since the integrand is bounded away from zero we obtain

$$2\varepsilon \geqq c \int_{-\infty}^{\infty} dH_n(x) = cH_n(\infty), \qquad n \geqq n_0(\varepsilon),$$

where $c > 0$ depends only on $t_0$. Thus $(H_n(\infty))$ is bounded.

Further, by virtue of $(1 + x^2)/x^2 \geqq 1$ we see that

$$2\varepsilon \geqq \int_{|x| \geqq 1/t_0} \left(1 - \frac{\sin t_0 x}{t_0 x}\right) dH_n(x) \geqq \frac{1}{2}\int_{|x| \geqq 1/t_0} dH_n(x).$$

This means that the sequence $(H_n)$ is tight. Hence, by virtue of Prop. 2.4.12, it is relatively compact.

Now we are able to select a subsequence $(H_{n'})$ such that $H_{n'} \xrightarrow{c} H$, say. Then the extended Helly-Bray theorem tells us that $\int_{-\infty}^{\infty} K(t, x)\, dH_{n'}(x) \to \int_{-\infty}^{\infty} K(t, x)\, dH(x)$, and therefore the sequence $(a_{n'})$ must also converge to a certain limit $a$. This yields $f(a_{n'}, H_{n'}; t) \to f(a, H; t)$, and by assumption we have $f(a, H; t) = f(t)$. Finally, L. 7.2.2 tells us that $a$ and $H$ are independent of the chosen subsequence so that the assertions follow. □

Now we are in a position to prove the first one of the representation formulae for inf.div.c.f.

**Theorem 7.2.4** (Lévy-Hinčin canonical representation). *A function $f$ is an inf.div.c.f. iff it permits the representation*

$$\log f(t) = iat + \int_{-\infty}^{\infty} \left(e^{itx} - 1 - \frac{itx}{1+x^2}\right)\frac{1+x^2}{x^2}\, dH(x), \qquad t \in \mathbb{R}_1, \qquad (7.2.2)$$

*where $a \in \mathbb{R}_1$ and $H$ is a non-decreasing, left-continuous, and bounded function on $\mathbb{R}_1$ such that $H(-\infty) = 0$. This representation is unique.*

Proof. *If-part*: We see that $f(t)^{1/n} = f(a/n, H/n; t)$ which is a c.f. by L. 7.2.1. Hence, $f$ is inf.div.

*Only-if part*: Let $F = ({}^nF)^{*n}$ be the d.f. corresponding to $f$. Then we put

$$a_n = n \int_{-\infty}^{\infty} \frac{y}{1+y^2} \, d({}^nF(y)) , \qquad H_n(x) = n \int_{-\infty}^{x} \frac{y^2}{1+y^2} \, d({}^nF(y))$$

and verify

$$f(a_n, H_n; t) = e^{n({}^nf(t)-1)} .$$

By virtue of Cor. 7.1.5 we have $f(a_n, H_n; t) \to f(t)$, so that (7.2.2) follows from L. 7.2.3.

Finally, the *uniqueness* of the representation (7.2.2) has been proved in L. 7.2.2. □

Since the representation (7.2.2) is unique we denote the corresponding inf.div.d.f. $F$ by $F = \mathcal{H}(a, H)$. The function $H$ is said to be the Lévy-Hinčin *spectral function of F*. Examples for the Lévy-Hinčin representation of several inf.div.d.f. will be given at the end of the presten section.

The proof of the only-if part of Th. 7.2.4 also permits the following conclusion.

**Corollary 7.2.5.** *Let $F = \mathcal{H}(a, H)$ be inf.div. Then we have, as $n \to \infty$,*

$$n \int_{-\infty}^{\infty} \frac{y}{1+y^2} \, d({}^nF(y)) \to a , \qquad n \int_{-\infty}^{x} \frac{y^2}{1+y^2} \, d({}^nF(y)) \xrightarrow{c} H(x) . \qquad \square$$

Next we want to modify the Lévy-Hinčin representation so that we obtain a formula which will often be useful for our purposes.

**Theorem 7.2.6** (Lévy canonical representation). *A function $f$ is an inf.div.c.f. iff it permits the representation*

$$\log f(t) = iat - \frac{\gamma^2}{2} t^2 + \int_{-\infty}^{-0} k(t, u) \, dM(u) + \int_{+0}^{\infty} k(t, u) \, dN(u) \tag{7.2.3}$$

*where $a$ and $\gamma^2 \geqq 0$ are real constants and*

$$k(t, u) := e^{itu} - 1 - \frac{itu}{1+u^2} \cdot$$

*Moreover, $M \geqq 0$ and $N \leqq 0$ are non-decreasing, left-continuos functions on* $(-\infty, 0)$ *and* $(0, \infty)$, *resp., such that*

$$M(-\infty) = N(\infty) = 0 , \qquad \int_{-1}^{-0} u^2 \, dM(u) + \int_{+0}^{1} u^2 \, dN(u) < \infty .$$

*The representation* (7.2.3) *is unique.*

Proof. Our proof amounts to showing that (7.2.3) is equivalent to (7.2.2); then all assertions follow from Th. 7.2.4. — If (7.2.2) is given we put

$$\left.\begin{aligned} &M(u) = \int_{-\infty}^{u} \frac{1+x^2}{x^2} \, dH(x) , \quad u < 0; \quad N(u) = -\int_{u}^{\infty} \frac{1+x^2}{x^2} \, dH(x) , \quad u > 0 , \\ &\gamma^2 = H(+0) - H(-0) \geqq 0 \end{aligned}\right\} \tag{7.2.4}$$

and obtain (7.2.3). On the other hand, if (7.2.3) is given, then we introduce the function

$$H(x) = \begin{cases} \int\limits_{-\infty}^{x} \frac{u^2}{1+u^2} \, \mathrm{d}M(u), & x < 0, \\ H(-0) + \gamma^2 + \int\limits_{+0}^{x} \frac{u^2}{1+u^2} \, \mathrm{d}N(u), & x > 0 \end{cases} \tag{7.2.5}$$

which leads us to the representation (7.2.2). Further, the functions $M$, $N$, and $H$ introduced in (7.2.4) and (7.2.5) have the properties specified in Th. 7.2.6 and 7.2.4, resp. □

An inf.div.d.f. $F$ the c.f. $f$ of which has the representation (7.2.3) will be denoted by $F = \mathscr{L}(a, \gamma^2, M, N)$. The functions $M$ and $N$ are respectively called *the left and the right Lévy spectral function of* $F$; $\gamma^2$ is said to be its Gaussian parameter. Note that the constant $a$ in (7.2.3) is the same as in (7.2.2).

**Corollary 7.2.7.** *Let* $F = \mathscr{L}(a, \gamma^2, M, N)$ *be inf.div. Then we have, as* $n \to \infty$,

$$n \int\limits_{-\infty}^{\infty} \frac{y}{1+y^2} \, \mathrm{d}\left({}^nF(y)\right) \to a,$$

$$n\,{}^nF \xrightarrow[(-\infty,0)]{} M, \qquad n({}^nF - 1) \xrightarrow[(0,\infty)]{} N,$$

$$\gamma^2 = \lim_{\varepsilon \to +0} \overline{\lim_{n \to \infty}}\, n \int\limits_{|y|<\varepsilon} y^2 \, \mathrm{d}\left({}^nF(y)\right) = \lim_{\varepsilon \to +0} \underline{\lim_{n \to \infty}}\, n \int\limits_{|y|<\varepsilon} y^2 \, \mathrm{d}\left({}^nF(y)\right).$$

Proof. The first three relations follow immediately from Cor. 7.2.5 by virtue of (7.2.4) and the extended Helly-Bray theorem. Further, if $\pm\varepsilon$ are continuity points of $H$, then

$$\gamma^2 = H(+0) - H(-0) = \lim_{\varepsilon \to +0} \lim_{n \to \infty} n \int\limits_{|y|<\varepsilon} \frac{y^2}{1+y^2} \, \mathrm{d}\left(F^n(y)\right),$$

and the inequalities

$$\frac{y^2}{1+\varepsilon^2} \leqq \frac{y^2}{1+y^2} \leqq y^2, \qquad |y| \leqq \varepsilon,$$

lead us to the last of the asserted relations. □

Next we give several further conclusions of the canonical representations the proofs of which are obvious.

**Corollary 7.2.8.** *Let* $F_k = \mathscr{H}(a_k, H_k) = \mathscr{L}(a_k, \gamma_k^2, M_k, N_k)$, $k = 1, 2$, *be inf.div. Then* $F = F_1 * F_2$ *is also inf.div. and we have*

$$F = \mathscr{H}(a_1 + a_2, H_1 + H_2) = \mathscr{L}(a_1 + a_2, \gamma_1^2 + \gamma_2^2, M_1 + M_2, N_1 + N_2). \quad \square$$

**Remark 7.2.9.** An inf.div.d.f. $F = \mathscr{L}(a, \gamma^2, M, N)$ permits the decomposition

$$F = \Phi_{a,\gamma} * F_M * F_N$$

where $\Phi_{a,\gamma}$ is a — possibly degenerate — normal d.f. and $F_M = \mathscr{L}(0, 0, M, 0)$, $F_N = \mathscr{L}(0, 0, 0, N)$ are inf.div. If $\gamma^2 > 0$ we say that $\Phi_{a,\gamma}$ is the *normal component* of $F$. □

**Corollary 7.2.10.** *If $f$ is inf.div., then so are $|f|$ and $f^\eta$, $\eta > 0$. The corresponding d.f. have the representations* $\mathscr{L}\left(0, \gamma^2, \frac{M(u) - N(-u+0)}{2}, \frac{N(u) - M(-u+0)}{2}\right)$ *and* $\mathscr{L}(\eta a, \eta\gamma^2, \eta M, \eta N)$, *resp. Further, $|f|$ has the representation*

$$\log |f(t)| = -\frac{\gamma^2}{2} t^2 - \int_{-\infty}^{-0} (1 - \cos tu)\, \mathrm{d}M(u) - \int_{+0}^{\infty} (1 - \cos tu)\, \mathrm{d}N(u)\,. \qquad \square$$

**Corollary 7.2.11.** *If $F = \mathscr{L}(a, \gamma^2, M, N)$ is inf.div., then so is $F'(x) = F(bx + c)$, $b > 0$, and we have $F' = \mathscr{L}\left(a', \gamma^2/b^2, M(bu), N(bu)\right)$ for a certain $a' \in R_1$.*

Proof. We insert the Lévy formula (7.2.3) for $f$ in the obvious relation

$$\log f'(t) = -it\frac{c}{b} + \log f\left(\frac{t}{b}\right).$$

Thus we obtain

$$\log f'(t) = it\frac{a-c}{b} - \frac{\gamma^2}{2b^2} t^2 + \int_{-\infty}^{-0} \left(e^{itu} - 1 - \frac{itu}{1 + (bu)^2}\right) \mathrm{d}M(bu) +$$

$$+ \int_{+0}^{\infty} (\ldots)\, \mathrm{d}N(bu) =$$

$$= it\left[\frac{a-c}{b} + \int_{-\infty}^{-0} \frac{u^3(b^2 - 1)}{(1 + u^2)\left(1 + (bu)^2\right)}\, \mathrm{d}M(bu) + \int_{+0}^{\infty} \ldots\, \mathrm{d}N(bu)\right]$$

$$-\frac{\gamma^2}{2b^2} t^2 + \int_{-\infty}^{-0} k(t, u)\, \mathrm{d}M(bu) + \int_{+0}^{\infty} k(t, u)\, \mathrm{d}N(bu)\,.$$

Since the integrals in brackets converge, the proof is complete because the Lévy representation is unique $\square$

To close this section we give some examples for the canonical representation of inf.div.d.f. which can be verified by inspection.

**Example 7.2.12.** (i) For the normal and the degenerate d.f. $\Phi_{\mu,\sigma}$, $\sigma^2 \geqq 0$, we have $\Phi_{\mu,\sigma} = \mathscr{H}(\mu, \sigma^2 U_0) = \mathscr{L}(\mu, \sigma^2, 0, 0)$.

(ii) For Cauchy d.f. we see that $F = \mathscr{H}\left(a, \frac{b}{2} + \frac{b}{\pi} \arctan x\right) = \mathscr{L}\left(a, 0, \frac{b}{\pi |u|}, -\frac{b}{\pi u}\right)$.

(iii) The Poisson d.f. $P_\lambda$ has the representation

$$P_\lambda = \mathscr{H}\left(\frac{\lambda}{2}, \frac{\lambda}{2} U_1\right) = \mathscr{L}\left(\frac{\lambda}{2}, 0, 0, \lambda(U_1 - 1)\right).$$

(iv) Let $G$ be a monotone function with Fourier-Stieltjes transform $g$ and suppose $G(-\infty) = 0$ and $0 \leqq G(\infty) < \infty$. Then $f = e^{\lambda(g - G(\infty))}$ is an inf.div.c.f. (see Ex. 7.1.3), and for the corresponding d.f. $F$ we have $F = \mathscr{H}(a, H) = \mathscr{L}(a, 0, M, N)$ where

$$a = \lambda \int_{-\infty}^{\infty} \frac{y}{1 + y^2}\, \mathrm{d}G(y),\ H(x) = \lambda \int_{-\infty}^{x} \frac{y^2}{1 + y^2}\, \mathrm{d}G(y),\ M = \lambda G \text{ and } N = \lambda(G - G(\infty)). \quad \square$$

## 7.3. Analytic properties of infinitely divisible characteristic functions

Now we present several conclusions which we derive from the Lévy canonical representation (7.2.3); all of them are related to analytic or boundary inf.div.c.f.

Throughout this section we denote by $M$ and $N$ respectively left and right Lévy spectral functions; the characteristic properties of such functions are specified in Th. 7.2.6. Further, for complex $z$ and real $u$ we define the integral functions

$$k_1(z, u) := e^{izu} - 1 - izu$$

and

$$k(z, u) := e^{izu} - 1 - \frac{izu}{1 + u^2} = k_1(z, u) + \frac{izu^3}{1 + u^2}.$$

Some elementary properties of these functions which will be needed in the following are collected together in App. B.4.

Clearly, if $M(u)$ $\big(N(u)\big)$ is a left (right) spectral function, then $-M(-u+0)$ $\big(-N(-u+0)\big)$ is a right (left) spectral function; therefore it suffices to formulate our results for one kind of spectral functions only.

**Lemma 7.3.1.** *For fixed $\eta > 0$, the function*

$$\psi_\eta(z) := \int_{-\eta}^{-0} k_1(z, u)\, dM(u)$$

*is an integral function and it satisfies*

$$\psi_\eta(z) = o(|z|^2) \quad \text{as} \quad |z| \to \infty, \qquad z \in S(-\infty, 0].$$

Proof. We consider the sequence $(\psi_{\eta,n})$ of integral functions defined by

$$\psi_{\eta,n}(z) = \int_{-\eta}^{-1/n} k_1(z, u)\, dM(u).$$

Now we choose an arbitrary real number $r > 0$. Then we obtain for all $|z| \leqq r$ and $m \leqq n$, by virtue of (B. 4.1),

$$|\psi_{\eta,n}(z) - \psi_{\eta,m}(z)| \leqq \frac{|z|^2}{2} \int_{-1/m}^{-1/n} e^{|zu|}\, u^2\, dM(u) \leqq \frac{r^2}{2}\, e^{r/m} \int_{-1/m}^{-1/n} u^2\, dM(u).$$

This bound tends to 0 as $n, m \to \infty$ since $\int_{-1}^{-0} u^2\, dM(u) < \infty$; hence, the sequence $(\psi_{\eta,n})$ converges uniformly in $|z| \leqq r$ so that $\psi_\eta$ is analytic in $|z| < r$ by Weierstrass' convergence theorem. Since $r$ is arbitrary we see that $\psi_\eta$ is an integral function. Further, for $z \in S(-\infty, 0]$, in view of (B. 4.2) we get for any $\varepsilon$, $0 < \varepsilon < \eta$,

$$\left|\frac{\psi_\eta(z)}{z^2}\right| \leqq \int_{-\eta}^{-\varepsilon} \left|\frac{k_1(z, u)}{z^2}\right| dM(u) + \frac{1}{2} \int_{-\varepsilon}^{-0} u^2\, dM(u) \to 0;$$

the limit is to be taken first as $|z| \to \infty$ and then as $\varepsilon \to +0$. □

**Proposition 7.3.2.** *The function*

$$g_M(z) := \int_{-\infty}^{-0} k(z, u)\, dM(u) \tag{7.3.1}$$

*is analytic, in the half plane $S(-\infty, 0)$, and it satisfies*

$$g_M(z) = o(|z|^2) \quad as \quad |z| \to \infty\,, \qquad z \in S(-\infty, 0]\,. \tag{7.3.2}$$

Proof. Let $-\eta < 0$ be an arbitrary continuity point of $M$. Then simple computations show that

$$g_M(z) = \int_{-\infty}^{-\eta} e^{izu}\, dM(u) + \psi_\eta(z) - M(-\eta) + izA_\eta \tag{7.3.3}$$

where

$$A_\eta = \int_{-\eta}^{-0} \frac{u^3}{1+u^2}\, dM(u) - \int_{-\infty}^{-\eta} \frac{u}{1+u^2}\, dM(u)\,.$$

The first summand in (7.3.3) is — but for a non-negative factor — a c.f. and in view of Re $(izu) = -u \operatorname{Im} z \leqq 0$ it is analytic in $S(-\infty, 0)$. Hence, all assertions follow from L. 7.3.1. □

**Corollary 7.3.3.** *Let $F = \mathcal{L}(a, \gamma^2, M, N)$ be inf.div. with c.f. $f$. Then we have*

$$\gamma^2 = -2 \lim_{t\to\infty} \frac{\log f(t)}{t^2} = -2 \lim_{t\to\infty} \frac{\log |f(t)|}{t^2}\,.$$

Proof. The estimate (7.3.2) and the analogous one for $g_N$ (valid in $z \in S[0, \infty)$) hold true together for real $z = t$. Therefore we obtain from the Lévy canonical representation (7.2.3), as $t \to \infty$, $\log f(t) = -(\gamma^2/2)t^2 + o(t^2)$; the second equality follows now from the fact that $\gamma^2$ is real. □

**Corollary 7.3.4.** *The function $g_M(t)$ is continuable to the strip $S(0, \bar{s})$, $0 < \bar{s} \leqq \infty$, iff $M(u) = O(e^{su})$, $u \to -\infty$, for all $s$ with $0 < s < \bar{s}$.*
*In either case the representation* (7.3.1) *holds in $S(-\infty, \bar{s})$.*
Proof. We know from Prop. 7.3.2 that the function (7.3.1) is analytic in $S(-\infty, 0)$. In formula (7.3.3) it is decomposed into a c.f. (but for a factor) and a function which is integral by L. 7.3.1.; by Prop. 4.3.8 the c.f. is an a.c.f. whose strip of regularity contains the strip $S(0, \bar{s})$ iff $M$ has the specified property so that the first assertion is shown.

Now it is also clear from (7.3.3) that $g_M$ — if it is continuable to $S(0, \bar{s})$ — has the representation (7.3.1). □

Before combining the results just obtained to an assertion on inf.div.c.f. analytic in a strip we need the following extension of Prop. 7.1.4.

**Proposition 7.3.5.** *Assume that the inf.div.c.f. $f$ is continuable to a strip $S(\underline{s}, \bar{s})$, $-\infty \leqq \underline{s} < \bar{s} \leqq \infty$. Then $f$ has no zeros within this strip.*

Proof. Since $f = ({}^n\!f)^n$ is continuable to $S(\underline{s}, \bar{s})$ the same is true for ${}^n\!f$ by Rajkov's Prop. 4.6.2. If $f$ should have a zero of a certain order $k$ at $z_0 \in S(\underline{s}, \bar{s})$, then ${}^n\!f$ would have a singularity at $z_0$ for $n > k$ which is impossible. □

**Theorem 7.3.6.** *Let $F = \mathcal{L}(a, \gamma^2, M, N)$ be inf.div. with c.f. $f$. Then $f$ is continuable to a strip $S(\underline{s}, \bar{s})$, $-\infty \leqq \underline{s} \leqq 0 \leqq \bar{s} \leqq \infty$, $\underline{s} < \bar{s}$, iff, as $u \to \infty$,*

$$M(-u) = O(e^{-su}) \quad \textit{for all} \quad s < \bar{s}\,,$$

$$N(u) = O(e^{su}) \qquad \textit{for all} \quad s > \underline{s}\,.$$

*In either case the representation*

$$\log f(z) = iaz - \frac{\gamma^2}{2} z^2 + \int_{-\infty}^{-0} k(z, u) \, dM(u) + \int_{+0}^{\infty} k(z, u) \, dN(u) \tag{7.3.4}$$

*holds true for all* $z \in S(\underline{s}, \bar{s})$.

Proof. To begin with, we make use of the decomposition

$$f(t) = \exp\left\{iat - \frac{\gamma^2}{2} t^2\right\} f_M(t) \, f_N(t) \, ,$$

cf. Rem. 7.2.9, where $f_M = \exp(g_M)$ and $f_N = \exp(g_N)$. Now the if-part follows immediately from Prop. 7.3.2 and Cor. 7.3.4 which also show that the representation (7.3.4) is valid in $S(\underline{s}, \bar{s})$.

Conversely, if $f$ is continuable to $S(\underline{s}, \bar{s})$ then, by Prop. 4.6.2, the same is true for both $f_M$ and $f_N$. Since in this strip there are no zeros of $f_M$ and $f_N$, the functions $g_M$ and $g_N$ are also continuable to $S(\underline{s}, \bar{s})$. Thus the assertions on $M$ and $N$ follow from Cor. 7.3.4 and its counterpart valid for $N$. □

The representation (7.3.4) suggests to investigate which conditions the parameters $a$ and $\gamma^2$ and the spectral functions $M$ and $N$ must satisfy in order that $F$ be a one-sided d.f. To this end we prove some auxiliary results which will be stated in the following two lemmas; they contain also assertions which are needed in later sections.

**Lemma 7.3.7.** (i) *If* $M(u_0) > 0$, $-\infty < u_0 < 0$, *then*

$$e^{su_0} \int_{-\infty}^{-0} k(is, u) \, dM(u) \to \infty \, , \qquad s \to \infty \, .$$

(ii) *If* $-\infty < u_1 < 0$, *then*

$$e^{su_1} \int_{u_1-0}^{-0} k(is, u) \, dM(u) \to 0 \; M(u_1 + 0) - M(u_1), \qquad s \to \infty \, .$$

Proof. (i) Since $M$ is left-continuous there exist numbers $u'$ and $u''$ such that $u' < u'' < u_0$ and $M(u'') - M(u') > 0$. Then we obtain, by virtue of (B. 4.5),

$$e^{su_0} \int_{-\infty}^{-0} k(is, u) \, dM(u) \geqq e^{su_0} \int_{u'}^{u''} k(is, u) \, dM(u) \, .$$

Further, for $u' < u < u''$ and sufficiently large $s$ we get

$$k(is, u) = e^{-su} - 1 + \frac{su}{1 + u^2} \geqq \frac{1}{2} e^{-su} \geqq \frac{1}{2} e^{-su''} \, .$$

Now the assertion follows in view of

$$e^{su_0} \int_{-\infty}^{-0} k(is, u) \, dM(u) \geqq \tfrac{1}{2} e^{s(u_0 - u'')} [M(u'') - M(u')] \to \infty \, , \qquad s \to \infty \, .$$

(ii) For any $\varepsilon$, $0 < \varepsilon < |u_1|$ we have, by virtue of $0 < k(is, u) \leqq e^{-su}$, $su < 0$,

$$(M(u_1 + 0) - M(u_1)) \leqq e^{su_1} \int_{u_1-0}^{-0} k(is, u) \, dM(u) + o(1) \leqq e^{su_1} [e^{-su_1} \int_{u_1-0}^{u_1+\varepsilon} dM(u) + o(e^{-su_1})] \, .$$

Letting first $s \to \infty$ and then $\varepsilon \to 0$ the assertion is shown. □

**Lemma 7.3.8.** (i) *We have*

$$\frac{1}{s}\int_{+0}^{\infty} k(\mathrm{i}s, u)\,\mathrm{d}N(u) = O(1), \qquad s \to \infty, \quad \textit{iff} \quad I := \int_{+0}^{1} u\,\mathrm{d}N(u) < \infty.$$

*In either case,*

$$\frac{1}{s}\int_{+0}^{\infty} k(\mathrm{i}s, u)\,\mathrm{d}N(u) = \int_{+0}^{\infty} \frac{u}{1+u^2}\,\mathrm{d}N(u) + o(1), \qquad s \to \infty. \tag{7.3.5}$$

(ii) *Assume that $I < \infty$. Then*

$$\int_{+0}^{\infty} (\mathrm{e}^{-su} - 1)\,\mathrm{d}N(u) = O(1), \qquad s \to \infty, \quad \textit{iff} \quad N(+0) > -\infty.$$

*In either case we have*

$$\int_{+0}^{\infty} (\mathrm{e}^{-su} - 1)\,\mathrm{d}N(u) = N(+0) + o(1), \qquad s \to \infty.$$

Proof. (i) Let $I < \infty$. Then we get, by virtue of $1 - \mathrm{e}^{-x} \leqq x$, $x \geqq 0$, for any $\varepsilon > 0$,

$$\left|\frac{1}{s}\int_{+0}^{\infty} k(\mathrm{i}s, u)\,\mathrm{d}N(u) - \int_{+0}^{\infty} \frac{u}{1+u^2}\,\mathrm{d}N(u)\right| = \frac{1}{s}\int_{+0}^{\infty} (1 - \mathrm{e}^{-su})\,\mathrm{d}N(u)$$

$$\leqq \int_{+0}^{\varepsilon} u\,\mathrm{d}N(u) + \int_{\varepsilon}^{\infty} \frac{1 - \mathrm{e}^{-su}}{s}\,\mathrm{d}N(u) \to 0;$$

the limit is to be taken first as $s \to \infty$ and then $\varepsilon \to 0$. Hence, the if-part and the formula (7.3.5) have been shown.

Suppose now $1/s \int_{+0}^{\infty} k(\mathrm{i}s, u)\,\mathrm{d}N(u) = O(1)$. Then it can easily be derived that we also have $1/s \int_{+0}^{1} k_1(\mathrm{i}s, u)\,\mathrm{d}N(u) = O(1)$. Accordingly, using (B. 4.3) and (B. 4.4), we obtain

$$\frac{1}{s}\int_{+0}^{1} k_1(\mathrm{i}s, u)\,\mathrm{d}N(u) \geqq \int_{1/s}^{1} \frac{k_1(\mathrm{i}s, u)}{su}\,u\,\mathrm{d}N(u) \geqq \frac{1}{\mathrm{e}}\int_{1/s}^{1} u\,\mathrm{d}N(u) \geqq 0$$

whence $I < \infty$ follows as $s \to \infty$

(ii) If $N(+0) > -\infty$, then we get immediately for any $\varepsilon > 0$

$$\left|\int_{+0}^{\infty} (\mathrm{e}^{-su} - 1)\,\mathrm{d}N(u) - N(+0)\right| \leqq \int_{+0}^{\varepsilon} \mathrm{d}N(u) + \int_{\varepsilon}^{\infty} \mathrm{e}^{-su}\,\mathrm{d}N(u) \to 0,$$

where first $s \to \infty$ and then $\varepsilon \to 0$. — Conversely, if $\int_{+0}^{\infty} (\mathrm{e}^{-su} - 1)\,\mathrm{d}N(u) = O(1)$, then we conclude

$$\int_{+0}^{\infty} (\mathrm{e}^{-su} - 1)\,\mathrm{d}N(u) \leqq \int_{1/s}^{\infty} (\mathrm{e}^{-su} - 1)\,\mathrm{d}N(u) \leqq \left(1 - \frac{1}{\mathrm{e}}\right) N\left(\frac{1}{s}\right) \leqq 0. \qquad \square$$

Now we turn to the characterization of one-sided inf.div.d.f.; we formulate the result only for d.f. bounded to the left.

**Proposition 7.3.9.** *An inf.div.d.f. $F = \mathcal{L}(a, \gamma^2, M, N)$ is bounded to the left iff*

$$\gamma^2 = 0, \qquad M = 0, \qquad \int_{+0}^{1} u \, dN(u) < \infty .$$

*In either case we have*

$$l_F := \operatorname{lext}(F) = a - \int_{+0}^{\infty} \frac{u}{1+u^2} \, dN(u)$$

*and $f$ is continuable to $S(0, \infty)$ with the representation*

$$\log f(z) = i l_F z + \int_{+0}^{\infty} (e^{izu} - 1) \, dN(u), \qquad z \in S[0, \infty) .$$

Proof. Under the assumption of the if-part we see from Th. 7.3.6 that $f$ is continuable to a half plane $S(\underline{s}, \infty)$ where $\underline{s} \leqq 0$ imoreover, (7.3.4) implies

$$\frac{\log f(is)}{s} = -a + \frac{1}{s} \int_{+0}^{\infty} k(is, u) \, dN(u), \qquad s > 0 .$$

Applying now L. 7.3.8, (i), we obtain

$$\frac{\log f(is)}{s} = -a + \int_{+0}^{\infty} \frac{u}{1+u^2} \, dN(u) + o(1), \qquad s \to \infty,$$

whence it follows that $F$ is bounded to the left, see Cor. 4.4.2; from this also the formula for $l_F$ follows. Now the representation of $\log f(z)$ is an easy consequence of (7.3.4).

Assume now that $F$ is bounded to the left. Then we see from Prop. 4.2.2 that $f$ is continuable at least to $S(0, \infty)$. Hence we have (7.3.4) in $S(0, \infty)$. By Prop. 4.2.8 $f$ satisfies

$$\lim_{s \to \infty} \frac{\log f(is)}{s^2} = 0 . \tag{7.3.6}$$

Now we apply Prop. 7.3.2 to $N$ and see that

$$g_N(z) := \int_{+0}^{\infty} k(z, u) \, dN(u) = o(|z|^2), \qquad z \in S[0, \infty), \qquad |z| \to \infty .$$

Thus we obtain from (7.3.4)

$$\frac{\log f(is)}{s^2} = \frac{\gamma^2}{2} \left(1 + o(1)\right) + \frac{1}{s^2} \int_{-\infty}^{-0} k(is, u) \, dM(u), \qquad s \to \infty .$$

Combining this with (7.3.6) we conclude from L. 7.3.7, (i), that necessarily $M = 0$; then we also see that $\gamma^2 = 0$. So we find from (7.3.4)

$$\frac{\log f(is)}{s} = -a + \frac{1}{s} \int_{+0}^{\infty} k(is, u) \, dN(u) .$$

The left side is bounded by virtue of Prop. 4.2.8 so that L. 7.3.8, (i), yields $\int_{+0}^{1} u \, dN(u) < \infty$. □

**Corollary 7.3.10.** *An inf.div.d.f. $F$ is a finite d.f. iff it is degenerated.*

For, Prop. 7.3.9 tells us that $F$ is bounded in either direction iff $\gamma^2 = 0$, $M = 0$, and $N = 0$. □

## 7.4. General convergence theorems and compactness conditions

This section presents convergence and compactness properties of inf.div.d.f. They will be applied in part III. For all our statements we use the Lévy-Hinčin representation. Clearly, it is also possible to reformulate them in terms of the Lévy representation.

First we show that complete limits of sequences of inf.div.d.f. are also inf.div.; more precisely, we have the following assertion.

**Proposition 7.4.1.** *Let* $F_n = \mathcal{K}(a_n, H_n)$, $n \geqq 1$, *be inf.div.d.f. Then* $F_n \xrightarrow{c} F$ *holds iff there exist a constant* $a \in \mathbb{R}_1$ *and a spectral function* $H$ *such that* $a_n \to a$ *and* $H_n \xrightarrow{c} H$. *In either case,* $F$ *is inf.div. and has the representation* $F = \mathcal{K}(a, H)$.

Proof. If $a_n \to a$, $H_n \xrightarrow{c} H$, then we can apply the extended Helly-Bray theorem to $(H_n)$ and see (since $K(t, x)$ is continuous and bounded in $x$) that

$$f_n(t) = \exp \{ia_n t + \int_{-\infty}^{\infty} K(t, x) \, dH_n(x)\} \to f(a, H; t) .$$

Accordingly, $F_n \xrightarrow{c} F = \mathcal{K}(a, H)$.

Conversely, if $F_n = \mathcal{K}(a_n, H_n) \xrightarrow{c} F$, then $f_n(t) = f(a_n, H_n; t) \to f(t)$, and L. 7.2.3 tells us that $a_n \to a$ and $H_n \xrightarrow{c} H$. □

Prop. 7.4.1 implies the following compactness criterion. It will be formulated in terms of tightness; concerning the interrelation between these notions see Th. 2.4.4 and Prop. 2.4.12.

**Corollary 7.4.2.** *A set* $\boldsymbol{F}$ *of inf.div.d.f.* $F = \mathcal{K}(a, H)$ *is tight iff the corresponding set* $\boldsymbol{H}$ *is tight and, moreover,*

$$\sup_{\boldsymbol{F}} H(\infty) < \infty , \qquad \sup_{\boldsymbol{F}} |a| < \infty . \tag{7.4.1}$$

Proof. *If part*: The first assumption (7.4.1) implies that the set $\boldsymbol{H}$ is uniformly bounded so that Prop. 2.4.12 can be applied to it. Thus we come to know that $\boldsymbol{H}$ is relatively compact. Now Helly's selection theorem tells us that there exists a sequence $(H_n)$, $H_n \in \boldsymbol{H}$, and a function $H$ such that we have $H_n \xrightarrow{c} H$. Now we choose a subsequence $(H_{n'})$ of $(H_n)$ such that also $a_{n'} \to a$, $a \in \mathbb{R}_1$, is guaranteed. Hence Prop. 7.4.1 implies $F_{n'} = \mathcal{K}(a_{n'}, H_{n'}) \xrightarrow{c} F = \mathcal{K}(a, H)$ so that $\boldsymbol{F}$ is relatively compact; hence, it is also tight.

*Only-if part*: Choose an arbitrary sequence of pairs $((a_n, H_n))$ such that $F_n = \mathcal{K}(a_n, H_n) \in \boldsymbol{F}$. By assumption, $(F_n)$ contains a subsequence $(F_{n'})$ satisfying $F_{n'} \xrightarrow{c} F$ for a certain d.f. $F$. Accordingly, from Prop. 7.4.1 we see that $F = \mathcal{K}(a, H)$, say, and that $a_{n'} \to a$, $H_{n'} \xrightarrow{c} H$. Now the assertions follow by an indirect argument. □

Next we wish to deal with weakly converging sequences of inf. div.d.f. the limit function $F$ of which may be defective. Here we require that probability mass "escapes" in at most one direction; for definiteness we suppose that $F(-\infty) = 0$ is guaranteed.

We start with the following general lemma. In view of the continuity theorem it is trivial if the assumed convergences are complete.

**Lemma 7.4.3.** *Let* $(F_{1n})$ *and* $(F_{2n})$ *be two sequences of arbitrary d.f. and assume* $F_{kn} \xrightarrow{w} F_k$ *where* $F_k(-\infty) = 0$, $k = 1, 2$. *Then* $F_{1n} * F_{2n} \xrightarrow{w} F_1 * F_2$.

Proof. Let $G$ stand for an arbitrary non-decreasing function satisfying $G(-\infty) = 0$, $G(\infty) \leqq 1$. Then we introduce, for fixed $y_0$, a corresponding d.f. $\hat{G}$ defined by

$$\hat{G}(y) = \begin{cases} G(y), & y \leqq y_0, \\ 1, & y > y_0. \end{cases}$$

Clearly, our assumptions ensure that $\hat{F}_{kn} \xrightarrow{c} \hat{F}_k$, $k = 1, 2$, and the continuity theorem tells us that $\hat{F}_{1n} * \hat{F}_{2n} \xrightarrow{c} \hat{F}_1 * \hat{F}_2$.

Now let $x \in R_1$ be fixed. Choose $y_0$ so that $x - y_0 < y_0$, and suppose that $x - y_0$ is a continuity point of all functions occurring in the course of the present proof. Then we consider two functions $G_1, G_2$ with the properties of $G$. Thus we obtain, by definition of the corresponding d.f. $\hat{G}_1$ and $\hat{G}_2$,

$$\hat{G}_1 * \hat{G}_2(x) = \int_{-\infty}^{x-y_0+0} + \int_{x-y_0+0}^{y_0-0} + \int_{y_0-0}^{\infty} \hat{G}_1(x-y)\,\hat{G}_2(dy)$$
$$\leqq G_2(x - y_0) + G_1 * G_2(x) + G_1(x - y_0).$$

Further, by virtue of $G \leqq \hat{G}$ we see that

$$G_1 * G_2(x) \leqq \int_{-\infty}^{\infty} \hat{G}_1(x-y)\,G_2(dy) = \int_{-\infty}^{\infty} G_2(x-y)\,\hat{G}_1(dy) \leqq \hat{G}_1 * \hat{G}_2(x).$$

Note that $\hat{G}_1 * G_2$ is commutative since either $G_2 = 0$ or $G_2/G_2(\infty)$ is a d.f. so that Cor. 1.4.3 applies.

Using the above estimates we obtain, by assumption and by virtue of Prop. 2.1.3,

$$F_1 * F_2(x) \leqq \hat{F}_1 * \hat{F}_2(x) \leqq \varliminf_{n\to\infty} \hat{F}_{1n} * \hat{F}_{2n}(x)$$
$$\leqq \varliminf_{n\to\infty} \big(F_{1n} * F_{2n}(x) + F_{1n}(x - y_0) + F_{2n}(x - y_0)\big)$$
$$= \varliminf_{n\to\infty} F_{1n} * F_{2n}(x) + F_1(x - y_0) + F_2(x - y_0),$$
$$\varlimsup_{n\to\infty} F_{1n} * F_{2n}(x) \leqq \varlimsup_{n\to\infty} \hat{F}_{1n} * \hat{F}_{2n}(x) \leqq \hat{F}_1 * \hat{F}_2(x+0)$$
$$\leqq F_1 * F_2(x+0) + F_1(x - y_0) + F_2(x - y_0).$$

Finally, on letting $y_0 \to \infty$ we get, using the assumption $F_k(-\infty) = 0$,

$$F_1 * F_2(x) \leqq \varliminf_{n\to\infty} F_{1n} * F_{2n}(x), \qquad \varlimsup_{n\to\infty} F_{1n} * F_{2n}(x) \leqq F_1 * F_2(x+0);$$

hence, the assertion follows from Prop. 2.1.3. □

Now we are able to prove the following weak convergence theorem.

**Proposition 7.4.4.** *Let $F_n = \mathcal{H}(a_n, H_n)$, $n \geqq 1$, be inf.div.d.f. and suppose (7.4.1) for the set $\boldsymbol{F} = \{F_n : n \geqq 1\}$.*

(i) *We have $F_n \xrightarrow{w} F$, $F(-\infty) = 0$, iff $H_n \xrightarrow{w} H$, $H(-\infty) = 0$, and, moreover, $H_n(\infty) \to \delta$, $a_n \to a$. In either case, $F$ has the representation $F = e^{H(\infty)-\delta}\,\mathcal{H}(a, H)$.*

(ii) *We have $F_n \xrightarrow{w} F$, $F(\infty) = 1$, iff $H_n \xrightarrow{w} H$, $H_n(\infty) \to H(\infty)$, and $a_n \to a$. In this case $F = 1 - e^{-H(-\infty)}\,\big(1 - \mathcal{H}(a, H)\big)$.*

Proof. The basic idea of the proof is an appropriate decomposition of the d.f. $F_n$.

*If-part of* (i): By assumption there exists a monotone function $H$, $H(-\infty) = 0$, such that $H_n \xrightarrow{w} H$. Let $x_0 > 0$ be a continuity point of $H$. Then we put

$$H_n^l(x) = \begin{cases} H_n(x), & x \leq x_0, \\ H_n(x_0), & x > x_0, \end{cases} \qquad H_n^r(x) = \begin{cases} 0, & x \leq x_0, \\ H_n(x) - H_n(x_0), & x > x_0, \end{cases}$$

$$a_n^l = a_n - \int_{x_0}^{\infty} \frac{1}{x}\,\mathrm{d}H_n(x), \qquad a_n^r = \int_{x_0}^{\infty} \frac{1}{x}\,\mathrm{d}H_n(x),$$

$$F_n^l = \mathcal{H}(a_n^l, H_n^l), \qquad F_n^r = \mathcal{H}(a_n^r, H_n^r).$$

From Cor. 7.2.8 we see that

$$F_n = F_n^l * F_n^r.$$

Further, using the limits $H$ and $a$ we introduce $H^l$, $H^r$; $a^l$, $a^r$; $F^l$, $F^r$ in a similar way.

First we consider the sequence $(F_n^l)$. Clearly, our assumptions imply that $H_n^l \xrightarrow{c} H^l$ and, moreover,

$$a_n^l = a_n - \int_{x_0}^{\infty} \frac{1}{x}\,\mathrm{d}H_n(x) \to a - \int_{x_0}^{\infty} \frac{1}{x}\,\mathrm{d}H(x) = a^l,$$

cf. Rem. 2.2.7. Hence, by virtue of Prop. 7.4.1 we have $F_n^l = \mathcal{H}(a_n^l, H_n^l) \xrightarrow{c} \mathcal{H}(a^l, H^l) = F^l$.

Next we deal with the sequence $(F_n^r)$. Put

$$G_n(x) = \int_{x_0}^{x} \frac{1+y^2}{y^2}\,\mathrm{d}H_n(y), \qquad G(x) = \int_{x_0}^{x} \frac{1+y^2}{y^2}\,\mathrm{d}H(y), \qquad x > x_0.$$

Then Ex. 7.2.12 (iv), and Ex. 7.1.3 tell us that $F_n^r$ and $F^r$ have respectively the representations

$$F_n^r = \mathrm{e}^{-G_n(\infty)} \sum_{k=0}^{\infty} \frac{G_n^{*k}}{k!}, \qquad F^r = \mathrm{e}^{-G(\infty)} \sum_{k=0}^{\infty} \frac{G^{*k}}{k!}. \tag{7.4.2}$$

By assumption we have $H_n \xrightarrow{w} H$ and $H_n(\infty) \to \delta$; thus we get, under consideration of Rem. 2.2.7,

$$G_n(\infty) = H_n(\infty) - H_n(x_0) + \int_{x_0}^{\infty} \frac{\mathrm{d}H_n(y)}{y^2} \to \delta - H(x_0) + \int_{x_0}^{\infty} \frac{\mathrm{d}H(y)}{y^2}$$
$$= \delta - H(\infty) + G(\infty).$$

Further, using the Helly-Bray theorem we obtain $G_n \xrightarrow{w} G$ so that for all $k \geq 0$, apply ing L. 7.4.3 $k$ times,

$$G_n^{*k} \xrightarrow{w} G^{*k}.$$

Now we see, by a dominated convergence argument, that in (7.4.2) the limit $n \to \infty$ can be carried out under the sum sign; hence

$$F_n^r \xrightarrow{w} \mathrm{e}^{-(\delta - H(\infty) + G(\infty))} \sum_{k=0}^{\infty} \frac{G^{*k}}{k!} = \mathrm{e}^{H(\infty)-\delta}\, F^r.$$

Finally, we again use L. 7.4.3 to conclude that

$$F_n = F_n^l * F_n^r \xrightarrow{w} e^{H(\infty)-\delta} F^l * F^r = e^{H(\infty)-\delta}\, \mathcal{H}(a, H)\,.$$

This terminates the proof of the if-part of (i).

The *if-part of* (ii) follows from the if-part of (i) by setting

$$\tilde{F}_n(x) = 1 - F_n(-x+0) = \mathcal{H}\big(-a_n, H_n(\infty) - H_n(-x+0)\big)\,.$$

*Only-if part of* (i): The supposition (7.4.1) makes it possible to choose a subsequence $(n')$ such that, for a certain monotone function $H$ and for some numbers $\delta$ and $a$,

$$H_{n'} \xrightarrow{w} H\,, \qquad H_{n'}(\infty) \to \delta\,, \qquad a_{n'} \to a\,. \tag{7.4.3}$$

Using the same notations and the same arguments as before, we see that

$$H_{n'}^l \xrightarrow{w} H^l\,, \qquad H_{n'}^l(\infty) \to H^l(\infty)\,, \qquad a_{n'}^l \to a^l\,.$$

Thus we can apply the if-part of (ii) and obtain

$$F_{n'}^l \xrightarrow{w} F^l\,, \qquad F^l(\infty) = 1\,.$$

Further, from (7.4.2) we see that $F_{n'}^r(-0) = 0$ and $F_{n'}^r(+0) = e^{-G_{n'}(\infty)}$ where, by virtue of (7.4.1),

$$G_{n'}(\infty) = \int_{x_0}^{\infty} \frac{1+y^2}{y^2}\, dH_{n'}(y) \leqq \frac{1+x_0^2}{x_0^2} H_{n'}(\infty) \leqq C < \infty\,.$$

Accordingly, for an arbitrary continuity point $x$ of $F_{n'}^l$ we get

$$F_{n'}(x) = F_{n'}^l * F_{n'}^r(x) \geqq \int_{-0}^{+0} F_{n'}^l(x-y)\, dF_{n'}^r(y) \geqq F_{n'}^l(x)\, e^{-C}\,,$$

and on letting first $n' \to \infty$ and then $x \to -\infty$ we obtain $F^l(-\infty) = 0$. This means that $F_{n'}^l \xrightarrow{c} F^l$, and Prop. 7.4.1 tells us that $H_{n'}^l \xrightarrow{c} H^l$ whence $H^l(-\infty) = 0$ follows. Clearly, $H^r(-\infty) = 0$ and therefore $H(-\infty) = H^l(-\infty) + H^r(-\infty) = 0$.

Now we see from (7.4.3) that all suppositions of the if-part of (i) are satisfied according to which

$$F_{n'} \xrightarrow{w} e^{H(\infty)-\delta}\, \mathcal{H}(a, H) = F$$

where the latter equality follows by the assumption $F_n \xrightarrow{w} F$.

It remains to show that the limits $H$, $\delta$, and $a$ are independent of the chosen subsequence $(n')$. Indeed, assuming

$$e^{\bar{H}(\infty)-\bar{\delta}}\, \mathcal{H}(\bar{a}, \bar{H}) = e^{H(\infty)-\delta}\, \mathcal{H}(a, H)$$

we obtain, on letting $x \to \infty$, $\bar{H}(\infty) - \bar{\delta} = H(\infty) - \delta$, and as a consequence, $\mathcal{H}(\bar{a}, \bar{H}) = \mathcal{H}(a, H)$. Now L. 7.2.2 tells us that $\bar{a} = a$ and $\bar{H} = H$. Hence, $\bar{H}(\infty) = H(\infty)$ so that $\bar{\delta} = \delta$ follows. This terminates the proof of the only-if part of (i).

The *only-if part of* (ii) is now immediate. □

It is obvious that Prop. 7.4.4 leads to a modification of Cor. 7.4.2. To formulate it we introduce the following notion.

A set $G$ of non-decreasing and bounded functions $G$ is said to be *left-tight* if

$$\lim_{x\to-\infty} \sup_{G} \big(G(x) - G(-\infty)\big) = \lim_{x\to-\infty} \sup_{G} \int_{-\infty}^{x} dG(y) = 0\,.$$

In analogy to the results of Sect. 2.4 we see that a set $\boldsymbol{G}$ of non-decreasing and uniformly bounded functions is left-tight iff for every sequence $(G_n)$, $G_n \in \boldsymbol{G}$, satisfying $G_n \xrightarrow{w} G$ we have $G_n(-\infty) \to G(-\infty)$.

**Corollary 7.4.5.** *Let $\boldsymbol{F}$ be a set of inf.div.d.f. $F = \mathcal{H}(a, H)$ and suppose* (7.4.1). *Then $\boldsymbol{F}$ is left-tight iff $\boldsymbol{H}$ is left-tight.*

The proof is similar to that of Cor. 7.4.2 where we apply Prop. 7.4.4 instead of Prop. 7.4.1. □

Clearly, all our left-sided assertions have right-sided analoga; the notion right-tight can be defined in an obvious manner. Note that for right-sided assertions the supposition $\sup_{\boldsymbol{F}} H(\infty) < \infty$ cannot be dropped; this is easily seen by checking the above proofs.

## 7.5. Analytic properties of infinitely divisible distribution functions

Applying the general explanations on analytic d.f. given in Chap. 6, we now wish to study the analytic character of inf.div.d.f. We remind the reader of the notations $l_\nu(F)$ and $\chi(F)$, see Sect. 6.1.

First we consider inf.div.d.f. with $\gamma^2 > 0$.

**Proposition 7.5.1.** *Let $F = \mathcal{L}(a, \gamma^2, M, N)$ be inf.div. and assume $\gamma^2 > 0$. Then $F$ is an integral d.f. of order $\varrho \leqq 2$. If $\varrho = 2$, then the type $\tau$ of $F$ is not greater than $1/(2\gamma^2)$. Further, if $F$ is symmetric, then $\varrho = 2$ and $\tau = 1/(2\gamma^2)$.*

Proof. According to Cor. 7.3.3 we have for the c.f. $f$ of $F$

$$-\frac{\log |f(t)|}{t^2} = \frac{\gamma^2}{2}[1 + o(1)], \qquad t \to \infty .$$

This yields $\chi(F) = 2$ and $l_2(F) = \gamma^2/2$. Now Cor. 6.1.12 shows that $F$ is integral while the estimates for $\varrho$ and $\tau$ follow from Prop. 6.4.4 and 6.4.5, resp. Further, if $F$ is symmetric, then $F \in \boldsymbol{P}$ since $f$ has no zeros, cf. Def. 6.2.1. Then the assertions follow from Suppl. 6.6.3 and 6.6.4. □

Now we turn to the case $\gamma^2 = 0$. We formulate three conditions in terms of the left spectral function $M$ which are sufficient for $F$ to be analytic; the reformulation of these conditions to conditions in terms of $N$ will be left to the reader.

**Proposition 7.5.2.** *Let $F = \mathcal{L}(a, 0, M, N)$ be inf.div. Assume that, for some $\nu \geqq 1$,*

$$l_\nu^{(1)} := \varliminf_{t\to\infty} \int_{-1}^{-0} \frac{1 - \cos tu}{t^\nu}\, dM(u) > 0 . \tag{7.5.1}$$

*Then $F$ is analytic at least in the strip $S(l_1^{(1)})$. Further, if* (7.5.1) *holds true for some $\nu > 1$, then $F$ is integral and of finite order $\varrho \leqq 1 + \dfrac{1}{\nu - 1}$. If $\varrho = 1 + \dfrac{1}{\nu - 1}$, then $F$ has a finite type $\tau$.*

Proof. First of all, for every $\nu$ we get by virtue of Cor. 7.2.10

$$l_\nu(F) = \lim_{\overline{t\to\infty}} \frac{1}{t^\nu}\left[\int_{-\infty}^{-0} (1 - \cos tu)\,\mathrm{d}M(u) + \int_{+0}^{\infty} (1 - \cos tu)\,\mathrm{d}N(u)\right]$$

$$\geqq \lim_{\overline{t\to\infty}} \int_{-1}^{-0} \frac{1 - \cos tu}{t^\nu}\,\mathrm{d}M(u) = l_\nu^{(1)}.$$

Accordingly, if (7.5.1) is satisfied for some $\nu \geqq 1$, then we have $l_1(F) \geqq l_\nu(F) \geqq l_\nu^{(1)} > 0$. Now Th. 6.1.11 shows that $F$ is analytic at least in $S(l_1)$ which contains $S(l_1^{(1)})$. Further, if $l_\nu^{(1)} > 0$ for some $\nu > 1$, then we have $l_\nu(F) > 0$ and $\chi(F) \geqq \nu > 1$, cf. (6.1.1). Therefore, Cor. 6.1.12 tells us that $F$ is integral. Finally, the assertions concerning the order and the type of $F$ now follow from Prop. 6.4.4 and 6.4.5. □

**Remark 7.5.3.** (i) In the above proposition, the case $l_1^{(1)} = \infty$ is admitted; then $F$ is integral but possibly of infinite order.

(ii) In view of Prop. 7.3.2 we have $l_\nu^{(1)} = 0$ for $\nu \geqq 2$; therefore the upper bound for $\varrho$ is always greater than 2.

(iii) If $\varrho = 1 + \dfrac{1}{\nu - 1}$, then $\tau \leqq (\nu-1)\,(l_\nu^{(1)}\nu^\nu)^{-1/(\nu-1)}$. This follows from Prop. 6.4.5.

(iv) If we check condition (7.5.1) for the right spectral function $N$ and find that it is true for $\nu' > \nu$, then we obtain $\varrho \leqq 1 + \dfrac{1}{\nu' - 1} < 1 + \dfrac{1}{\nu - 1}$ and a better upper bound for $\varrho$ is reached. □

**Proposition 7.5.4.** *Let $F = \mathscr{L}(a, 0, M, N)$ be inf.div. Assume that, for some $\eta > 1$ and a certain $\nu \geqq 1$,*

$$l_\nu^{(2)} := \lim_{\overline{u\to -0}} |u|^\nu\,[M(u) - M(\eta u)] > 0\,. \tag{7.5.2}$$

*Then $F$ is analytic in a strip $S(s)$, $s > 0$. If (7.5.2) holds true for some $\nu > 1$, then $F$ is integral of order $\varrho \leqq 1 + \dfrac{1}{\nu - 1}$; if $\varrho = 1 + \dfrac{1}{\nu - 1}$, then $F$ has a finite type $\tau$.*

Proof. We put $c = \pi\dfrac{\eta - 1}{\eta + 1}$ so that $0 < c < \pi$. Now we get for sufficiently large $t > 0$

$$\int_{-1}^{-0} (1 - \cos tx)\,\mathrm{d}M(x) \geqq \int_{-\frac{\pi-c}{t}}^{-\frac{\pi+c}{t}} (1 - \cos tx)\,\mathrm{d}M(x)\,.$$

Since in the interval of integration we have $-2\pi < -(\pi + c) \leqq tx \leqq -(\pi - c) < 0$ there exists a $\delta > 0$ such that $1 - \cos tx \geqq \delta$. Hence,

$$\int_{-1}^{-0} (1 - \cos tx)\,\mathrm{d}M(x) \geqq \delta\left[M\left(-\frac{\pi - c}{t}\right) - M\left(-\frac{\pi + c}{t}\right)\right].$$

Setting $u = -\frac{\pi - c}{t}$ we have $-\frac{\pi + c}{t} = -\frac{\pi - c}{t}\frac{\pi + c}{\pi - c} = u\eta$ and thus we obtain

$$\frac{1}{t^\nu}\int_{-1}^{-0}(1 - \cos tx)\,\mathrm{d}M(x) \geqq \frac{\delta}{(\pi - c)^\nu}|u|^\nu[M(u) - M(\eta u)]\,.$$

Now $t \to \infty$ is equivalent to $u \to -0$ so that we conclude from (7.5.2) that $l_\nu^{(1)} \geqq \delta(\pi - c)^{-\nu} l_\nu^{(2)} > 0$. In this way, all assertions follow from Prop. 7.5.2. □

We remark that the strip $S(s)$ and the upper bound for the type of $F$ can be quantified more precisely by virtue of

$$\delta(\pi - c)^{-\nu} \leqq \left(\frac{\eta + 1}{2\pi}\right)^\nu\left(1 - \cos\frac{2\pi}{\eta + 1}\right).$$

**Proposition 7.5.5.** *Let $F = \mathcal{L}(a, 0, M, N)$ be inf.div. Assume that, for a certain $\nu \geqq 1$,*

$$0 < c \leqq |u|^\nu M(u) \leqq C < \infty\,, \qquad u_0 < u < 0\,.$$

*Then all assertions of Prop. 7.5.4 remain true.*

Proof. We choose $\eta = (2C/c)^{1/\nu} > 1$ and obtain for $u > u_0$

$$|u|^\nu[M(u) - M(\eta u)] = |u|^\nu M(u) - \frac{1}{\eta^\nu}|\eta u|^\nu M(\eta u) \geqq c - \frac{c}{2C}C = \frac{c}{2} > 0\,.$$

Thus the suppositions of Prop. 7.5.4 are satisfied whence the assertion follows. □

## 7.6. Continuation and uniqueness theorems

In Sect. 1.5 we introduced our general concept of uniqueness theorems for d.f. Now we wish to turn to the very first problem in this field, which was posed by A. N. Kolmogorov in the fifties. Namely, he asked if an inf.div.d.f. which agrees with the normal d.f. on a half axis coincides with the normal d.f. everywhere. — Afterwards, Kolmogorov's problem fell into oblivion; but in the course of proving a conjecture of V. M. Zolotarev concerning the central limit theorem (see Th. 10.4.1) it proved necessary to solve this problem. The affirmative answer runs as follow.

**Theorem 7.6.1.** *Let $F$ be an inf.div.d.f. such that*

$$F(x) = \Phi(x)\,, \qquad x < x_0\,,$$

*for a certain $x_0 \in \mathbb{R}_1$. Then $F = \Phi$.* □

We need not give a proof of this result since it is a special case of the recent Prop. 4.7.3. Besides, there does already exist the following generalization of Th. 7:6.1.

**Proposition 7.6.2.** *Let $F_k = \mathcal{L}(a_k, \gamma_k^2, M_k, N_k)$, $k = 1, 2$, be inf.div. such that $F_1(x) = O(e^{sx})$, $x \to -\infty$, for all $s > 0$. If, moreover,*

$$F_1(x) = F_2(x) > 0\,, \qquad x < x_0\,,$$

*for some $x_0 \in \mathbb{R}_1$, then $F_1 = F_2$.*

The original proof of this result, which is due to Ibragimov (1977), is rather comprehensive so that we confine ourselves to giving only a short outline of it. At first, one applies Nevanlinna's value distribution theory (in particular, Nevanlinna's second main theorem) to show that $M_1 = M_2 =: M$ and $\gamma_1^2 = \gamma_2^2 =: \gamma^2$. Accordingly, $F_k = \mathscr{L}(0, \gamma^2, M, 0) * \mathscr{L}(a_k, 0, 0, N_k)$. Then the theory of Phragmén and Lindelöf can be used to prove that in this decomposition the second components also coincide. □

Note that the supposition $F_1(x) = O(e^{sx})$, $x \to -\infty$, $s > 0$, occurring in Prop. 7.6.2 means that the c.f. $f_1$ of $F_1$ is continuable to the upper half-plane $S(0, \infty)$, see Prop. 4.3.8. The following example shows that this supposition cannot be dropped in general.

**Example 7.6.3.** For $\lambda = \log 2$ we put

$$F(x) = \begin{cases} e^{\lambda x}, & x \leqq 0, \\ 1, & x > 0, \end{cases}$$

$$G_1(x) = 1 - F(-x), \qquad G_2(x) = P_{2\lambda}(x),$$

where $P$ stands for a Poisson distribution. Clearly, $F$, $G_1$, and $G_2$ are inf.div.; the c.f. $g_k$ are analytic in $S(0, \infty)$ while $f$ has a pole at $z = i\lambda \in S(0, \infty)$. Now we introduce $F_k = F * G_k$, $k = 1, 2$, so that the c.f. $f_k = fg_k$ also have poles at $z = i\lambda$. Further, we have $F_1 \neq F_2$ but it can easily be checked that

$$F_1(x) = F_2(x) = \tfrac{1}{2} e^{\lambda x}, \qquad x < 0. \qquad \square$$

In spite of Ex. 7.6.3, we are able to prove the following continuation theorem without the analyticity of $f_1$. — The special subclass of inf.div.d.f. occurring in it will play a particular role in Sect. 13.3.

**Proposition 7.6.4.** *Let $F_k = \mathscr{L}(a_k, \gamma_k^2, c_k |u|^{-\alpha_k}, N_k)$, $k = 1, 2$, be inf.div. where $c_k \geqq 0$ and $0 < \alpha_k < 2$. Assume that*

$$F_1(x) = F_2(x) > 0, \qquad x < x_0,$$

*for a certain $x_0 \in \mathbb{R}_1$. Then $F_1 = F_2$.*

Proof. To begin with, we write $F_k = F_{1k} * F_{2k}$ where

$$F_{1k} = \mathscr{L}(0, 0, c_k |u|^{-\alpha_k}, 0), \qquad F_{2k} = \mathscr{L}(a_k, \gamma_k^2, 0, N_k).$$

Now we anticipate a result concerning stable c.f. Namely, in Sect. 8.3 we will see that the c.f. $f_{1k}$ can be continued from the half axis $\{t > 0\}$ to a function $f_{1k}^+$ analytic in $S(0, \infty)$. If we check the formula (8.3.1) we find that this continuation has the property

$$f_{1k}^+(is) = \varphi_k(s)\, e^{-ic_k b(\alpha_k) s^{\alpha_k}}, \qquad s > 0,$$

no matter if $\alpha_k \neq 1$ or $\alpha_k = 1$. Here $b(\alpha_k) > 0$ is a certain constant and $\varphi_k$ is a real function such that $\varphi_k(s) > 0$ for all $s > 0$. Further, by Th. 7.3.6 the c.f. $f_{2k}$ is continuable from the whole real axis to $S(0, \infty)$. Accordingly, $f_k = f_{1k} f_{2k}$ can be continued from $\{t > 0\}$ to a function $f_k^+$ which is analytic in $S(0, \infty)$, and we have, for $s > 0$,

$$f_1^+(is) - f_2^+(is) = \psi_1(s)\, e^{-ic_1 b(\alpha_1) s^{\alpha_1}} - \psi_2(s)\, e^{-ic_2 b(\alpha_2) s^{\alpha_2}}.$$

Here we put $\psi_k = \varphi_k f_{2k}$ so that $\psi_k(s) > 0$ for all $s > 0$.

On the other hand, by hypothesis we have

$$f_1(t) - f_2(t) = \int_{x_0}^{\infty} e^{itx}\, d[F_1(x) - F_2(x)];$$

hence, this function is continuable to $S(0, \infty)$ and it approaches real values on the imaginary axis. Since this continuation must coincide with the continuation of $f_1 - f_2$

from $\{t > 0\}$ to $S(0, \infty)$ we get $\operatorname{Im}\left(f_1^+(is) - f_2^+(is)\right) = 0$ and therefore, for all $s > 0$,

$$\psi_1(s) \sin [c_1 b(\alpha_1) s^{\alpha_1}] = \psi_2(s) \sin [c_2 b(\alpha_2) s^{\alpha_2}] . \tag{7.6.1}$$

By virtue of $\psi_k(s) > 0$ and $b(\alpha_k) > 0$ we can conclude that exactly one of the following cases occurs.

(i) $c_1 = c_2 = 0$. Then $M_1 = M_2 = 0$, so that the $f_k$ are continuable to $S(0, \infty)$ by Th. 7.3.6. Now the assertion follows from Prop. 7.6.2.

(ii) $c_1 \neq 0$ and $c_2 \neq 0$. Then we denote by $s_1 > 0$ and $s_2 > s_1$ the smallest zeros of the function $\sin [c_1 b(\alpha_1) s^{\alpha_1}]$. Now (7.6.1) tells us that

$$c_1 b(\alpha_1) s_j^{\alpha_1} = c_2 b(\alpha_2) s_j^{\alpha_2}, \qquad j = 1, 2 .$$

This yields $(s_2/s_1)^{\alpha_1} = (s_2/s_1)^{\alpha_2}$ whence $\alpha_1 = \alpha_2$ follows; then we also conclude $c_1 = c_2$. Accordingly, from (7.6.1) we get $\psi_1(s) = \psi_2(s)$ for all $s > 0$. This means that $f_1^+(is) = f_2^+(is)$, $s > 0$, and therefore $f_1(t) = f_2(t)$ for $t > 0$. □

A further step we wish to take is to conclude only the coincidence of certain elements occurring in the Lévy representation of inf.div.d.f. The next result shows that such a conclusion can be drawn from weak asymptotic assumptions.

**Proposition 7.6.5.** *Let* $F_k = \mathcal{L}(a_k, \gamma_k^2, M_k, N_k)$, $k = 1, 2$, *be inf.div. Assume that* lext $(F_1) = -\infty$ *and* lext $(M_1) > -\infty$. *If, moreover, there exist constants* $c$ *and* $C$ *such that*

$$0 < c \leqq \frac{F_2(x)}{F_1(x)} \leqq C < \infty, \qquad x < x_0, \tag{7.6.2}$$

*for some* $x_0 \in R_1$, *then* $M_1 = M_2$ *and* $\gamma_1^2 = \gamma_2^2$.

Proof. Since $l_1 :=$ lext $(M_1) > -\infty$ the c.f. $f_1$ is continuable to $S(0, \infty)$, see Th. 7.3.6. Now (7.6.2) and Prop. 4.4.5 tell us that $f_2$ is also continuable to $S(0, \infty)$ and that

$$v(s) := \log f_2(is) - \log f_1(is) = O(1), \qquad s \to \infty . \tag{7.6.3}$$

On the other hand, using the Lévy representation and applying Prop. 7.3.2 to $N_k$ we get, as $s \to \infty$,

$$v(s) = o(s^2) + \frac{\gamma_2^2 - \gamma_1^2}{2} s^2 + \int_{-\infty}^{-0} k(is, u) \, d[M_2(u) - M_1(u)] . \tag{7.6.4}$$

Now assume that $M_2(u_0) > 0$ for a certain $u_0 < l_1 =$ lext $(M_1) \leqq 0$. Then we apply L. 7.3.7, (ii), to $M_1$ and see that

$$e^{su_0} \int_{-\infty}^{-0} k(is, u) \, dM_1(u) \to 0, \qquad s \to \infty .$$

Thus we obtain from (7.6.3) and (7.6.4)

$$o(1) = e^{su_0} v(s) = o(1) + e^{su_0} \int_{-\infty}^{-0} k(is, u) \, dM_2(u), \qquad s \to \infty,$$

so that L. 7.3.7, (i), when applied to $M_2$, yields a contradiction. Accordingly, lext $(M_2) \geqq l_1$ and we have $M_1 = M_2 = 0$, provided that $l_1 = 0$.

Otherwise, if $l_1 < 0$ let $u_1$ be an arbitrary continuity point of both $M_1$ and $M_2$ such that $l_1 < u_1 < 0$. Then we define the function

$$\overline{M}(u) = M_2(u) - M_1(u), \qquad u \leqq u_1 .$$

Clearly, $\overline{M}$ is of bounded variation on $(-\infty, u_1)$ and we have $\overline{M}(u) = 0$, $u \leqq l_1$. Now we again make use of (7.6.3), (7.6.4), and L. 7.3.7, (ii), and obtain, as $s \to \infty$,

$$o(1) = e^{su_1} v(s) = o(1) + e^{su_1} \int_{l_1}^{u_1} k(is, u)\, d\overline{M}(u)$$

whence it easily follows that

$$e^{su_1} \int_{l_1}^{u_1} e^{-su}\, d\overline{M}(u) = o(1)\,, \qquad s \to \infty\,.$$

Next we introduce the integral function $\overline{m}(z) = \int_{l_1}^{u_1} e^{izu}\, d\overline{M}(u)$ and see that

$$\overline{m}(is) = \int_{l_1}^{u_1} e^{-su}\, d\overline{M}(u) = o(e^{-su_1})\,, \qquad s \to \infty;$$

therefore

$$\overline{\lim_{s\to\infty}} \frac{\log |\overline{m}(is)|}{s} \leqq -u_1\,.$$

Now Prop. C. 2.2.1 tells us that $\overline{M}(u) = 0$, $u \leqq u_1$, whence $M_1(u) = M_2(u)$ for $u \leqq u_1$ follows. Since $u_1 < 0$ is arbitrary we obtain $M_1 = M_2$. — Finally, from (7.6.3) and (7.6.4) we now conclude

$$o(1) = \frac{v(s)}{s^2} = o(1) + \frac{\gamma_2^2 - \gamma_1^2}{2}\,, \qquad s \to \infty\,,$$

and therefore $\gamma_1^2 = \gamma_2^2$. □

Next we prove three corollaries to Prop. 7.6.5 which supplement also the statement given in Suppl. 7.9.18. The first of them gives some additional information; the other ones are uniqueness theorems for ind.div.d.f.

**Corollary 7.6.6.** *Assume that all conditions of Prop. 7.6.5 be satisfied. Then we have*

$$I_1 := \int_{+0}^{1} u\, dN_1(u) < \infty \quad \textit{iff} \quad I_2 := \int_{+0}^{1} u\, dN_2(u) < \infty\,,$$

*and in either case*

$$a_1 - \int_{+0}^{\infty} \frac{u}{1+u^2}\, dN_1(u) = a_2 - \int_{+0}^{\infty} \frac{u}{1+u^2}\, dN_2(u)\,. \tag{7.6.5}$$

*Further, if $I_1, I_2 < \infty$, then*

$$N_1(+0) > -\infty \quad \textit{iff} \quad N_2(+0) > -\infty\,,$$

*and in either case we have*

$$\log c \leqq N_2(+0) - N_1(+0) \leqq \log C\,. \tag{7.6.6}$$

Proof. From Prop. 7.6.5 we have $M_1 = M_2$ and $\gamma_1^2 = \gamma_2^2$. Accordingly, it follows by virtue of (7.6.3) that, as $s \to \infty$,

$$o(1) = \frac{v(s)}{s} = -(a_2 - a_1) + \frac{1}{s} \int_{+0}^{\infty} k(is, u)\, d[N_2(u) - N_1(u)]\,.$$

Hence, the first assertion and (7.6.5) can immediately be concluded from L. 7.3.8, (i). Further, if $I_1$, $I_2 < \infty$, then we have in view of (7.6.3) and (7.6.5)

$$O(1) = v(s) = \int_{+0}^{\infty} (e^{-su} - 1) \, d[N_2(u) - N_1(u)] \,, \qquad s \to \infty \,.$$

So L. 7.3.8, (ii), tells us that $N_1(+0) > -\infty$ is equivalent to $N_2(+0) > -\infty$ and that, if this is the case,

$$v(s) = N_2(+0) - N_1(+0) + o(1) \,, \qquad s \to \infty \,.$$

Finally, from (7.6.2) and Prop. 4.4.5 it follows that for any $\varepsilon$, $0 < \varepsilon < c$, and sufficiently large $s$

$$\log (c - \varepsilon) \leqq v(s) = N_2(+0) - N_1(+0) + o(1) \leqq \log (C + \varepsilon)$$

thus showing (7.6.6). □

**Corollary 7.6.7.** *Let* $F_k = \mathcal{L}(a_k, \gamma_k^2, M_k, N_k)$, $k = 1, 2$, *be inf.div. Assume, further,* lext $(F_1) = -\infty$, lext $(M_1) > -\infty$, *and* $N_1 = 0$. *Then the condition*

$$\lim_{x \to -\infty} \frac{F_2(x)}{F_1(x)} = 1 \tag{7.6.7}$$

*implies* $F_1 = F_2$.

Proof. Clearly, (7.6.7) implies (7.6.2) where $c$ and $C$ can be chosen arbitrarily close to 1. Now Prop. 7.6.5 tells us that $M_1 = M_2$ and $\gamma_1^2 = \gamma_2^2$, while (7.6.6) can be used to obtain $N_2(+0) = 0$ so that $N_2 = 0$. Finally, (7.6.5) now yields $a_1 = a_2$. □

**Corollary 7.6.8.** *Let* $F_k = \mathcal{L}(a_k, \gamma_k^2, M_k, N_k)$, $k = 1, 2$, *be inf.div. such that* lext $(M_1) > -\infty$. *Assume, further, that at least one of the following conditions be satisfied.*

(i) $\quad \gamma_1^2 > 0 \,,$

(ii) $\quad \varliminf_{t \to \infty} \int_{-1}^{-0} \frac{1 - \cos tu}{t} \, dM_1(u) > 0 \,.$

*If* (7.6.2) *holds true and if there exists a bounded sequence* $(x_n)$ *of real numbers such that*

$$F_1(x_n) = F_2(x_n) \,, \qquad n \geqq 1 \,, \tag{7.6.8}$$

*then* $F_1 = F_2$.

Proof. By Prop. 7.3.9, both (i) and (ii) imply lext $(F_1) = -\infty$. Therefore all assumptions of Prop. 7.6.5 are in force whence $M_1 = M_2$ and $\gamma_1^2 = \gamma_2^2$ follow. This means that the conditions (i) and/or (ii) hold true also for $\gamma_2^2$ and/or $M_2$. Now Prop. 7.5.1 and/or 7.5.2 show that both $F_1$ and $F_2$ are analytic d.f. Hence, the assertion follows from (7.6.8) by virtue of the uniqueness theorem for analytic functions. □

So far we have exclusively dealt with uniqueness theorems in the suppositions of which there occur relations between $F_1$ and $F_2$ on a half-axis $(-\infty, x_0)$. To close this section we wish to prove uniqueness theorems in the suppositions of which there occur relations between $F_1$ and $F_2$ only on a bounded set $S \subset R_1$. The following example shows some difficulties we have to expect in this context.

**Example 7.6.9.** Let $F_1$ be a non-degenerate (not necessarily inf.div.) d.f. Then we will construct an inf.div.d.f. $F_2 \neq F_1$ which coincides with $F_1$ on a certain interval $(c, d)$.

To this end, choose $c < d$ such that $0 < F_1(c) < 1$ and $F_1(d) \leqq F_1(c)\,[1 - \log F_1(c)]$; this is always possible. Now we define the function

$$G_1(x) = \begin{cases} 0\,, & x \leqq d - c\,, \\ \dfrac{F_1(x - d + 2c) - F_1(c)}{-F_1(c)\log F_1(c)}\,, & d - c < x \leqq 2(d - c)\,, \\ 1\,, & x > 2(d - c)\,; \end{cases}$$

then our assumptions ensure that $G_1$ is a d.f. Further, we put $\lambda = -\log F_1(c)$ and introduce the compound Poisson d.f.

$$G_2 = \mathrm{e}^{-\lambda} \sum_{k=0}^{\infty} \frac{\lambda^k}{k!} G_1^{*k}\,.$$

Now the choice of $G_1$ implies lext $(G_1^{*k}) \geqq k(d - c)$, and therefore

$$G_2(x) = \begin{cases} 0\,, & x \leqq 0\,, \\ F_1(c)\,, & 0 < x \leqq d - c\,, \\ F_1(x - d + 2c)\,, & d - c < x \leqq 2(d - c)\,. \end{cases}$$

So it comes out that the inf.div.d.f. $F_2(x) := G_2(x + d - 2c)$ coincides with $F_1(x)$ for $c < x < d$. Note that $c >$ lext $(F_1)$ and also $c > 2c - d =$ lext $(F_2)$. □

This example indicates that we cannot expect uniqueness theorems in the suppositions of which there occur only relations between $F_1$ and $F_2$ on sets $S$ which are bounded away from lext $(F_k)$ (or rext $(F_k)$, resp.). Therefore we have to ensure that lext $(F_k) = \inf \{x : x \in S\}$ which also implies that lext $(F_1) =$ lext $(F_2)$ must be assumed. Without loss of generality we suppose lext $(F_k) = 0$.

**Proposition 7.6.10.** *Let* $F_k = \mathcal{L}(a_k, 0, 0, N_k)$, $k = 1, 2$, *be inf.div. such that* lext $(F_k) = 0$. *Then we have, for some* $x_0 > 0$,

$$F_1(x) = F_2(x)\,, \qquad 0 < x < x_0 \tag{7.6.9}$$

*iff*

$$N_1(u) = N_2(u)\,, \qquad 0 < u < x_0\,. \tag{7.6.10}$$

Proof. Putting $l_F = 0$ in Prop. 7.3.9, we obtain

$$f_k(t) = \exp\Big(\int_{+0}^{\infty} (\mathrm{e}^{itu} - 1)\,\mathrm{d}N_k(u)\Big)\,, \qquad k = 1, 2\,. \tag{7.6.11}$$

*If-part.* We define two sequences of spectral functions by

$$N_k^{(n)}(u) := \begin{cases} N_k(1/n)\,, & 0 < u \leqq 1/n\,, \\ N_k(u)\,, & u > 1/n\,, \end{cases} \qquad k = 1, 2\,,$$

so that by assumption (7.6.10)

$$N_1^{(n)}(u) = N_2^{(n)}(u)\,, \qquad 0 < u < x_0\,, \qquad n \geqq n_0 > x_0^{-1}\,.$$

Using the abbreviations

$$\lambda_{nk} = -N_k(1/n)\,, \qquad k = 1, 2, \qquad n \geqq n_0\,,$$

we introduce the d.f.

$$G_k^{(n)}(u) := \begin{cases} 0\,, & u \leqq 0\,, \\ \lambda_{nk}^{-1}\big(N_k^{(n)}(u) + \lambda_{nk}\big)\,, & u > 0\,, \end{cases} \qquad n \geqq n_0\,,$$

($k = 1, 2$) which obviously coincide on $(-\infty, x_0)$. Now from (7.6.11) it is clear that

$$f_k^{(n)}(t) := \exp\,(\lambda_{nk} \int_0^\infty (e^{itu} - 1)\, dG_k^{(n)}(u))$$

are c.f. of compound Poisson d.f. (see Ex. 7.1.3)

$$F_k^{(n)} = e^{-\lambda_{nk}} \sum_{j=0}^\infty \frac{\lambda_{nk}^j}{j!} (G_k^{(n)})^{*j} \qquad k = 1, 2\,.$$

By construction we clearly have

$$F_1^{(n)}(x) = F_2^{(n)}(x)\,, \qquad x < x_1\,, \qquad n \geqq n_0\,.$$

Since $f_k^{(n)} \to f_k$, as $n \to \infty$, it follows that $F_k^{(n)} \xrightarrow{c} F_k$ ($k = 1, 2$) whence the desired result follows.

*Only-if part*: We denote by ${}^nF_k$ the "roots" which satisfy $({}^nF_k)^{*n} = F_k$, $k = 1, 2$. Then Prop. 4.7.7 tells us, by virtue of (7.6.9), that ${}^nF_1(x) = {}^nF_2(x)$ for $0 < x < x_0$ and all $n \geqq 1$. Since $N_k(u) = \lim_{n\to\infty} n[{}^nF_k(u) - 1]$ for all continuity points $u$ of $N_k$ (see Cor. 7.2.7) the assertion is shown. □

Prop. 7.6.10 yields the following uniqueness theorem. Its proof is immediate since the assumption $N(u_0) = 0$ ensures that there is only one continuation of $N$ to $(u_0, \infty)$.

**Corollary 7.6.11.** *Let $F_k = \mathcal{L}(a_k, 0, 0, N_k)$, $k = 1, 2$, be inf. div. Assume that* lext $(F_1)$ $= 0$ *and $N_1(u_0) = 0$ for some $u_0 > 0$. Then*

$$F_1(x) = F_2(x)\,, \qquad 0 \leqq x < u_0\,,$$

*implies $F_1 = F_2$.* □

**Example 7.6.12.** Let $F$ be a lattice inf.div.d.f. with atoms $x_k = k$ and jumps $p_k$, $k \geqq 0$, such that $0 < p_0 < 1$. If

$$p_1 = -p_0 \log p_0\,, \tag{7.6.12}$$

then $F = P_\lambda$ where $P_\lambda$ is a Poisson d.f. with $\lambda = -\log p_0$.

Indeed, for any $\lambda > 0$ the Poisson d.f. $P_\lambda$ has the properties lext $(P_\lambda) = 0$ and $N_\lambda(u) = 0$ for all $u > 1$. Further, the jumps of $P_\lambda$ at $x_0 = 0$ and $x_1 = 1$ are given by $e^{-\lambda}$ and $\lambda\, e^{-\lambda}$, resp. Now (7.6.12) means that $F(x) = P_\lambda(x)$ for $0 \leqq x < 2$ where $\lambda = -\log p_0$. So Cor. 7.6.11 shows the assertion. □

Next we prove that the supposition (7.6.12) can essentially be weakened without impairing the conclusion drawn from it. We start with the following lemma.

**Lemma 7.6.13.** *Let $F$ be a lattice inf.div.d.f. with atoms $x_k = k$, $k \geqq 0$, and jumps $p_k$ such that $0 < p_0 < 1$. Then there exists a lattice d.f. $G$ with atoms $x_k = k$, $k \geqq 1$, and jumps $q_k$ such that*

$$F = e^{-\lambda} \sum_{k=0}^\infty \frac{\lambda^k}{k!} G^{*k}\,, \tag{7.6.13}$$

*where $\lambda = -\log p_0$, i.e. $F$ is a compound Poisson d.f. with c.f.*

$$f(t) = \exp\,\{\lambda \sum_{k=1}^\infty (e^{itk} - 1)\, q_k\}\,.$$

Proof. First of all, because of lext $(F) = 0$ we see from Prop. 7.3.9 that

$$\log f(t) = \int_{+0}^\infty (e^{itu} - 1)\, dN(u)\,.$$

Further, since $F$ is lattice, Prop. 3.6.2 shows that $|f(2\pi)| = 1$ so that

$$0 = \log |f(2\pi)| = \int_{+0}^{\infty} (\cos 2\pi u - 1) \, \mathrm{d}N(u) .$$

By virtue of $\cos x - 1 \leqq 0$ it comes out that $N = \text{const.}$ in $(k, k+1)$, $k \geqq 0$. Accordingly, $N$ is a step function with jumps $\lambda q_k$ at $k \geqq 1$ where $\lambda = |N(+0)|$ so that $\sum_{k=1}^{\infty} q_k = 1$. Now we see that $f$ and hence $F$ have the asserted form. Finally, since $G^{*0}(+0) = 1$ and $G^{*k}(+0) = 0$, $k \geqq 1$, from (7.6.13) we obtain $p_0 = F(+0) = \mathrm{e}^{-\lambda}$ whence $\lambda = -\log p_0$ follows. This terminates the proof. □

**Proposition 7.6.14.** *Let $F$ be a lattice inf.div.d.f. with atoms $x_k = k$, $k \geqq 0$, and jumps $p_k$ such that $0 < p_0 < 1$. If, for a certain $n \geqq 1$,*

$$\sum_{k=0}^{n} p_k \geqq p_0 \sum_{k=0}^{n} \frac{(-\log p_0)^k}{k!} \tag{7.6.14}$$

*then $F = P_\lambda$ with $\lambda = -\log p_0$.*

Proof. Denote by $G$ the d.f. occurring in L. 7.6.13. Then we see that

$$G^{*k}(n+0) = \begin{cases} q_1^n , & k = n , \\ 0 , & k > n , \end{cases}$$

while, of course, $G^{*k}(n+0) \leqq 1$ for $k < n$. Accordingly, if $q_1 < 1$, then we get from (7.6.13)

$$\sum_{k=0}^{n} p_k = F(n+0) < \mathrm{e}^{-\lambda} \sum_{k=0}^{n} \frac{\lambda^k}{k!}$$

which contradicts (7.6.14) because of $\lambda = -\log p_0$. Hence, $q_1 = 1$ and $q_k = 0$, $k > 1$, so that $\log f(t) = \lambda(\mathrm{e}^{it} - 1)$ and therefore $F = P_\lambda$. □

In this way, there appears an extremal property of the Poisson d.f. Namely, in the class of all lattice inf.div.d.f. which are concentrated on $\{k: k \geqq 0\}$ the Poisson d.f. is the only one which satisfied (7.6.14) for every $n \geqq 1$, and in this case even the equality sign is valid.

## 7.7. Asymptotic properties

Let us consider a non-degenerate inf.div.d.f. $F = \mathscr{L}(a, \gamma^2, M, N)$ such that $\operatorname{lext}(F) = -\infty$. The aim of the present section is to link asymptotic properties of the left tail of $F$ with certain properties of the left spectral function $M$. Corresponding results for the right tail can be obtained as simple consequences.

In analogy to the notion $\operatorname{lext}(F)$ and $\operatorname{rext}(F)$ introduced in Sect. 1.1 we now write

$$l_M := \operatorname{lext}(M) = \sup \{x \leqq 0 \colon M(x) = 0\}$$

$$r_N := \operatorname{rext}(N) = \inf \{x \geqq 0 \colon N(x) = 0\}$$

with the obvious meaning that $l_M = -\infty$ if $M > 0$ and $r_N = \infty$ if $N < 0$. Note that $l_M = 0$ iff $M = 0$, and $r_N = 0$ iff $N = 0$.

**Proposition 7.7.1.** *Let $F$ be an inf.div.d.f. with* lext $(F) = -\infty$, *then*

$$\lim_{x\to\infty} \frac{x \log x}{\log F(-x)} = l_M\,, \qquad -\infty \leqq l_M \leqq 0\,,$$

*that is, the limit exists finitely or infinitely.*

Before proving this result we give two simple consequences the proof of the first of which is trivial.

**Corollary 7.7.2.** *Let* lext $(F) = -\infty$. *Then the following three statements are equivalent.*

(i) $\quad M = 0;$

(ii) $$\lim_{x\to\infty} \frac{x \log x}{\log F(-x)} = 0\,;$$

(iii) *There is a sequence $(x_n)$, $x_n \to \infty$, such that*

$$\lim_{n\to\infty} \frac{x_n \log x_n}{\log F(-x_n)} = 0\,.$$ □

**Corollary 7.7.3.** *For* $x_0 = \max\{-l_M, r_N\}$ *we have* – if $F$ is non-degenerate –

$$\lim_{x\to\infty} \frac{x \log x}{\log T_F(x)} = -x_0\,, \qquad 0 \leqq x_0 \leqq \infty\,.$$

Proof. Assuming first rext $(F) < \infty$ we have lext $(F) = -\infty$ and see from Prop. 7.3.9 that $r_N = 0$. Thus the assertion follows directly from Prop. 7.7.1. On the other hand, if lext $(F) = -\infty$, rext $(F) = \infty$ we write, as $x \to \infty$, $\big(1 + o(1)\big) \max\big(\log F(-x), \log\big(1 - F(x)\big)\big) = \log T_F(x)$. Thus

$$\frac{x \log x}{\log T_F(x)} = \big(1 + o(1)\big) \min\left(\frac{x \log x}{\log F(-x)}, \quad \frac{x \log x}{\log\big(1 - F(x)\big)}\right).$$

Now our assertion easily follows from Prop. 7.7.1 and the corresponding right-sided relation given by

$$\lim_{x\to\infty} \frac{x \log x}{\log\big(1 - F(x)\big)} = -r_N\,, \qquad 0 \leqq r_N \leqq \infty\,.$$ □

Proof of Proposition 7.7.1. It suffices to prove that

$$\underline{\lim}_{x\to\infty} \frac{x \log x}{\log F(-x)} \geqq l_M \tag{7.7.1}$$

and

$$\overline{\lim}_{x\to\infty} \frac{x \log x}{\log F(-x)} \leqq l_M\,. \tag{7.7.2}$$

First we show (7.7.1). The case $l_M = -\infty$ is obvious. Let us now assume $-\infty < l_M \leqq 0$. Then Th. 7.3.6 tells us that

$$\log f(z) = iza - \tfrac{1}{2}\gamma^2 z^2 + g_M(z) + g_N(z) \tag{7.7.3}$$

where

$$g_M(z) = \int_{-\infty}^{-0} k(z, x)\, dM(x)\,, \qquad g_N(z) = \int_{+0}^{\infty} k(z, x)\, dN(x)\,, \qquad z \in S[0, \infty)\,.$$

If $l_M = 0$, then $g_M = 0$. Now we apply Prop. 7.3.2 to $N$ and see that

$$\log f(is) = -sa + \tfrac{1}{2}\gamma^2 s^2 + g_N(is) < c_1 s^2 \tag{7.7.4}$$

for every $c_1 > \frac{1}{2}\gamma^2$ and sufficiently large $s > 0$. Putting $h(s) = c_1 s$, $y = 1/2$, and applying Prop. 4.4.3 we get

$$F(-x) \leqq \exp\left(-\frac{x^2}{4c_1}\right) \tag{7.7.5}$$

for $x$ sufficiently large whence (7.7.1) follows with $l_M = 0$. If $-\infty < l_M < 0$, then L. 7.3.7 (ii), implies $g_M(is) \leqq c_2 \exp(-l_M s)$ for certain $c_2 > 0$ and large $s$. Using (7.7.3) and (7.7.4) we obtain

$$\log f(is) \leqq c_1 s^2 + c_2 \, e^{-l_M s} \leqq e^{-l' s}$$

for every fixed $l'$, $-\infty < l' < l_M$, and sufficiently large $s$. Now we put $h(s) = e^{-l's}$ and apply again Prop. 4.4.3 so that

$$F(-x) \leqq \exp\left(\frac{(1-y)\,x}{l'}(\log x + \log y)\right)$$

for $0 < y < 1$ and sufficiently large $x$. Therefore we have

$$\varliminf_{x\to\infty} \frac{x \log x}{\log F(-x)} \geqq \frac{l'}{1-y}.$$

This proves (7.7.1) as $y \to 0$ and $l' \to l_M$.

Now we turn to the proof of (7.7.2). Since the case $l_M = 0$ is trivial we may assume $-\infty \leqq l_M < 0$ and choose some $l'$, $l_M < l' < 0$. Further we introduce i.i.d.r.v. $X_k$, $k = 1, 2, \ldots, n$, subject to the common d.f. ${}^nF$. Then

$$F(nl') = ({}^nF)^{*n}(nl') = \mathsf{P}\Big(\sum_{k=1}^{n} X_k < nl'\Big) \geqq \mathsf{P}\Big(\bigcap_{k=1}^{n} \{X_k < l'\}\Big) = \big({}^nF(l')\big)^n .$$

Accordingly,

$$\log F(nl') \geqq n \log {}^nF(l') = n \log\big(n\,{}^nF(l')\big) - n \log n . \tag{7.7.6}$$

Let us now suppose that $l'$ is a continuity point of $M$; then we see from Cor. 7.2.7 that

$$n\,{}^nF(l') \to M(l'), \qquad n \to \infty .$$

Moreover, $l' > l_M$ implies $\delta := \frac{1}{2} M(l') > 0$ and thus, by (7.7.6),

$$\log F(nl') \geqq n \log \delta - n \log n$$

for sufficiently large $n$. Next we choose an arbitrary $x > 0$ and determine the natural number $n \geqq 1$ such that $-(n-1)\,l' < x \leqq -nl'$. Then we obtain, as $x \to \infty$,

$$\begin{aligned}\frac{\log F(-x)}{x \log x} &\geqq \frac{\log F(nl')}{-l'(n-1)\log\big(-l'(n-1)\big)} \\ &= \big(1+o(1)\big)\frac{\log F(nl')}{-l'n \log n} \geqq \frac{\big(1+o(1)\big)}{-l'}\left(\frac{\log \delta}{\log n} - 1\right) = \frac{\big(1+o(1)\big)}{l'}.\end{aligned}$$

Hence we get

$$\varlimsup_{x\to\infty} \frac{x \log x}{\log F(-x)} \leqq l' .$$

On letting $l' \to l_M$ the relation (7.7.2) follows. □

**Example 7.7.4.** Let $P_\lambda = \mathcal{L}(a_\lambda, 0, 0, N_\lambda)$ be a Poisson d.f. for some $\lambda > 0$, cf. Ex. 7.2.12 (iii); then we have rext $(N_\lambda) = 1$, and Cor. 7.7.3 yields

$$\lim_{x \to \infty} \frac{\log(1 - P_\lambda(x))}{x \log x} = -1 .$$

We note that a more precise relation is given by

$$\frac{\log(1 - P_\lambda(x))}{x \log x} = -1 + \frac{1 + \log \lambda}{\log x} + O\left(\frac{1}{x}\right), \qquad x \to \infty . \tag{7.7.7}$$

Indeed, the remainder term of the series $e^{-\lambda} \sum_0^\infty \frac{\lambda^k}{k!} = 1$ admits the estimate

$$e^{-\lambda} \frac{\lambda^{[x+1]}}{[x+1]!} \leqq 1 - P_\lambda(x+0) \leqq \frac{\lambda^{[x+1]}}{[x+1]!}, \qquad x \geqq 0 .$$

Now (7.7.7) can be verified using the Stirling formula, cf. App. B.5. □

Let us now concentrate on inf.div.d.f. $F$ satisfying $l_M = -\infty$. We are interested in asymptotic relations between $M(x)$ and $F(x)$ as $x$ tends to $-\infty$.

**Definition 7.7.5.** A non-decreasing function $\psi$: $[1, \infty) \to [0, \infty)$ belongs to the class $C$ if

$$\psi(x_1 + x_2) \leqq c_0 \psi(x_1)\, \psi(x_2) \tag{7.7.8}$$

for some $c_0 \geqq 1$ and all $x_1, x_2 \geqq 1$. □

**Example 7.7.6.** (i) $\psi(x) = x^\eta \in C$, $\eta > 0$. To see this we write

$$(x_1 + x_2)^\eta \leqq 2^\eta (\max\{x_1, x_2\})^\eta \leqq 2^\eta x_1^\eta x_2^\eta , \qquad x_1, x_2 \geqq 1 ,$$

i.e. (7.7.8) is satisfied with $c_0 = 2^\eta$.

(ii) $\psi(x) = \log^\eta (1 + x) \in C$, $\eta > 0$, where $c_0 = (2/\log 2)^\eta$.

(iii) $\psi(x) = \exp(\tau x^\eta) \in C$, $\tau > 0$, $0 < \eta \leqq 1$, where $c_0 = 1$.

The proof of (ii) and (iii) will be left to the reader. □

Note that from $\psi_1, \psi_2 \in C$ it follows that $\psi = \psi_1 \psi_2 \in C$.

**Lemma 7.7.7.** *For every $\psi \in C$ there exist $c_1, c_2 > 0$ such that*

$$\psi(x) \leqq c_1 e^{c_2 x} , \qquad x \geqq 1 .$$

Proof. By induction we obtain, for every $x \geqq 1$,

$$\begin{aligned}\psi(x) &\leqq \psi([x] + 1) \leqq c_0 \psi([x])\, \psi(1) \leqq c_0^{[x]} \psi(1)^{[x]+1} \\ &\leqq c_0^x \psi(1)^{x+1} = \psi(1) \exp\left(\log\left(c_0 \psi(1)\right) x\right) .\end{aligned}$$ □

**Proposition 7.7.8.** *Let $F = \mathcal{L}(a, \gamma^2, M, N)$ be inf.div. with* lext $(M) = -\infty$ *and suppose $\psi \in C$. Then*

$$\int_{-\infty}^{-1} \psi(-x)\, M(dx) < \infty \tag{7.7.9}$$

*is satisfied iff*

$$\int_{-\infty}^{-1} \psi(-x)\, F(dx) < \infty . \tag{7.7.10}$$

Proof. (i) First we prove this assertion under the additional assumption that $F$ is a compound Poisson d.f.

$$F = e^{-\lambda} \sum_{0}^{\infty} \frac{\lambda^n}{n!} G^{*n} \tag{7.7.11}$$

where the d.f. $G$ satisfies rext $(G) \leqq -1$.

Let us suppose that (7.7.10) is true. Then by (7.7.11) and $\psi \geqq 0$ we have

$$\infty > \int_{-\infty}^{-1} \psi(-x)\, F(dx) \geqq e^{-\lambda}\lambda \int_{-\infty}^{-1} \psi(-x)\, G(dx)$$

whence (7.7.9) follows since $M = \lambda G$.

Conversely, we now suppose (7.7.9) and introduce i.i.d.r.v. $Y_k$, $k \geqq 1$, subject to $G$. By virtue of (7.7.11) and by $\mathsf{P}(Y_k \leqq -1) = 1$ we see that

$$0 \leqq \int_{-\infty}^{-1} \psi(-x)\, F(dx) = e^{-\lambda} \sum_{n=1}^{\infty} \mathsf{E}\psi\Big(-\sum_{k=1}^{n} Y_k\Big) \frac{\lambda^n}{n!}.$$

Applying (7.7.8) inductively we get

$$\mathsf{E}\psi\Big(-\sum_{k=1}^{n} Y_k\Big) \leqq c_0^{n-1} (\mathsf{E}\psi(-Y_1))^n \leqq (c_0 \mathsf{E}\psi(-Y_1))^n .$$

Thus we have

$$\int_{-\infty}^{-1} \psi(-x)\, F(dx) \leqq e^{-\lambda} \sum_{n=1}^{\infty} \frac{1}{n!} (c_0 \lambda \mathsf{E}\psi(-Y_1))^n < \infty .$$

(ii) We are now in a position to prove the general case. We put

$$M_1(x) = \begin{cases} M(x), & x \leqq -1, \\ M(-1), & -1 < x < 0, \end{cases}$$

$$M_2(x) = \begin{cases} 0, & x \leqq -1, \\ M(x) - M(-1), & -1 < x < 0, \end{cases}$$

so that $M = M_1 + M_2$. Moreover, we write $F = F_1 * F_2$ where $F_1 = \mathscr{L}(a_1, 0, M_1, 0)$ is — for a suitably chosen $a_1 \in \mathbb{R}_1$ — a compound Poisson d.f. and $F_2 = \mathscr{L}(a_2, \gamma^2, M_2, N)$. In view of lext $(M_2) \geqq -1$ and by Prop. 7.7.1 we get $F_2(x) = O(e^{-c|x|})$, $x \to -\infty$, for all $c > 0$; hence and by L. 7.7.7 we find

$$\int_{-\infty}^{-1} \psi(-x)\, F_2(dx) \leqq c_1 \int_{-\infty}^{-1} e^{c_1|x|}\, F_2(dx) < \infty . \tag{7.7.12}$$

Now we introduce i.r.v. $X_1$, $X_2$ subject to $F_1$, $F_2$, resp. For an arbitrary r.v. $X$ we set $X^- = \min\{X, -1\}$.

Let us now suppose (7.7.9). Then part (i) of the proof yields $\mathsf{E}\psi(-X_2^-) < \infty$. Accordingly,

$$\begin{aligned} \int_{-\infty}^{-1} \psi(-x)\, F(dx) &\leqq \mathsf{E}\psi(-(X_1 + X_2)^-) \leqq \mathsf{E}\psi(-X_1^- - X_2^-) \\ &\leqq c_0 \mathsf{E}\psi(-X_1^-)\, \mathsf{E}\psi(-X_2^-) < \infty . \end{aligned}$$

Note that the second inequality is true since $\psi$ is non-decreasing.

Conversely, let (7.7.10) be satisfied. For some $x_0 > 0$ we put $F'(x) = F(x + x_0)$, $F_2'(x) = F_2(x + x_0)$, $F_1' = F_1$ and get

$$\int_{-\infty}^{-1} \psi(-x)\, F'(\mathrm{d}x) = \int_{-\infty}^{-1+x_0} \psi(-u + x_0)\, F(\mathrm{d}u) = \int_{-\infty}^{-1} \dots + \int_{-1}^{-1+x_0} \dots$$
$$\leqq c_0\psi([x_0] + 1) \int_{-\infty}^{-1} \psi(-u)\, F(\mathrm{d}u) + x_0\psi(1 + x_0) < \infty$$

so that the assumption (7.7.10) is also true for $F'$. Hence we may assume without loss of generality that $C := F_2(0) > 0$. Then

$$0 \leqq \int_{-\infty}^{-1} \psi(-x_1)\, F_1(\mathrm{d}x_1) \leqq \frac{1}{C} \int_{-\infty}^{0} \int_{-\infty}^{-1} \psi(-x_1 - x_2)\, F_1(\mathrm{d}x_1)\, F_2(\mathrm{d}x_2) := I\,.$$

Applying now Fubini's theorem and noting that $0 \leqq -1 - x_1$ we get

$$I \leqq \frac{1}{C} \int_{-\infty}^{-1} \int_{-\infty}^{-1-x_1} \psi(-x_1 - x_2)\, F_2(\mathrm{d}x_2)\, F_1(\mathrm{d}x_1) = \frac{1}{C} \int_{-\infty}^{-1} \psi(-x)\, F(\mathrm{d}x) < \infty$$

for the latter equality cf. Prop. 1.3.10. This implies $\mathsf{E}\psi(-X_1) < \infty$. Finally, we use part (i) and arrive at the required relation (7.7.9). The proof is complete. □

A special choice of $\psi$ leads to the following assertions.

**Corollary 7.7.9.** (i) *For any $\eta > 0$ we have*

$$\int_{-\infty}^{-1} |x|^\eta \, \mathrm{d}M(x) < \infty \quad \textit{iff} \quad \int_{-\infty}^{-1} |x|^\eta \, \mathrm{d}F(x) < \infty\,.$$

(ii) *For any $s > 0$ we have*

$$\int_{-\infty}^{-1} \mathrm{e}^{-sx} \, \mathrm{d}M(x) < \infty \quad \textit{iff} \quad \int_{-\infty}^{-1} \mathrm{e}^{-sx} \, \mathrm{d}F(x) < \infty\,.$$

□

## 7.8. Power mixtures

The method of mixing represents an essential tool for constructing special classes of In the present section we only treat power mixtures.

Let us consider some d.f. $V$ with lext $(V) \geqq 0$. According to Prop. 4.2.2 the corresponding c.f. $v$ is continuable to $S(0, \infty)$. Further, let $G$ be a d.f. with c.f. $g$. To define power mixtures we distinguish two cases.

(i) Let $G$ be arbitrary and assume that $V$ is lattice with atoms $x_k = k \geqq 0$ and jumps $p_k \geqq 0$. Making use of Prop. 3.1.12 we may introduce the c.f.

$$f(t) = \sum_{k=0}^{\infty} g^k(t)\, p_k = \int_{-0}^{\infty} g^y(t)\, \mathrm{d}V(y) \tag{7.8.1}$$

which belongs to the d.f. $F = \sum_{0}^{\infty} G^{*k} p_k$.

(ii) Let $G$ be inf.div. Then $g^y$, $y \geqq 0$, is a c.f., the corresponding d.f. will be denoted by $G_y$. We introduce the c.f.

$$f(t) = \int_{-0}^{\infty} g^y(t)\, \mathrm{d}V(y)\,, \tag{7.8.2}$$

cf. Suppl. 3.8.8. Its d.f. is given by

$$F(x) = \int_{-0}^{\infty} G_y(x)\, V(\mathrm{d}y)\,.$$

In both cases the d.f. $F$ is called the *power mixture of $G$ with respect to $V$*. The integrals in (7.8.1) and (7.8.2) can be rewritten as

$$f(t) = \int_{-0}^{\infty} \exp\left(iy(-i \log g(t))\right) dV(y) .$$

Here we suppose that $g(t) \neq 0$ and use the principal branch of $\log g(t)$ satisfying $\log g(0) = 0$. Putting $0^\circ = 1$ we see that

$$f(t) = \begin{cases} v(-i \log g(t)) , & g(t) \neq 0 , \\ V(+0) , & g(t) = 0 . \end{cases} \tag{7.8.3}$$

If $V = U_1$ is degenerate then $f$ in (ii) is inf. div. But note that $F$ is no always inf. div. if $V$ is not t inf.div. This is seen by the following example.

**Example 7.8.1.** Let $V$ be the d.f. with atoms $x_1 = 1$, $x_2 = 2$, and jumps

$$p_1 = \exp(-\pi^2/2)\left(1 - \exp(-\pi^2/2)\right) > 0 , \qquad p_2 = 1 - p_1 > 0 .$$

Further we put $G = \Phi_{1,1}$. Then (7.8.1) implies

$$f(t) = p_1 \exp(it - \tfrac{1}{2} t^2) + p_2 \exp(2it - t^2) .$$

Now we easily verify that $f(\pi) = 0$ and, by Prop. 7.1.4, $F$ is not inf.div. □

Next we state a result which is true for both case (i) and case (ii).

**Proposition 7.8.2.** *Let $F$ be the power mixture of $G$ with respect to $V$. If $V$ is inf. div. then so is $F$. In this case, ${}^nF$ is the power mixture of $G$ with respect to ${}^nV$ for all $n \geqq 2$.*

Proof. By assumption we have $V = ({}^nV)^{*n}$ for certain d.f. ${}^nV$ and $n \geqq 2$. In view of (7.8.3) we have

$$f(t) = v(-i \log g(t)) = ({}^nv(-i \log g(t))^n = ({}^nf(t))^n .$$

Since ${}^nv(-i \log g(t))$ is the c.f. of the power mixture of $G$ with respect to ${}^nV$ all assertions are shown. □

We now list three particularly interesting classes of inf.div. power mixtures which use special inf.div.d.f. $V$; the first two belong to case (i) while the third belongs to case (ii).

**Example 7.8.3.** *Compound Poisson d.f.* It is clearly a power mixture where $V$ is a Poisson d.f. with parameter $\lambda > 0$ and $G$ is arbitrary (cf. Ex. 7.1.3). □

**Example 7.8.4.** *Compound negative binomial d.f.* Let $V$ be a negative binomial d.f. with parameters $\eta > 0$ and $p$, $0 < p < 1$. The c.f. $v$ is given by

$$v(t) = \left(\frac{1-p}{1-p\,e^{it}}\right)^{\eta}$$

(cf. Table). Further, let $G$ be an arbitrary d.f. Then from (7.8.3) we get the c.f. of a power mixture

$$f(t) = \left(\frac{1-p}{1-pg(t)}\right)^{\eta} .$$

It is called *compound negative binomial d.f.* In particular, for $\eta = 1$ the d.f. $V$ is a geometric d.f. and $F$ is then called *compound geometric d.f.* □

**Example 7.8.5.** *Compound gamma d.f.* Suppose $V$ is a gamma *d.f.* with parameters $\eta > 0, \lambda > 0$. Then

$$v(t) = \left(1 + \frac{it}{\lambda}\right)^{-\eta}.$$

For an arbitrary inf.div.d.f. $G$ we now obtain the c.f. of a compound gamma d.f.

$$f(t) = \left(1 + \frac{1}{\lambda}\log g(t)\right)^{-\eta}.$$

In particular, for $\eta = 1$ this is the c.f. of a *compound exponential d.f.* □

Next we deal with representations of inf.div. power mixtures of type (ii).

**Lemma 7.8.6.** *Let $G = \mathcal{H}(a_G, H_G)$ be inf.div. and put $G_y = \mathcal{H}(ya_G, yH_G)$, $y \geqq 0$. Then we have the two relations*

$$\frac{1}{y} T_{G_y}(x) = O(1), \qquad y \to +0, \qquad x > 0, \tag{7.8.4}$$

$$\frac{1}{y}\int_{-\infty}^{\infty} \frac{x}{1+x^2}\,\mathrm{d}G_y(x) = O(1), \qquad y \to +0. \tag{7.8.5}$$

Proof. For some sequence $y_n \to +0$ we introduce inf.div.c.f.

$$f_n(t) = \exp\left(\frac{1}{y_n}\left(g^{y_n}(t) - 1\right)\right), \qquad n \geqq 1,$$

which belong to the d.f. $F_n = \mathcal{H}(a_n, H_n)$ where (see Ex. 7.2.12 (iv))

$$a_n = \frac{1}{y_n}\int_{-\infty}^{\infty} \frac{x}{1+x^2}\,\mathrm{d}G_{y_n}(x),$$

$$H_n(x) = \frac{1}{y_n}\int_{-\infty}^{x} \frac{u^2}{1+u^2}\,\mathrm{d}G_{y_n}(u).$$

By the mean value theorem we have $f_n \to g$, as $n \to \infty$, and by Prop. 7.4.1 we see that $a_n \to a_G$. Accordingly, (7.8.5) follows by an indirect argument.

On the other hand, Prop. 7.4.1 also implies $H_n \xrightarrow{c} H_G$ so that, for every $x > 0$

$$\infty > H_G(\infty) \geqq \overline{\lim_{n\to\infty}} \frac{1}{y_n}\int_{|u|\geqq x} \frac{u^2}{1+u^2}\,\mathrm{d}G_{y_n}(u) \geqq \frac{x^2}{1+x^2}\overline{\lim_{n\to\infty}}\frac{1}{y_n} T_{G_{y_n}}(x)$$

whence (7.8.4) follows. □

Before formulating the main result we emphasize that it suffices to consider the case $l_V := \operatorname{lext}(V) = 0$. Indeed, if $l_V > 0$, then we write the power mixture (7.8.2) in the form

$$f(t) = \int_{l_V-0}^{\infty} g^y(t)\,\mathrm{d}V(y) = g^{l_V}(t)\int_{-0}^{\infty} g^y(t)\,\mathrm{d}\hat{V}(y),$$

where the inf.div.d.f. $\hat{V}(x) = V(x + l_V)$ satisfies $\operatorname{lext}(\hat{V}) = 0$.

**Proposition 7.8.7.** *Let $G$ and $V = \mathcal{L}(a_V, 0, 0, N_V)$ be inf.div. and assume $l_V = 0$. Then the power mixture $F$ of $G$ with respect to $V$ is inf.div. and admits the representation $F = \mathcal{L}(a_F, 0, M_F, N_F)$ where*

$$a_F = \int_{+0}^{\infty} \int_{-\infty}^{\infty} \frac{x}{1 + x^2} \, dG_y(x) \, N_V(dy) ,$$

$$M_F(x) = \int_{+0}^{\infty} G_y(x) \, N_V(dy) , \qquad x < 0 ,$$

$$N_F(x) = \int_{+0}^{\infty} \big(G_y(x) - 1\big) \, N_V(dy) , \qquad x > 0 .$$

*All three integrals exist finitely.*

For the proof we need the following statement.

**Lemma 7.8.8.** *Under the conditions of Prop. 7.8.7 we have*

$$\log f(t) = \int_{+0}^{\infty} \big(g^y(t) - 1\big) \, dN_V(y) . \tag{7.8.6}$$

Proof. From Prop. 7.8.2 we get the representation

$${}^n f(t) = \int_{-0}^{\infty} g^y(t) \, d\big({}^n V(y)\big) .$$

Accordingly, we can write

$$n\big({}^n f(t) - 1\big) = \int_{+0}^{\infty} \big(g^y(t) - 1\big) \, n \, d\big({}^n V(y)\big)$$

$$= \int_{+0}^{\infty} \left(g^y(t) - 1 - \frac{y \log g(t)}{1 + y^2}\right) \frac{1 + y^2}{y^2} \, dH_n(y)$$

$$+ \log g(t) \int_{+0}^{\infty} \frac{y}{1 + y^2} \, n \, d\big({}^n V(y)\big) = I_n^{(1)} + I_n^{(2)} ,$$

say; here we put $H_n(x) = 0$, $x \leq 0$, and

$$H_n(x) = \int_{+0}^{x} \frac{u^2}{1 + u^2} \, n \, d\big({}^n V(u)\big) , \qquad x > 0 .$$

From Cor. 7.1.5 we obtain $n\big({}^n f(t) - 1\big) \to \log f(t)$. On the other hand, by Cor. 7.2.5 we have $I_n^{(2)} \to a_V \log g(t)$. But Cor. 7.2.5 also implies that $H_n \xrightarrow{c} H_V$ so that, using the extended Helly-Bray theorem,

$$\lim_{n \to \infty} I_n^{(1)} = \int_{+0}^{\infty} \left(g^y(t) - 1 - \frac{y \log g(t)}{1 + y^2}\right) \frac{1 + y^2}{y^2} \, dH_V(y)$$

$$= \int_{+0}^{\infty} \big(g^y(t) - 1\big) \, dN_V(y) - \log g(t) \int_{+0}^{\infty} \frac{y}{1 + y^2} \, dN_V(y) .$$

Since $l_V = 0$ the latter integral equals $a_V$, see Prop. 7.3.9. Combining these results we arrive at the required formula. □

Proof of proposition 7.8.7. First we show that the integrals exist. Put $\psi(y) = \min\{1, y\}$, $y > 0$, then from (7.8.4) it follows that

$$\int_{+0}^{\infty} T_{G_y}(x)\, N_V(\mathrm{d}y) \leqq C \int_{+0}^{\infty} \psi(y)\, N_V(\mathrm{d}y)\,, \qquad x > 0\,.$$

Remembering that $l_V = 0$ and using Prop. 7.3.9 we see that the latter integral is finite.
In a similar way we see from (7.8.5) that

$$\int_{+0}^{\infty} \left| \int_{-\infty}^{\infty} \frac{x}{1+x^2}\, \mathrm{d}G_y(x) \right| N_V(\mathrm{d}y) \leqq C \int_{+0}^{\infty} \psi(y)\, N_V(\mathrm{d}y) < \infty$$

Next we conclude from L. 7.8.8 that

$$\log f(t) = \int_{+0}^{\infty} \int_{-\infty}^{\infty} \mathrm{e}^{itx}\, \mathrm{d}_x\big(G_y(x) - U_0(x)\big)\, N_V(\mathrm{d}y)\,.$$

As in Sect. 7.2 we set $k(t, x) = \mathrm{e}^{itx} - 1 - \dfrac{itx}{1+x^2}$. Then it is easily verified by the definition of $a_F$ that

$$\log f(t) = ita_F + \int_{+0}^{\infty} \int_{-\infty}^{\infty} k(t, x)\, \mathrm{d}_x\big(G_y(x) - U_0(x)\big)\, N_V(\mathrm{d}y)\,.$$

Introducing the function $\hat{N}_V(y) = \int_{+0}^{y} \psi(t)\, N_V(\mathrm{d}t)$, $y > 0$, and using $k(t, 0) = 0$ we obtain

$$\log f(t) = ita_F + \int_{+0}^{\infty} \int_{|x|>0} k(t, x)\, \mathrm{d}_x \left[\frac{G_y(x) - U_0(x)}{\psi(y)}\right] \hat{N}_V(\mathrm{d}y)\,.$$

The function $\hat{N}_V$ is bounded and monotone; $\dfrac{1}{\psi(y)}\big(G_y(x) - U_0(x)\big)$ is bounded in $y$ by L. 7.8.6, it is monotone for $x > 0$ and $x < 0$, resp. Hence we are able to apply Prop. A.4 so that

$$\log f(t) = ita_F + \int_{|x|>0} k(t, x)\, \mathrm{d}_x \int_{+0}^{\infty} \big(G_y(x) - U_0(x)\big)\, N_V(\mathrm{d}y)\,.$$

Finally, the uniqueness of the Lévy representation for inf.div.c.f. completes the proof of the asserted representation. □

We now consider power mixtures of type (i). Here we use the fact that an inf.div. lattice d.f. $V$ with lext $(V) = 0$ is a compound Poisson d.f. with

$$\log v(t) = \sum_{k=1}^{\infty} (\mathrm{e}^{itk} - 1)\, q_k$$

(see L. 7.6.13), that is, the corresponding spectral function $N_V$ is also lattice with atoms $x_k = k \geqq 1$ and jumps $q_k \geqq 0$.

**Corollary 7.8.9.** *Let $G$ be arbitrary and assume that $V$ is lattice and inf.div. with* lext $(V = 0$. *If $q_k$, $k \geqq 1$, are the jumps of $N_V$, then the power mixture $F$ of $G$ with respect to $V$ can be written as $F = \mathcal{L}(a_F, 0, M_F, N_F)$ where*

$$a_F = \sum_{k=1}^{\infty} \int_{-\infty}^{\infty} \frac{x}{1+x^2} \, \mathrm{d}G^{*k}(x) \, q_k ,$$

$$M_F(x) = \sum_{k=1}^{\infty} G^{*k}(x) \, q_k , \qquad x < 0 ,$$

$$N_F(x) = \sum_{k=1}^{\infty} \left(G^{*k}(x) - 1\right) q_k , \qquad x > 0 .$$

Obviously, for the special case $V = P_\lambda$, $\lambda > 0$, this statement leads to the formulae of Ex. 7.2.12, (iv).

The proof is quite similar to the proof of Prop. 7.8.7. Note that $\sum_{k=1}^{\infty} q_k = -N(1) < \infty$ implies the convergence of the above sums. Moreover, (7.8.6) can be rewritten as

$$\log f(t) = \sum_{k=1}^{\infty} \left(g^k(t) - 1\right) q_k . \qquad \square$$

## 7.9. Supplement

The proofs of the results contained in 7.9.1.—7.9.4. can be found in standard books such as Lukacs (1970).

**7.9.1.** *A d.f. $F$ is inf.div. iff it is the complete limit of a sequence of compound Poisson d.f.* (de Finetti).

**7.9.2.** *A d.f. $F$ is inf.div. iff it is the complete limit of a sequence of finite convolutions of Poisson type d.f.*

**7.9.3.** *A d.f. $F$ is inf.div. if every component of $F$ is decomposable* (Hinčin). The converse is not true; for an example see Lukacs (1970).

**7.9.4.** *A d.f. $F$ with finite variance is inf.div. iff there exist a constant $a \in R_1$ and a non-decreasing and bounded function $K$ such that the c.f. $f$ of $F$ permits the representation*

$$\log f(t) = \mathrm{i}at + \int_{-\infty}^{\infty} \left(\mathrm{e}^{\mathrm{i}tu} - 1 - \mathrm{i}tu\right) \frac{\mathrm{d}K(u)}{u^2} ;$$

*this representation is unique* (Kolmogorov).

**7.9.5.** In connection with Cor. 7.2.10, it is noteworthy that there exist c.f. $f$ which are not inf.div. although $|f|$ is inf.div. (see Lukacs (1970)).

**7.9.6.** Prop. 7.4.1 can be reformulated in terms of the Lévy representation as follows (for a detailed proof see Gnedenko and Kolmogorov (1959)).

*Let* $F_n = \mathcal{L}(a_n, \gamma_n^2, M_n, N_n)$, $n \geqq 1$, *be inf.div. Then we have* $F_n \xrightarrow{c} F$ *for a certain d.f.* $F$ *iff there exist constants* $a \in \mathbb{R}_1$ *and* $\gamma^2 \geqq 0$, *and spectral functions* $M$ *and* $N$ *such that*

$$a_n \to a\,, \qquad M_n \xrightarrow[(-\infty,0)]{} M\,, \qquad N_n \xrightarrow[(0,\infty)]{} N\,,$$

$$\lim_{\varepsilon\to+0} \underline{\lim}_{n\to\infty} \Big( \int_{-\varepsilon}^{-0} u^2 \, dM_n(u) + \gamma_n^2 + \int_{+0}^{\varepsilon} u^2 \, dN_n(u) \Big) = \lim_{\varepsilon\to+0} \overline{\lim}_{n\to\infty} (\ldots) = \gamma^2\,.$$

*In either case,* $F = \mathcal{L}(a, \gamma^2, M, N)$ *is inf.div.*

**7.9.7.** An easy consequence of Cor. 7.6.11 runs as follows. *Let* $P_\lambda$ *be a Poisson d.f. with atoms* $x_k = k$, $k \geqq 0$ *and* $F = \mathcal{L}(a, 0, 0, N)$ *an inf.div. d. f. satisfying* lext $(F) = 0$. *If* $x_1 > 1$ *and* $F(x) = P_\lambda(x)$ *for* $x \leqq x_1$, *then* $F = P_\lambda$ (Rossberg, Jesiak, and Siegel (1981 a)).

**7.9.8.** *Let* $X \sim F = \mathcal{L}(a, \gamma^2, M, N)$ *be inf.div. and set, for* $h > 0$,

$$\psi(h) = \frac{1}{h^2} \int\limits_{0<u<h} u^2 \, d[N(u) - M(-u)] + \int\limits_{u\geqq h} d[N(u) - M(-u)]\,.$$

*Then we have for the concentration function* $Q$ *of* $X$

$$Q(X, h) \leqq \left(1 + \frac{\gamma^2}{\eta h^2}\right)^{-\eta/2} \left(\frac{\psi(h)}{c_1(1-\eta)}\right)^{\eta-1/2},$$

*where* $0 < \eta < 1$ *is arbitrary and* $c_1 = 33{,}0659 \ldots$

As consequences we obtain the estimates

$$Q(X, h) \leqq \left(\frac{\gamma^2}{h^2} + \frac{\psi(h)}{c_1}\right)^{-1/2}, \qquad Q(X, h) \leqq c_2 \left(1 + \frac{\gamma^2}{h^2} + \frac{\psi(h)}{c_1}\right)^{-1/2}$$

where $c_2 = 1{,}2019 \ldots$ (Siegel (1981 b)). Further estimates for the concentration function of inf.div.r.v. can be found in Petrov (1975).

**7.9.9.** *Let* $F = \mathcal{L}(a, \gamma^2, M, N)$ *be inf.div. and let* $h\colon (0, \infty) \to (0, \infty)$ *be continuous and strictly increasing. Further, assume that for every* $\eta$, $0 < \eta < 1$,

$$h^{-1}(\eta y)/h^{-1}(y) = O\left(\frac{1}{y}\right), \qquad y \to y_0 + 0\,,$$

*where* $h^{-1}$ *denotes the inverse function of* $h$ *and* $y_0 = h^{-1}(+0)$. *Then the condition*

$$\underline{\lim}_{x\to\infty} \frac{-\log M(-x)}{x h(x)} = c\,, \qquad 0 < c \leqq \infty\,,$$

*is equivalent to*

$$\underline{\lim}_{x\to\infty} \frac{-\log F(-x)}{x h(\log x)} = c\,, \qquad 0 < c \leqq \infty$$

(Antonov (1980 b)).

**7.9.10.** *Let* $V$ *be a d.f. satisfying* lext $(V) \geqq 0$. *Then the mixture of exponential c.f.*

$$f(t) = \int_0^\infty \frac{y}{y - it} \, dV(y)$$

*is an inf.div.c.f.* (Steutel (1967)).

Applying arguments similar to those in the proof of the above assertion it can be proved that mixtures of geometric c.f. are inf.div.: *If the d.f. V is concentrated on* $[0, 1]$, *then the c.f.*

$$f(t) = \int_0^1 \frac{1-y}{1-y\,e^{it}}\,dV(y)$$

*is inf.div.*

**7.9.11.** *Let* $F = \mathcal{L}(a, \gamma^2, M, N)$ *be inf.div. Then the following assertions are equivalent:*

(i) *F is a compound Poisson d.f. (but for a shift),*
(ii) $\gamma^2 = 0$, $M(-0) < \infty$, $N(+0) > -\infty$,
(iii) *F is non-continuous.*

The equivalence (i) $\Leftrightarrow$ (ii) is obvious by Ex. 7.1.3 and Ex. 7.2.12 (iv). The relation (ii) $\Leftrightarrow$ (iii) was first proved by Doeblin, see Petrov (1975).

**7.9.12.** The following general problems concerning the unique determination of inf.div.d.f. are due to Zolotarev (1978).

**Problem.** Let $F$ be inf.div. Characterize sets $S = S(F) \subset R_1$ such that the values $F(x)$, $x \in S$, define $F$ uniquely.

**Problem.** Let $S \subset R_1$ be arbitrary. Characterize the class $\boldsymbol{F} = \boldsymbol{F}(S)$ of inf.div.d.f. which are uniquely determined provided that their values $F(x)$, $x \in S$, are given. In particular, find conditions such that $\boldsymbol{F}(S) \neq \emptyset$.

**7.9.13.** According to Ibragimov's theorem 7.6.2, all inf.div.d.f. $F > 0$ whose c.f. are continuable to the upper half plane belong to $\boldsymbol{F}((-\infty, x_0))$ for every $x_0 \in R_1$. A full characterization of this class in terms of domains of partial attractions can be found in Rossberg (1979).

**Problem.** Give a purely analytic characterization of the class $\boldsymbol{F}\big((-\infty, x_0)\big)$.

**7.9.14.** Ibragimov's theorem 7.6.2 has been modified by Rossberg (1982) and Jesiak (1978) in two different directions. Rossberg's result has been formulated as Prop. 4.7.2, and Jesiak's runs as follows.

**Proposition 7.9.1.** *Let F be some d.f. with c.f.f. Assume*

(i) $F(x) = O(e^{sx})$, $x \to -\infty$, *for all* $s > 0$,

(ii) $\lim\limits_{s\to\infty} \dfrac{\log f(is)}{s \log s} = \infty$,

(iii) $n(f, R) = O(\log R)$, $R \to \infty$,

*where* $n(f, R)$ *denotes the number of zeros of f in the semi-disk* $|z| \leqq R$, $\operatorname{Im} z \geqq 0$.

*If G is a d.f. such that* $n(g, R) = O(\log R)$, $R \to \infty$, *then* $G(x) = F(x)$, $x < x_0$ *implies* $G = F$.

Note that neither of the results just mentioned implies Ibragimov's theorem but both together do so.

**7.9.15.** A full generalization of Ibragimov's theorem has been given by Ostrovskij (1979); it concerns certain functions of bounded variation (cf. Suppl. 4.10.10).

**7.9.16.** Prop. 7.4.6 can be generalized in the following way.
*Let* $F_k = \mathscr{L}(a_k, 0, M_k, N_k)$, $k = 1, 2$, *be inf.div. such that*

$$M_k(u) = \sum_{j=1}^{m_k} d_{kj} |u|^{-\alpha_{kj}}, \qquad u < 0,$$

*for certain real numbers* $d_{kj} > 0$ *and* $0 < \alpha_{kj} < 2$. *Then* $F_1(x) = F_2(x)$, $x < x_0$, *implies* $F_1 = F_2$ (Bischoff (1982); it can easily be seen that the restriction $0 < \alpha_{kj} < 1$ made there is superfluous; cf. our proof of Prop. 7.6.4.)

**7.9.17. Problem.** Let $F_k = \mathscr{L}(a_k, \gamma_k^2, M_k, N_k)$, $k = 1, 2$, be inf.div. and assume lext $(F_1) = -\infty$. Then the question arises of whether $F_1(x) = F_2(x)$, $x < x_0$, implies $M_1 = M_2$ and $\gamma_1^2 = \gamma_2^2$. More generally, find asymptotic relations between the left tails of $F_1$ and $F_2$ which permit these conclusions.

For partial answers to this problem see Prop. 7.6.5 and Suppl. 7.9.18.

**7.9.18.** Prop. 7.6.5 also permits the following modifications.
*Let* $F_k = \mathscr{L}(a_k, \gamma_k^2, M_k, N_k)$, $k = 1, 2$, *be inf.div. such that* lext $(F_1) = -\infty$ *and* $F_1(x) = O(e^{sx})$, $x \to -\infty$, *for all* $s > 0$. *Assume, further, that at least one of the following two conditions holds true.*

(i) $|M_1(u) - M_2(u)| \leqq C_0\, e^{-u^2}$, $u \leqq u_0 < 0$,

(ii) $M_1(u) \leqq M_2(u)$ *or* $M_2(u) \leqq M_1(u)$, $u \leqq u_0 < 0$.

*Then* (7.6.2) *implies* $M_1 = M_2$ *and* $\gamma_1^2 = \gamma_2^2$; see Rossberg, Jesiak, and Siegel (1981).

Here the Phragmén-Lindelöf theorem Prop. C.1.13 is essentially used. The given statement also admits the Cor. 7.6.6—7.6.8.

7.9.19. In view of Prop. 7.6.10 it is of interest to characterize the subclass of inf.div. d.f. $F = \mathscr{L}(a, 0, 0, N)$ satisfying lext $(F) = 0$, $0 <$ rext $(N) = r_N < \infty$, because they are uniquely defined by their values on $(0, r_N)$. This problem is solved by Cor. 7.7.3.

## Notes and bibliographical data

Today, infinite divisibility is a vast field, and it is hardly possible to give a comprehensive survey. But many standard books contain material concerning these d.f. As examples we mention Gnedenko and Kolmogorov (1959), Lukacs (1970), and Petrov (1975). Steutel (1979) points out applications of inf.div.d.f. in different fields. Here we only give some references for recent results.

L.7.4.3. was stated in Rossberg and Siegel (1975) Prop. 7.4.4 is due to Titov (1981 a); he gives another proof which also shows that the assumption (7.4.1) is superfluous.

The results of Sect. 7.5 follow Jesiak (1978, 1979 a, b).

Th. 7.6.1 as the first continuation theorem for inf.div.d.f. has been proved by Rossberg (1974); its generalization (Prop. 7.6.2) is due to Ibragimov (1977). Prop. 7.6.4 is also due to Rossberg (1979), and the statements 7.6.5 to 7.6.8 have been shown by Jesiak (1978, 1981). Prop. 7.6.10 is contained in Rossberg, Jesiak, and Siegel (1981 a).

Many results — or parts of them — contained in Sect. 7.7 are due to various authors, we mention Kruglov (1970, 1974 a, b), Steutel (1974), Antonov (1980 b), Esseen (1981), and Kruglov and Antonov (1982). Prop. 7.7.8. is a counterpart to a two-sided result proved by Kruglov (1970).

Prop. 7.8.2 is a well-known fact, see Feller (1971).

# Chapter 8. Stable distributions

In the present chapter we investigate an important subclass of inf.div.d.f. which plays a central role in the summation theory for i.i.d.r.v.

In Sect. 8.1, a canonical representation for a stable c.f. $f$ will be derived; for this purpose we successfully apply the theory of regularly varying functions. In particular, formulas for $|f|$ and for the Lévy spectral functions $M$ and $N$ can easily be concluded.

Most of the results contained in Sect. 8.2 and 8.3 are well-known; they will be applied in Sect. 8.4 to prove several recent uniqueness theorems for stable d.f.

## 8.1. Basic properties. Canonical representation

**Definition 8.1.1.** A d.f. $F$ is said to be *stable* if for every $n \geqq 2$ there exist constants $a_n \in R_1$ and $b_n > 0$ such that

$$F(x) = F^{*n}(b_n x + a_n) . \tag{8.1.1}$$

The corresponding c.f. $f$ is then also called stable. □

We see that a d.f. $F$ is stable iff, for every collection $X_0, X_1, \ldots, X_n$ of i.i.d.r.v., $X_k \sim F$, $0 \leqq k \leqq n$, $n \geqq 2$, there exist constants $a_n$ and $b_n > 0$ such that

$$X_0 \sim \frac{1}{b_n}(X_1 + \ldots + X_n - a_n) .$$

Further, a c.f. $f$ is stable iff there exist constants $c_n$ and $b_n > 0$ such that, for every $n \geqq 2$,

$$f(t) = e^{itc_n} f^n\left(\frac{t}{b_n}\right) \tag{8.1.2}$$

where the $b_n$ are the same as in (8.1.1) while $c_n = -a_n/b_n$.

It is clear that a stable d.f. is inf.div. where the "roots" ${}^nF$ are distinguished by a particular property, namely that they are of the same type as $F$. Next we wish to derive step by step the special properties of stable c.f. which characterize the class of stable c.f. within the class of all inf.div.c.f. — In all that follows we exclude degenerate laws from consideration though they are trivially stable.

**Lemma 8.1.2.** *Let $f$ be a stable c.f. Then*

(i) *for the constants $b_n$ occurring in* (8.1.2) *we have*

$$b_n \to \infty , \qquad b_n/b_{n+1} \to 1 , \quad \text{as} \quad n \to \infty ,$$

(ii) *the function $|f(t)|$ is non-increasing for $t > 0$.*

Proof. (i) We carry an indirect proof and assume $b_{n'} \to b < \infty$ for some subsequence $(n')$ of natural numbers. Then we choose $t_0$ such that $|f(t_0)| \neq 1$ and obtain, using (8.1.2),

$$0 = \lim_{n' \to \infty} |f(t_0)|^{n'} = \lim_{n' \to \infty} |f(t_0 b_{n'})| = |f(t_0 b)| > 0;$$

the inequality follows from the fact that inf.div.c.f. have no zeros by Prop. 7.1.4. This contradiction yields $b_n \to \infty$.

Next we assume that $b_{n'}/b_{n'+1} \to b < 1$ for some subsequence $(n')$. Then we get for all $t$, using (8.1.2) twice and letting $n' \to \infty$,

$$|f(t)| = \left|f\left(\frac{t}{b_{n'+1}}\right)\right|^{n'+1} = \left|f\left(t\frac{b_{n'}}{b_{n'+1}}\right)\right| \cdot \left|f\left(\frac{t}{b_{n'+1}}\right)\right| \to |f(tb)| .$$

Therefore $|f(t)| = |f(tb)|$, and by induction we get $|f(t)| = |f(tb^k)|$ for all $k \geqq 1$. Thus choosing $t_0$ as before we obtain the contradiction

$$1 = |f(0)| = \lim_{k \to \infty} |f(t_0 b^k)| = |f(t_0)| \neq 1;$$

similarly we can exclude the case $b_{n'}/b_{n'+1} \to b > 1$. Accordingly, $b_n/b_{n+1} \to 1$.

(ii) Let $0 < t_1 < t_2$. Then, for given $n$, we put $m_n = \max\{j: b_j \leqq b_n(t_2/t_1)\}$. From $b_{m_n} \leqq b_n(t_2/t_1) < b_{m_n+1}$ and by virtue of (i) it follows then that $b_{m_n}/b_n \to t_2/t_1$ as $n \to \infty$. Further, since $t_2/t_1 > 1$ we have $m_n \geqq n$. This yields

$$|f(t_1)| = \left|f\left(\frac{t_1}{b_n}\right)\right|^n \geqq \left|f\left(\frac{t_1}{b_n}\right)\right|^{m_n} = \left|f\left(t_1\frac{b_{m_n}}{b_n}\right)\right| \to |f(t_2)|$$

because c.f. are continuous. □

**Proposition 8.1.3.** *Let $f$ be a non-degenerate stable c.f. Then there exist constants $b > 0$ and $\alpha$, $0 < \alpha \leqq 2$, such that*

$$|f(t)| = e^{-b|t|^\alpha} . \tag{8.1.3}$$

Proof. By L. 8.1.2 (ii), the function $H(t) := \log |f(t)|$ is monotone for $t > 0$. Further the formula (8.1.2) yields

$$\lim_{n \to \infty} \frac{1}{n} H(b_n t) = \lim_{n \to \infty} \frac{1}{n} \log |f(b_n t)| = \log |f(t)| .$$

Thus the properties of regularly varying functions tell us that

$$\log |f(t)| = bt^\alpha , \qquad t \geqq 0 ,$$

for some real $b$ and $\alpha$, cf. Cor. 2.5.5. Since $\log |f(t)| < 0$ for certain $t$ we get $b < 0$; since $\log |f(t)|$ is non-increasing by L. 8.1.2 (ii), $\alpha > 0$ follows. Further, in view of $|f(-t)| = |f(t)|$ we obtain

$$|f(t)| = e^{-b|t|^\alpha} \quad \text{for some} \quad b > 0, \alpha > 0 .$$

Finally, assume $\alpha > 2$. Then we have

$$|f(t)| = 1 + o(t^2) , \qquad t \to 0 ,$$

and Cor. 3.1.10 implies that $|f| = 1$. Since this case has been excluded we get $\alpha \leqq 2$. □

Obviously, for a given non-degenerate stable d.f. $F$ with c.f. $f$ the parameter $\alpha$ in (8.1.3) is uniquely determined. It is called the *characteristic exponent* of $F$ and of $f$.

**Corollary 8.1.4.** *Let $f$ be stable with characteristic exponent $\alpha$.*

(i) *For the constants $b_n$ in (8.1.2) we have $b_n = n^{1/\alpha}$.*

(ii) *If $\alpha = 2$, then $f$ is a normal c.f.*

Proof. (i) is an easy consequence of (8.1.2) and (8.1.3).

(ii) The symmetrization $F^s$ of $F$ has the c.f.

$$f^s(t) = |f(t)|^2 = e^{-2bt^2},$$

so that $F^s$ is a normal d.f. Therefore, by Cramér's theorem 4.6.4, $F$ is also normal. □

Now we wish to derive the special form which Lévy's canonical representation takes for stable c.f. By virtue of Cor. 8.1.4 (ii), it suffices to consider the case $\alpha < 2$.

**Proposition 8.1.5.** *Let $F = \mathscr{L}(a, \gamma^2, M, N)$ be stable with characteristic exponent $\alpha < 2$ Then*

$$\gamma^2 = 0\,, \qquad M(u) = d_1\,|u|^{-\alpha}\,, \qquad N(u) = -d_2 u^{-\alpha} \tag{8.1.4}$$

*for certain constants $d_1, d_2 \geqq 0$ such that $d_1 + d_2 > 0$.*

Proof. Firstly, Cor. 7.3.3 tells us that

$$\gamma^2 = -2 \lim_{t\to\infty} \frac{\log|f(t)|}{t^2}\,.$$

Accordingly, $\gamma^2 = 0$ by Prop. 8.1.3. Further, from $F(x) = F^{*n}(b_n x + a_n)$ we conclude, by virtue of Cor. 7.2.8 and 7.2.11, that for a certain $a' \in R_1$

$$\mathscr{L}(a, 0, M, N) = \mathscr{L}\big(a', 0, nM(b_n u), nN(b_n u)\big)\,, \qquad n \geqq 1\,.$$

Since the Lévy representation is unique we get for all $n \geqq 1$

$$M(u) = nM(b_n u)\,, \qquad u < 0; \qquad N(u) = nN(b_n u)\,, \qquad u > 0\,.$$

Now Cor. 2.5.5 tells us that either $M = 0$ or $M(u) = d_1\,|u|^r$ for some $d_1 > 0$ and a certain real $r$. Moreover, for all $u < 0$ and $n \geqq 1$ we obtain, by virtue of $b_n = n^{1/\alpha}$,

$$d_1\,|u|^r = M(u) = nM(n^{1/\alpha}u) = nd_1 n^{r/\alpha}\,|u|^r\,,$$

whence $r = -\alpha$ follows. In the same manner se see that $N(u) = -d_2 u^{-\alpha}$ for some $d_2 \geqq 0$. Finally, the case $d_1 + d_2 = 0$ is impossible since degenerate laws have been excluded once for all. □

The special form of the spectral functions $M$ and $N$ makes it possible to give an explicit canonical representation for stable c.f. which will be derived later in this section.

For this purpose we first show that the parameter $b$ occurring in (8.1.3) can be expressed in terms of $\alpha$ and the constants $d_1$ and $d_2$.

**Lemma 8.1.6.** *Let $f$ be stable with $\alpha < 2$. Then*

$$b = \alpha(d_1 + d_2) \int_0^\infty (1 - \cos x)/x^{1+\alpha}\, dx\,.$$

Proof. We insert (8.1.4) in Lévy's canonical formula and obtain by (8.1.3), for all $t > 0$,

$$-bt^\alpha = \log|f(t)| = d_1\alpha \int_0^\infty \frac{\cos tu - 1}{u^{1+\alpha}}\, du + d_2\alpha \int_0^\infty \frac{\cos tu - 1}{u^{1+\alpha}}\, du$$

$$= -\alpha(d_1 + d_2)\, t^\alpha \int_0^\infty (1 - \cos x)/x^{1+\alpha}\, dx\,. \qquad \square$$

Next we give an explicit formula for $\arg f(t)$.

**Proposition 8.1.7.** *Let $f$ be stable with $\alpha < 2$. Then*

$$\arg f(t) = at - \alpha(d_1 - d_2)\,|t|^\alpha \operatorname{sgn} t \cdot \psi_\alpha(t) \tag{8.1.5}$$

*for some $a \in R_1$ where*

$$\psi_\alpha(t) := \begin{cases} \int\limits_0^\infty \sin x/x^{1+\alpha}\,dx\,, & \text{if } 0 < \alpha < 1\,, \\ -\log|t|\,, & \text{if } \alpha = 1\,, \\ \int\limits_0^\infty (\sin x - x)/x^{1+\alpha}\,dx\,, & \text{if } 1 < \alpha < 2\,. \end{cases}$$

Proof. It is enough to prove (8.1.5) for $t > 0$ in view of $\arg f(-t) = -\arg f(t)$. — According to (8.1.4) and Lévy's formula (7.2.3) we have for $t > 0$

$$\arg f(t) = at - \alpha(d_1 - d_2)\int\limits_0^\infty \left(\sin tu - \frac{tu}{1+u^2}\right)\frac{du}{u^{1+\alpha}}\,. \tag{8.1.6}$$

Assume now $\alpha < 1$. Then the integral $\int\limits_0^\infty \frac{u}{1+u^2}\frac{du}{u^{1+\alpha}}$ exists and we have for a certain $a' \in R_1$

$$\arg f(t) = a't - \alpha(d_1 - d_2)\int\limits_0^\infty \frac{\sin tu}{u^{1+\alpha}}\,du = a't - \alpha(d_1 - d_2)\,t^\alpha\,\psi_\alpha(t)\,.$$

If $\alpha > 1$ we make use of $u/(1+u^2) = u - u^3/(1+u^2)$. Since the integral

$$\int\limits_0^\infty \frac{u^3}{1+u^2}\frac{du}{u^{1+\alpha}}$$

exists we get from (8.1.6) for some $a''$

$$\arg f(t) = a''t - \alpha(d_1 - d_2)\,t^\alpha \int\limits_0^\infty (\sin x - x)/x^{1+\alpha}\,dx$$

whence the assertion follows from the definition of $\psi_\alpha$.

If $\alpha = 1$, then we first note that for any $\varepsilon > 0$ and fixed $t > 0$

$$\int\limits_\varepsilon^\infty \left(\sin tu - \frac{tu}{1+u^2}\right)\frac{du}{u^2} = \int\limits_\varepsilon^{\varepsilon/t} \frac{\sin tu}{u^2}\,du + \int\limits_{\varepsilon/t}^\infty \frac{\sin tu}{u^2}\,du - \int\limits_\varepsilon^\infty \frac{t\,du}{u(1+u^2)}$$

$$= \int\limits_\varepsilon^{\varepsilon/t} \frac{\sin tu}{u^2}\,du + t\int\limits_\varepsilon^\infty \left(\frac{\sin u}{u^2} - \frac{1}{u(1+u^2)}\right)du\,.$$

One sees that the latter integral tends to a finite limit as $\varepsilon \to +0$ while

$$\lim_{\varepsilon\to+0} \int\limits_\varepsilon^{\varepsilon/t} \frac{\sin tu}{u^2}\,du = t \lim_{\varepsilon\to+0} \int\limits_\varepsilon^{\varepsilon/t} \frac{\sin tu}{tu}\frac{du}{u} = t \lim_{\varepsilon\to+0} \int\limits_\varepsilon^{\varepsilon/t} \frac{du}{u} = -t\log t\,.$$

In this way we get by virtue of (8.1.6), for some $a''' \in R_1$,

$$\arg f(t) = a'''t - (d_1 - d_2)\,t(-\log t)\,.$$

□

Before formulating the main result of the present section we prove the following purely analytic formulas.

**Lemma 8.1.8.** *For $0 < \alpha < 1$ it holds*

$$\arg \int_0^\infty (e^{ix} - 1)/x^{1+\alpha}\, dx = -\frac{\pi\alpha}{2},$$

*while for $1 < \alpha < 2$*

$$\arg \int_0^\infty (e^{ix} - 1 - ix)/x^{1+\alpha}\, dx = -\frac{\pi\alpha}{2}.$$

Proof. Let $0 < r < R < \infty$. We make use of the Cauchy theorem on contour integration taking as the contour of integration the intervals $(r, R)$ and $(ir, iR)$ and the circular arcs of radii $r$ and $R$ with center at the origin. Integrating $g(z) = (e^{iz} - 1)/z^{1+\alpha}$ on this contour, it is not hard to show that the integrals on the circular arcs approach 0 as $r \to +0$ and $R \to \infty$. Accordingly, for $\alpha < 1$ we find

$$\arg \int_0^\infty \frac{e^{ix} - 1}{x^{1+\alpha}}\, dx = \arg \int_0^\infty \frac{e^{-y} - 1}{(iy)^{1+\alpha}}\, d(iy) = \arg i^{-\alpha} = -\frac{\pi\alpha}{2};$$

the second assertion follows analogously. □

**Theorem 8.1.9.** *A non-degenerate d.f. $F$ with c.f. $f$ is stable iff there exist real constants $\alpha$, $\beta$, $a$, and $b$ with $0 < \alpha \leqq 2$, $|\beta| \leqq 1$, and $b > 0$, such that*

$$\log f(t) = iat - b\,|t|^\alpha\,[1 + i\beta \operatorname{sgn} t \cdot \omega_\alpha(t)] \tag{8.1.7}$$

*where*

$$\omega_\alpha(t) := \begin{cases} \tan \dfrac{\pi\alpha}{2} & \text{if } \alpha \neq 1, \\ -\dfrac{2}{\pi} \log |t| & \text{if } \alpha = 1. \end{cases}$$

*In either case, $\alpha$ is the characteristic exponent of $F$.*

Proof. Let $f$ be stable with characteristic exponent $\alpha$.

If $\alpha = 2$, then $f$ is normal and it has indeed the representation (8.1.7); note that in this case the parameter $\beta$ is not specified. — To treat the case $\alpha < 2$ we first conclude from (8.1.3) and (8.1.5) that

$$\log f(t) = iat - b\,|t|^\alpha \left[1 + i\frac{\alpha}{b}\,(d_1 - d_2) \operatorname{sgn} t \cdot \psi_\alpha(t)\right];$$

in the following we put

$$\beta = \frac{d_1 - d_2}{d_1 + d_2}. \tag{8.1.8}$$

If $0 < \alpha < 1$, then L. 8.1.6 tells us that

$$\frac{\alpha}{b}(d_1 - d_2)\,\psi_\alpha(t) = \beta\,\frac{\int_0^\infty \sin x/x^{1+\alpha}\,\mathrm{d}x}{\int_0^\infty (1 - \cos x)/x^{1+\alpha}\,\mathrm{d}x}$$

$$= \beta\,\frac{\operatorname{Im}\int_0^\infty (\mathrm{e}^{\mathrm{i}x} - 1)/x^{1+\alpha}\,\mathrm{d}x}{-\operatorname{Re}\int_0^\infty (\mathrm{e}^{\mathrm{i}x} - 1)/x^{1+\alpha}\,\mathrm{d}x}$$

$$= -\beta\tan\left(\arg\int_0^\infty \frac{\mathrm{e}^{\mathrm{i}x} - 1}{x^{1+\alpha}}\,\mathrm{d}x\right) = \beta\tan\frac{\pi\alpha}{2}\,;$$

the latter equality is due to L. 8.1.8. Hence, (8.1.7) follows. The cases $1 < \alpha < 2$ and $\alpha = 1$ can be treated in a similar way, for $\alpha = 1$ we note that $\int_0^\infty (1 - \cos x)/x^2\,\mathrm{d}x = \pi/2$.

To prove the converse statement we put $b_n = n^{1/\alpha}$ and

$$c_n = \begin{cases} a(1 - n/b_n) & \text{if } \alpha \neq 1\,, \\ -\dfrac{2}{\pi}\beta b \log n & \text{if } \alpha = 1\,. \end{cases}$$

Then it can easily be checked that the c.f. $f$ given by (8.1.7) satisfies (8.1.2) so that $f$ is stable. □

**Remark 8.1.10.** (i) The parameters $a$ and $b$ in (8.1.7) are respectively location and scale parameters, so that every pair $\alpha$, $\beta$ satisfying $0 < \alpha \leqq 2$ and $|\beta| \leqq 1$ defines a type of stable d.f. — Note that the parameter $a$ is not the same as in the corresponding Lévy representation of $F$, cf. the proof of Prop. 8.1.7.

(ii) But for the case $\alpha = 2$, it is clear from the uniqueness theorem for c.f. that the four parameters $\alpha$, $\beta$, $a$, and $b$ in (8.1.7) are uniquely determined by $F$. Therefore, we will briefly denote a stable d.f. $F$ by $F = \mathscr{S}(\alpha, \beta; a, b)$; we formally include the case $\alpha = 2$ although in this case $\beta$ is not specified. □

**Example 8.1.11.** (i) For $F = \mathscr{S}(1, 0; 0, 1)$ we obtain $f(t) = \mathrm{e}^{-|t|}$ so that $F$ is the Cauchy d.f.

(ii) For $F = \mathscr{S}(\frac{1}{2}, -1; 0, 1)$ we have $f(t) = \mathrm{e}^{-\sqrt{|t|}\,[1 - \mathrm{i}\,\operatorname{sgn} t]}$, and $F$ has the density

$$p(x) = \begin{cases} 0\,, & x < 0\,, \\ \dfrac{1}{\sqrt{2\pi}}\,\mathrm{e}^{-1/2x}\,x^{-3/2}\,, & x > 0\,. \end{cases}$$ □

The canonical representation (8.1.7) is basic to the investigation of stable d.f. For example we collect together a number of simple properties of stable d.f.

**Proposition 8.1.12.** *Let $F = \mathscr{S}(\alpha, \beta; a, b)$ be stable. Then the following assertions are true.*

(i) $F_-(x) = 1 - F(x)$ *is also stable where* $F_- = \mathscr{S}(\alpha, -\beta; -a, b)$.

(ii) $F'(x) = F(\mu x + \nu)$, $\mu > 0$, *is also stable. We have* $F' = \mathcal{S}(\alpha, \beta; a', b')$ *where* $b' = b/\mu^\alpha$ *and*

$$a' = \begin{cases} \dfrac{a-\nu}{\mu} & \text{if} \quad \alpha \neq 1, \\ \dfrac{a - \nu - \dfrac{2}{\pi}\beta \log \mu}{\mu} & \text{if} \quad \alpha = 1. \end{cases}$$

(iii) *$F$ is symmetric with respect to $x_0$ iff $a = x_0$ and either $\alpha = 2$ or $\beta = 0$.*

(iv) *$F$ is a one-sided d.f. iff $\alpha < 1$ and $|\beta| = 1$. In this case we have* lext $(F) = a$ *if* $\beta = -1$ *and* rext $(F) = a$ *if* $\beta = 1$.

Proof. (i) and (ii) follow immediately from (8.1.7).

(iii) By Cor. 3.2.6, $F$ is symmetric with respect to $x_0$ iff $f(t)\,\mathrm{e}^{-itx_0}$ is real. Now (8.1.7) proves the assertion.

(iv) According to Prop. 7.3.9, $F$ is bounded to the left iff $\gamma^2 = 0$, $M = 0$, and $\int_{+0}^{1} u \, \mathrm{d}N(u) < \infty$. In view of (8.1.4) these requirements are equivalent to $d_1 = 0$ and $\alpha < 1$. Further, $d_1 = 0$ is equivalent to $\beta = -1$ by (8.1.8). In this case we obtain from Prop. 4.2.8

$$\text{lext}\,(F) = -\lim_{s\to\infty} \frac{\log f(is)}{s} = a\,.$$

The other assertion can be proved by considering a d.f. $F$ bounded to the right. □

**Proposition 8.1.13.** (i) *Let $F_k = \mathcal{S}(\alpha_k, \beta_k; a_k, b_k)$, $k = 1, 2$, be stable such that $\alpha_1 = \alpha_2 = \alpha$. Then $F = F_1 * F_2$ is also stable and we have $F = \mathcal{S}\big(\alpha, (\beta_1 b_1 + \beta_2 b_2)/(b_1 + b_2); a_1 + a_2, b_1 + b_2\big)$.*

(ii) *Every stable d.f. $F = \mathcal{S}(\alpha, \beta; a, b)$ permits the decomposition $F = F_1 * F_2$ where $F_k = \mathcal{S}\big(\alpha, (-1)^k; a/2, b/2\,[1 + (-1)^k \beta]\big)$, $k = 1, 2$.*

Proof. Using (8.1.7), (i) can easily be checked while (ii) follows immediately from (i). □

## 8.2. Series expansions and asymptotic formulae

From (8.1.3) we see that the c.f. of a non-degenerate stable d.f. $F$ is absolutely integrable on $\mathbb{R}_1$ so that $F$ is absolutely continuous by the Fourier inversion theorem 3.2.2. But except for the normal density, the Cauchy density, and the density given in Ex. 8.1.11 (ii), no more stable densities are known which can be expressed by means of "elementary" functions. Therefore it is very useful to know series expansions for stable densities or, at least, asymptotic formulas for small or large $x$. The present section contains a survey on such formulas; they will be applied in Sect. 8.3 and 8.4.

We omit most of the proofs since these can be found in the literature; for example we refer the reader to Skorohod (1954) or Ibragimov and Linnik (1965). To apply the formulas given there we note that, by virtue of Prop. 8.1.12 (ii), a stable density $p$ corresponding to $F = \mathcal{S}(\alpha, \beta; a, b)$ can be written as

$$p(x) = b^{-1/\alpha} q\big(b^{-1/\alpha}(x - a')\big)$$

where $q$ is the density of $G = \mathcal{T}(\alpha, \beta; 0, 1)$ while

$$a' = \begin{cases} a & \text{if } \alpha \neq 1, \\ a - \frac{2}{\pi}\beta b \log b & \text{if } \alpha = 1. \end{cases}$$

Throughout this section we denote by $p$ a stable density corresponding to $F = \mathcal{S}(\alpha, \beta; a, b)$ and we exclude the trivial case $\alpha = 2$ in which $p$ is normal. Further, for $\alpha \neq 1$ we write

$$b_0 = b\sqrt{1 + \beta^2 \tan^2 \frac{\pi\alpha}{2}}, \qquad \beta_0 = \arctan\left(\beta \tan \frac{\pi\alpha}{2}\right).$$

**Proposition 8.2.1.** (i) *If $\alpha \neq 1$, then we have for all $N \geqq 1$*

$$p(x) = \sum_{k=1}^{N} \frac{C_k}{(x - a)^{1+\alpha k}} + O\left(\frac{1}{x^{1+\alpha(N+1)}}\right), \qquad x \to \infty,$$

*where*

$$C_k = \frac{(-1)^{k+1}}{\pi k!} b_0^k \Gamma(1 + \alpha k) \sin k \left[\frac{\pi\alpha}{2} - \beta_0\right].$$

*For $\alpha < 1$ the corresponding series converges for all $x > a$.*

(ii) *If $\alpha = 1$, then we have for all $N \geqq 1$*

$$p\left(x - \frac{2}{\pi}\beta b \log (x - a)\right) = \sum_{k=1}^{N} \frac{C_k}{(x - a)^{1+k}} + O\left(\frac{1}{x^{2+N}}\right), \qquad x \to \infty,$$

*where*

$$C_k = \frac{b^k}{\pi k!} \operatorname{Im} \int_0^\infty e^{-t} t^k \left(i - i\beta + \frac{2}{\pi}\beta \log t\right)^k dt. \qquad \square$$

Next we slightly modify this result for $\alpha = 1$.

**Corollary 8.2.2.** *If $\alpha = 1$, then*

$$p(x) = \frac{C_1}{(x - a)^2}\left[1 - \frac{4}{\pi}\beta b \frac{\log x}{x}\right] + \frac{C_2}{(x - a)^3} + o\left(\frac{1}{x^3}\right), \qquad x \to \infty,$$

*where the $C_k$ are as in Prop. 8.2.1 (ii).*

Proof. Put $h(y) = y - (2/\pi)\beta b \log (y - a)$, $y > a$. We have $h(y) \to \infty$ as $y \to \infty$, and $h$ is strictly increasing for $y > y_0 = \max\{a, a + (2/\pi)\beta b\}$. Hence, for large $x$ we can define a function $y = y(x)$ by means of

$$y(x) - \frac{2}{\pi}\beta b \log \big(y(x) - a\big) = x, \qquad y(x) > y_0. \tag{8.2.1}$$

We easily see that $x/y(x) \to 1$ as $x \to \infty$. This implies

$$y - a = x - a + \frac{2}{\pi}\beta b \log (y - a) = x\left[1 + O\left(\frac{\log x}{x}\right)\right]$$

whence it follows that

$$\log (y - a) = \log x + O\left(\frac{\log x}{x}\right) = \log x \left[1 + O\left(\frac{1}{x}\right)\right].$$

Accordingly,

$$\frac{\log(y-a)}{y-a}=\frac{\log x}{x}\,\frac{1+O\left(\frac{1}{x}\right)}{1+O\left(\frac{\log x}{x}\right)}=\frac{\log x}{x}\left[1+O\left(\frac{\log x}{x}\right)\right];$$

thus we get from (8.2.1)

$$\frac{x-a}{y-a}=1-\frac{2}{\pi}\beta b\,\frac{\log(y-a)}{y-a}=1-\frac{2}{\pi}\beta b\,\frac{\log x}{x}+o\left(\frac{1}{x}\right). \tag{8.2.2}$$

Now we obtain from Prop. 8.2.1 (ii), because of (8.2.1)

$$\begin{aligned}p(x)&=p\left(y-\frac{2}{\pi}\beta b\log(y-a)\right)=\frac{C_1}{(y-a)^2}+\frac{C_2}{(y-a)^3}+o\left(\frac{1}{y^3}\right)\\&=\frac{C_1}{(x-a)^2}\left(\frac{x-a}{y-a}\right)^2+\frac{C_2}{(x-a)^3}\left(\frac{x-a}{y-a}\right)^3+o\left(\frac{1}{x^3}\right);\end{aligned}$$

thus the assertion follows by inserting (8.2.2). □

**Corollary 8.2.3.** *Let $C_k$, $k \geqq 1$, be the coefficients occurring in Prop.* 8.2.1.

(i) *If $\beta < 1$, then $C_1 > 0$.*
(ii) *If $\beta = 1$, then $C_k = 0$ for all $k \geqq 1$.*

Proof. (i) Assume $\alpha \neq 1$, $\beta < 1$. Then it is not hard to verify that the definition of $\beta_0$ implies

$$0<\frac{\pi\alpha}{2}-\beta_0<\pi. \tag{8.2.3}$$

Accordingly, $C_1=\frac{1}{\pi}b_0\Gamma(1+\alpha)\sin\left(\frac{\alpha\pi}{2}-\beta_0\right)>0$. – If $\alpha = 1$ and $\beta < 1$, then $C_1=\frac{q}{\pi}(1-\beta)>0$.

(ii) can be shown by inspection. □

With regard to Cor. 8.2.3 (ii), it is desirable to have more precise asymptotic formulae for stable densities with $\beta = 1$. We quote such formulae only in a form which will be needed later. (Note that for $\beta = 1$ and $\alpha < 1$ we have $p(x) = 0$, $x > a$; cf. Prop. 8.1.12 (iv).)

**Proposition 8.2.4.** *Suppose $\beta = 1$.*

(i) *If $\alpha > 1$, then we have for a certain constant $B(\alpha) > 0$*

$$p(x)=\exp\{-B(\alpha)\,b^{1/(1-\alpha)}(x-a)^{\alpha/(\alpha-1)}+O(\log x)\},\qquad x\to\infty.$$

(ii) *If $\alpha = 1$, then*

$$p(x)=\exp\left\{-\frac{2b}{\pi e}\exp\left[\frac{\pi}{2b}(x-a)+o(1)\right]\right\},\qquad x\to\infty.$$ □

We summarize the results formulated so far in the following way.

**Corollary 8.2.5.** *Let $p$ be a stable density with $\alpha < 2$. Then there exists a constant $C > 0$ such that, as $x \to \infty$,*

$$p(x) = \begin{cases} \dfrac{C + o(1)}{x^{1+\alpha}} & \text{if } \beta < 1\,, \\ \exp\{-(C + o(1))\, x^{\alpha/(\alpha-1)}\} & \text{if } \beta = 1, \alpha > 1\,, \\ \exp\left\{-\,(C + o(1)) \exp\left(\dfrac{\pi}{2b}\, x\right)\right\} & \text{if } \beta = 1, \alpha = 1\,. \end{cases}$$ □

Now we are going to deal with asymptotic formulae for small $x$.

**Proposition 8.2.6.** (i) *If $\alpha \neq 1$, then we have for all $N \geqq 1$*

$$p(x) = \sum_{k=1}^{N} c_k (x - a)^{k-1} + O((x - a)^N)\,, \qquad x \to a \pm 0\,,$$

*where*

$$c_k = \frac{(-1)^{k+1}}{\pi k!}\, b_0^{-k/\alpha}\, \Gamma\left(1 + \frac{k}{\alpha}\right) \sin k \left[\frac{\pi}{2} - \frac{\beta_0}{\alpha}\right].$$

*For $\alpha > 1$ the corresponding series converges for all $x$.*

(ii) *If $\alpha = 1$ and $\beta \neq 0$, then we have for all $x$*

$$p(x) = \sum_{k=0}^{\infty} c_k \left(x - a + \frac{2}{\pi}\beta b \log b\right)^k$$

*where*

$$c_k = \frac{(\operatorname{sgn} \beta)^k}{\pi k!\, b^{k+1}} \int_0^\infty t^{k-(2/\pi)|\beta|} \sin t(1 + |\beta|)\, dt\,.$$ □

The formulae just given contain less information if $\alpha < 1$ and $|\beta| = 1$ since then $c_k = 0$ for all $k \geqq 1$. The following gives a better estimate of the speed of convergence of $\alpha < 1$ and $\beta = -1$; note that in this case $p(x) = 0$, $x < a$, by Prop. 8.1.12 (iv). — The case $\beta = 1$ can be treated by using Prop. 8.1.12 (i).

**Proposition 8.2.7.** *Suppose $\alpha < 1$ and $\beta = -1$. Then we have for a constant $B(\alpha) > 0$*

$$p(x) = \exp\{-B(\alpha)\, b^{1/(1-\alpha)} (x - a)^{\alpha/(\alpha-1)} + O(\log x)\}\,, \qquad x \to a + 0\,.$$ □

## 8.3. Analytic properties

First we make some remarks on the analytic character of stable c.f. — Except for the c.f. of the normal and the degenerate d.f., a stable c.f. $f$ cannot be an a.c.f. This is easily seen from the canonical representation (8.1.7), but follows also from Prop. 4.3.8 and the results of the previous section. Indeed, at least one of the tails $F(x)$ or $1 - F(-x)$ decreases only as a power of $x$.

From Prop. 4.3.8 we also know that $f$ is a boundary c.f. iff exactly one of these tails decreases at least exponentially; this is equivalent to $|\beta| = 1$, cf. Cor. 8.2.5. For example, if $\beta = -1$, then the function

$$\psi(z) = \begin{cases} \exp\left\{iaz - bz^\alpha \left[1 - i \tan \dfrac{\pi\alpha}{2}\right]\right\} & \text{if } \alpha \neq 1\,, \\ \exp\left\{iaz - bz \left[1 + i\, \dfrac{2}{\pi} \log z\right]\right\} & \text{if } \alpha = 1\,, \end{cases}$$

is analytic in the upper half-plane and represents the continuation of $f$. This can be seen from (8.1.7) by inspection.

Nevertheless, all stable c.f. have a further interesting analytic property; namely, they are continuable from the half-axis $\{t > 0\}$ (resp. $\{t < 0\}$) to a function analytic in the whole complex plane cut from 0 to $\infty$ in an appropriate way. These continuations are given by

$$\psi_\pm(z) = \begin{cases} \exp\left\{iaz - b(\pm z)^\alpha \left[1 \pm i\beta \tan \frac{\pi\alpha}{2}\right]\right\} & \text{if} \quad \alpha \neq 1, \\ \exp\left\{iaz - b(\pm z)\left[1 \pm i\,\frac{2}{\pi}\beta \log(\pm z)\right]\right\} & \text{if} \quad \alpha = 1. \end{cases} \tag{8.3.1}$$

Indeed, both $\psi_+$ and $\psi_-$ are analytic in the cut plane and they coincide with $f(t)$ for $t > 0$ (resp. $t < 0$), provided that the cut does not touch (but at $t = 0$) the line $\{t > 0\}$ (resp. $\{t < 0\}$). This fact has already been used in the proof of Prop. 7.6.4.

Now we are going to investigate the analytic character of stable d.f. The general considerations on a.d.f. outlined in Chap. 6 yield certain assertions about the analyticity of stable d.f. But if we make use of the series expansions for stable densities stated in the previous section, we obtain better results.

**Proposition 8.3.1.** *Let $F = \mathscr{S}(\alpha, \beta; a, b)$ be stable with density $p$.*

(i) *If $\alpha > 1$, then $p$ is an integral density of order $\varrho = \alpha/(\alpha-1)$ and type $\tau = (\alpha-1) \times (b_0\alpha^\alpha)^{1/(1-\alpha)}$.*

(ii) *If $\alpha = 1$ and $\beta \neq 0$, then $p$ is an integral density. — If $\alpha = 1$ and $\beta = 0$, then $p$ is analytic in the strip $S(b)$ and has poles at $z_\pm = a \pm ib$.*

(iii) *If $\alpha < 1$, then $p$ permits the representation*

$$p(x) = \begin{cases} \dfrac{1}{x-a}\, p_+\big((x-a)^{-\alpha}\big), & x > a, \\ \dfrac{1}{a-x}\, p_-\big((a-x)^{-\alpha}\big), & x < a, \end{cases}$$

*where $p_\pm$ are integral functions. If $|\beta| < 1$, then $p_\pm$ have order $\varrho = 1/(1-\alpha)$ and type $\tau = (1-\alpha)(b_0\alpha^\alpha)^{1/(1-\alpha)}$. If $\beta = 1$ ($\beta = -1$), then $p_+ = 0$ ($p_- = 0$) while $p_-(p_+)$ has the same order and type as before.*

Sketch of the proof. (i), (ii) If $\alpha > 1$ or $\alpha = 1, \beta \neq 0$, then $p$ has a series representation which is given in Prop. 8.2.6. Using Prop. B. 3.3 and the Stirling formula B. 5, one can show that this series represents an integral function of the stated growth. — If $\alpha = 1$ and $\beta = 0$, then $p(x) = \dfrac{1}{\pi}\dfrac{b}{b^2 + (x-a)^2}$; see Ex. 8.1.11 (i).

(iii) As before we conclude that the functions

$$p_\pm(z) = \sum_{k=1}^{\infty} \frac{(-1)^{k+1}\,\Gamma(1+\alpha k)}{\pi k!}\, b_0^k \sin k\left[\frac{\pi\alpha}{2} \mp \beta_0\right] z^k$$

are integral functions of order and type as asserted; by virtue of Prop. 8.2.1 (i), we see that $p$ has the desired form. □

**Remark 8.3.2.** (i) If $\alpha > 1$, then $F = \mathscr{S}(\alpha, \beta; a, b)$ is an integral d.f. of order $\varrho = \alpha/(\alpha-1)$. Thus we could apply Prop. 6.4.5 (ii), to obtain

$$\tau \leq (\alpha - 1)\,(b\alpha^\alpha)^{1/(1-\alpha)} =: \tau_0\,.$$

Now we see that the actual type $\tau$ of $F$, given in Prop. 8.3.1 (i), may be less than $\tau_0$ since $b_0 = b\sqrt{1 + \beta^2 \tan^2(\pi\alpha)/2} \geqq b$ (cf. Rem. 6.4.6). The equation $\tau = \tau_0$ is satisfied iff $\beta = 0$ or $\alpha = 2$. This is the case in which $F$ is symmetric, see Prop. 8.1.12 (iii).

(ii) If $\alpha = 1$ and $\beta \neq 0$, then $p$ has infinite order. A proof of this fact can be found in ZOLOTAREV (1983). □

The next corollary contains the results of Prop. 8.3.1 in a simple form which is convenient for later purposes.

**Corollary 8.3.3.** *Let $F = \mathscr{S}(\alpha, \beta; a, b)$ be stable with density $p$. If $\alpha \geqq 1$, then $p$ is analytic at least in a strip $S(s)$, $s > 0$. If $\alpha < 1$, then $p$ is analytic in both $\operatorname{Re} z > a$ and $\operatorname{Re} z < a$ while $z = a$ is a singularity of $p$.* □

Also, we have the following.

**Corollary 8.3.4.** *If $p^{(k)}(0) = 0$ for all $k \geqq 0$, then either $p(x) = 0$ in a certain interval $0 < x < x_0$ or $\alpha < 1$, $\beta = -1$, and $a = 0$.*

Proof. Assume that $p$ is analytic in a neighbourhood of the origin 0. Then from $p^{(k)}(0) = 0$, $k \geqq 0$, it follows from the uniqueness theorem for analytic functions that $p(x) = 0$, $|x| < x_0$, for a certain $x_0$. — Suppose now that 0 is a singularity of $p$. Then Cor. 8.3.3 tells us that necessarily $\alpha < 1$ and $a = 0$. Now we obtain from Prop. 8.2.6 (i),

$$p(0) = \frac{1}{\pi} b_0^{-1/\alpha} \Gamma\left(1 + \frac{1}{\alpha}\right) \sin\left[\frac{\pi}{2} - \frac{\beta_0}{\alpha}\right];$$

and the definition of $\beta_0$ implies $|\beta_0| \leqq \pi\alpha/2$. Since by assumption $p^{(0)}(0) = p(0) = 0$, we get $|\beta_0| = \pi\alpha/2$ whence $|\beta| = 1$ follows. Finally, if $\beta = 1$, then $p(x) = 0$ for all $x > 0$ by Prop. 8.1.12 (iv). □

## 8.4. Uniqueness theorems

The results of Sect. 8.2 and 8.3 form the basis for the uniqueness theorems for stable d.f. to which we are going now. Our main result is the following continuation theorem the proof of which will be postponed to the end of this section.

**Theorem 8.4.1.** *Let $F_1$ and $F_2$ be stable d.f., and let $S = \{x_k : k \geqq 1\}$ be an infinite real point set. If*

$$0 < F_1(x_k) = F_2(x_k) < 1\,, \qquad k \geqq 1\,,$$

*then $F_1 = F_2$.*

Now we wish to prove some uniqueness theorems for stable densities. They will provide us the tools for the proof of the above theorem.

We start with statements in the suppositions of which there occur asymptotic relations for $x \to \infty$. By $p_j$, $j = 1, 2$, we denote stable densities belonging to $F_j = \mathscr{S}(\alpha_j, \beta_j; a_j, b_j)$, and $C_{jk}$, $k \geqq 1$, are the corresponding coefficients occurring in Prop. 8.2.1. — If $\alpha_j < 1$ we exclude the case $\beta_j = 1$ since then $p_j(x) = 0$ for $x > a_j$; cf. Prop. 8.1.12 (iv).

**Proposition 8.4.2.** *Assume that, as $x \to \infty$*

$$\frac{p_1(x)}{p_2(x)} = o(x^\varepsilon)\,, \qquad \frac{p_2(x)}{p_1(x)} = o(x^\varepsilon)\,, \quad \text{for all} \quad \varepsilon > 0\,. \tag{8.4.1}$$

*Then* $\alpha_1 = \alpha_2 = \alpha$, *say. Further, if* $p_1 \neq p_2$, *then* $\alpha < 2$ *and* $\beta_1, \beta_2 < 1$, *and we have*

$$\frac{p_1(x)}{p_2(x)} = \frac{C_{11}}{C_{21}} + o(1)\,, \qquad x \to \infty\,. \tag{8.4.2}$$

Proof. First suppose $\alpha_1 \neq \alpha_2$. Then it is an immediate consequence of Cor. 8.2.5 that either $p_1/p_2$ or $p_2/p_1$ increases at least as a power of $x$. This contradicts (8.4.1) so that $\alpha_1 = \alpha_2 = \alpha$ follows.

Assume $\alpha = 2$. Then $p_1$ and $p_2$ are normal densities and from (8.4.1) it follows that $p_1 = p_2$ by an indirect argument. Let now $\alpha < 2$. Then we see from Cor. 8.2.5 and (8.4.1) that either $\beta_1 = \beta_2 = 1$ (and then $\alpha \geqq 1$) or $\beta_1, \beta_2 < 1$. In the first case we make use of Prop. 8.2.4 to conclude first $b_1 = b_2$ and then $a_1 = a_2$. Hence, $p_1 = p_2$.

Accordingly, if $p_1 \neq p_2$, then we have $\alpha < 2$ and $\beta_1, \beta_2 < 1$. Now Cor. 8.2.3 (i), tells us that then $C_{j1} > 0$, and (8.4.2) is an immediate consequence of Prop. 8.2.1. □

**Lemma 8.4.3.** *Assume* $\alpha_1 = \alpha_2 = \alpha < 2$ *and* $\beta_1, \beta_2 < 1$. *If*

$$C_{11} = C_{21}\,, \qquad C_{12} = C_{22}\,, \tag{8.4.3}$$

*then* $\beta_1 = \beta_2$ *and* $b_1 = b_2$.

Proof. According to Cor. 8.2.3 (i), we have $C_{j1} > 0$. Therefore, (8.4.3) yields

$$C_{12}/C_{11}^2 = C_{22}/C_{21}^2\,. \tag{8.4.4}$$

Assume now $\alpha \neq 1$. Then we express (8.4.4) by means of the formulas given in Prop. 8.2.1 (i). Thus we see, by virtue of $\sin 2x = 2 \sin x \cos x$, that

$$\cot\left[\frac{\pi\alpha}{2} - \beta_{01}\right] = \cot\left[\frac{\pi\alpha}{2} - \beta_{02}\right].$$

Since $0 < (\pi\alpha/2) - \beta_{0j} < \pi$ (cf. (8.2.3)) we get $\beta_{01} = \beta_{02}$ whence we conclude that $\beta_1 = \beta_2$. Now $b_1 = b_2$ follows from $C_{11} = C_{21}$.

If $\alpha = 1$, then (8.4.4) and the formulae given in Prop. 8.2.1 (ii), lead us after some calculation to

$$\beta_1/(1 - \beta_1) = \beta_2/(1 - \beta_2)\,.$$

This yields $\beta_1 = \beta_2$, and $b_1 = b_2$ follows again from $C_{11} = C_{21}$. □

**Proposition 8.4.4.** *Assume that*

$$R(x) := \frac{p_1(x)}{p_2(x)} - 1 = o(x^{-\max\{1,\alpha_1,\alpha_2\}})\,, \qquad x \to \infty\,. \tag{8.4.5}$$

*Then we have* $p_1 = p_2$.

Proof. First of all, (8.4.5) implies (8.4.1); thus Prop. 8.4.2 tells us that $\alpha_1 = \alpha_2 = \alpha$, say. Moreover, it is enough to consider the case $\alpha < 2$ and $\beta_1, \beta_2 < 1$ since otherwise $p_1 = p_2$ by virtue of this proposition (using an indirect argument).

Now we see from (8.4.5) and (8.4.2) that

$$C_{11} = C_{21} =: C_1\,. \tag{8.4.6}$$

where $C_1 > 0$ by Cor. 8.2.3 (i). Further, in view of

$$p_1(x) - p_2(x) = p_2(x)\, R(x)\,,$$

we obtain from Cor. 8.2.5 and (8.4.5) that, as $x \to \infty$,

$$x^{2+\alpha}\big(p_1(x) - p_2(x)\big) = x^{2+\alpha} O(x^{-1-\alpha})\, o(x^{-1}) = o(1) \tag{8.4.7}$$

and also

$$x^{1+2\alpha}(p_1(x) - p_2(x)) = x^{1+2\alpha}O(x^{-1-\alpha})\, o(x^{-\alpha}) = o(1)\,. \tag{8.4.8}$$

In the following we put, for given $\alpha$,

$$\Delta(x) = \frac{1}{(x-a_1)^{1+\alpha}} - \frac{1}{(x-a_2)^{1+\alpha}} = (1+\alpha)\int_{a_1}^{a_2} \frac{du}{(x-u)^{2+\alpha}}$$

so that

$$x^{2+\alpha}\Delta(x) = (1+\alpha)(a_2 - a_1) + o(1)\,, \qquad x \to \infty\,. \tag{8.4.9}$$

Assume now $\alpha > 1$. Then we obtain from Prop. 8.2.1 (i),

$$p_1(x) - p_2(x) = C_1\Delta(x) + \frac{C_{21}}{(x-a_1)^{1+2\alpha}} - \frac{C_{22}}{(x-a_2)^{1+2\alpha}} + o\left(\frac{1}{x^{1+2\alpha}}\right). \tag{8.4.10}$$

Combining (8.4.7), (8.4.9), and (8.4.10) we get

$$o(1) = x^{2+\alpha}(p_1(x) - p_2(x)) = C_1(1+\alpha)(a_2 - a_1) + o(1) \tag{8.4.11}$$

whence $a_1 = a_2$ follows. Accordingly, $\Delta = 0$, and from (8.4.8) and (8.4.10) we now conclude that

$$o(1) = x^{1+2\alpha}(p_1(x) - p_2(x)) = C_{21} - C_{22} + o(1)\,, \tag{8.4.12}$$

and therefore

$$C_{12} = C_{22}\,. \tag{8.4.13}$$

Because of (8.4.6) and (8.4.13) we may apply L. 8.4.3 and see that $\beta_1 = \beta_2$ and $b_1 = b_2$. This means that $p_1 = p_2$.

Let now $\alpha < 1$. Then we again start from (8.4.10) but conclude in another succession than before. Namely, from (8.4.8), (8.4.9), and (8.4.10) we first get (8.4.12). This yields (8.4.13), and by virtue of (8.4.6) and L. 8.4.3 we again obtain $\beta_1 = \beta_2$ and $b_1 = b_2$. Accordingly, $C_{1k} = C_{2k}$ for all $k \geqq 1$; so Prop. 8.2.1 (i), tells us that

$$p_1(x) - p_2(x) = C_1\,\Delta(x)\,[1 + o(1)]\,. \tag{8.4.14}$$

Now we make use of (8.4.7) and (8.4.9) to conclude (8.4.11) so that $a_1 = a_2$. This yields $p_1 = p_2$.

Assume, finally, that $\alpha = 1$. Then Cor. 8.2.2 tells us that

$$p_1(x) - p_2(x) = C_1\left[\Delta(x) - \frac{4}{\pi}\frac{\log x}{x}\left(\frac{\beta_1 b_1}{(x-a_1)^2} - \frac{\beta_2 b_2}{(x-a_2)^2}\right)\right] + O(x^{-3}).$$

Multiplying this equation with $x^3$ and letting $x \to \infty$, we see from (8.4.7) and (8.4.9) that

$$o(1) = O(1) - \frac{4}{\pi}C_1(\beta_1 b_1 - \beta_2 b_2 + o(1))\log x\,.$$

So we get necessarily $\beta_1 b_1 = \beta_2 b_2$. Accordingly, since $C_{j1} = 1/\pi\,(b_j - \beta_j b_j)$ we obtain by virtue of (8.4.6) first $b_1 = b_2$ and then $\beta_1 = \beta_2$. Hence, $C_{1k} = C_{2k}$ for all $k \geqq 1$, and from Cor. 8.2.2 we again conclude (8.4.14) whence $a_1 = a_2$ follows as before. Therefore, $p_1 = p_2$. □

Prop. 8.4.4 immediately yields the following.

**Corollary 8.4.5.** *If*

$$\frac{p_1(x)}{p_2(x)} = 1 + o\left(\frac{1}{x^{2-\varepsilon}}\right), \qquad x \to \infty, \quad \textit{for all} \quad \varepsilon > 0,$$

*then* $p_1 = p_2$. □

Now we are going to prove a uniqueness theorem for stable densities in the supposition of which there occurs an asymptotic relation for small $x$. As before, $p_j$ denote stable densities belonging to $F_j = \mathscr{S}(\alpha_j, \beta_j; a_j, b_j)$; we exclude the case that $p_j(x) = 0$ in a certain interval $0 < x < x_0$.

**Proposition 8.4.6.** *If*

$$\frac{p_1(x)}{p_2(x)} = 1 + o(x^n), \qquad x \to +0, \quad \textit{for all} \quad n \geqq 1, \tag{8.4.15}$$

*then* $p_1 = p_2$.

Proof. First of all, from (8.4.15) it follows that

$$p_1(x) - p_2(x) = o(x^n), \qquad x \to +0, \qquad n \geqq 1. \tag{8.4.16}$$

Next we conclude from Cor. 8.3.3 that exactly one of the following three cases can occur.

(i) The origin 0 is a singularity of both $p_1$ and $p_2$,

(ii) 0 is a singularity of $p_1$ while $p_2$ is analytic in a certain neighbourhood of 0 (or conversely),

(iii) both $p_1$ and $p_2$ are analytic in a neighbourhood of 0.

(i) In this case we see from Cor. 8.3.3 that $\alpha_1, \alpha_2 < 1$ and $a_1 = a_2 = 0$.

If $\beta_1 = 1$, then $p_1(x) = 0$, $x > 0$, by Prop. 8.1.12 (iv), which has been excluded.

Let $\beta_1 = -1$. Then we see from Prop. 8.2.7 that $p_1$ decreases exponentially as $x \to +0$, and by (8.4.16) $p_2$ must have the same behaviour. Accordingly, $\beta_2 = -1$. Now it is an immediate consequence of Prop. 8.2.7 and (8.4.16) that $\alpha_1 = \alpha_2$ and $b_1 = b_2$, whence $p_1 = p_2$ follows.

In the same manner we can treat the cases $|\beta_2| = 1$.

Assume now $|\beta_j| < 1$, $j = 1, 2$. Then we have, for all $N \geqq 1$,

$$p_j(x) = \sum_{k=1}^{N} c_{jk} x^{k-1} + O(x^N), \qquad x \to +0, \tag{8.4.17}$$

where the $c_{jk}$ are as in Prop. 8.2.6 (i). Clearly, from (8.4.16) it follows that

$$c_{1k} = c_{2k}, \quad \text{for all} \quad k \geqq 1. \tag{8.4.18}$$

Next we apply the Stirling formula to conclude

$$\varlimsup_{k \to \infty} \frac{k \log k}{-\log |c_{jk}|} = \frac{\alpha_j}{\alpha_j - 1};$$

thus (8.4.18) implies $\alpha_1 = \alpha_2$. Finally, we use the same method as in the proof of L. 8.4.3 to obtain $\beta_1 = \beta_2$ and $b_1 = b_2$ from (8.4.18) for $k = 1, 2$. This again yields $p_1 = p_2$.

(ii) In this case we have $\alpha_1 < 1$ and $a_1 = 0$ so that $p_1$ satisfies (8.4.17) for all $N \geqq 1$, where the $c_{1k}$ are as in Prop. 8.2.6 (i). On the other hand, since $p_2$ is analytic in some neighbourhood of 0 we have, for a certain $x_0 > 0$ and for certain coefficients $c_{2k}$, $k \geqq 1$,

$$p_2(x) = \sum_{k=1}^{\infty} c_{2k} x^{k-1}, \qquad |x| < x_0. \tag{8.4.19}$$

Now (8.4.16) again implies (8.4.18).

Assume $|\beta_1| = 1$. Then $c_{1k} = 0$ for all $k \geqq 1$ so that, by virtue of (8.4.18) and (8.4.19), $p_2(x) = 0$ for $|x| < x_0$, which has been excluded.

Let $|\beta_1| < 1$. Then we apply the Stirling formula to compute

$$l := \overline{\lim_{k\to\infty}} |c_{1k}|^{1/k} = \infty .$$

On the other hand, because of (8.4.18) we can connect this limit with the radius of convergence $\varrho$ of the series in (8.4.19); namely, we have

$$l = \overline{\lim_{k\to\infty}} |c_{2k}|^{1/k} = \frac{1}{\varrho} .$$

Since $\varrho \geqq x_0$ we get $l \leqq 1/x_0$, a contradiction. Hence, case (ii) is not possible.

(iii) In this case we can also expand $p_1$ in a series analogous to (8.4.19) with coefficients $c_{1k}$ where $x_0$ must be chosen sufficiently small. Now (8.4.16) again yields (8.4.18) so that $p_1(x) = p_2(x)$ for $|x| < x_0$. Obviously, the singularities of $p_1$ and $p_2$, if any, must coincide. Since the $p_j$ have at most one singularity we see that $p_1$ and $p_2$ coincide at least on a half-axis. Now Cor. 8.4.5 and its counterpart valid for $x \to -\infty$ prove the assertion in this case. □

Before proving our main result, which was formulated at the very beginning, we verify the following "smoothness" properties for stable densities.

**Lemma 8.4.7.** *Let $p_1$ and $p_2$ be stable densities.*

(i) *For every $\eta \in R_1$ there exists the (possibly infinite) limit*

$$l_1 := \lim_{x\to\infty} x^{\eta} \left(\frac{p_1(x)}{p_2(x)} - 1\right),$$

*unless $p_2(x) = 0$ on a certain half-axis $x > x_0$.*

(ii) *For every $n \geqq 1$ there exists the (possibly infinite) limit*

$$l_2 := \lim_{x\to+0} x^{-n} \left(\frac{p_1(x)}{p_2(x)} - 1\right),$$

*unless $p_2(x) = 0$ in a certain interval $0 < x < x_2$.*

Proof. (i) is an immediate consequence of Cor. 8.2.5.

(ii) If $p_1(x) = 0$ in a certain interval $0 < x < x_1$, then $l_2 = -\infty$. Otherwise, let $n \geqq 1$ be fixed. Then we consider the ratio

$$\frac{p_1(x) - p_2(x)}{x^n p_2(x)}$$

and try to apply de l'Hospital's rule $k$ times for an appropriate $k \geqq 0$. Obviously, the only case which yields difficulties occurs when all derivatives of both the nominator and the denominator tend to 0 as $x \to +0$. This happens iff $p_2^{(k)}(0) = 0$ and $p_1^{(k)}(0) = 0$ for all $k \geqq 0$. Since the case $p_j(x) = 0$, $0 < x < x_j$, has been excluded, we see from Cor. 8.3.4 that then $\alpha_j < 1$, $\beta_j = -1$, and $a_j = 0$. Hence, we can apply the asymptotic formula given in Prop. 8.2.7 to verify that the limit $l_2$ exists where $|l_2| \leqq \infty$. □

Proof of Theorem 8.4.1. We distinguish the following cases.
(i) $S$ contains an unbounded subsequence $(x_{k'})$,
(ii) $S$ contains a converging subsequence $(x_{k'})$.

Without loss of generality we may assume that either $x_{k'} \to \infty$ or $x_{k'} \to +0$; otherwise we make use of Prop. 8.1.12 and replace $F_j(x)$ by $1 - F_j(-x)$ or $F_j(x - \lim x_{k'})$. — In the following let $p_j$ be the density of $F_j$.

(i) Since we assume $0 < F_2(x_k) < 1$ it is not possible that $p_2(x) = 0$ in a certain half-axis $x > x_0$. Thus we can apply L. 8.4.7 (i), according to which the limit $l_1$ for $\eta = 2 - \varepsilon$ exists, where $\varepsilon > 0$ is arbitrary. On the other hand, by assumption and using Rolle's theorem we find a sequence $(y_k)$, $y_k \to \infty$, such that $p_1(y_k) = p_2(y_k)$, $k \geqq 1$. This means that $l_1 = 0$; accordingly, the supposition of Cor. 8.4.5 is satisfied. Applying this corollary we get $p_1 = p_2$.

(ii) As before we see that L. 8.4.7 (ii), can be applied which tells us that the limit $l_2$ exists for every $n \geqq 1$, and the same reasoning as in part (i) yields $l_2 = 0$. Hence, the supposition of Prop. 8.4.6 is satisfied whence $p_1 = p_2$ follows. □

## 8.5. Supplement

8.5.1. *A d.f. $F$ is stable iff its type is closed under convolution.* In other words, $F$ is stable iff for all constants $a_1$, $a_2$ and $b_1, b_2 > 0$ there exist $a$ and $b > 0$ such that

$$F(b_1x + a_1) * F(b_2x + a_2) = F(bx + a) .$$

This property is often used to define stable d.f.

8.5.2. *All stable d.f. are unimodal.* This has been proved by Yamazato even for a broader class of inf.div.d.f.; for a proof see Lukacs (1983).

8.5.3. *Let $F_1$ and $F_2$ be convolutions of stable d.f. If $F_1(x) = F_2(x)$, $x < x_0$, for a certain $x_0$ such that $F_1(x_0) > 0$, then $F_1 = F_2$. Moreover, the components of $F_1$ and $F_2$ are uniquely determined except for the location parameters* (Bischoff (1982)).

8.5.4. **Problem** (Ibragimov): Let $F$ be stable and $G$ inf.div. Assume $G(x) = F(x)$, $x < x_0$, for a certain $x_0 \in \mathbb{R}_1$. Does it then follow that $G = F$?

The answer is affirmative if $\alpha = 2$ or $1 \leqq \alpha < 2$ and $\beta = -1$. Indeed, in these cases we have $F > 0$, and the c.f. $f$ of $F$ is continuable to the upper half-plane so that Ibragimov's theorem 7.6.2 applies. However, the answer is negative if $\alpha < 2$ and $\beta > -1$. This is an immediate consequence of the following general result due to Ostrovskij (1981).

**Proposition 8.5.1.** *Let $F$ be a d.f. which is absolutely continuous on $(-\infty, 0)$. Assume that its derivative $p$ satisfies*

$$0 < C_1 \leqq p(x)\,|x^{1+\alpha}|\,H(|x|) \leqq C_2 < \infty\,, \qquad x < 0\,,$$

*where $\alpha > 0$ is a constant and $H$ stands for a function of regular variation with exponent $r = 0$. Then there exists an inf.div.d.f. $G \neq F$ such that $G(x) = F(x)$, $x < x_0$, for some $x_0 \in \mathbb{R}_1$.* □

This result yields the following.

**Corollary 8.5.2.** *Let $F = \mathcal{S}(\alpha, \beta; a, b)$ such that $\alpha < 2$ and $\beta > -1$. Then there exists an inf.div.d.f. $G \neq F$ satisfying $G(x) = F(x)$, $x < x_0$, for a certain $x_0 \in \mathbb{R}_1$.* □

**8.5.5. Problem** (Ostrovskij): Let $F$ be as in Cor. 8.5.2. Characterize the class of all inf.div.d.f. $G \neq F$ satisfying $G(x) = F(x)$, $x < x_0$, for a certain $x_0 \in R_1$.

The following preliminary answer was given by Ostrovskij (1981) himself.

**Proposition 8.5.3.** *Let $F$ be as in Cor. 8.5.2 and let $G \neq F$ be an inf.div. d.f. which satisfies $G(x) = F(x)$, $x < x_0$. Then there exist inf.div.d.f. $G_1$ and $G_2$ such that* rext $(G_1) \leq$ lext $(G_2)$ *and* $G = G_1 * G_2$. □

This proposition, Prop. 8.5.1, and Cor. 8.5.2 can all be carried over to the multivariate case, see Ostrovskij (1982).

**8.5.6. Problem** (Zolotarev): Since a stable d.f. $F$ depends only on four real parameters, it might be possible to determine it uniquely by means of its values $F(x_k)$ given on a finite point set appropriately chosen.

**8.5.7.** The uniqueness theorems for stable d.f. proved in Rossberg and Jesiak (1978) and Zolotarev (1978) (they are special cases of Th. 8.4.1) were carried over to "quasi-stable" d.f. by Antonov (1980a) and (1981). For the definition and basic properties of this class see Lukas (1983); there these d.f. are called *sumi-stable.*

## Notes and bibliographical data

The results contained in Sect. 8.1 to 8.3 inclusive are due to various authors. They can be found in standard books such as Ibragimov and Linnik (1965) and Lukacs (1970) where further references are also given.

The initial steps to prove Th. 8.4.1 were made by Rossberg and Jesiak (1978) and Zolotarev (1978). The final version stated here was obtained by Jesiak (1979c). The uniqueness theorems given in Sect. 8.4. are due to Jesiak (1985).

The first monograph to exclusively deal with stable d.f. has recently been written by Zolotarev (1983). In this outstanding work, the author gives not only a deep theoretical analysis of the class of stable d.f., but also a survey of results which are of practical interest.

# Chapter 9. The normal distribution

The normal distribution has already been considered in previous chapters. We remind the reader, for instance, of the results given in Sect. 4.6 and 4.7; further we have investigated this d.f. as integral, inf.div., and stable d.f., see Chap. 6–8.

In the present chapter we discuss some asymptotic properties of d.f. which uniquely characterize the normal d.f. In detail, Sect. 9.2 and 9.3 are devoted to the unique determination of this d.f. in the set of inf.div.d.f. by means of two-sided and one-sided asymptotic conditions, resp. To prove these results we make use of the powerful relations given in Sect. 7.7. In Sect. 9.4 we will show that an arbitrary d.f. $F$ coincides with a normal d.f. when both the tail function and the c.f. $f$ converge to zero sufficiently quickly for increasing arguments; this is a probabilistic modification of a theorem of Hardy.

## 9.1. Elementary properties

Let us consider a normal d.f. $\Phi_{\mu,\sigma}$ with density

$$p_{\mu,\sigma}(x) = \frac{1}{\sqrt{2\pi}\,\sigma} \exp\left(-\frac{1}{2}\left(\frac{x-\mu}{\sigma}\right)^2\right), \qquad x \in \mathbb{R}_1\,, \tag{9.1.1}$$

for some $\sigma > 0$, $\mu \in \mathbb{R}_1$, and c.f.

$$\varphi(t) = \varphi_{\mu,\sigma}(t) = \exp\left(i\mu t - \tfrac{1}{2}\sigma^2 t^2\right).$$

Now let $X$ be a r.v. subject to $\Phi_{\mu,\sigma}$; then we wish to determine all moments of $X$. First we find the differential equation

$$\varphi'(t) = (i - \sigma^2 t)\,\varphi(t)$$

and, by induction,

$$\varphi^{(n+1)}(t) = -n\sigma^2\varphi^{(n-1)}(t) + (i\mu - \sigma^2 t)\,\varphi^{(n)}(t)\,, \qquad n \geqq 1\,.$$

Setting $t = 0$ and using Prop. 3.5.1 we obtain

$$\mathsf{E}X = \mu\,, \qquad \mathsf{Var}\,X = \mathsf{E}X^2 - \mu^2 = \sigma^2\,.$$

Putting $\mathsf{E}X^0 = 1$ we further get

$$\mathsf{E}X^{n+1} = n\sigma^2\mathsf{E}X^{n-1} + \mu\mathsf{E}X^n \quad \text{for each} \quad n \geqq 1\,.$$

In particular, in the case that $\mu = 0$ this implies

$$\mathsf{E}X^n = \begin{cases} \dfrac{n!\sigma^n}{2^{n/2}\left(\dfrac{n}{2}\right)!}, & \text{if } n \text{ is even}, \\ 0, & \text{if } n \text{ is odd}. \end{cases}$$

Next we turn to expansions of $\Phi_{\mu,\sigma}$. First we trivially see from (9.1.1) that

$$p_{\mu,\sigma}(x) = \frac{1}{\sqrt{2\pi}\,\sigma} \sum_{k=0}^{\infty} \left(-\frac{1}{2}\right)^k \left(\frac{x-\mu}{\sigma}\right)^{2k} \frac{1}{k!}, \qquad x \in \mathbb{R}_1 .$$

We integrate over $[\mu, x]$ and note that $\Phi_{\mu,\sigma}(\mu) = 1/2$; then

$$\Phi_{\mu,\sigma}(x) = \frac{1}{2} + \frac{1}{\sqrt{2\pi}\,\sigma} \sum_{k=0}^{\infty} \left(-\frac{1}{2}\right)^k \frac{(x-\mu)^{2k+1}}{k!\sigma^{2k}(2k+1)}, \qquad x \in \mathbb{R}_1 .$$

For large $x$ we have

$$\Phi_{\mu,\sigma}(x) = \frac{\sigma}{|x|\sqrt{2\pi}} \exp\left(-\frac{1}{2}\left(\frac{x-\mu}{\sigma}\right)^2\right)(1 + o(1)), \qquad x \to -\infty, \tag{9.1.2}$$

which immediately follows from de l'Hospital's rule. A similar relation can be given for $1 - \Phi_{\mu,\sigma}(x)$, as $x \to \infty$.

## 9.2. Characterizations by two-sided tail conditions

We start with two asymptotic relations for the normal d.f. $\Phi_{\mu,\sigma}$. First we conclude from (9.1.2) that

$$\lim_{x\to\infty} \frac{x^2}{\log \Phi_{\mu,\sigma}(-x)} = \lim_{x\to\infty} \frac{x^2}{\log\left(1 - \Phi_{\mu,\sigma}(x)\right)} = -2\sigma^2 . \tag{9.2.1}$$

Next we note that, for any d.f. $F$ and its tail function $T_F$,

$$\max\{F(-x), 1 - F(x)\} \leq T_F(x) \leq 2\max\{F(-x), 1 - F(x)\}, \qquad x > 0 .$$

Thus, setting $F = \Phi_{\mu,\sigma}$ we find from (9.2.1)

$$\lim_{x\to\infty} \frac{x^2}{\log T_{\Phi_{\mu,\sigma}}(x)} = -2\sigma^2 . \tag{9.2.2}$$

Turning to an arbitrary inf.div.d.f. $F = \mathcal{L}(a_F, \gamma_F^2, M_F, N_F)$, it is tempting to ask what can be concluded from (9.2.2) if we replace $\Phi_{\mu,\sigma}$ by $F$. We will see that even still weaker asymptotic conditions on $T_F$ ensure that $F$ is normal.

**Proposition 9.2.1.** *Suppose that $F$ is a non-degenerate inf.div. d.f. Then the following three conditions are equivalent.*

(i) *$F$ is normal;*

(ii) $$\lim_{x\to\infty} \frac{x \log x}{\log T_F(x)} = 0; \tag{9.2.3}$$

(iii) *There is a sequence $(x_n)$, $x_n \to \infty$, such that*

$$\lim_{n\to\infty} \frac{x \log x_n}{\log T_F(x_n)} = 0 . \tag{9.2.4}$$

Proof. Clearly, $F$ is normal iff $M_F = N_F = 0$, i.e., iff

$$x_0 := \max\{-\operatorname{lext}(M_F), \operatorname{rext}(N_F)\} = 0 .$$

Using this fact our statement immediately follows from Cor. 7.7.3. □

Another unique determination of a normal d.f. can be expressed by integrals.

**Proposition 9.2.2.** *Suppose that $F$ is a non-degenerate inf.div.d.f. Then $F$ is normal iff*

$$\int_{-\infty}^{\infty} \exp\left(\eta\,|x| \log(|x|+1)\right) \mathrm{d}F(x) < \infty\,, \quad \textit{for all} \quad \eta > 0\,. \tag{9.2.5}$$

Proof. Let us assume $F = \Phi_{\mu,\sigma}$. Then the integral in (9.2.5) can be written as

$$\frac{1}{\sqrt{2\pi}\,\sigma} \int_{-\infty}^{\infty} \exp\left(-\frac{1}{2}\left(\frac{x-\mu}{\sigma}\right)^2 + \eta\,|x| \log(|x|+1)\right) \mathrm{d}x$$

which trivially converges for all $\eta > 0$.

For the converse assume (9.2.5). Then

$$T_F(r) \exp\left(\eta r \log(r+1)\right) \leqq \int_{|x| \geqq r} \exp\left(\eta\,|x| \log(|x|+1)\right) \mathrm{d}F(x) \to 0\,,$$

as $r \to \infty$ for all $\eta > 0$. Switching to the logarithm, we now verify that

$$-\frac{1}{\eta} \leqq \overline{\lim_{r\to\infty}} \frac{r \log(r+1)}{\log T_F(r)} \leqq 0$$

whence (9.2.4) follows. Using Prop. 9.2.1 we arrive at the desired result. □

Next we state another consequence of Prop. 9.2.1.

**Proposition 9.2.3.** *If $F$ is a non-degenerate inf.div.d.f., then only the following two cases are possible: Either $F$ is normal or we have, for some $c > 0$, $x_0 > 0$,*

$$T_F(x) \geqq \mathrm{e}^{-cx \log x}\,, \qquad x \geqq x_0\,. \tag{9.2.6}$$

Proof. Assume that $F$ is not normal. Then by Prop. 9.2.1, (iii), we see that

$$\overline{\lim_{x\to\infty}} \frac{x \log x}{\log T_F(x)} < 0\,.$$

This relation implies (9.2.6) which was to be proved. □

By virtue of (9.2.2) and (9.2.6) we can restate the above result.

**Corollary 9.2.4.** *If $F$ is an arbitrary d.f. satisfying*

$$\exp(-\varepsilon x^2) \leqq T_F(x) \leqq \exp(-\eta x \log x)\,, \qquad x \geqq x_0\,,$$

*for every fixed pair $\varepsilon > 0$, $\eta > 0$, and sufficiently large $x_0 > 0$, then $F$ cannot be an inf.div.d.f.*

The above results make it obvious that in the class of all inf.div.d.f. a normal d.f. is distinguished by a pronounced asymptotic tail property.

## 9.3. Characterizations by one-sided tail conditions

Let $F = \mathscr{L}(a_F, \gamma_F^2, M_F, N_F)$ be an inf.div.d.f. Contrary to the subject of the preceding section it is now our aim to obtain left-sided asymptotic tail conditions of $F$ which uniquely characterize a normal d.f. Clearly, they must be stronger than the conditions for $T_F$ given above. Here we exclude the case $\operatorname{lext}(F) > -\infty$.

First we consider the condition

$$\lim_{x\to\infty} \frac{x \log x}{\log F(-x)} = 0 \tag{9.3.1}$$

which corresponds to (9.2.3). From Cor. 7.7.2 we see that (9.3.1) is necessary and sufficient for $M_F = 0$. The next step is to determine the component $\gamma_F^2$ by an asymptotic property, which is sharper than (9.3.1).

**Proposition 9.3.1.** *Let $F$ be a non-degenerate inf.div.d.f. satisfying $M_F = 0$ and* lext $(F) = -\infty$. *Then*

$$\lim_{x\to\infty} \frac{x^2}{\log F(-x)} = -2\gamma_F^2\,, \qquad \gamma_F \geqq 0\,, \tag{9.3.2}$$

*that is, the limit always exists finitely.*

Note that for $F = \Phi_{\mu,\sigma}$ formula (9.3.2) has already been stated as (9.2.1).

Proof. It is enough to prove the two inequalities

$$\overline{\lim_{x\to\infty}} \frac{x^2}{\log F(-x)} \leqq -2\gamma_F^2 \tag{9.3.3}$$

and

$$\underline{\lim_{x\to\infty}} \frac{x^2}{\log F(-x)} \geqq -2\gamma_F^2 \tag{9.3.4}$$

for $F = \mathcal{L}(a_F, \gamma_F^2, 0, N_F)$.

To prove (9.3.3) it sufficies to assume $\gamma_F > 0$. Then $F$ can be written as $F = \Phi_{0,\gamma_F} * F_N$ where $F_N = \mathcal{L}(a_F, 0, 0, N_F)$. Next we select a number $y_0 \in R_1$ with $c_0 := F_N(y_0) > 0$ so that

$$F(-x) \geqq \int_{-\infty}^{y_0} \Phi_{0,\gamma_F}(-x-y)\, \mathrm{d}F_N(y) \geqq c_0 \Phi_{0,\gamma_F}(-x-y_0)\,.$$

Using (9.1.2) we obtain, as $x \to \infty$,

$$F(-x) \geqq \left[\frac{c_0\gamma_F}{|x+y_0| \sqrt{2\pi}}\,(1+o(1))\right] \exp\left(-\frac{1}{2\gamma_F^2}(x+y_0)^2\right)$$
$$= \exp\left(-\frac{x^2}{2\gamma_F^2}\,(1+o(1))\right),$$

This relation terminates the proof of (9.3.3).

Now we show (9.3.4). In view of the assumption lext $(M) = 0$ we may proceed as in the proof of Prop. 7.7.1 so that (7.7.5) follows. Hence

$$\underline{\lim_{x\to\infty}} \frac{x^2}{\log F(-x)} \geqq -4c_1$$

for every $c_1 > \frac{1}{2}\gamma_F^2$. Now we obtain the inequality (9.3.4) by letting $c_1 \to \frac{1}{2}\gamma_F^2$. □

Prop. 9.3.1 directly implies the next result.

**Corollary 9.3.2.** *For a non-degenerate inf.div.d.f. $F$ satisfying* lext $(F) = 0$ *the following three conditions are equivalent.*

(i) $M_F = 0$ *and* $\gamma_F = 0$;

(ii) $$\lim_{x\to\infty} \frac{x^2}{\log F(-x)} = 0;$$

(iii) *There is a sequence $(x_n)$, $x_n \to \infty$, such that*

$$\lim_{n\to\infty} \frac{x_n^2}{\log F(-x_n)} = 0. \qquad \square$$

The relation (9.3.2) leads to a connection between a normal d.f. and an arbitrary inf.div.d.f.

**Proposition 9.3.3.** *Let $F = \mathcal{L}(a_F, \gamma_F^2, M_F, N_F)$ be an inf.div.d.f. satisfying* lext $(F) = -\infty$ *and let $\mu \in R_1$, $\sigma > 0$ be certain numbers. Then we have the representation*

$$F = \mathcal{L}(a_F, \sigma^2, 0, N_F) \tag{9.3.5}$$

*iff there exist two sequences $(x_n)$, $(y_n)$ tending to infinity such that, as $n \to \infty$,*

$$\frac{F(-x_n)}{\Phi_{\mu,\sigma}(-x_n)} \leqq \exp\left(o(x_n^2)\right), \tag{9.3.6}$$

$$\frac{\Phi_{\mu,\sigma}(-y_n)}{F(-y_n)} \leqq \exp\left(o(y_n^2)\right). \tag{9.3.7}$$

Proof. Let us first assume (9.3.5), i.e. we have $M_F = 0$, $\gamma_F = \sigma > 0$. From (9.3.2) and (9.2.1) we find

$$\Delta := \lim_{x\to\infty} \frac{1}{x^2}\left(\log F(-x) - \log \Phi_{\mu,\sigma}(-x)\right) = 0.$$

This relation is equivalent to

$$\max\left\{\frac{F(-x)}{\Phi_{\mu,\sigma}(-x)}, \frac{\Phi_{\mu,\sigma}(-x)}{F(-x)}\right\} = \exp\left(o(x^2)\right), \qquad x \to \infty,$$

so that (9.3.6) and (9.3.7) follow for all sequences $(x_n)$, $(y_n)$ tending to infinity.

Now we, conversely, assume (9.3.6) and (9.3.7). We carry out an indirect proof in several steps. Suppose first that $M_F \neq 0$, i.e. $l_M < 0$. Then Prop. 7.7.1 implies

$$\lim_{x\to\infty} \frac{1}{x^2} \log F(-x) = 0.$$

Accordingly, (9.2.1) leads to $\Delta = 1/2\sigma^2 > 0$ which contradicts (9.3.6). Hence $M_F = 0$.

Now we are able to make use of (9.3.2) according to which

$$\Delta = \frac{1}{2\sigma^2} - \frac{1}{2\gamma_F^2}.$$

If $\gamma_F < \sigma$ then $\Delta < 0$ and we have a contradiction to (9.3.7). On the other hand, if $\gamma_F > \sigma$ then $\Delta > 0$ which contradicts (9.3.6). This proves $\gamma_F = \sigma$ which completes the proof. $\square$

The above conditions guarantee the representation (9.3.5) where $a_F$, $N_F$ remain undefined. Now we introduce the d.f. $F_N = \mathcal{L}(a_F - \mu, 0, 0, N_F)$ and rewrite (9.3.5) as

$$F = \Phi_{\mu,\sigma} * F_N.$$

Our aim is to find more restrictive tail conditions implying

$$l := \operatorname{lext}(F_N) = 0\,.$$

We remind the reader of Prop. 7.3.9 which links $a_F - \mu$ to $N_F$.

**Proposition 9.3.4.** *Let* $F = \mathscr{L}(a_F, \gamma_F^2, M_F, N_F)$ *be an inf.div.d.f. satisfying* lext $(F) = -\infty$, *and let* $\mu \in R_1$, $\sigma > 0$ *be certain numbers. Then we have* (9.3.5) *where* lext $(F_N) = 0$ *iff there exist two sequences* $(x_n)$, $(y_n)$ *tending to infinity such that, as* $n \to \infty$,

$$\frac{F(-x_n)}{\Phi_{\mu,\sigma}(-x_n)} \leqq \exp\big(o(x_n)\big) \tag{9.3.8}$$

*and*

$$\frac{\Phi_{\mu,\sigma}(-y_n)}{F(-y_n)} \leqq \exp\big(o(y_n)\big)\,. \tag{9.3.9}$$

Proof. Assume first that (9.3.8) and (9.3.9) are satisfied. Then the assumptions of Prop. 9.3.3 are also true so that (9.3.5) follows. It remains to show $l=0$. Let us assume $l>0$, then from (9.1.2) we conclude that

$$\begin{aligned} F(-x) &= \int_{l-0}^{\infty} \Phi_{\mu,\sigma}(-x-y)\, \mathrm{d}F_N(y) \leqq \Phi_{\mu,\sigma}(-x-l) \\ &= \Phi_{\mu,\sigma}(-x) \exp\left(-\frac{lx}{\sigma^2} + O(1)\right) \end{aligned}$$

which contradicts (9.3.9). On the other hand, if $-\infty \leqq l < 0$ then we choose a number $x_0 \in (l, 0)$ so that $c_1 := F_N(x_0) > 0$, and we trivially obtain $F(-x) \geqq c_1\Phi_{\mu,\sigma}(-x - x_0)$ whence for another number $c_2 > 0$ and sufficiently large $x$ it follows that

$$F(-x) \geqq \Phi_{\mu,\sigma}(-x)\, c_2 \exp\left(-\frac{xx_0}{\sigma^2}\right).$$

This contradicts (9.3.8) since $x_0 < 0$. Thus $l = 0$.

Now we conversely assume (9.3.5) and $l = 0$. Then

$$F(-x) = \int_{-0}^{\infty} \Phi_{\mu,\sigma}(-x-y)\, \mathrm{d}F_N(y) \leqq \Phi_{\mu,\sigma}(-x)$$

which leads to (9.3.8). On the other hand, for arbitrary $\varepsilon > 0$ we put $c_3 = \int_{-0}^{\varepsilon} \mathrm{d}F_N(y) > 0$ and obtain

$$F(-x) \geqq \int_{-0}^{\varepsilon} \Phi_{\mu,\sigma}(-x-y)\, \mathrm{d}F_N(y) \geqq c_3\Phi_{\mu,\sigma}(-x-\varepsilon)\,.$$

Applying again (9.1.2) we can verify (9.3.9). The proof is complete. □

We have obtained conditions under which $M_F = 0$ and $\gamma_F = \sigma$. Now we are in a position to give necessary and sufficient conditions under which even $N_F = 0$ and $a_F = \mu$ are fulfilled so that $F = \Phi_{\mu,\sigma}$.

**Proposition 9.3.5.** *Let* $F$ *be an inf.div.d.f. satisfying* lext $(F) = -\infty$, *and let* $\mu \in R_1$, $\sigma > 0$ *be certain numbers. Then* $F = \Phi_{\mu,\sigma}$ *iff there exist two sequences* $(x_n)$, $(y_n)$ *tending to infinity such that the two relations* (9.3.8) *and*

$$F(-y_n) \geqq \Phi_{\mu,\sigma}(-y_n)\big(1 + o(1)\big)\,, \qquad n \to \infty\,, \tag{9.3.10}$$

*are satisfied.*

Proof. It is enough to prove the if-part. Let us assume (9.3.8) and (9.3.10). Then Prop. 9.3.4 yields (9.3.5) where lext $(F_N) = 0$. To finish the proof we show that $F_N$ is degenerate, i.e., $r = \text{rext}\,(F_N) = 0$ is also true. If this were wrong, then we could choose $x_0 \in (0, r)$ and write

$$F(-x) \leqq F_N(x_0)\, \Phi_{\mu,\sigma}(-x) + \Phi_{\mu,\sigma}(-x - x_0)$$
$$= F_N(x_0)\, \Phi_{\mu,\sigma}(-x)\, \big(1 + o(1)\big)\,.$$

This relation contradicts (9.3.10) since $F_N(x_0) < 1$. □

Now we get a further two consequences; the first is connected with the Kolmogorov original conjecture (cf. Th. 7.6.1) and the second was partially formulated as Cor. 7.6.7.

**Corollary 9.3.6.** *Let $F$ be an inf.div.d.f. and $S$ an unbounded set of real numbers. If for some $\sigma > 0$, $\mu \in R_1$ the equation*

$$F(x) = \Phi_{\mu,\sigma}(x)\,, \qquad x \in S\,,$$

*is satisfied, then $F = \Phi_{\mu,\sigma}$.*

This relation immediately follows from Prop. 9.3.5. □

**Corollary 9.3.7.** *Let $F$ be an inf.div.d.f. and $c$ some positive number. If we further assume that*

$$\lim_{x\to\infty} \frac{F(-x)}{\Phi_{\mu,\sigma}(-x)} = c\,, \tag{9.3.11}$$

*then either $c = 1$ and $F = \Phi_{\mu,\sigma}$, or we have $0 < c < 1$ and $F$ admits the representation* (9.3.5) *where* lext $(F_N) = 0$ *and* $c = F_N(+0)$.

Proof. Suppose first $c \geqq 1$, then the conditions (9.3.8) and (9.3.10) are fulfilled whence $F = \Phi_{\mu,\sigma}$ and $c = 1$ follow, that is, the case $1 < c < \infty$ is impossible.

Suppose now $0 < c < 1$. By Prop. 9.3.4 we get (9.3.5) and lext $(F_N) = 0$. Thus, we may write

$$0 \leqq F(x) - F_N(+0)\, \Phi_{\mu,\sigma}(x) = \int_{+0}^{\infty} \Phi_{\mu,\sigma}(x - y)\, \mathrm{d}F_N(y)$$
$$= \int_{+0}^{\varepsilon} + \int_{\varepsilon}^{\infty} \leqq \big(F_N(\varepsilon) - F_N(+0)\big)\, \Phi_{\mu,\sigma}(x) + \Phi_{\mu,\sigma}(x - \varepsilon)$$
$$= \big(F_N(\varepsilon) - F_N(+0)\big)\, \Phi_{\mu,\sigma}(x)\, \big(1 + o(1)\big)\,, \quad \text{as} \quad x \to -\infty\,,$$

(for the latter equality cf. (9.1.2)). The above estimates prove (9.3.11) for $c = F_N(+0)$ since $F_N(\varepsilon) - F_N(+0)$ can be made arbitrarily small. □

**Example 9.3.8.** Let $P_\lambda$ be the Poisson d.f. with parameter $\lambda > 0$ and set, for certain $\sigma > 0$, $\mu \in R_1$,

$$F = \Phi_{\mu,\sigma} * P_\lambda\,;$$

then we show that

$$F(-x) = \mathrm{e}^{-\lambda}\, \Phi_{\mu,\sigma}(-x) \left(1 + \lambda \exp\left(\frac{x}{\sigma^2}\right) \big(1 + o(1)\big)\right), \tag{9.3.12}$$

as $x \to \infty$, that is, (9.3.11) is valid with $c = e^{-\lambda}$. Indeed, we proceed as in the proof of Cor. 9.3.7 and obtain

$$0 \leqq F(-x) - e^{-\lambda} \Phi_{\mu,\sigma}(-x) = e^{-\lambda} \sum_{1}^{\infty} \Phi_{\mu,\sigma}(-x - k) \frac{\lambda^k}{k!}$$

$$\leqq \lambda e^{-\lambda} \Phi_{\mu,\sigma}(-x) \left[\frac{\Phi_{\mu,\sigma}(-x-1)}{\Phi_{\mu,\sigma}(-x)}\right] (1 + o(1))$$

whence (9.3.12) follows. □

**Remark 9.3.9.** If the assumptions of Cor. 9.3.6. hold for a bounded set $S$ of real numbers, then we cannot draw the conclusion that $F = \Phi_{\mu,\sigma}$ without additional assumptions, see Ex. 7.6.9. □

## 9.4. A probabilistic version of a theorem of Hardy

It was N. Wiener who remarked that a non-trivial pair of functions $p$ and its Fourier transform $f$ cannot both be very small at infinity, in other words if $p$ and $f$ both converge to zero very quickly, then $p = f = 0$.

**Example 9.4.1.** We consider the extreme case in which $f(x) = p(x) = 0$, $|x| \geqq x_0 > 0$ Then

$$f(z) = \frac{1}{\sqrt{2\pi}} \int_{-x_0}^{x_0} e^{-izx} p(x) \, dx$$

is integral in view of the same argument as we used in the proof of Prop. 4.2.1. Hence from the assumption $f(x) = 0$, $|x| \geqq x_0$ we clearly have $f = 0$. □

The precise statement of Wiener's remark is contained in a theorem of Hardy. We give a probabilistic reformulation in which it takes the form of a characterization theorem for the normal d.f. Note that the assumption (9.4.1) on the c.f. implies (by the Fourier inversion theorem Th. 3.2.2) that the density $p$ exists; further in view of the reflection property a c.f. can have only an even number of zeros or infinitely many.

**Proposition 9.4.2.** *Let the c.f. $f$ of the density $p$ satisfy*

$$f(t) = O(|t|^m e^{-bt^2}), \qquad |t| \to \infty, \tag{9.4.1}$$

*and assume*

$$p(x) = O(|x|^n e^{-ax^2}), \qquad |x| \to \infty, \tag{9.4.2}$$

*where $m$ and $n$ are non-negative integers and $a > 0$, $b > 0$, $4ab \geqq 1$. Then $4ab = 1$ and*

$$f(t) = e^{-bt^2} P(t) \tag{9.4.3}$$

*where $P$ stands for a polynomial with an even degree not higher than* $\min\{m, n\}$. *If $m = 0$ or $n = 0$ then* $f(t) = \exp(-bt^2)$, *i.e., $p$ is the density of a normal d.f.*

Note that in the case of the normal d.f. $\Phi_{0,\sigma}$ we have $1/a = 2\sigma^2$, $b = \sigma^2/2$ so that $4ab = 1$. It is also the statement that $4ab > 1$ is impossible that confirms Wiener's remark.

Proof. By Prop. 4.3.8 it is clear from (9.4.2) that the c.f. $f$ is integral. We are going to estimate its growth. By assumption (9.4.2) we have, for a certain $C > 0$ and $z = t + is$,

$$|f(z)| = \left| \int_{-\infty}^{\infty} e^{izu}\, p(u)\, du \right| \leqq \int_{-1}^{1} e^{|zu|}\, p(u)\, du + C \int_{-\infty}^{\infty} |u|^n\, e^{-au^2}\, e^{|zu|}\, du .$$

It is enough to consider the growth of the latter term for which we can write

$$2C \left(\frac{d}{d\,|z|}\right)^n \int_0^{\infty} \exp\,(-au^2 + |z|\, u)\, du$$

$$\leqq 2C \left(\frac{d}{d\,|z|}\right)^n \exp\left(\frac{|z|^2}{4a}\right) \int_{-\infty}^{\infty} \exp\left(-a\left(u - \frac{|z|}{2a}\right)^2\right) du .$$

Thus the integral on the right does not depend on $|z|$ so that

$$f(z) = O\left(|z|^n \exp\left(\frac{|z|^2}{4a}\right)\right), \qquad |z| \to \infty . \tag{9.4.4}$$

Now we introduce the even integral function

$$g_1(z) := f(z) + f(-z) = \sum_{k=0}^{\infty} a_{2k} z^{2k} , \tag{9.4.5}$$

which has obviously no higher growth than $f$. Moreover, from (9.4.1) we obtain

$$g_1(t) = O(|t|^m\, e^{-bt^2}) , \qquad |t| \to \infty . \tag{9.4.6}$$

Putting $w = z^2$ we now write

$$g_1(z) = g_2(w) := \sum_{j=0}^{\infty} a_{2j} w^j ;$$

from (9.4.4) we find for this integral function the estimate

$$g_2(w) = O\left(|w|^{n/2} \exp\left(\frac{|w|}{4a}\right)\right), \qquad |w| \to \infty . \tag{9.4.7}$$

From (9.4.6) we see that

$$g_2(r) = O(r^{m/2}\, e^{-br}) , \qquad r > 0 . \tag{9.4.8}$$

Now we apply the theory of Phragmén and Lindelöf where we write $P_k$ for polynomials with degree not higher than $k$ which are not necessarily the same. Prop. C.2.4.5 (with $\tau = 1/4a$) rules out the case $4ab > 1$ (so that $4ab = 1$ by assumption) and tells us that

$$g_2(w) = \exp\,(-w/4a)\, P_q(w)$$

with $q = \min\,\{n/2, m/2\}$. Hence

$$g_1(z) = \exp\,(-z^2/4a)\, P_{2q}(z) .$$

Next we consider the odd function

$$h_1(z) := f(z) - f(-z);$$

then the even function $zh_1(z)$ can be treated like $g_1$ with $m + 1$ instead of $m$; accordingly $h_1(z) = \exp\,(-z^2/4a)\, P_{2q}(z)$ follows. Thus we obtain

$$f(z) = \big(g_1(z) + h_1(z)\big)/2 = \exp\,(-z^2/4a)\, P_{2q}(z) . \qquad \square$$

Example 9.4.3. C.f. of the form (9.4.3) with $p = 1$ occur. This is shown by the c.f. $f(t) = (1 - t^2) \exp(-t^2/2)$. It corresponds to the density $p(x) = (x^2/\sqrt{2\pi}) \exp(-x^2/2)$ as is seen by writing

$$f(t) = -\frac{d^2}{dt^2}(e^{-t^2/2}) = -\frac{d^2}{dt^2}\frac{1}{\sqrt{2\pi}}\int_{-\infty}^{\infty} e^{itx-(x^2/2)}\,dx = \int_{-\infty}^{\infty} e^{itx}\,p(x)\,dx\,. \quad \square$$

## 9.5. Supplement

**9.5.1.** *Let* $G = \mathcal{L}(a_G, \gamma_G^2, M_G, N_G)$ *be inf.div. and* $V$ *a d.f. with* lext $(V) \geqq 0$. *Further, let* $F$ *be the power mixture of* $G$ *with respect to* $V$ *(formula* (7.8.2)).

(i) *If* $\gamma_G > 0$ *and we have, as* $x \to \infty$,

$$\big(1 + o(1)\big) \leqq \frac{F(-x)}{\Phi(-x)} \leqq e^{o(x)}\,,$$

*then* $F = \Phi$. *Moreover, there is a number* $\sigma > 0$ *such that* $G = \Phi_{0,\sigma}$ *and* $V = U_{1/\sigma}$ *(degenerate d.f.)* (Bethmann (1984)).

(ii) $F$ *has a normal component* $\Phi_{0,\sigma_F}$ *iff* $\gamma_G > 0$ *and* lext $(V) > 0$. *In either case we have* $\sigma_F^2 =$ lext $(V)\,\gamma_G^2$. *Note that this result is true even if* $F$ *is not inf.div.* (Siegel (1983)).

A simple consequence of (ii) is the following assertion. *The relation* $F = \Phi$ *follows iff there is a constant* $\sigma > 0$ *such that* $G = \Phi_{0,\sigma}$ *and* $V = U_{1/\sigma}$ (Kruglov (1976b)).

**9.5.2.** *Let* $F$ *be an inf.div.d.f. satisfying* lext $(F) = -\infty$, *and let there exists a sequence* $(x_n)$, $x_n \to \infty$, *such that*

$$\frac{\log F(-x_n)}{x_n^2} \to c\,, \qquad 0 < c < \infty\,.$$

*Assume, further, that* $F(x) = \Phi_{\mu,\sigma}(x)$, $x \in S$, *for some set* $S$ *of real numbers with a finite limit point. Then* $F = \Phi_{\mu,\sigma}$. — For the proof we apply Prop. 9.3.1 and use the fact that an inf.div.d.f. $F$ with $\gamma_F > 0$ is analytic, cf. Prop. 7.5.1.

**9.5.3.** Answering a question of Rossberg it was Siegel (1980) who generalized Th. 7.6.1 in the following way where (. , .) stands for the scalar product. *Let* $H$ *be a separable Hilbertspace, let* $F$ *stand for an inf.div. distribution with values in* $H$, *and introduce a half space* $H = \{x \in H : (x, x_0) < c\}$ *for some* $c \in R_1$; *moreover, let* $\varphi$ *be a normal distribution on* $H$ *with* $\varphi(H) < 0$. *If* $F(A) = \varphi(A)$ *for all Borel sets* $A \subseteq H$ *then* $F = \varphi_0$. The proof makes use of the method of projections which permits to reduce the problem to Th. 7.6.1.

**Problem.** Does the above statement remain true if we replace $H$ by a narrower angle?

## Notes and bibliographical data

Prop. 9.2.2 is due to Kruglov (1970), and Prop. 9.2.3 can be found in Steutel (1974). Sect. 9.3 contains an extension on the main results of the paper Rossberg and Siegel (1981), but Cor. 9.3.7 Contains an earlier result of Riedel (1976). Prop. 9.4.2 is a probabilistic reformulation of a theorem of Hardy contained in Kawata (1972), Th. 8.1.1.

# III. Limit theorems for sums of independent random variables

## Chapter 10. Basic ideas and concepts

The main part of probability theory investigates random mass phenomena, i.e. events and processes in which a large number of r.v. is involved. A very successful method of tackling problems of this kind consists in the following: Instead of considering a finite, large number $n$ of r.v. we pass to the limit as $n \to \infty$ and state theorems on the asymptotic behaviour of the probabilities under consideration. The most famous examples of such limit theorems are provided by the *law of large numbers* (LLN) and the *central limit theorem* (CLT). The first tells us under which conditions the well known stability of the arithmetic means occurs such that we obtain results which are nearly independent of chance; we have stated an elegant example in Th. 3.7.1. Very general formulations are given in Sect. 11.4. On the other hand, the different versions of the CLT contain conditions under which the normalized sums are subject to d.f. which completely converge to a normal d.f. A very well-known example for such a result is given in Th. 3.7.2; more general results are given in Chap. 12.

Relations of these two types have been the main object of research in probability theory for more than 200 years. But there exists a third classic limit law which long stood isolated — the theorem of Poisson, which we stated in Th. 3.7.3. We will see later that there is a deep reason for this isolation; it is connected with the simple triangular array which we had to introduce in Sect. 3.7 before we could prove Th. 3.7.3.

The present chapter explains how the general limit problems arose from the above classic directions of research. Further we acquaint the reader with several basic notions, such as infinite smallness, accompanying distributions, and restricted convergence.

In connection with the latter we explain the general plan of proofs of theorems involving the assumption of restricted convergence.

After studying Chap. 10 the reader may skip Chap. 11 and 12 and may simply look up the few results from here which are needed when reading Chap. 13.

### 10.1. The general limit problem

Let there be a sequence $(X_n)$ of i.r.v. Let $(a_n)$ and $(b_n)$, $b_n > 0$, be two sequences of real numbers and consider the normed and centred sums

$$S_n = \frac{1}{b_n}(X_1 + X_2 + \dots + X_n - a_n) \sim F_n . \tag{10.1.1}$$

The oldest limit problem for $(S_n)$ is the question of under which conditions does the convergence

$$\mathsf{P}(|S_n| \geqq \varepsilon) \to 0 , \qquad n \to \infty , \tag{10.1.2}$$

take place for all $\varepsilon > 0$; the classic theory investigated this convergence only in the special case $b_n = n$. Results of this kind are called LLN. As mentioned above, a simple example is given in Th. 3.7.1.

Clearly, the LLN does not contain much information about the deviations between $S_n$ and 0 for large $n$. For instance, by means of the LLN it is not possible to estimate the probabilities

$$\mathsf{P}(\alpha \leqq \sum_{k=1}^{n} X_k \leqq \beta)$$

for all numbers $\alpha < \beta$ even if $n$ is large. A useful solution to this problem is connected with the well-known CLT. We say that the sequence $(X_n)$ of i.r.v. is subject to the CLT, if there exist two sequences $(a_n)$ and $(b_n)$, $b_n > 0$, and two numbers $\mu \in \mathbb{R}_1$, $\sigma > 0$, such that

$$F_n \xrightarrow{c} \Phi_{\mu,\sigma}, \qquad n \to \infty. \tag{10.1.3}$$

We stated a classic example of the CLT in Th. 3.7.2. In this case $b_n := \sqrt{n \operatorname{Var} X_1}$ and $a_n := n\mathsf{E}X_1$.

Now we turn to the general problem of finding the class of all possible limit d.f. for $(F_n)$. Historically the first step towards the solution of this problem concerned the case in which the i.r.v. $X_k$ are identically distributed. It will be briefly treated in Chap. 13. There we will see that the corresponding set $\boldsymbol{S}$ of limit d.f. coincides with the set of all stable laws. Note that the above mentioned Th. 3.7.2 yields $\Phi_{\mu,\sigma} \in \boldsymbol{S}$; in view of this result, in Chap. 13 we can focus our attention on the case that the second moment of $X_1$ does not exist.

The next problem to be considered is a generalization of the one above in the sense that the r.v. $X_k$ are not necessarily identically distributed. Then an additional restriction on the r.v. $X_k$ has to be imposed; it means that the role of a single summand becomes vanishingly small as $n \to \infty$. This condition will be treated in a general form in the next section; it implies, for instance, $b_n \to \infty$. In this model the class of limit laws is usually denoted by $\boldsymbol{L}$ (Lévy); for details we refer to Petrov (1975). In particular, it can be proved that every non-degenerate limit law $F \in \boldsymbol{L}$ is absolutely continuous and inf.div. Further, we trivially see that $\boldsymbol{S} \subseteqq \boldsymbol{L}$. This model is not treated in the present book especially since results involving r-convergence are still lacking.

As early as in the 19th century the theory of summation of r.v. knew not only the degenerate and the normal d.f. as limit d.f., but also a third type, the Poisson d.f. $P_\lambda$; the corresponding classic result due to Poisson is given in Th. 3.7.3. Since $P_\lambda$ is not continuous, it is clear that $P_\lambda \notin \boldsymbol{L}$ for all $\lambda > 0$, i.e., we cannot interprete it as a limit d.f. for the sums introduced in (10.1.1). This shows that it might be reasonable to generalize the model under consideration. Let us define i.r.v.

$$X_{nk} = X_k/b_n; \qquad k = 1, 2, \dots, n; \qquad n = 1, 2, \dots,$$

and set $A_n = a_n/b_n$. Then the sums (10.1.1) can be written as

$$S_n = \sum_{k=1}^{n} X_{nk} - A_n.$$

We now observe that we are really dealing with a special case of the following more general model.

**Definition 10.1.1.** Let $(k_n)$ be a certain sequence of natural numbers. Then the double sequence of row-wise i.r.v.

$$\begin{array}{l} X_{11}, X_{12}, \dots, X_{1k_1} \\ X_{21}, X_{22}, \dots, X_{2k_2} \\ \dots \qquad \dots \\ X_{n1}, X_{n2}, \dots, X_{nk_n} \\ \dots \qquad \dots \end{array}$$

is called *triangular array of i.r.v.* □

The d.f. of $X_{nk}$ will be denoted by $F_{nk}$. Further, for some sequence $(A_n)$ of real numbers we introduce the sums

$$S_n = \sum_{k=1}^{k_n} X_{nk} - A_n \sim F_n . \tag{10.1.4}$$

Now we ask what can be said about the limit behaviour of $(F_n)$.

First of all, it is easy to see that any d.f. $F$ can appear as the limit d.f. of such a sequence. Indeed, assume $F_{n1} = F$ while $\mathsf{P}(X_{nk} = 0) = 1$ for $k > 1$ and every $n$. Then $F_n = F$ follows so that $F_n \xrightarrow{c} F$ is trivially satisfied. This fact leads to the following general problems.

(P1) Determine the set of all limit laws for the sequence $(F_n)$ in (10.1.4) by imposing appropriate additional conditions on the triangular array $(X_{nk})$.

(P2) Find necessary and sufficient conditions such that the sequence $(F_n)$ completely converges to a given limit d.f. $F$.

In the theory of summation of i.r.v. it is also often of interest to present conditions which are only necessary or only sufficient. For an example, we will study compactness conditions which are necessary while statements involving sufficient conditions are given in Sect. 3.7.

Two important special cases of (P2) are the questions of under which assumptions the LLN and the CLT, resp., hold true for the triangular array $(X_{nk})$ of i.r.v. We state some results for the LLN in Sect. 11.4 while the CLT is studied in detail in Chap. 12.

In Sect. 3.7 we have obtained the Poisson d.f. as a limit d.f. To this end, we have considered a very special triangular array of i.r.v. A more general model will be useful in proving Prop. 10.3.2; it concerns identically distributed summands. This proposition also tells us that the corresponding set of limit d.f. is contained in the set $\boldsymbol{G}$ of all inf.div.d.f. We note that the converse is trivially also valid. In fact, if $F$ is an arbitrary inf.div.d.f., then we may write $F = ({}^nF)^{*n}$ for all $n \geqq 2$. Now we choose i.i.d.r.v. $X_{nk} \sim {}^nF$, $k = 1, 2, \dots, n := k_n$, so that $F_n = ({}^nF)^{*n} \xrightarrow{c} F$ follows.

In the next section we will introduce another mild restriction on the summands $X_{nk}$. We will see later that the corresponding set of limit d.f. also coincides with the set $\boldsymbol{G}$.

## 10.2. Infinite smallness

We consider a triangular array defined as above and introduce an important condition; it implies intuitively that no summand is likely to receive a preponderating influence on the sums $S_n$. In all that follows the expressions $\max\limits_k$, $\sum\limits_k$, $\bigcup\limits_k$, mean $\max\limits_{1 \leqq k \leqq k_n}$, $\sum\limits_{k=1}^{k_n}$, $\bigcup\limits_{k=1}^{k_n}$.

**Definition 10.2.1.** The triangular array $(X_{nk})$ of i.r.v. satisfies the condition of *infinite smallness*, if

$$\lim_{n\to\infty} \max_k \mathsf{P}(|X_{nk}| \geqq \varepsilon) = 0\,, \quad \text{for all} \quad \varepsilon > 0\,.$$

The r.v. $X_{nk}$ are then also called *infinitely small.* □

First of all, we see that infinite smallness is intimately connected with $k_n \to \infty$ as $n \to \infty$. Indeed, let us suppose that there is a subsequence $(k_{n'})$ such that $k_{n'} \leqq C$ for a number $C$ and all $n'$. Then

$$\mathsf{P}(|\sum_k X_{n'k}| \geqq \varepsilon) \leqq \mathsf{P}\left(\bigcup_k \left\{|X_{n'k}| \geqq \frac{\varepsilon}{k_{n'}}\right\}\right)$$

$$\leqq \sum_k \mathsf{P}\left(|X_{n'k}| \geqq \frac{\varepsilon}{k_{n'}}\right) \leqq C \max_k \mathsf{P}\left(|X_{n'k}| \geqq \frac{\varepsilon}{C}\right) \to 0$$

as $n' \to \infty$, so that for this subsequence only a degenerate limit d.f. is possible.

In the rest of this section we prove equivalent reformulations and several consequences of infinite smallness. They can be skipped at first reading and studied as soon as they are needed.

Before giving a first lemma we remind the reader of the function $w(x) = \min\{x^2, 1\}$.

**Lemma 10.2.2.** *The following statements are equivalent:*

(i) *The infinite smallness is satisfied.*

(ii) $$\max_k \int_{-\infty}^{\infty} w(x)\, \mathrm{d}F_{nk}(x) \to 0\,, \qquad n \to \infty;$$

(iii) $$\max_k |f_{nk}(t) - 1| \to 0\,, \qquad n \to \infty\,,$$

*uniformly on any bounded interval.*

Proof. We show the implications (ii) ⇒ (i) ⇒ (iii) ⇒ (ii).

First we suppose (ii) and make use of the inequality

$$\max \mathsf{P}\,(|X_{nk}| \geqq \varepsilon) \leqq \frac{1}{\varepsilon^2} \max_k \int_{-\infty}^{\infty} w(x)\, \mathrm{d}F_{nk}(x)\,, \qquad 0 < \varepsilon < 1\,,$$

so that (i) follows.

(i) ⇒ (iii) We fix $T > 0$, then for every $|t| \leqq T$ and arbitrary $\varepsilon > 0$ we find from Prop. B.4.1 that

$$\max_k |f_{nk}(t) - 1| \leqq \max_k \Big|\int_{|x|<\varepsilon} (\mathrm{e}^{\mathrm{i}tx} - 1)\, \mathrm{d}F_{nk}(x)\Big| + \max_k \Big|\int_{|x|\geqq\varepsilon} (\mathrm{e}^{\mathrm{i}tx} - 1)\, \mathrm{d}F_{nk}(x)\Big|$$

$$\leqq T\varepsilon + 2 \max_k \mathsf{P}(|X_{nk}| \geqq \varepsilon)\,.$$

(iii) ⇒ (ii) We need the relation

$$w(x) \leqq 1 - \max\{0, 1 - |x|\} = \frac{2}{\pi} \int_{-\infty}^{\infty} (1 - \mathrm{e}^{\mathrm{i}tx})\, \frac{\sin^2 t/2}{t^2}\, \mathrm{d}t\,,$$

see Table. Interchanging the order of integration we have

$$\int_{-\infty}^{\infty} w(x)\, \mathrm{d}F_{nk}(x) \leqq \frac{1}{2\pi} \int_{|t|\leqq T} |f_{nk}(t) - 1|\, \mathrm{d}t + \frac{4}{\pi} \int_{|t|>T} \frac{\sin^2 t/2}{t^2}\, \mathrm{d}t\,.$$

Now by (iii) we see that the right side is arbitrarily small if we choose first $T$ and then $n$ sufficiently large. □

We denote by $m_{nk}$ a median of $X_{nk}$ and introduce the numbers

$$a_{nk} = \int_{|x|<\tau} x \, \mathrm{d}F_{nk}(x) \,, \quad \text{for fixed} \quad \tau > 0 \,.$$

**Lemma 10.2.3.** *Infinite smallness implies*

$$\max_k |m_{nk}| \to 0 \,, \qquad n \to \infty \,, \tag{10.2.1}$$

*and*

$$\max_k |a_{nk}| \to 0 \,, \qquad n \to \infty \,. \tag{10.2.2}$$

*Further, the differences $X_{nk} - m_{nk}$ and $X_{nk} - a_{nk}$ are also infinitely small.*

Proof. Our assumption implies $\min_k \mathsf{P}(|X_{nk}| \leqq \varepsilon) > 1/2$ for all $\varepsilon > 0$ and $n$ large enough; thus

$$F_{nk}(-\varepsilon) < 1/2 < F_{nk}(\varepsilon) \,.$$

Hence $m_{nk} \in (-\varepsilon, \varepsilon)$ for all $k$ which proves (10.2.1).

The relation (10.2.2) follows from

$$|a_{nk}| \leqq \varepsilon + \int_{\varepsilon \leqq |x| < \tau} |x| \, \mathrm{d}F_{nk}(x) \leqq \varepsilon + \tau \mathsf{P}(|X_{nk}| \geqq \varepsilon) \,.$$

Now we use $|m_{nk}| < \varepsilon/2$ and obtain

$$\max_k \mathsf{P}(|X_{nk} - m_{nk}| \geqq \varepsilon) \leqq \max_k \mathsf{P}(|X_{nk}| \geqq \varepsilon/2) \to 0 \,, \qquad n \to \infty \,.$$

The corresponding assertion for the r.v. $X_{nk} - a_{nk}$ can be shown in just the same way. □

Next we introduce the d.f. $F'_{nk}(x) = F_{nk}(x + a_{nk})$ and denote by $f'_{nk}$ the c.f. of $F'_{nk}$.

**Lemma 10.2.4.** *Infinite smallness implies*

$$\max_{|t| \leq T} |f'_{nk}(t) - 1| \leqq c_1 \int_{-\infty}^{\infty} w(x) \, \mathrm{d}F'_{nk}(x) \tag{10.2.3}$$

*for any fixed $T > 0$, sufficiently large $n$, and a certain constant $c_1$ only depending on $\tau$, $T$.*

Proof. For the sake of brevity we set $a = a_{nk}$, $F = F_{nk}$. The definition of the number $a$ obviously implies

$$\int_{|x|<\tau} (x - a) \, \mathrm{d}F(x) = a \int_{|x| \geqq \tau} \mathrm{d}F(x) \,. \tag{10.2.4}$$

Further, we have

$$I := |f'_{nk}(t) - 1| = \left| \int_{-\infty}^{\infty} (\mathrm{e}^{\mathrm{i}t(x-a)} - 1) \, \mathrm{d}F(x) \right|$$

$$\leqq 2 \int_{|x| \geqq \tau} \mathrm{d}F(x) + \left| \int_{|x|<\tau} t(x - a) \, \mathrm{d}F(x) \right| + \int_{|x|<\tau} |\mathrm{e}^{\mathrm{i}t(x-a)} - 1 - \mathrm{i}t(x - a)| \, \mathrm{d}F(x) \,.$$

Using (10.2.4) and Prop. B.4.1 we get, for $|t| \leq T$,

$$I \leqq (2 + |a| \, T) \int_{|x| \geqq \tau} \mathrm{d}F(x) + T^2/2 \int_{|x|<\tau} (x - a)^2 \, \mathrm{d}F(x) \,. \tag{10.2.5}$$

By (10.2.2) we have $|a| \leqq \tau/2$ for $n$ large enough. Hence

$$w(x-a) = \min\{(x-a)^2, 1\} \geqq \min\{\tau^2/4, 1\} \geqq \left(\frac{4}{\tau^2}+1\right)^{-1}, \qquad |x| \geqq \tau. \tag{10.2.6}$$

On the other hand, we have, for $|a| \leqq \tau/2$ and $|x| < \tau$,

$$1 > 4(x-a)^2/(9\tau^2)$$

so that

$$w(x-a) \geqq (x-a)^2 \min\{1, 4/(9\tau^2)\} \geqq (x-a)^2 \left(1 + (3\tau/2)^2\right)^{-1}, \qquad |x| < \tau \tag{10.2.7}$$

Combining (10.2.5)—(10.2.7) we finally obtain

$$\max_{|t|\leqq T} |f'_{nk}(t) - 1|$$
$$\leqq \left[\left(2+\frac{T\tau}{2}\right)\left(\frac{4}{\tau^2}+1\right) + \frac{T^2}{2}\left(1+\left(\frac{3\tau}{2}\right)^2\right)\right] \int_{-\infty}^{\infty} w(x-a)\,\mathrm{d}F(x). \qquad \square$$

**Lemma 10.2.5.** *Infinite smallnes implies*

$$\tfrac{1}{8} \int_{-\infty}^{\infty} w(x-m_{nk})\,\mathrm{d}F_{nk}(x) \leqq \int_{-\infty}^{\infty} w(x)\,\mathrm{d}F'_{nk}(x) \leqq c_2 \int_{-\infty}^{\infty} w(x-m_{nk})\,\mathrm{d}F_{nk}(x)$$

*for sufficiently large $n$ and a certain constant $c_2$ only depending on $\tau$.*

Proof. The first inequality easily follows from Cor. 1.4.8. Next we set $a = a_{nk}$, $m = m_{nk}$, $F = F_{nk}$ and write

$$I := \int_{-\infty}^{\infty} w(x-a)\,\mathrm{d}F(x) \leqq \int_{|x|<\tau} (x-a)^2\,\mathrm{d}F(x) + \int_{|x|\geqq\tau} \mathrm{d}F(x).$$

The inequality $(x-a)^2 \leqq (x-m)^2 + 2(m-a)(x-a)$ is obviously true both for $m \leqq a$ and $m > a$. Accordingly,

$$I \leqq \int_{|x|<\tau} (x-m)^2\,\mathrm{d}F(x) + 2(m-a)\int_{|x|<\tau} (x-a)\,\mathrm{d}F(x) + \int_{|x|\geqq\tau} \mathrm{d}F(x).$$

By (10.2.4) we further obtain

$$I \leqq \int_{|x|<\tau} (x-m)^2\,\mathrm{d}F(x) + \left(1 + 2|a|(|m|+|a|)\right)\int_{|x|\geqq\tau} \mathrm{d}F(x).$$

The inequality (10.2.6) and (10.2.7) rest basicly on the fact that $|a| < \tau/2$ for sufficiently large $n$. Since $|m| < \tau/2$ by L. 10.2.3 we see that these inequalities hold also true on replacing $a$ by $m$. Thus

$$I \leqq \left[\left(1+\left(\frac{3\tau}{2}\right)^2\right) + (1+\tau^2)\left(\frac{4}{\tau^2}+1\right)\right] \int_{-\infty}^{\infty} w(x-m)\,\mathrm{d}F(x).$$

This completes the proof of the second inequality. $\square$

## 10.3. The method of accompanying distributions

Let us consider a triangular array $(X_{nk})$, $k = 1, 2, \ldots, k_n$, of i.r.v. Then for suitably chosen real numbers $c_{nk}$ we introduce another triangular array $(\tilde{X}_{nk})$, $k = 1, 2, \ldots, k_n$, of i.r.v. where $\tilde{X}_{nk}$ are subject to the inf.div.d.f. $\tilde{F}_{nk}$ with c.f.

$$\tilde{f}_{nk}(t) = \exp\left(f_{nk}(t)\, e^{-ic_{nk}t} - 1 + ic_{nk}t\right). \tag{10.3.1}$$

It is obvious that $\tilde{F}_{nk}(x - c_{nk})$ are compound Poisson d.f.

**Definition 10.3.1.** Let $F_n$ be the d.f. of the sum

$$S_n = X_{n1} + X_{n2} + \dots + X_{nk_n} - A_n\,, \qquad n = 1, 2, \dots,$$

for some real numbers $A_n$ and introduce the sum

$$\tilde{S}_n = \tilde{X}_{n1} + \tilde{X}_{n2} + \dots + \tilde{X}_{nk_n} - A_n\,, \qquad n = 1, 2, \dots$$

Then the d.f. $\tilde{F}_n$ of $\tilde{S}_n$ are called the *accompanying d.f. of* $F_n$. □

In modern representations of the theory of summation of i.r.v. there is apparently no other access to the main results than via the accompanying d.f. It is the point of the method that we can establish a relation between $(F_n)$ and $(\tilde{F}_n)$. In this way the asymptotic behaviour of $(F_n)$ can be studied by considering that of $(\tilde{F}_n)$. Since the d.f. $\tilde{F}_n$ are inf.div. we are then able to apply the powerful tools which we have at hand for sequences of inf.div.d.f. This idea is due to Bavli and Gnedenko and was first elaborated in detail by Gnedenko and Kolmogorov (1959).

To explain the method of accompanying d.f. we will present here two statements for a special triangular array by specifying the d.f. $F_{nk}$; namely we put $F_{nk} = V_n$ for all $k$. Then we will see that the simple choice of $c_{nk} = 0$ for all $n$, $k$ is possible so that the proofs are particularly simple. In Chap. 11 we treat more general models; then the constants $c_{nk}$ will be chosen as medians $m_{nk}$ of $X_{nk}$ or as the numbers $a_{nk}$ introduced in the previous section. Note that under certain additional assumptions on the moments of $X_{nk}$ we can set $c_{nk} = \mathsf{E}X_{nk}$, cf. Sect. 12.2.

**Proposition 10.3.2.** *Put $A_n = 0$ and $c_{nk} = 0$ for all $n$, $k$, and assume that the r.v. $X_{nk}$ are i.d. with the common d.f. $V_n$ for $k = 1, 2, \dots, k_n$. Assume, moreover, $k_n \to \infty$ as $n \to \infty$. Then*

$$F_n \xrightarrow{c} F\,, \qquad n \to \infty\,, \tag{10.3.2}$$

*holds iff*

$$\tilde{F}_n \xrightarrow{c} F\,, \qquad n \to \infty\,, \tag{10.3.3}$$

*is satisfied. In either case $F$ is inf.div.*

Proof. First we suppose (10.3.3) or, in other words, for the corresponding c.f. we have $\tilde{f}_n = e^{k_n(v_n - 1)} \to f$ where $v_n$ stands for the c.f. of $V_n$. Then from $k_n \to \infty$ we necessarily obtain $v_n(t) \to 1$ for all $t$. This convergence is uniform on any finite interval $I$ so that $v_n(t) \neq 0$ for all $t \in I$ and sufficiently large $n$. Thus we get, as $n \to \infty$,

$$\begin{aligned} \log f_n(t) &= k_n \log\left[1 + \left(v_n(t) - 1\right)\right] \\ &= k\left(v_n(t) - 1\right)\left(1 + o(1)\right) = \log \tilde{f}(t)\left(1 + o(1)\right). \end{aligned} \tag{10.3.4}$$

Hence, $f_n \to f$, i.e., $F_n \xrightarrow{c} F$ holds.

Now we suppose (10.3.2) so that $f_n = v_n^{k_n} \to f$. Since $f$ is a c.f. there exists $t_0 > 0$ such that $f(t) \neq 0$, $|t| \leqq t_0$. Thus we have $v_n(t) \neq 0$ for $|t| \leqq t_0$ and sufficiently large $n$, and we may write

$$\log f_n(t) = k_n \log v_n(t) \to \log f(t)\,, \qquad |t| \leqq t_0\,.$$

Applying again $k_n \to \infty$ we obtain $v_n(t) \to 1$ on this interval. Clearly, $\hat{f} = 1$ is an analytic c.f. which is trivially defined by its values on an arbitrary interval. Therefore, we conclude from Prop. 3.4.5 that $v_n(t) \to 1$ for all $t$. Thus (10.3.4) is again true for all $t$ so that $\tilde{f}_n \to f$, i.e., $\tilde{F}_n \xrightarrow{c} F$.

Finally, Prop. 7.4.1 tells us that $F$ is inf.div. □

**Proposition 10.3.3.** *Under the same assumptions as in Prop. 10.3.2 the sequence $(F_n)$ is relatively compact iff so is $(\tilde{F}_n)$.*

Proof. Let $(F_n)$ be relatively compact and choose an arbitrary subsequence $(\tilde{F}_{n'})$ of $(\tilde{F}_n)$. By assumption there is a subsequence $(n'')$ of $(n')$ such that $F_{n''} \xrightarrow{c} F$. Now Prop. 10.3.2 can be applied to $(F_{n''})$ so that $\tilde{F}_{n''} \to F$, i.e., $(\tilde{F}_n)$ is relatively compact by Def. 2.4.1. The converse assertion can be shown similarly. □

## 10.4. The first central limit theorem involving restricted convergence

The concept of r-convergence on sets $S \subseteq R_1$ has already been defined in Chap. 2; it generalizes weak convergence, which is only defined for $S = R_1$. It is now in order to explain its origin. We state a result which was conjectured by Zolotarev and emphasize that it represents a CLT of a new type asserting that the r-convergence assumed on $S = (-\infty, \tau)$ continues to $R_1$ and that the limit function is the normal d.f.

**Theorem 10.4.1.** *Consider a sequence $(X_k)$ of i.i.d.r.v. with common d.f. $V$ and write $F_n$ for the d.f. of the sums*

$$S_n = \frac{1}{b_n}(X_1 + X_2 + \dots + X_n - a_n), \qquad n \geqq 1 .$$

*Then the r-convergence*

$$F_n \xrightarrow[(-\infty,\tau)]{} \Phi, \qquad n \to \infty, \tag{10.4.1}$$

*where $\tau \in R_1$ is arbitrary, implies the complete convergence*

$$F_n \xrightarrow{c} \Phi, \qquad n \to \infty. \tag{10.4.2}$$

Sketch of the proof. By virtue of Helly's selection theorem there exists a subsequence $(F_{n'})$ and a monotone function $F$ such that $F_{n'} \xrightarrow{w} F$. By assumption (10.4.1), $F$ obviously satisfies the condition

$$F(x) = \Phi(x), \qquad x < \tau . \tag{10.4.3}$$

Now we are faced with two very different problems.

(i) *Compactness problem.* Is the sequence $(F_n)$ relatively compact, in other words, do we have $F(\infty) = 1$? The answer is, in fact, affirmative due to the fact that the summands $X_k$ are i.i.d.r.v. Namely, for the limit function $F$ we can obtain the set (8.1.1) of functional equations at least on certain half axes, i.e., we have (see formula (13.1.3))

$$F(x) = F^{*n}(B_n x + A_n), \qquad n \geqq 2, \qquad x < \tau, \tag{10.4.4}$$

for certain constants $B_n > 0$, $A_n \in \mathbb{R}_1$. Thus we conclude that

$$0 < F(\tau) = F^{*n}(x_\tau) \leqq F^{*n}(\infty) \leqq (F(\infty))^n$$

where the number $x_\tau$ is defined by $B_n x_\tau + A_n = \tau$. Assuming now that $F(\infty) < 1$ and letting $n \to \infty$ we arrive at a contradiction, whence $F(\infty) = 1$ follows.

Next we show that $b_n \to \infty$. For this purpose we fix $h > 0$ and $x \in \mathbb{R}_1$ such that $x + h < \tau$. Using the concentration function $Q$ we then have

$$F_n(x+h) - F_n(x-h) \leqq \sup_y \left(F_n(y+h) - F_n(y-h)\right)$$
$$= Q(S_n, h) = Q\left(\sum_{k=1}^{n} X_k, hb_n\right).$$

We carry an indirect proof and assume that there exists a bounded subsequence $(b_{n'})$, $0 < b_{n'} < C$ for some $C > 0$. Then we obtain

$$0 < \Phi(x+h) - \Phi(x-h) \leqq \overline{\lim_{n'\to\infty}}\, Q\left(\sum_{k=1}^{n'} X_k, hC\right).$$

The right side vanishes by Cor. 5.4.5 so that we get a contradiction.

Accordingly, $b_n \to \infty$ so that the summands $X_k/b_n$ are infinitely small. Now Hinčin's theorem (Th. 11.2.3) tells us that $F$ is inf.div.

(ii) *Continuation problem.* Does an inf.div.d.f. $F$ satisfying (10.4.3) coincide with $\Phi$? This purely analytic question was answered in the affirmative in Th. 7.6.1. Hence $F = \Phi$.

Thus we see that an arbitrary weakly converging subsequence $(F_{n'})$ has the limit function $\Phi$ whence the assertion follows by virtue of Cor. 2.2.3. □

Obviously we have proved a little more in (i); in fact, any limit function $F$ of $(F_n)$ is non-defective and satisfies the relation (10.4.4) — an important property of a stable d.f. — at least on a half axis, and (10.4.3) is not needed for this conclusion. Thus it is possible to explain the details of the above proof by proving the more general Prop. 13.3.1.

Zolotarev's conjecture was remarkable since it was later possible to find quite a number of results of this new type; moreover, the proof of Th. 10.4.1 stimulated further research in two different directions. Indeed, it leads first to a compactness problem and then to a continuation problem. The first gave rise to further investigations of the relative compactness of sets of d.f., that is, of a subject which we treat on several occasions in the present book. The latter proved to be the origin of continuation theory and of a generalized setting of problems of this kind which we call characterization problems (see Sect. 1.5). The sketch of the proof given above serves us as a model for most of the proofs of our results involving r-convergence.

We are now going to compare Th. 10.4.1 with a well-known version of the CLT for i.i.d.r.v. according to which the complete convergence (10.4.2) holds true iff

$$\lim_{x\to\infty} \frac{x^2 T_V(x)}{\int_{|u|\leqq x} u^2 \, dV(u)} = 0, \tag{10.4.5}$$

see Th. 13.2.7. Intuitively, this result asserts that the CLT holds iff the tail function $T_V(x)$ tends to 0 sufficiently quickly as $x \to \infty$. This is a two-sided asymptotic property of $V$. From Th. 10.4.1 we now see that we can conclude (10.4.5) from r-convergence on $(-\infty, \tau)$ only, i.e., from a one-sided assumption on $F_n$. This is a consequence of the fact that the d.f. $F_n$ are convolutions.

Now we explain the general structure of theorems involving r-convergence; here we return to the notation of sums in (10.1.4) and note that (10.1.1) is merely a particular case of it. The limit theorems for sums of i.r.v. known so far have, in general, the structure

$$(F_n \xrightarrow{c} F) \Leftrightarrow (C_1) \tag{10.4.6}$$

where $(C_1)$ stands for a certain complex of conditions. Usually, it includes conditions on the d.f. of the individual summands. Relation (10.4.5) is an example, which concerns i.i.d.r.v.; for more general cases we refer the reader to Prop. 11.2.4 and Prop. 11.2.5.

It is the purpose of the theory of sums involving r-convergence to state results of the type

$$(F_n \xrightarrow{c} F) \Leftrightarrow (F_n \xrightarrow[S]{} F) \quad \text{and} \quad (C_2)\,. \tag{10.4.7}$$

Here $(C_2)$ stands for a complex of conditions which are of particular interest when they are distinctly weaker than $(C_1)$. An important example is Th. 10.4.1 where $(C_2)$ is omitted.

Such theorems lead us to a new insight into the meaning of the conditions $(C_1)$ occurring in (10.4.6). In several cases the reader will precisely recognize which part of $(C_1)$ can be dropped if r-convergence on a certain set is added. We remark that there exist also results in which the r-convergence occurring in (10.4.7) is somewhat modified.

## 10.5. Supplement

**10.5.1.** Clearly, it follows from Th. 10.4.1 that the assumption of r-convergence (10.4.1) implies the relation (10.4.5). It is well known that the relation (10.4.5) implies, in its turn, the existence of all absolute moments $\mathsf{E}\,|X_1|^\delta$, $0 < \delta < 2$, but $\mathsf{Var}\,X_1 < \infty$ does not follow. On the other hand, using Th. 6, Sect. 35, in Gnedenko and Kolmogorov (1959) on normal attraction it is easily seen that $\mathsf{Var}\,X_1 = \sigma^2 < \infty$ is granted if (10.4.1) holds with $b_n = \sqrt{n}$.

**10.5.2.** The fact formulated in 10.5.1 was applied to the theory of moderate deviations in Amosova (1979); the important thing to notice in the following result is that (10.5.1) represents a purely one-sided assumption; here $a_n = 0$, $b_n = \sigma\sqrt{n}$, and $\sigma$ stands for a certain constant. It is noteworthy that (10.5.1) implies, among others, that $\mathsf{Var}\,X_1$ exists and equals $\sigma^2$.

*We have, for an arbitrary number $c > 0$,*

$$1 - F_n(x\sigma\sqrt{n}) = \big(1 - \Phi(x)\big)\,\big(1 + o(1)\big)\,, \quad as \quad n \to \infty\,, \tag{10.5.1}$$

*uniformly in the intervals $0 \leq x \leq c\sqrt{\log n}$ iff the following conditions are satisfied.*

(i) $\mathsf{P}(X_1 \geq x) = o\big(x^{-(c^2+2)}(\log x)^{(c^2+1)/2}\big)\,, \quad x \to \infty;$

(ii) $\log x \int\limits_{|y| \geq x} y^2\,\mathrm{d}V(y) \to 0\,, \quad x \to \infty;$

(iii) $\mathsf{E}X_1 = 0\,, \quad \mathsf{Var}\,X_1 = \sigma^2\,.$

For an application of Th. 10.4.1 to narrower zones see Amosova (1981).

In Petrov (1975) we find an elegant theorem (Th. 12, Sect. VII, 3) on large deviations where the assumption (10.5.1) is made uniformly on the intervals $0 \leq x \leq cn^\beta$ for some fixed $\beta$, $0 < \beta < 1/2$. This result is nearly in the spirit of the theory involving r-convergence, the only drawback being that $\operatorname{Var} X_1 < \infty$ is also assumed, i.e., a two-sided assumption on $V$ is made. This can easily be remedied since Amosova's theorem guarantees the existence of $\operatorname{Var} X_1$.

**10.5.3.** Assume that the d.f. $V$ has density $v$ so that the d.f. $(F_n)$ have also densities $v_n$ for all $n \geqq 1$. It is well known that in this case (10.4.5) does not imply the convergence

$$v_n(x) \to \frac{1}{\sqrt{2\pi}} e^{-x^2/2}, \qquad x \in R_1. \tag{10.5.2}$$

For additional assumptions which guarantee (10.5.2) see FELLER (1971) and PETROV (1975).

**Problem** (PADITZ). Which conclusions are possible if (10.5.2) holds true only on $S = (-\infty, \tau)$; in particular, does the r-convergence of $(v_n)$ assumed on $(-\infty, \tau)$ continue to the whole line?

**10.5.4.** A peculiar generalization of Th. 10.4.1 was obtained by RIEDEL (1977); here, for the first time, convergence of $F_n$ is assumed not in any point. *Let*

$$\underline{F}(x) = \underline{\lim}_{n\to\infty} F_n(x), \qquad \overline{F}(x) = \overline{\lim}_{n\to\infty} F_n(x), \qquad x \in R_1.$$

*Then the assumption*

$$\lim_{x\to-\infty} \frac{\underline{F}(x)}{\Phi(x)} = \lim_{x\to-\infty} \frac{\overline{F}(x)}{\Phi(x)} = 1$$

*implies* $F_n \xrightarrow{c} \Phi$.

For a generalization see Th. 12.4.1.

## Notes and bibliographical data

The assertions of Sect. 10.2 represent important tools for the theory of summation of i.r.v. They are well-known results or slight modifications of them. The results of Sect. 10.3 are also well known; for more information on the method of accompanying laws see GNEDENKO and KOLMOGOROV (1959) and KRUGLOV (1976a).

Th. 10.4.1 is due to ROSSBERG and SIEGEL (1975).

# Chapter 11. Limit theorems for triangular arrays

In this chapter we study fundamental limit relation for sums of i.r.v. forming a triangular array $(X_{nk})$. The authors' purpose was to give an introduction to well-known standard results on this subject. Moreover, a number of the present statements has not been treated in books or monographs so far.

Sect. 11.1 is devoted to the problem of compactness of sums without the condition of infinite smallness, while in Sect. 11.2 we demonstrate the role of this assumption in connection with the infinite divisibility of the limit d.f. Extensions of these results when only weak convergence is granted and representations for the defective limit d.f. are given in Sect. 11.3. In the later sections we treat special complete convergences, namely to the degenerate d.f. (LLN), the Poisson d.f. and analytic d.f. Following the essential ideas of this book, in Sect. 11.6 and partially also in Sect. 11.5 we concentrate on limit relations under the assumption of r-convergence.

Throughout this chapter we adhere to the notation $X_{nk} \sim F_{nk}$ and $S_n = \sum_k X_{nk} - A_n \sim F_n$ for certain real numbers $A_n$; moreover we write $a_{nk} = \int_{|x|<\tau} x \, \mathrm{d}F_{nk}(x)$ for fixed $\tau > 0$, $F'_{nk}(x) = F(x + a_{nk})$, and denote by $m_{nk}$ a median of $X_{nk}$.

## 11.1. Compactness of sums

In Sect. 10.3 we studied the method of accompanying d.f. for a special triangular array. Now we will see that this method can also be applied to the general case. In the present section the special choice $c_{nk} = m_{nk}$ for the centering constants occurring in (10.3.1) will prove to be convenient. Thus the c.f. $\tilde{f}_n$ of the accompanying d.f. $\tilde{F}_n$ can be written as

$$\log \tilde{f}_n(t) = \mathrm{i}ta_n + \sum_k \int_{-\infty}^{\infty} \left( \mathrm{e}^{\mathrm{i}tx} - 1 - \frac{\mathrm{i}tx}{1 + x^2} \right) \mathrm{d}F_{nk}(x + m_{nk}) \tag{11.1.1}$$

with

$$a_n = \sum_k \left( m_{nk} + \int_{-\infty}^{\infty} \frac{x}{1 + x^2} \, \mathrm{d}F_{nk}(x + m_{nk}) \right) - A_n . \tag{11.1.2}$$

Putting

$$H_n(x) = \sum_k \int_{-\infty}^{x} \frac{y^2}{1 + y^2} \, \mathrm{d}F_{nk}(y + m_{nk}) , \tag{11.1.3}$$

we can briefly write $\tilde{F}_n = \mathcal{H}(a_n, H_n)$.

Before formulating the main result of this section we remind the reader of the function $w(x) = \min\{1, x^2\}$ and the concentration function $Q$, see Chap. 5.

**Theorem 11.1.1.** *The following assertions are equivalent.*

(i) $(F_n)$ *is shift compact.*

(ii) $(\tilde{F}_n)$ *is shift compact.*

(iii) *The two conditions*

$$\sup_n \sum_k \int_{-\infty}^{\infty} w(x)\, \mathrm{d}F_{nk}(x + m_{nk}) < \infty \tag{11.1.4}$$

*and*

$$\lim_{x\to\infty} \sup_n \sum_k \mathsf{P}(|X_{nk} - m_{nk}| \geqq x) = 0 \tag{11.1.5}$$

*are satisfied.*

(iv) *The two conditions* (11.1.5) *and*

$$\inf_n Q(S_n, h) > 0\,, \quad \textit{for some} \quad h > 0\,, \tag{11.1.6}$$

*are satisfied.*

For the proof we need two lemmas.

**Lemma 11.1.2.** *The condition* (11.1.6) *implies* (11.1.4).

Proof. Let us assume (11.1.6). Then we easily see from Prop. 5.1.4 (ii), that (11.1.6) is true in fact for every $h > 0$. In particular we put $h = 1$ and conclude from Th. 5.4.1 that

$$\sup_n \sum_k \int_{-\infty}^{\infty} w(x)\, \mathrm{d}F_{nk}^{s}(x) \leqq d_0^2 \big(\inf_n Q(S_n, 1)\big)^{-2} < \infty\,.$$

Applying now the first inequality in Cor. 1.4.8 we arrive at the desired condition (11.1.4). □

**Lemma 11.1.3.** *Let* $\eta_0 \in (0, 1)$; *then there exists a constant* $c_0 > 0$ *only depending on* $\eta_0$ *such that*

$$1 - \eta \geqq -c_0 \log \eta \quad \textit{for all} \quad \eta \in [\eta_0, 1]\,.$$

Proof. Since the function log is strictly concave all its cords lie under its graph; thus

$$\frac{1-\eta}{1-\eta_0} \log \eta_0 \leqq \log \eta\,, \qquad \eta \in [\eta_0, 1]$$

whence the assertion follows with $c_0 = \dfrac{1-\eta_0}{|\log \eta_0|}$. □

Proof of Theorem 11.1.1. (i) ⇒ (iv) The shift compactness of $(F_n)$ obviously implies (11.1.6), see Prop. 5.2.1. On the other hand, by Prop. 2.4.11 we have relative compactness, i.e., also the tightness of $(F_n^s)$. Now Prop. 3.3.3 tells us that the corresponding set of c.f. $f_n^s(t) = \prod_k |f_{nk}(t)|^2$ is equicontinuous at 0; that is, for $\varepsilon > 0$, there exists $\delta(\varepsilon) > 0$ such that, for all $n$ and $|t| \leqq \delta$,

$$\begin{aligned} \varepsilon > |\log f_n^s(t)| &= -\sum_k \log |f_{nk}(t)|^2 = \sum_k \log\big(1 - (1 - |f_{nk}(t)|^2)\big) \\ &\geqq \sum_k (1 - |f_{nk}(t)|^2)\,. \end{aligned}$$

Using the Fubini theorem and the relation $1 - \sin(\delta x)/\delta x \geqq 1/10$ for $|x| \geqq 1/\delta$, we obtain

$$\varepsilon > \frac{1}{2\delta} \sum_k \int_{|t| \leqq \delta} (1 - |f_{nk}(t)|^2)\, dt = \sum_k \int_{-\infty}^{\infty} \left(1 - \frac{\sin \delta x}{\delta x}\right) dF_{nk}^s(x)$$
$$\geqq \frac{1}{10} \sum_k \int_{|x| \geqq 1/\delta} dF_{nk}^s(x)$$

whence (11.1.5) follows by virtue of Cor. 1.4.9.

(iv) $\Rightarrow$ (iii) is obvious because of L. 11.1.2.

(iii) $\Rightarrow$ (ii) We write $\tilde{F}_n = \mathscr{H}(a_n, H_n)$ where $a_n$, $H_n$ are given by (11.1.2) and (11.1.3), resp. By $y^2/(1 + y^2) \leqq w(y)$ and in view of the assumption (11.1.4) we have

$$\sup_n H_n(\infty) \leqq \sup_n \sum_k \int_{-\infty}^{\infty} w(y)\, dF_{nk}(y + m_{nk}) < \infty .$$

We further obtain from (11.1.5) that

$$\sup_n \int_{|y| \geqq x} dH_n(y) \leqq \sup_n \sum_k \mathsf{P}(|X_{nk} - m_{nk}| \geqq x) \to 0 , \qquad x \to \infty ,$$

i.e., $(H_n)$ is tight. Now we are ready to apply Cor. 7.4.2 so that the relative compactness of $(\tilde{F}_n)$ follows for suitably chosen centering constants $A_n$ hence $(\tilde{F}_n)$ is shift compact.

(ii) $\Rightarrow$ (i) We assume the shift compactness of $(\tilde{F}_n)$. Then, by Prop. 3.3.6, the set $\{|\tilde{f}_n(t)|\}$ is equicontinuous at 0. Since $\log x$ is continuous at $x = 1$ we get the same assertion for the set $\{\log |\tilde{f}_n(t)|\}$. Thus we have in view of (11.1.1), for $\varepsilon > 0$ and all $n$,

$$\varepsilon > -\log |\tilde{f}_n(t)| = \sum_k \left(1 - \operatorname{Re}\left(f_{nk}(t)\, e^{-im_{nk}t}\right)\right)$$
$$\geqq \sum_k (1 - |f_{nk}(t)|) , \qquad |t| \leqq \delta(\varepsilon) . \tag{11.1.7}$$

Now we see from (11.1.7) that $1 - |f_{nk}(t)| \leqq \varepsilon$. Accordingly, L. 11.1.3 can be applied with $\eta = |f_{nk}(t)|$ and $\eta_0 = 1 - \varepsilon$ so that, by (11.1.7),

$$\varepsilon \geqq -c_0 \sum_k \log |f_{nk}(t)| = -c_0 \log |f_n(t)| , \qquad |t| \leqq \delta .$$

This means that $\{\log |f_n(t)|\}$ is equicontinuous at 0 and, as a consequence, $\{|f_n(t)|\}$ is also equicontinuous at 0. Hence Prop. 3.3.6 applies and entails the desired condition (i). □

**Example 11.1.4.** Let us be given i.r.v. $X_{nk}$, $k = 1, 2, \ldots, n$, having the distribution

$$\mathsf{P}(X_{nk} = 0) = 1 - \frac{1}{n}, \qquad \mathsf{P}(X_{nk} = 1) = \frac{1}{2n}, \qquad \mathsf{P}(X_{nk} = n) = \frac{1}{2n},$$
$$n = 2, 3, \ldots$$

Then $m_{nk} = 0$ and we obtain, for $n \geqq x > 0$,

$$\sum_{k=1}^{n} \int_{-\infty}^{\infty} w(y)\, dF_{nk}(y) = \tfrac{1}{2} + \sum_{k=1}^{n} \mathsf{P}(|X_{nk}| \geqq x) = 1 .$$

This proves (11.1.4) while (11.1.5) is violated, i.e., $(F_n)$ is not shift compact. Finally, we note that the condition (11.1.6) is true; this can be shown by L. 11.2.1 (see also Ex. 11.3.6). □

For the following we note that $(F_n)$ is shift compact iff $(\overline{F}_n)$ is shift compact where

$$\overline{F}_n := \mathsf{P}(\sum_k X_{nk} < x) .$$

**Corollary 11.1.5.** *Let $(F_n)$ be shift compact and $\tau > 0$ be fixed. Then the shift constants of $(\overline{F}_n)$ can be chosen as*

$$A_n = A_n(\tau) = \sum_k (m_{nk} + \int_{|y|<\tau} y \, \mathrm{d}F_{nk}(y + m_{nk})) . \tag{11.1.8}$$

Proof. Let $(A_n')$ be some sequence of shift constants, i.e., $(\overline{F}_n(x + A_n'))$ is relatively compact. Accordingly, Prop. 2.4.7 tells us that it is enough to show that $(A_n(\tau) - A_n')$ is bounded for every $\tau > 0$.

To prove this we fix $\hat{\tau} > 0$ to be determined below. Then we introduce truncated r.v.

$$X_{nk}^{\hat{\tau}} = \begin{cases} X_{nk} - m_{nk} , & \text{for } |X_{nk} - m_{nk}| < \hat{\tau} , \\ 0 , & \text{otherwise} , \end{cases}$$

and random events $B_n^{\hat{\tau}} = \bigcap_k \{|X_{nk} - m_{nk}| < \hat{\tau}\}$. Then

$$\mathsf{P}(|S_n| \geqq x) \leqq \mathsf{P}(\overline{B}_n^{\hat{\tau}}) + \mathsf{P}(\{|S_n| \geqq x\} \cap B_n^{\hat{\tau}}) =: I_1 + I_2 .$$

By Th. 11.1.1, our assumption implies (11.1.5). Hence, we have

$$I_1 = \mathsf{P}(\bigcup_k \{|X_{nk} - m_{nk}| \geqq \hat{\tau}\}) \leqq \sum_k \mathsf{P}(|X_{nk} - m_{nk}| \geqq \hat{\tau}) \leqq 1/4$$

for a certain $\hat{\tau} > 0$, chosen sufficiently large, and all $n$. On the other hand, we use (11.1.8) for $\tau = \hat{\tau}$ and obtain $S_n = \sum_k (X_{nk} - m_{nk} - \mathsf{E}X_{nk}^{\hat{\tau}})$ and

$$\{\{|S_n| \geqq x\} \cap B_n^{\hat{\tau}}\} \subseteqq \{|\sum_k (X_{nk}^{\hat{\tau}} - \mathsf{E}X_{nk}^{\hat{\tau}})| \geqq x\}$$

because of $\mathsf{E}X_{nk}^{\hat{\tau}} = \int_{|y|<\tau} y \, \mathrm{d}F_{nk}^{\hat{\tau}}(y + m_{nk})$. Now we see from the Čebyšev inequality that

$$I_2 \leqq \mathsf{P}(|\sum_k (X_{nk}^{\hat{\tau}} - \mathsf{E}X_{nk}^{\hat{\tau}})| \geqq x) \leqq \frac{1}{x^2} \sum_k \mathsf{Var}\, X_{nk}^{\hat{\tau}}$$

$$\leqq \frac{1}{x^2} \sum_k \mathsf{E}(X_{nk}^{\hat{\tau}})^2 \leqq \frac{1}{x^2} \hat{\tau}^2 \sum_k \mathsf{E}w(X_{nk} - m_{nk}) .$$

The latter inequality is true since $y^2 \leqq \hat{\tau}^2 w(y)$, $|y| \leqq \hat{\tau}$. By Th. 11.1.1, the assumption of shift compactness implies (11.1.4). Hence there exists a constant $c$ such that $I_2 \leqq \hat{\tau}^2 c x^{-2} \leqq 1/4$ for all $n$ and sufficiently large $x$.

Combining the above estimates we obtain $\mathsf{P}(|S_n| \geqq x) \leqq I_1 + I_2 \leqq 1/2$ so that

$$\inf_n \mathsf{P}\big(-x + A_n(\hat{\tau}) \leqq \sum_k X_{nk} < x + A_n(\hat{\tau})\big) > 0 .$$

Now we are in a position to apply L. 5.2.2 (with $a_n' = A_n$); accordingly, $A_n(\hat{\tau}) - A_n$ is bounded and $(A_n(\hat{\tau}))$ is a sequence of shift constants.

To finish the proof we have to show that $(A_n(\tau'') - A_n(\tau'))$ is bounded for every pair $\tau', \tau''$, $0 < \tau' < \tau''$; see Prop. 2.4.7. Indeed, we have

$$|x| \leqq \frac{\max\{1, \tau''\}}{\min\{1, \tau'\}} w(x) , \qquad \tau' \leqq |x| \leqq \tau'' ,$$

and, as a consequence,

$$|A_n(\tau'') - A_n(\tau')| = |\sum_k \int_{\tau' \leq |x| \leq \tau''} x \, dF_{nk}(x + m_{nk})|$$

$$\leq \frac{\max\{1, \tau''\}}{\min\{1, \tau'\}} \sum_k \int_{-\infty}^{\infty} w(x) \, dF_{nk}(x + m_{nk}) .$$

The sum on the right is bounded because of (11.1.4). □

We now consider mutually i.r.v. $X_1, X_2, \ldots$ Then Th. 11.1.1 and Cor. 11.1.5 can be written in a more elegant form.

**Proposition 11.1.6.** *Let* $X_1, X_2, \ldots$ *be i.r.v.,* $X_k \sim V_k$, *and* $S_n = \sum_{k=1}^{n} X_n - A_n \sim F_n$. *Further, let* $m_k$ *be a median of* $X_k$. *Then the following assertions are equivalent:*

(i) $(F_n)$ *is shift compact.*

(ii) $\sum_{k=1}^{\infty} \int_{-\infty}^{\infty} w(x) \, dV_k(x + m_k) < \infty .$

(iii) $\inf_n Q(S_n, h) > 0$, *for some* $h > 0$.

*The shift constants can be chosen as*

$$A_n = \sum_{k=1}^{n} \Big(m_k + \int_{|x|<\tau} x \, dV_k(x + m_k)\Big) .$$

Proof. (i) ⇒ (iii) is obvious since Prop. 5.2.1 tells us that (i) implies the tightness of the sequence $(Q(X_n, h))$.

(iii) ⇒ (ii) is a special case of L. 11.1.2 for $F_{nk} = V_k$.

(ii) ⇒ (i) In view of Th. 11.1.1 it suffices to show that

$$\lim_{x \to \infty} \sum_{k=1}^{\infty} \mathsf{P}(|X_k - m_k| \geq x) = 0 . \tag{11.1.9}$$

In fact, we have

$$\sum_{k=1}^{\infty} \mathsf{P}(|X_k - m_k| \geq x_0) \leq w(x_0)^{-1} \sum_{k=1}^{\infty} \int_{-\infty}^{\infty} w(y) \, dV_k(y + m_k) < \infty$$

for fixed $x_0 > 0$ since $w$ is non-decreasing for $x > 0$. Now we note that the convergence of the sum on the left of (11.1.9) is uniform on $(x_0, \infty)$. Hence we may carry out the limit under the sum sign which proves (11.1.9).

The representation of $A_n$ can be obtained from Cor. 11.1.5. □

Finally, we turn to one-sided compactness conditions. We will formulate them in terms of the right- or the left-tightness of $(F_n)$, which has been defined in Sect. 7.4.

**Proposition 11.1.7.** *Let* $A_n$ *be given by* (11.1.8) *for some* $\tau > 0$ *and assume that* (11.1.4) *is true. Then* $(F_n)$ *is left-tight iff*

$$\lim_{x \to -\infty} \sup_n \sum_k \mathsf{P}(X_{nk} - m_{nk} \leq x) = 0 . \tag{11.1.10}$$

Proof. Without loss of generality we set $m_{nk} = 0$ for all $n, k$ so that, by (11.1.8), $A_n = \sum_k a_{nk}$. Here $a_{nk} = \int_{|x|<\tau} x \, dF_{nk}(x)$. Moreover, we introduce one-sided truncated r. v.

$$X_{nk}^{\tau} = \begin{cases} X_{nk}, & \text{for } X_{nk} < \tau, \\ \tau, & \text{for } X_{nk} \geq \tau, \end{cases} \qquad X_{nk}^{\tau} \sim F_{nk}^{\tau} .$$

For the corresponding medians we have $m_{nk}^\tau = m_{nk} = 0$. Further,

$$a_{nk}^\tau = \int_{|x|<\tau} x \, \mathrm{d}F_{nk}^\tau(x) = a_{nk} \, .$$

Hence the shift constants $A_n$ given in (11.1.8) are the same for the r.v. $X_{nk}^\tau$ as for $X_{nk}$. Thus we may apply the above results to

$$S_n^\tau := \sum_k (X_{nk}^\tau - a_{nk}) \sim F_n^\tau \, .$$

Assume now that (11.1.10) is satisfied. Then we obtain

$$\lim_{x\to\infty} \sup_n \sum_k \mathsf{P}(|X_{nk}^\tau| \geqq x) = 0 \, .$$

On the other hand, we derive from (11.1.4) that

$$\sup_n \sum_k \int_{-\infty}^{\infty} w(x) \, \mathrm{d}F_{nk}^\tau(x) \leqq \sup_n \sum_k \int_{-\infty}^{\infty} w(x) \, \mathrm{d}F_{nk}(x) < \infty \, .$$

Thus we can apply Th. 11.1.1 and Prop. 11.1.5 to the sequence $(F_n^\tau)$ and get the relative compactness of $(F_n^\tau)$. Therefore $(F_n^\tau)$ is tight. Let $\varepsilon > 0$, then for all $n$ and sufficiently large $x$ we have

$$\begin{aligned} \varepsilon &> \mathsf{P}(|\sum_k (X_{nk}^\tau - a_{nk})| \geqq x) \geqq \mathsf{P}\big(\sum_k (X_{nk}^\tau - a_{nk}) \leqq -x\big) \\ &\geqq \mathsf{P}\big(\sum_k (X_{nk} - a_{nk}) \leqq -x\big) = \mathsf{P}(S_n \leqq -x) \end{aligned}$$

whence the required left-tightness of $(F_n)$ follows.

Conversely, let us assume that $(F_n)$ is left-tight. First we will show the left-tightness of $(F_n^\tau)$. We introduce random events $C_n^\tau = \bigcap_k \{X_{nk} < \tau\}$. By $m_{nk} = 0$ we have

$$\eta := \mathsf{P}(X_{nk} < \tau) \geqq 1/2 =: \eta_0 \tag{11.1.11}$$

and, by L. 11.1.3 there exists a constant $c_0 > 0$ such that

$$\begin{aligned} \mathsf{P}(C_n^\tau) &= \prod_k \mathsf{P}(X_{nk} < \tau) = \exp\big(\sum_k \log \mathsf{P}(X_{nk} < \tau)\big) \\ &\geqq \exp\left(-\frac{1}{c_0} \sum_k \mathsf{P}(X_{nk} \geqq \tau)\right). \end{aligned}$$

In view of the assumption (11.1.4) there exists a constant $c_1$ depending on $\tau$ such that

$$\sum_k \mathsf{P}(X_{nk} \geqq \tau) \leqq \max\left\{1, \frac{1}{\tau^2}\right\} \sum_k \int_{-\infty}^{\infty} w(x) \, \mathrm{d}F_{nk}(x) \leqq c_1 \, .$$

Hence, $\mathsf{P}(C_n^\tau) \geqq \exp(-c_1/c_0) =: c_2 > 0$.

Next we easily see by definition of $X_{nk}^\tau$ that for $x + a_{nk} < \tau$

$$\mathsf{P}(X_{nk} - a_{nk} \leqq x, X_{nk} < \tau) = \mathsf{P}(X_{nk} - a_{nk} \leqq x) = \mathsf{P}(X_{nk}^\tau - a_{nk} \leqq x) \, .$$

This implies

$$\mathsf{P}(X_{nk} - a_{nk} \leqq x \mid X_{nk} < \tau) \geqq \mathsf{P}(X_{nk}^\tau - a_{nk} \leqq x)$$

for any $x \in \mathbb{R}^1$, and, by induction,

$$\mathsf{P}(S_n \leqq x \mid C_n^\tau) \geqq \mathsf{P}(S_n^\tau \leqq x) \, .$$

So we get

$$\mathsf{P}(S_n \leqq x) \geqq \mathsf{P}(C_n^\tau)\, \mathsf{P}(S_n \leqq x \mid C_n^\tau) \geqq c_2 \mathsf{P}(S_n^{\tau'} \leqq x) .$$

Now the assumption of left-tightness of $(F_n)$ implies the left-tightness of $(F_n^\tau)$.

On the other hand, the r.v. $X_{nk}^\tau$ trivially satisfy (11.1.10) to the right, i.e.,

$$\lim_{x\to\infty} \sup_n \sum_k \mathsf{P}(X_{nk}^\tau \geqq x) = 0 .$$

Hence we can apply the first part of the present proof to the right. Thus $(F_n^\tau)$ is also right-tight. Left- and right-tightness of $(F_n^\tau)$ imply tightness. Accordingly, $(F_n^\tau)$ is relatively compact. Applying again Th. 11.1.1 to the sequence $(F_n^\tau)$ we see that (11.1.5) is valid for the r.v. $X_{nk}^\tau$. But this relation implies the required condition (11.1.10). □

## 11.2. Complete convergence

Now we will study the convergence of $(F_n)$ in a triangular array of i.r.v. The method of accompanying d.f. will play a crucial role. But in contrast to the preceding section we now use the centering constants

$$c_{nk} = a_{nk} = \int_{|x|<\tau} x \, \mathrm{d}F_{nk}(x) \text{ for some } \tau > 0 .$$

So we get from (10.3.1)

$$\log \tilde{f}_n(t) = ita_n + \sum_k \int_{-\infty}^{\infty} \left( \mathrm{e}^{itx} - 1 - \frac{itx}{1+x^2} \right) \mathrm{d}F'_{nk}(x) \tag{11.2.1}$$

with

$$a_n = \sum_k \left( a_{nk} + \int_{-\infty}^{\infty} \frac{x}{1+x^2} \, \mathrm{d}F'_{nk}(x) \right) - A_n . \tag{11.2.2}$$

Hence we may write $\tilde{F}_n = \mathcal{H}(a_n, H_n)$ where

$$H_n(x) = \sum_k \int_{-\infty}^{x} \frac{y^2}{1+y^2} \mathrm{d}F'_{nk}(y) . \tag{11.2.3}$$

First we prove an important lemma.

**Lemma 11.2.1.** *Suppose infinite smallness of the triangular array $(X_{nk})$ of i.r.v. Then the following three assertions are equivalent:*

(i) $\inf_n Q(S_n, h) > 0$, *for some* $h > 0$;

(ii) $\inf_n Q(\tilde{S}_n, h) > 0$, *for some* $h > 0$;

(iii) $\sup_n \sum_k \int_{-\infty}^{\infty} w(x) \, \mathrm{d}F'_{nk}(x) < \infty$.

*In all these cases we have the convergence*

$$\lim_{n\to\infty} \frac{f_n(t)}{\tilde{f}_n(t)} = 1; \tag{11.2.4}$$

*it is uniform on any finite interval.*

Proof. (i) $\Rightarrow$ (iii) By L. 11.1.2, the assumption (i) yields

$$\sup_n \sum_k \int_{-\infty}^{\infty} w(x)\, \mathrm{d}F_{nk}(x + m_{nk}) < \infty .$$

This relation, in its turn, implies (iii) because of L. 10.2.5.

(iii) $\Rightarrow$ (ii) We fix some $T > 0$. By L. 10.2.4, the assumption (iii) yields

$$\sum_k |f'_{nk}(t) - 1| \leqq c_1 \sum_k \int_{-\infty}^{\infty} w(x)\, \mathrm{d}F'_{nk}(x) \leqq c_2 , \qquad |t| \leqq T , \tag{11.2.5}$$

for all $n$ and certain numbers $c_1$, $c_2$ only depending on $\tau$, $T$. Next we make use of (11.2.1), (11.2.2), (11.2.5), and get

$$-\log |\tilde{f}_n(t)| = -\sum_k (\operatorname{Re} f'_{nk}(t) - 1) \leqq \sum_k |f'_{nk}(t) - 1| \leqq c_2 , \qquad |t| \leqq T .$$

Hence, $|\tilde{f}_n(t)| \geqq \mathrm{e}^{-c_2} > 0$. We now apply Prop. 5.3.3 and obtain the required assertion from

$$Q\left(\tilde{S}_n, \frac{1}{T}\right) \geqq \frac{1}{5T} \int_{|t| \leqq T} |\tilde{f}_n(t)|^2\, \mathrm{d}t \geqq \frac{2}{5} \mathrm{e}^{-2c_2} > 0 .$$

(ii) $\Rightarrow$ (i) First we see from infinite smallness and L. 10.2.3 that the differences $X_{nk} - a_{nk} \sim F'_{nk}$ are also infinitely small. Thus L. 10.2.2 implies

$$\max_k |f'_{nk}(t) - 1| \to 0 , \qquad n \to \infty , \tag{11.2.6}$$

uniformly on $[-T, T]$. Since $\log(1 + \alpha) = \alpha + o(|\alpha|)$, $\alpha \to 0$, and in view of (11.2.6) there is an integer $n_0(T)$ such that $\left(\text{with } \alpha = |f'_{nk}(t)| - 1\right)$

$$\log |f'_{nk}(t)| = -\left(1 - |f'_{nk}(t)|\right)\left(1 + o(1)\right) \geqq -2\left(1 - |f'_{nk}(t)|\right) ,$$

for all $k$, $|t| \leqq T$, and $n \geqq n_0$. Accordingly,

$$\begin{aligned} |f_n(t)|^2 &= \prod_k |f'_{nk}(t)|^2 = \exp\left(2 \sum_k \log |f'_{nk}(t)|\right) \\ &\geqq \exp\left(-4 \sum_k (1 - |f'_{nk}(t)|)\right) \geqq \exp\left(-4 \sum_k (1 - \operatorname{Re} f'_{nk}(t))\right) = |\tilde{f}_n(t)|^4 . \end{aligned}$$

Next we introduce r.v. $Y_n \sim \tilde{F}_n^{*4}$. Using first Prop. 5.3.3 and then Cor. 5.3.5, we get

$$\begin{aligned} Q(S_n, 1/T) &\geqq \frac{1}{5T} \int_{|t| \leqq T} |f_n(t)|^2\, \mathrm{d}t \\ &\geqq \frac{11}{60}\left[\frac{12}{11} \frac{1}{T} \int_{|t| \leqq T} |\tilde{f}_n(t)|^4\, \mathrm{d}t\right] \geqq \frac{11}{60} Q(Y_n, 1/T) , \qquad n \geqq n_0 . \end{aligned}$$

Applying the first inequality in Prop. 5.1.3 (v), twice ($\eta_1 = \eta_2 = 1/2$) we further conclude that

$$Q(Y_n, 1/T) \geqq Q\left(\tilde{S}_n, 1/(4T)\right)^4 .$$

It is only now that we make use of the assumption (ii) where we put $h = (4T)^{-1}$. Combining the above estimates we conclude from it the relation

$$\inf_{n \geqq n_0} (QS_n, 1/T) > 0 .$$

Now the desired assertion (i) is obvious.

Finally we have to show (11.2.4). Expanding the logarithm we see that

$$|\log (1 + \alpha) - \alpha| = \tfrac{1}{2} |\alpha|^2 (1 + o(1)), \quad \text{as} \quad |\alpha| \to 0 .$$

Hence, using (11.2.6) we find, for large $n$ and all $k$

$$|\log f'_{nk}(t) - (f'_{nk}(t) - 1)| = \tfrac{1}{2} |f'_{nk}(t) - 1|^2 (1 + o(1)) \leqq |f'_{nk}(t) - 1|^2 .$$

By the definitions of $f'_{nk}$, $\tilde{f}_n$, $a_n$, and by (11.2.5) we then have

$$|\log f_n(t) - \log \tilde{f}_n(t)| = |\sum_k (\log f_{nk}(t) - ita_{nk} - (f'_{nk}(t) - 1)|$$

$$= |\sum_k (\log f'_{nk}(t) - (f'_{nk}(t) - 1))| \leqq \sum_k |f'_{nk}(t) - 1|^2$$

$$\leqq \max_k |f'_{nk}(t) - 1| \sum_k |f'_{nk}(t) - 1| \leqq c_2 \max_k |f'_{nk}(t) - 1| .$$

By virtue of (11.2.6) we obtain

$$\log f_n(t) - \log \tilde{f}_n(t) \to 0 , \qquad n \to \infty ,$$

uniformly in $[-T, T]$ so that the relation (11.2.4) is obvious. □

Now we turn to the main results of the present section.

**Theorem 11.2.2** (Bavli-Gnedenko). *Let us suppose infinite smallness of the triangular array $(X_{nk})$ of i.r.v. Then we have $F_n \xrightarrow{c} F$ iff $\tilde{F}_n \xrightarrow{c} F$.*

Proof. We assume $F_n \xrightarrow{c} F$. Then it is clear from Prop. 5.2.1 — applied to the sums $S_n$ — that (i) in L. 11.2.1 holds true. Thus we get (11.2.4) and therefore

$$|f_n(t) - \tilde{f}_n(t)| \leqq \left|\frac{f_n(t)}{\tilde{f}_n(t)} - 1\right| \to 0 , \qquad n \to \infty . \tag{11.2.7}$$

Note that this convergence also holds uniformly for any bounded interval. Now the continuity theorem for c.f. yields the required assertion. The converse statement follows similarly. □

**Theorem 11.2.3** (Hinčin). *Let us suppose infinite smallness of the triangular array $(X_{nk})$ of i.r.v. Then the set of d.f. which are complete limits of $(F_n)$ coincides with the set of all inf.div.d.f.*

Proof. Let the $X_{nk}$ be infinitely small and assume $F_n \xrightarrow{c} F$. Then Th. 11.2.2 tells us that we have for the accompanying d.f. $\tilde{F}_n \xrightarrow{c} F$. We know from Prop. 7.4.1 that the complete limit of inf.div.d.f. is also inf.div. Thus $F$ is inf.div. because the $\tilde{F}_n$ are inf.div.

For the converse we consider an arbitrary inf.div. $F = ({}^nF)^{*n}$, $n \geqq 2$. We introduce i.i.d.r.v. $X_{nk}$ subject to ${}^nF$, $k = 1, \dots, n$, so that we have a triangular array with $k_n = n$. Clearly, $S_n = X_{n1} + \dots + X_{nn} \sim F_n = F$. Accordingly, it remains to prove infinite smallness. We recall that inf.div.c.f. have no zeros. Thus we have ${}^nf(t) = f^{1/n}(t) \to 1$, $n \to \infty$. By Cor. 3.4.4 we see that this convergence is uniform in any bounded interval. So we get the condition (iii) in L. 10.2.2 which implies infinite smallness. □

In other words, we have proved that *the set of limit d.f. for sequences $(F_n)$ of d.f. of sums $S_n$ with independent and infinitely small summands $X_{nk}$ coincides with the set of inf.div.d.f.*

Clearly, Th. 11.2.3 gives a final solution to problem (P1) under the hypothesis of infinite smallness (cf. Sect. 10.1). Let us now turn to conditions guaranteeing the convergence of $(F_n)$ to a given inf.div.d.f., i.e., we wish to solve problem (P2). In the first step we apply the Lévy-Hinčin formula and write $\tilde{F}_n = \mathcal{H}(a_n, H_n)$ where $a_n$ and $H_n$ are given by (11.2.2) and (11.2.3), resp.

**Proposition 11.2.4.** *Let $F = \mathcal{H}(a, H)$ be some inf.div.d.f. and suppose infinite smallness of the $(X_{nk})$. Then $F_n \xrightarrow{c} F$ holds iff $a_n \to a$ and $H_n \xrightarrow{c} H$ are satisfied.*

Proof. By Th. 11.2.2 we have $F_n \xrightarrow{c} F$ iff $\tilde{F}_n \xrightarrow{c} F$. According to Prop. 7.4.1, the latter relation is equivalent to $a_n \to a$ and $H_n \xrightarrow{c} H$. □

Using the Lévy formula, one obtains a similar proposition.

**Proposition 11.2.5.** *Let $F = \mathcal{L}(a, \gamma^2, M, N)$ be some inf.div.d.f. and suppose infinite smallness of the $(X_{nk})$. Then $F_n \xrightarrow{c} F$ holds iff the following three conditions are satisfied*

$$\sum_k F_{nk}(x) \xrightarrow[(-\infty,0)]{} M(x), \qquad \sum_k \left(F_{nk}(x) - 1\right) \xrightarrow[(0,\infty)]{} N(x); \tag{11.2.8}$$

$$\begin{aligned}\gamma^2 &= \lim_{\varepsilon\to 0} \overline{\lim_{n\to\infty}} \sum_k \Big( \int_{|x|<\varepsilon} x^2 \,\mathrm{d}F_{nk}(x) - \big( \int_{|x|<\varepsilon} x \,\mathrm{d}F_{nk}(x)\big)^2\Big) \\ &= \lim_{\varepsilon\to\infty} \underline{\lim_{n\to\infty}} \sum_k \Big( \int_{|x|<\varepsilon} x^2 \,\mathrm{d}F_{nk}(x) - \big( \int_{|x|<\varepsilon} x \,\mathrm{d}F_{nk}(x)\big)^2\Big).\end{aligned} \tag{11.2.9}$$

*If $\pm\tau$ are continuity points of $N$, $M$, resp., then*

$$\sum_k \int_{|x|<\tau} x \,\mathrm{d}F_{nk}(x) - A_n \to a + \int_{+0}^{\infty} s(x) \,\mathrm{d}N(x) + \int_{-\infty}^{-0} s(x) \,\mathrm{d}M(x), \tag{11.2.10}$$

*where*

$$s(x) = \begin{cases} x^3/(1 + x^2), & \text{for } |x| < \tau, \\ -x/(1 + x^2), & \text{for } |x| \geqq \tau. \end{cases}$$

The proof of this statement is given in Petrov (1975), Chap. 4. □

## 11.3. Weak convergence

We adhere to the notation of the preceding section. First we state a condition on the i.r.v. $X_{nk}$ which ensures that weak convergence of $(F_n)$ also implies complete convergence. It is interesting to note that in this case infinite smallness is superfluous.

**Proposition 11.3.1.** *Let us suppose $F_n \xrightarrow{w} F \neq$ const. Then $F$ is non-defective and the convergence is complete iff*

$$\lim_{x\to\infty} \sup_n \sum_k \mathsf{P}(|X_{nk} - m_{nk}| \geqq x) = 0. \tag{11.3.1}$$

Proof. We know from Prop. 5.1.3, (i), that any concentration function $Q(X, h) > 0$ for all $h > 0$. Now we show by an indirect proof that the assumption $F \neq$ const. implies

$$\inf_n Q(S_n, h) > 0, \quad \text{for all} \quad h > 0. \tag{11.3.2}$$

Let $\inf_n Q(S_n, h_0) = 0$ for some $h_0 > 0$. Then we can find a subsequence $(S_{n'})$ which satisfies

$$\lim_{n'\to\infty} Q(S_{n'}, h_0) = 0. \tag{11.3.3}$$

Since $F \neq$ const. there exist two continuity points $x + h_0$, $x - h_0$ of $F$ such that $p := F(x + h_0) - F(x - h_0) > 0$. Thus,

$$Q(S_{n'}, h_0) \geqq F_{n'}(x + h_0) - F_{n'}(x - h_0) \to p\,, \qquad n' \to \infty\,.$$

This contradicts (11.3.3) which proves (11.3.2).

Now Th. 11.1.1 tells us that under the condition (11.3.2) the shift compactness of $(F_n)$ is equivalent to (11.3.1). Since $F \neq 0$ and $F \neq 1$ we obtain from Prop. 2.4.7 that the shift compactness is equivalent to the relative compactness of $(F_n)$. □

Now we turn to convergence to a defective limit function. Before giving the next result we remind the reader of the w*-convergence (see Def. 3.4.6) and of the accompanying d.f. $\tilde{F}_n$ (see formula (11.2.1)).

**Proposition 11.3.2.** *Suppose infinite smallness of the $(X_{nk})$ and let $F \neq$ const. be a non-decreasing and left-continuous function. Then $F_n \xrightarrow{w^*} F$ holds iff $\tilde{F}_n \xrightarrow{w^*} F$.*

Proof. Let us assume $F_n \xrightarrow{w^*} F$. Since $F \neq$ const. we get from the same arguments as in the proof of Prop. 11.3.1 that (11.3.2) is true. Now we apply L. 11.2.1 and obtain (11.2.4). Hence (11.2.7) follows uniformly on every bounded interval. Making use of the integral characteristic defined in (3.4.1) we have, by (11.2.7),

$$f_n^*(u) - \tilde{f}_n^*(u) = \int_0^u \left(f_n(t) - \tilde{f}_n(t)\right) \mathrm{d}t \to 0\,, \qquad n \to \infty\,, \qquad u > 0\,.$$

Let $f^*$ be the integral characteristic of $F$, then Prop. 3.4.7 implies $f_n^* \to f^*$. Thus

$$\tilde{f}_n^*(u) = \left(\tilde{f}_n^*(u) - f_n^*(u)\right) + f_n^*(u) \to f^*(u)\,, \qquad u > 0\,.$$

Applying again Prop. 3.4.7 we get $\tilde{F}_n \xrightarrow{w^*} F$. — The converse assertion can be proved similarly. □

Now we are interested in representations of defective limit d.f. Here, for the sake of simplicity, we assume $F(-\infty) = 0$, that is, a defect is only admitted as $x \to \infty$. It is clear that $F(-\infty) = 0$ can be equivalently expressed by the left-tightness of $(F_n)$, cf. Sect. 7.4. As in Sect. 11.2 we write $\tilde{F}_n = \mathcal{H}(a_n, H_n)$ where $a_n$, $H_n$ are given by (11.2.2) and (11.2.3), resp.

**Proposition 11.3.3.** *Suppose infinite smallness of the $(X_{nk})$ and let $(F_n)$ be left-tight. Moreover, let $(a_n)$ be bounded. Then all functions $F \neq 0$ which are weak limits of $(F_n)$ admit the representation $F = \varrho G$ for some non-defective inf.div.d.f. $G$ and a constant $\varrho$, $0 < \varrho = F(\infty) \leqq 1$.*

Proof. Let us assume $F_n \xrightarrow{w} F \neq 0$. Then as in the proof of Prop. 11.3.1 we get (11.3.2), i.e., condition (i) of L. 11.2.1 is satisfied. Accordingly, (iii) is also fulfilled. Thus we have, using L. 10.2.5,

$$\sup_n \sum_k \int_{-\infty}^{\infty} w(x)\, \mathrm{d}F_{nk}(x + m_{nk}) \leqq \sup_n \sum_k \int_{-\infty}^{\infty} w(x)\, \mathrm{d}F'_{nk}(x) < \infty\,. \tag{11.3.4}$$

Since $x^2/(1 + x^2) \leqq w(x)$ this implies $\sup_n H_n(\infty) < \infty$.

The assumption of left-tightness of $(F_n)$ implies $F(-\infty) = 0$. Now we see from Prop. 11.1.7[1]) that

$$\sup_n \sum_k \mathsf{P}(X_{nk} - m_{nk} < x) \to 0\,, \qquad x \to -\infty\,.$$

[1]) For the choice of the centering constants $A_n$ see the reformulation of Prop. 11.1.7 given in Suppl. 11.7.4.

In view of $x^2/(1+x^2) \leqq 1$ and since $a_{nk}$, $m_{nk}$ tend to zero as $n \to \infty$ (see L. 10.2.3) we have by (11.2.3)

$$H_n(x) \leqq \sum_k \mathsf{P}(X_{nk} - a_{nk} \leqq x) = \sum_k \mathsf{P}(X_{nk} - m_{nk} \leqq x - m_{nk} + a_{nk})$$
$$\leqq \sum_k \mathsf{P}(X_{nk} - m_{nk} \leqq x + 1) \tag{11.3.5}$$

for sufficiently large $n$. Hence the relation

$$\sup_n H_n(x) \to 0\,, \qquad x \to -\infty\,,$$

is also satisfied, that is, $(H_n)$ is left-tight.

Combining these two results and the boundedness of $(a_n)$ we are ready to apply Cor. 7.4.5 to the sequence $(\tilde{F}_n)$ according to which $(\tilde{F}_n)$ is also left-tight.

On the other hand, by Prop. 11.3.2, $F_n \xrightarrow{w} F$ implies $\tilde{F}_n \xrightarrow{w^*} F$. Thus $\tilde{F}_n \xrightarrow{w} F$ since $(\tilde{F}_n)$ is left-tight and $F(-\infty) = 0$. Finally, we apply Prop. 7.4.4 to the sequence $(\tilde{F}_n)$ so that the desired representation of $F$ follows. □

Clearly, Prop. 11.3.3 determines the set of limit functions of $(F_n)$ under the hypotheses of infinite smallness and left-tightness, i.e., we have treated problem (P1) formulated in Sect. 10.1. Next we turn to conditions implying weak convergence to a certain limit function, i.e., we solve problem (P2).

**Proposition 11.3.4.** *Suppose infinite smallness of the $(X_{nk})$ and let $(a_n)$ be bounded. Further, let $0 < \varrho \leqq 1$ be a constant and $G = \mathcal{H}(a, H)$ be some inf.div.d.f. Then $F_n \xrightarrow{w} F = \varrho G$ holds iff*

$$a_n \to a\,, \qquad n \to \infty\,, \tag{11.3.6}$$

*and*

$$H_n \xrightarrow{w} H\,, \qquad H_n(\infty) \to H(\infty) - \log \varrho\,, \qquad n \to \infty\,. \tag{11.3.7}$$

Proof. Assume first $F_n \xrightarrow{w} F$. Then we see from the above proof that $\tilde{F}_n \xrightarrow{w} F$; hence (11.3.6) and (11.3.7) follow from Prop. 7.4.4.

Conversely, assume that (11.3.6) and (11.3.7) are satisfied. Then by Prop. 7.4.4 we have $\tilde{F}_n \xrightarrow{w} F = \varrho G$. Now Prop. 11.3.2 tells us that $F_n \xrightarrow{w^*} F$.

To finish the proof it suffices to show that $(F_n)$ is left-tight. Indeed, in view of Cor. 7.4.5, the left-tightness of $(\tilde{F}_n)$ implies the left-tightness of $(H_n)$. Accordingly,

$$\sup_n \sum_k \mathsf{P}(X_{nk} - a_{nk} \leqq x) \leqq 2 \sup_n H_n(x) \to 0\,, \qquad x \to -\infty\,,$$

in view of $1 \leqq 2x^2/(1+x^2)$, $x \leqq -1$, and (11.2.3). Reasoning as in (11.3.5) we obtain

$$\sup_n \sum_k \mathsf{P}(X_{nk} - m_{nk} < x) \to 0\,, \qquad x \to -\infty\,.$$

Thus we can use Prop. 11.1.7 whence the left-tightness of $(F_n)$ follows. □

Putting $\varrho = 1$ it is obvious that Prop. 11.3.4 includes Prop. 11.2.4 as a special case. Next we give a reformulation of Prop. 11.3.4 in terms of the Lévy formula for inf.div.d.f.; it includes Prop. 11.2.5 as a special case.

**Proposition 11.3.5.** *Suppose infinite smallness of the $(X_{nk})$ and let $(a_n)$ be bounded. Further, let $\varrho$, $0 < \varrho \leqq 1$, be a constant and $G = \mathcal{L}(a, \gamma^2, M, N)$ be some inf.div.d.f. Then $F_n \xrightarrow{w} F = \varrho G$ holds iff we have (11.2.9), (11.2.10), and as $n \to \infty$*

$$\sum_k F_{nk}(x) \xrightarrow[(-\infty,0)]{} M(x)\,, \qquad \sum_k (F_{nk}(x) - 1) \xrightarrow[(0,\infty)]{} N(x) + \log \varrho\,. \tag{11.3.8}$$

The proof of this result which is based on Prop. 11.3.4 will be left to the reader. It can be carried out in the same way as the proof of Prop. 11.2.5. □

**Remark 11.3.6.** Prop. 11.3.5 also implies an explicit formula for the defect of the limit function $F$. Using the second relation in (11.3.8) and $\varrho = F(\infty)$ we get

$$-\log F(\infty) = \lim_{\tau\to\infty} \overline{\lim_{n\to\infty}} \sum_k \mathsf{P}(X_{nk} \geqq \tau)$$
$$= \lim_{\tau\to\infty} \underline{\lim_{n\to\infty}} \sum_k \mathsf{P}(X_{nk} \geqq \tau) .$$
□

**Example 11.3.7.** Let the *d.f.* of $X_{nk}$ be defined in just the same way as in Ex. 11.1.4 and choose $0 < \tau \leqq 1$. Further, we put $A_n = -1/4$. Then we easily verify $m_{nk} = 0$, $a_{nk} = 0$, $a_n = 0$ for all $k$, $n$, and

$$H_n(x) = \begin{cases} 0 , & x \leqq 1 , \\ 1/4 , & 1 < x \leqq n , \\ 1/4 + \dfrac{1}{2}\left(\dfrac{n^2}{1+n^2}\right), & x > n . \end{cases}$$

Accordingly,

$$H_n(x) \xrightarrow{w} H(x) = \begin{cases} 0 , & x \leqq 1 , \\ 1/4 , & x > 1 . \end{cases}$$

Moreover, we have

$$H_n(\infty) = \frac{1}{4} + \frac{1}{2}\left(\frac{n^2}{1+n^2}\right) \to \frac{3}{4} .$$

Now Prop. 11.3.4 implies $F_n \xrightarrow{w} F = \left(1/\sqrt{e}\right) P_{1/2}$ where $P_{1/2}$ is a Poisson d.f. with parameter $\lambda = 1/2$. □

## 11.4. The weak law of large numbers

The principal importance of the law of large numbers to probability theory has already been underlined in Sect. 3.7 and 10.1. Next we state the general notion of this law for the sums

$$S_n = X_{n1} + X_{n2} + \dots + X_{nk_n} - A_n \sim F_n .$$

**Definition 11.4.1.** We say that the triangular array $(X_{nk})$ obeys the *weak law of large numbers* (LLN) if there is a sequence of constants $(A_n)$ such that

$$\lim_{n\to\infty} \mathsf{P}(|S_n| \geqq \varepsilon) = 0 , \quad \text{for all} \quad \varepsilon > 0 . \quad \square \tag{11.4.1}$$

First we show a remarkable property of the medians.

**Lemma 11.4.2.** *If the triangular array $(X_{nk})$ of i.r.v. obeys the LLN, then infinite smallness holds for the triangular array $(X_{nk} - m_{nk})$.*

Proof. Let $f_{nk}$ be the c.f. of $X_{nk}$. By (11.4.1) we have

$$\exp(-\mathrm{i}A_n t) \prod_k f_{nk}(t) \to 1 , \qquad n \to \infty ,$$

uniformly on every bounded interval. Hence it follows that $\prod_k |f_{nk}(t)|^2 \to 1$. Now from

$$\prod_k |f_{nk}(t)|^2 \leqq \min_k |f_{nk}(t)|^2 \leqq 1$$

it is easily seen that $\max (1 - |f_{nk}(t)|^2) \to 0$. Since $f^s_{nk}(t) = |f_{nk}(t)|^2$ is the c.f. of $X^s_{nk}$ obtained from $X_{nk}$ by symmetrization we get from L. 10.2.2 that the array $(X^s_{nk})$ satisfies the condition of infinite smallness. This proves our assertion by virtue of the first inequality in Cor. 1.4.9. □

Now we wish to prove necessary and sufficient conditions for the LLN. Here we continue to use the notation $a_{nk} = \int_{|x|<\tau} x \, dF_{nk}(x)$ and $F'_{nk}(x) = F(x + a_{nk})$ for some $\tau > 0$.

**Theorem 11.4.3.** *Suppose infinite smallness of the triangular array $(X_{nk})$ of i.r.v. Then the* LLN *holds for $(X_{nk})$ iff*

$$\lim_{n\to\infty} \sum_k \int_{-\infty}^{\infty} w(x) \, dF'_{nk}(x) = 0 . \tag{11.4.2}$$

*In either case the shift constants $A_n$ can be chosen as*

$$A_n = \sum_k a_{nk} . \tag{11.4.3}$$

Proof. First we see from (11.4.1) that the LLN is equivalent to $F_n \xrightarrow{c} U_0$. The degenerate d.f. $U_0$ is inf.div. and admits the representation $U_0 = \mathcal{H}(0, 0)$. Thus we are able to apply Prop. 11.2.4 so that the LLN is equivalent to $H_n \xrightarrow{c} 0$ and $a_n \to 0$. The first convergence is equivalent to (11.4.2) because we have — in view of $x^2/(1 + x^2) \leqq w(x) \leqq 2x^2/(1 + x^2)$ and (11.2.3) —

$$H_n(\infty) \leqq \sum_k \int_{-\infty}^{\infty} w(x) \, dF'_{nk}(x) \leqq 2H_n(\infty) .$$

The relation $a_n \to 0$ is trivially satisfied if $A_n$ is suitably chosen. This completes the proof of the first assertion.

To prove the special choice of $A_n$ in (11.4.3) we make use of the representation $U_0 = \mathcal{L}(0, 0, 0, 0)$ and apply (11.2.10). □

Now we study the LLN where infinite smallness is not mentioned. Using medians $m_{nk}$ of $X_{nk}$ we obtain the following criterion.

**Theorem 11.4.4.** *The LLN holds for the triangular array $(X_{nk})$ of i.r.v. iff*

$$\lim_{n\to\infty} \sum_k \int_{-\infty}^{\infty} w(x) \, dF_{nk}(x + m_{nk}) = 0 . \tag{11.4.4}$$

*In either case the shift constants $A_n$ can be chosen as*

$$A_n = \sum_k \Big(m_{nk} + \int_{|x|<\tau} x \, dF_{nk}(x + m_{nk})\Big) . \tag{11.4.5}$$

Proof. Assume first the LLN. Then L. 11.4.2 implies infinite smallness for $(X_{nk} - m_{nk})$. Applying Th. 11.4.3 to the triangular array $(X_{nk} - m_{nk})$ we get

$$\lim_{n\to\infty} \sum_k \int_{-\infty}^{\infty} w(x) \, dF_{nk}(x + d_{nk}) = 0 \tag{11.4.6}$$

where

$$d_{nk} := m_{nk} + \int_{|x|<\tau} x \, dF_{nk}(x + m_{nk}) .$$

Now the first inequality in L. 10.2.5 implies (11.4.4).

Assume conversely (11.4.4). Then from L. 10.2.2 and from

$$\max_k \int_{-\infty}^{\infty} w(x)\, \mathrm{d}F_{nk}(x + m_{nk}) \leqq \sum_k \int_{-\infty}^{\infty} w(x)\, \mathrm{d}F_{nk}(x + m_{nk}) \to 0\,, \qquad n \to \infty\,,$$

we conclude infinite smallness for the r.v. $X_{nk} - m_{nk}$. Using the second relation in L. 10.2.5 we obtain (11.4.6). This completes the proof of the LLN in view of Th. 11.4.3.

Finally, we see that (11.4.5) easily follows from (11.4.3) applied to the r.v. $X'_{nk} = X_{nk} - m_{nk}$. □

The above results are well known. Now we turn to another version of the LLN. The basic question is the following. We suppose the conditions $F_n \xrightarrow{w} F$ and $F(+0) - F(0) > 0$ which are obviously necessary for the LLN. Then we are interested in what additional assumptions are needed to ensure that the LLN is satisfied.

**Proposition 11.4.5.** *Let there exist a number $\eta > 0$ such that*

$$\varliminf_{n\to\infty} \mathsf{P}(|S_n| \leqq \varepsilon) \geqq \eta\,, \quad \textit{for all} \quad \varepsilon > 0\,. \tag{11.4.7}$$

*Then the following assertions are equivalent:*

(i) *The LLN holds for the triangular array $(X_{nk})$ of i.r.v.*

(ii) *There exist numbers $b_n > 0$, $n \geqq 1$, such that $b_n \to 0$ and*

$$\limsup_{x\to\infty}\, \sup_n \sum_k \mathsf{P}(|X_{nk} - m_{nk}| \geqq b_n x) = 0\,.$$

(iii) $$\lim_{n\to\infty} \sum_k \mathsf{P}(|X_{nk} - m_{nk}| \geqq \varepsilon) = 0\,, \quad \textit{for all} \quad \varepsilon > 0\,.$$

Proof. (i) ⇒ (ii) The LLN tells us that there exist numbers $b_n > 0$ such that $b_n \to 0$ and $\mathsf{P}(|S_n| \leqq b_n) \geqq 1 - b_n$. Now we introduce another triangular array $(\overline{X}_{nk})$ by putting

$$(\overline{X}_{nk}, \overline{S}_n, \overline{m}_{nk}, \overline{A}_n) = b_n^{-1}(X_{nk}, S_n, m_{nk}, A_n)\,, \qquad \overline{S}_n \sim \overline{F}_n\,,$$

and verify that $\mathsf{P}(|\overline{S}_n| \leqq 1) \geqq 1 - b_n$. Hence $(\overline{F}_n)$ is tight and, accordingly, also relatively compact. Applying Th. 11.1.1 we get

$$\limsup_{x\to\infty}\, \sup_n \sum_k \mathsf{P}(|\overline{X}_{nk} - \overline{m}_{nk}| \geqq x) = 0$$

which leads to the desired relation (ii).

(ii) ⇒ (iii) By assumption we get, for all $\varepsilon > 0$, $\delta > 0$,

$$\delta > \sum_k \mathsf{P}(|X_{nk} - m_{nk}| \geqq x_0 b_n) \geqq \sum_k \mathsf{P}(|X_{nk} - m_{nk}| \geqq \varepsilon)\,;$$

the first inequality holds for all $n$ and sufficiently large (but fixed) $x_0$; the second holds for sufficiently large $n$. This proves (iii).

(iii) ⇒ (i) We select a subsequence $(F_{n'})$ which satisfies $F_{n'} \xrightarrow{w} F$ for a certain $F$. Then by (11.4.7) we have $F(+0) - F(0) \geqq \eta$, and so $F \neq \text{const}$. Moreover, it is obvious that (iii) implies the relation (11.3.1). Now we are ready to apply Prop. 11.3.1, i.e., $F$ is non-defective. On the other hand, it is easily seen from (iii) that infinite smallness is valid for $(X_{nk} - m_{nk})$. Thus $F$ is inf.div., $F = \mathcal{L}(a, \gamma^2, M, N)$ say. Next we use Prop. 11.2.5 and conclude from the assumption (iii) that $M = N = 0$, that is, $F$ is normal or degenerate. Since $F$ is not continuous at 0 we also obtain $\gamma^2 = a = 0$. Clearly, all above conclusions are independent of the special subsequence $(F_{n'})$ whence $F_n \xrightarrow{c} U_0$ follows by virtue of Cor. 2.2.3. □

**Corollary 11.4.6.** *Let us assume infinite smallness of the* $(X_{nk})$. *Moreover, assume* $F_n \xrightarrow{w} F$ *and* $F(x_0+0) - F(x_0) > 0$ *for some* $x_0 \in \mathbb{R}_1$. *Then* $F_n \xrightarrow{c} U_{x_0}$ *holds iff we have*

$$\lim_{n\to\infty} \sum_k \mathsf{P}(|X_{nk}| \geqq \varepsilon) = 0\,, \quad \textit{for all} \quad \varepsilon > 0\,.$$

## 11.5. The Poisson distribution

The aim of this section is to state conditions implying the convergence of $S_n \sim F_n$ to a Poisson distribution. We already know that the Poisson d.f. $P_\lambda$, $\lambda > 0$, is inf.div. and admits the representation $P_\lambda = \mathcal{H}(a_\lambda, H_\lambda)$ where $a_\lambda = \lambda/2$, and

$$H_\lambda(x) = \begin{cases} 0\,, & x \leqq 1\,, \\ \lambda/2\,, & x < 1\,, \end{cases}$$

see Ex. 7.2.12 (iii). Making Prop. 11.2.4 more specific it is not difficult to derive the following result.

**Proposition 11.5.1.** *Suppose infinite smallness of the* $(X_{nk})$. *Then* $F_n \xrightarrow{c} P_\lambda$ *holds for suitably chosen constants* $A_n$ *iff we have*

$$\lim_{n\to\infty} \sum_k \int_{|x-1|\geqq\varepsilon} w(x)\,\mathrm{d}F'_{nk}(x) = 0\,, \quad \textit{for all} \quad \varepsilon > 0\,. \tag{11.5.1}$$

*and*

$$\lim_{n\to\infty} \sum_k \mathsf{P}(|X_{nk} - 1| < \varepsilon) = \lambda\,, \quad \textit{for all} \quad \varepsilon > 0\,. \tag{11.5.2}$$

*If these two conditions are satisfied, then* $A_n$ *can be chosen as*

$$A_n = \sum_k \left( a_{nk} + \int_{-\infty}^{\infty} \frac{x}{1+x^2}\,\mathrm{d}F'_{nk}(x) \right) - \lambda/2\,. \quad \square \tag{11.5.3}$$

In the remaining part of this section we restrict our attention to non-negative lattice r.v. $X_{nk}$. Then Prop. 11.5.1 takes on an essentially simpler form. Here we put

$$p_{nk}(j) = \mathsf{P}(X_{nk} = j)\,, \qquad j = 0, 1, \ldots$$

**Proposition 11.5.2.** *Put* $A_n = 0$ *for all* $n$ *and suppose infinite smallness of the* $(X_{nk})$. *Further, let* $\lambda > 0$ *and* $\mathsf{P}(X_{nk} \in \{0, 1, \ldots\}) = 1$. *Then* $F_n \xrightarrow{c} P_\lambda$ *holds iff we have*

$$\sum_k \sum_{j=2}^{\infty} p_{nk}(j) \to 0\,, \qquad n \to \infty\,, \tag{11.5.4}$$

*and*

$$\sum_k p_{nk}(1) \to \lambda\,, \qquad n \to \infty\,. \tag{11.5.5}$$

Proof. For $0 < \tau \leqq 1$ we obviously obtain $a_{nk} = 0$ and $F'_{nk} = F_{nk}$. Now it is clear that (11.5.1) and (11.5.2) can be rewritten as (11.5.4) and (11.5.5), resp. Hence it remains to show that we may put $A_n = 0$ for all $n$. In view of Prop. 2.3.1 it suffices to prove that the constants $A_n$ defined in (11.5.3) tend to zero as $n \to \infty$. Indeed, since $F'_{nk} = F_{nk}$ and by (11.5.4), (11.5.5), we have

$$|A_n| = \left| \sum_k \left( a_{nk} + \int_{-\infty}^{\infty} \frac{x}{1+x^2}\,\mathrm{d}F'_{nk}(x) \right) - \lambda/2 \right|$$

$$\leqq \tfrac{1}{2} \left| \sum_k p_{nk}(1) - \lambda \right| + \sum_k \sum_{j=2}^{\infty} p_{nk}(j) \to 0\,, \qquad n \to \infty\,. \qquad \square$$

Note that Prop. 11.5.2 contains the classic theorem of Poisson, see Prop. 3.7.3.

Next we turn to limit theorems under the assumption of r-convergence. More precisely, if the r.v. $X_{nk}$ take on only non-negative integer values and $\mathsf{P}(S_n = 0) \to p_0$, $0 < p_0 < 1$, holds, then we are interested in conditions which must be added such that $F_n \xrightarrow{c} P_\lambda$ for $\lambda = -\log p_0$.

**Proposition 11.5.3.** *Put $A_n = 0$ for all $n$ and suppose infinite smallness. Further, let $\lambda > 0$ and $\mathsf{P}(X_{nk} \in \{0, 1, \dots\}) = 1$. Then $F_n \xrightarrow{c} P_\lambda$ holds iff the following conditions are satisfied:*

(i) $$\lim_{n\to\infty} \mathsf{P}(S_n = 0) = \mathrm{e}^{-\lambda}.$$

(ii) *There exists an integer $k \geqq 1$ such that*

$$\varliminf_{n\to\infty} \mathsf{P}(S_n \leqq k) \geqq \mathrm{e}^{-\lambda} \sum_{j=0}^{k} \frac{\lambda^j}{j!}.$$

Proof. The necessity of the above conditions is obvious. Let us now assume (i) and (ii). Then we can choose a subsequence $(F_{n'})$ which satisfies $F_{n'} \xrightarrow{w} F$ for a certain function $F$. Because of $p_0 = \mathrm{e}^{-\lambda} > 0$ we have $F \neq$ const. and Prop. 11.3.4 implies $F = \varrho G$ where $G$ is inf.div. and $0 < \varrho = F(\infty) \leqq 1$. Next we show by an indirect proof that $\varrho = 1$. If this were wrong, i.e., $0 < \varrho < 1$, then we could write

$$q_j := G(j+0) - G(j), \qquad j = 0, 1, \dots, \lambda := -\log p_0, \qquad \mu := -\log q_0.$$

Thus we have $\lambda = \mu - \log \varrho > \mu$ and, by assumption (ii),

$$\sum_{j=0}^{k} q_j = \frac{1}{\varrho} \lim_{n'\to\infty} \mathsf{P}(S_{n'} \leqq k) \geqq \frac{1}{\varrho} \mathrm{e}^{-\lambda} \sum_{j=0}^{k} \frac{\lambda^j}{j!}$$
$$= \mathrm{e}^{-\mu} \sum_{j=0}^{k} \frac{\lambda^j}{j!} > \mathrm{e}^{-\mu} \sum_{j=0}^{k} \frac{\mu^j}{j!}.$$

Now Prop. 7.6.14 tells us that such an inf.div.d.f. $G$ does not exists. Hence $\varrho = 1$ and $\lambda = \mu$. Prop. 7.6.14 also implies $F = P_\lambda$. This result is independent of the special subsequence $(F_{n'})$; this terminates the proof by virtue of Cor. 2.2.3. □

Note that this proof is not quite in accordance with was said in Sect. 10.4 since compactness and uniqueness are obtained in the same step.

Putting $k = 1$ and writing the above condition (ii) in a simplified manner, we get the following result.

**Corollary 11.5.4.** *Put $A_n = 0$ for all $n$, let $\mathsf{P}(X_{nk} \in \{0, 1, \dots\}) = 1$, and suppose infinite smallness. Then $F_n \xrightarrow{c} P_\lambda$ holds for some $\lambda > 0$ iff there are constants $p_0$, $p_1$, $0 < p_0 < 1$, such that*

$$\lim_{n\to\infty} \mathsf{P}(S_n = j) = p_j; \qquad j = 0, 1, \tag{11.5.6}$$

*and*

$$-p_0 \log p_0 \leqq p_1. \tag{11.5.7}$$

*In either case we have $\lambda = -\log p_0$ so that only the equality sign is possible in* (11.5.7). □

The special feature of the above result is that we only assume the convergence of the first and the second jump of $(F_n)$ to the corresponding jumps of a Poisson d.f. $P_\lambda$. It comes out that under the hypothesis of convergence to other jumps of $P_\lambda$ this assertion is not true. This is seen by the following example.

**Example 11.5.5.** We set $\mu = 7/48$, $\lambda = 1/2$, and for $n \geqq 2$, $k = 1, 2, \ldots, n$,

$$\mathsf{P}(X_{nk} = j) = \begin{cases} 1 - 1/2n\,, & j = 0\,,\\ \lambda^2/2n\,, & j = 2\,,\\ \lambda^3/6n\,, & j = 3\,,\\ (\lambda - \mu)/n\,, & j = n\,,\\ 0\,, & \text{otherwise}\,. \end{cases}$$

Using Prop. 11.3.5 it is not difficult to prove that

$$S_n = \sum_{k=1}^{n} X_{nk} \sim F_n \xrightarrow{\text{w}} F$$

where $F$ is lattice and defective with $F(\infty) = \mathrm{e}^{\mu-\lambda} < 1$. On the other hand, we can show the convergence

$$\lim_{n\to\infty} \mathsf{P}(S_n = j) = \mathrm{e}^{-\lambda} \frac{\lambda^j}{j!}\,; \qquad j = 0, 2, 3\,,$$

which corresponds to (11.5.6). □

## 11.6. Analytic limit distribution functions

Throughout this section we denote by $S$ a set of real numbers with a finite limit point. Our aim is to treat the convergence of $(F_n)$ under the hypothesis of r-convergence on $S$. The idea of the corresponding proofs is presented in Sect. 10.4. Namely, we first have to show the relative compactness of $(F_n)$; the next step is to prove the uniqueness of the limit d.f. $F$. To solve the latter problem we make use of the analyticity of $F$.

Our two basic assumptions, each of which ensures that $F$ is analytic, are

$$\lim_{\varepsilon\to+0} \varliminf_{n\to\infty} \sum_k \int_{|x|<\varepsilon} x^2 \,\mathrm{d}F'_{nk}(x) > 0\,, \tag{11.6.1}$$

and

$$\lim_{\varepsilon\to+0} \lim_{n\to\infty} \varepsilon^\nu \sum_k \mathsf{P}(\eta\varepsilon \leqq |X_{nk}| \leqq \varepsilon) > 0\,, \quad \text{for some} \quad \eta \in (0, 1)\,, \quad \nu \geqq 1\,. \tag{11.6.2}$$

Note that in (11.6.1) the limit $\varepsilon \to +0$ exists by monotonicity. Intuitively, the conditions (11.6.1) and (11.6.2) imply that the probability masses of the r.v. $X_{nk}$ are concentrated closely, but not too closely about the origin. In view of Prop. 11.2.4 the condition (11.6.1) guarantees that there exists a normal component $\gamma^2 = H(+0) - H(0)$ of the limit d.f. But the relation (11.6.2) in its turn implies a certain growth of the spectral functions $M$, $N$ in a neighbourhood of the origin, see (11.2.8). By simple examples we can show that these conditions are independent of each other.

**Proposition 11.6.1.** *Let $F$ be an inf.div.d.f. with a normal component and suppose infinite smallness. Then $F_n \xrightarrow{\text{c}} F$ holds iff we have the three conditions* (11.6.1),

$$F_n \underset{S}{\rightarrow} F\,, \qquad n \to \infty\,, \tag{11.6.3}$$

*and*

$$\sup_n \sum_k \mathsf{P}(|X_{nk}| \geqq x) \to 0\,, \qquad x \to \infty\,. \tag{11.6.4}$$

Proof. Let us first assume $F_n \xrightarrow{c} F$ where $F$ has a normal component. Then from Prop. 11.2.4 and Cor. 2.1.4 it follows that

$$\gamma^2 = H(+0) - H(0) = \lim_{\varepsilon \to +0} \varliminf_{n \to \infty} \sum_k \int_{|x| < \varepsilon} \frac{x^2}{1 + x^2} \, dF'_{nk}(x) > 0 .$$

Now the relation (11.6.1) is obvious. The r-convergence in (11.6.3) easily follows from $F_n \xrightarrow{c} F$ since $F$ is continuous. Next we apply L. 10.2.3 and get $\max_k |m_{nk}| \leqq x/2$ for sufficiently large $n$. Hence

$$\sup_n \sum_k \mathsf{P}(|X_{nk}| \geqq x) \leqq \sup_n \sum_k \mathsf{P}(|X_{nk} - m_{nk}| \geqq x/2) .$$

Using the relative compactness of $(F_n)$, we see from Th. 11.1.1 that the term on the right converges to zero. This proves (11.6.4).

Now we conversely assume that (11.6.1), (11.6.3), (11.6.4) are true. Then we choose a subsequence $(F_{n'})$ which satisfies $F_{n'} \xrightarrow{w} G$ for a certain function $G$.

*Compactness.* The r-convergence given by (11.6.3) and the assumed analyticity of $F$ imply $G \neq$ const. on $S$. Further, we obtain from Cor. 1.4.9 and from (11.6.4) that, as $x \to \infty$,

$$\sup_n \sum_k \mathsf{P}(|X_{nk} - m_{nk}| \geqq x) \leqq 4 \sup_n \sum_k \mathsf{P}(|X_{nk}| \geqq x/2) \to 0 . \tag{11.6.5}$$

Thus Prop. 11.3.1 applies to conclude that $G$ is non-defective. Moreover, $G$ is inf.div. since infinite smallness is satisfied, $G = \mathscr{H}(a, H)$ say.

*Continuation.* We define $H_n$ as in (11.2.3) and see from Prop. 11.2.4 that $H_{n'} \xrightarrow{c} H$. By Cor. 2.1.4 we then have

$$\begin{aligned} \gamma^2 = H(+0) - H(0) &= \lim_{\varepsilon \to +0} \varliminf_{n' \to \infty} \sum_k \int_{|x| < \varepsilon} \frac{x^2}{1 + x^2} \, dF'_{nk}(x) \\ &\geqq \lim_{\varepsilon \to +0} \varliminf_{n' \to \infty} \frac{1}{1 + \varepsilon^2} \sum_k \int_{|x| < \varepsilon} x^2 \, dF'_{n'k}(x) . \end{aligned}$$

Now it is clear that (11.6.1) implies $\gamma^2 > 0$, that is, $G$ is also analytic. Combining (11.6.3) and $F_{n'} \xrightarrow{c} G$ we observe that $F(x) = G(x)$, $x \in S$, and, by analytic continuation, $F$ coincides with $G$. This result is independent of the special subsequence $(F_{n'})$ whence $F_n \xrightarrow{c} F$ follows by virtue of Cor. 2.2.3. □

Next we replace the assumption (11.6.1) by (11.6.2).

**Proposition 11.6.2.** *Let $F$ be an inf.div. analytic d.f. and suppose infinite smallness. Moreover, let* (11.6.2) *be satisfied. Then $F_n \xrightarrow{c} F$ holds iff* (11.6.3) *and* (11.6.4) *are true.*

Proof. This proposition can be shown in just the same way as Prop. 11.6.1. In this case we make use of Prop. 7.5.4 and (11.6.2) to prove the analyticity of the limit d.f. □

We now give a sufficient condition for (11.6.1).

**Lemma 11.6.3.** *If for every sufficient small $\varepsilon > 0$ there exist a sequence $(\eta_n(\varepsilon))$ satisfying $0 < \eta_n(\varepsilon) < \varepsilon$ and*

$$\lim_{\varepsilon \to +0} \varliminf_{n \to \infty} \eta_n^2(\varepsilon) \sum_k \mathsf{P}(\eta_n(\varepsilon) \leqq |X_{nk} - a_{nk}| \leqq \varepsilon) > 0 ,$$

*then* (11.6.1) *holds true.*

Proof. Assuming the existence of such a sequence $\eta_n(\varepsilon)$ the assertion follows from

$$\sum_k \int_{|x|<\varepsilon} x^2 \, \mathrm{d}F'_{nk}(x) \geqq \sum_k \int_{\eta(\varepsilon)\leqq|x|\leqq\varepsilon} x^2 \, \mathrm{d}F'_{nk}(x)$$
$$\geqq \eta_n^2(\varepsilon) \sum_k \mathsf{P}\big(\eta_n(\varepsilon) \leqq |X_{nk} - a_{nk}| \leqq \varepsilon\big) . \qquad \square$$

**Remark 11.6.4.** It is possible to drop the assumption „$F$ is inf.div. with a normal component" in Prop. 11.6.1 or "$F$ is inf.div. and analytic" in Prop. 11.6.2. Then the corresponding statements can be reformulated in the following way.

(i) Suppose infinite smallness. Then $F_n \xrightarrow{c} F$ holds for some inf.div.d.f. with a normal component iff we have (11.6.1), (11.6.4), and $F_n \underset{S}{\to} \psi$ for a certain function $\psi \neq$ const. on $S$.

(ii) Suppose infinite smallness and let (11.6.2) be satisfied. Then $F_n \xrightarrow{c} F$ holds for some inf.div. and analytic d.f. iff we have (11.6.4) and $F_n \underset{S}{\to} \psi$ for a certain function $\psi \neq$ const. on $S$. $\square$

The results stated above get a significantly simpler form under the condition that $X_{nk} \sim F_{nk}$ are symmetric. It is then noteworthy, that r-convergence on an interval implies complete convergence without the additional assumptions (11.6.1) or (11.6.2).

**Proposition 11.6.5.** *Suppose infinite smallness and let the d.f. $F$ be inf.div. and analytic. Further, let $F_{nk}$ be symmetric. Then*

$$S_n = \sum_k X_{nk} \sim F_n \xrightarrow{c} F \,, \quad \text{as} \quad n \to \infty \,,$$

*holds iff* (11.6.4) *is satisfied and*

$$F_n \xrightarrow[(-x_0,\, x_0)]{} F \,, \quad \text{for some } x_0 > 0 \,. \tag{11.6.6}$$

Proof. The necessity of (11.6.4) and (11.6.6) is obvious. To prove the sufficiency we select a subsequence $(F_{n'})$ which satisfies $F_{n'} \xrightarrow{w} G$ for certain $G$.

*Compactness.* The analyticity of $F$ implies $F \neq$ const. on $S = (-x_0, x_0)$. Hence $G \neq$ const. According to Prop. 11.3.1 we further see that (11.6.4) implies $F_{n'} \xrightarrow{c} G$ and $G$ is non-defective.

*Continuation.* By our assumptions we easily verify that $G$ is symmetric and inf.div. The corresponding c.f. $g$ is real and $g(t) \neq 0$ holds true for all $t$, i.e., $g > 0$. Taking account of the set $\boldsymbol{P}$ (see Def. 6.2.1) we have $G \in \boldsymbol{P}$. On the other hand, we combine (11.6.6) with $F_{n'} \xrightarrow{c} G$ and get

$$G(x) = F(x) \,, \qquad x \in (-x_0, x_0) \,.$$

Now we observe that all suppositions of Prop. 6.5.1 are satisfied. Thus $G$ coincides with $F$; this completes the proof by virtue of Cor. 2.2.3. $\square$

## 11.7. Supplement

*Compactness of sums*

We put $S_n = \sum_k X_{nk} - A_n \sim F_n$, $a_{nk} = \int_{|x|<\tau} x \, \mathrm{d}F_{nk}(x)$, and $F'_{nk}(x) = F(x + a_{nk})$.

**11.7.1.** *Let the accompanying d.f. $\tilde{F}_n$ be defined as in* (11.1.1)–(11.1.3). *Then the sequence $(F_n)$ is relatively compact iff so is $(\tilde{F}_n)$.* — This statement is not hard to prove using Th. 11.1.1 and Cor. 11.1.5.

Now we turn to reformulations of results given in Sect. 11.1; note that we make here the additional assumption of infinite smallness.

**11.7.2.** *Let the accompanying d.f. $\tilde{F}_n$ be defined as in* (11.2.1)–(11.2.3) *and suppose infinite smallness of* $(X_{nk})$. *Then the following assertions are equivalent:*

(i) $(F_n)$ *is shift compact.*

(ii) $(\tilde{F}_n)$ *is shift compact.*

(iii)
$$\sup_n \sum_k \int_{-\infty}^{\infty} w(x)\, dF'_{nk}(x) < \infty\,, \tag{11.7.1}$$

*and*

$$\lim_{x\to\infty} \sup_n \sum_k \mathsf{P}(|X_{nk}| \geqq x) = 0\,. \tag{11.7.2}$$

(iv) *We have the relations* (11.7.2) *and*

$$\inf_n Q(S_n, h) > 0\,, \quad \textit{for some} \quad h > 0\,. \tag{11.7.3}$$

We can show this result in a similar way as was used for Th. 11.1.1, with the addition of L. 11.2.1. Note that a simple proof of the equivalence (i) $\Leftrightarrow$ (ii) $\Leftrightarrow$ (iii) can also be given using Helly's selection theorem, Cor. 7.4.2, and Prop. 11.2.4.

**11.7.3.** *Suppose infinite smallness. If* $(F_n)$ *is shift compact, then the shift constants may be chosen as*

$$A_n^1 = \sum_k \Big(a_{nk} + \int_{|x|<\tau} x\, dF'_{nk}(x)\Big)\,, \qquad n \geqq 1\,,$$

$$A_n^2 = \sum_k \left(a_{nk} + \int_{-\infty}^{\infty} \frac{x}{1+x^2}\, dF'_{nk}(x)\right), \qquad n \geqq 1\,,$$

$$A_n^3 = \sum_k a_{nk}\,, \qquad n \geqq 1\,.$$

Hints for the proof. That $A_n^1$ are shift constants can be proved in just the same way as the shift constants $A_n(\tau)$ in (11.1.8) on replacing $m_{nk}$ by $a_{nk}$.

To see that $A_n^2$ are shift constants we use (11.7.1) and verify that $(A_n^2 - A_n^1)$ is bounded. Then the assertion follows from Prop. 2.4.7. — Finally, the sequence $(A_n^3 - A_n^1)$ is also bounded. To prove this we again use (11.7.1) and show the estimate

$$\Big|\sum_k \int_{|x|<\tau} x\, dF'_{nk}(x)\Big| \leqq C$$

using arguments similar to the proof of theorem 6, Chap. 4 in Petrov (1975).

**11.7.4.** We formulate another version of Prop. 11.1.7. *Let us suppose infinite smallness and choose $A_n^1$ as in* 11.7.3. *Further, we assume that* (11.7.1) *is true. Then* $(F_n)$ *is left-tight iff we have*

$$\lim_{x\to-\infty} \sup_n \sum_k \mathsf{P}(X_{nk} \leqq x) = 0\,.$$

The proof of this assertion can be carried out as the proof of Prop. 11.1.7 on replacing $m_{nk}$ by $a_{nk}$. In this case we use the estimate

$$\mathsf{P}(X_{nk} - a_{nk} < \tau) \geqq 1/2 =: \eta_0$$

instead of (11.1.11), which easily follows from infinite smallness and $\max_k |a_{nk}| \to 0$.

*General convergence theorems*

**11.7.5.** Convergence of $(F_n)$ without infinite smallness. It is in view of such results that the assertions of Sect. 11.1 are of particular interest. We denote the Lévy metric of two d.f. $F_1, F_2$ by $L(F_1, F_2)$, see GNEDENKO and KOLMOGOROV (1959). Let $S_n = \sum_k X_{nk} \sim F_n$.

*Assume that $(F_n)$ is relatively compact. Then $F_n(x + A_n) \xrightarrow{c} F(x)$ holds for some d.f. $F$ and suitably chosen constants $A_n$ iff $A_n \to A$ for some $A \in \mathbb{R}_1$ and there exists a triangular array $(Y_{nk})$ of i.r.v., $k = 1, 2, \ldots, k_n$, $Y_{nk} \sim G_{nk}$, such that*

$$F = G_{n1} * G_{n2} * \ldots * G_{nk_n}\,, \quad \textit{for all } n\,,$$

$$\lim_{n\to\infty} \max_k L(F_{nk}, G_{nk}) = 0\,,$$

$$\lim_{n\to\infty} \sum_k L(\hat{F}_{nk}, G_{nk}) = 0; \qquad \hat{F}_{nk}(x) = F_{nk}\left(x + \frac{A_n}{k_n}\right)$$

(ZOLOTAREV (1970), BERGSTRÖM (1973)). — For more general results concerning sums of Hilbert space valued r.v. see KRUGLOV (1974). See also 11.7.8.

**11.7.6.** According to Prop. 11.2.4 and 11.2.5, the terms

$$C_n(\varepsilon) := \sum_k \int_{|x|<\varepsilon} x^2 \, \mathrm{d}F'_{nk}(x)\,,$$

$$D_n(\varepsilon) := \sum_k \Big( \int_{|x|<\varepsilon} x^2 \, \mathrm{d}F_{nk}(x) - \big( \int_{|x|<\varepsilon} x \, \mathrm{d}F_{nk}(x)\big)^2\Big)$$

play an outstanding role in the construction of the normal component of inf.div. limit d.f. Next we state the connection between $C_n$ and $D_n$.

(i) *Let us suppose infinite smallness. Further, let* (11.7.1) *and*

$$\lim_{\delta\to+0} \sup_n \sum_k \mathsf{P}(\varepsilon - \delta \leq |X_{nk}| \leq \varepsilon + \delta) = 0 \tag{11.7.4}$$

*be satisfied for some $\varepsilon$, $0 < \varepsilon < \tau$. Then $C_n(\varepsilon) - D_n(\varepsilon) \to 0$, $n \to \infty$. Note that* (11.7.1) *and* (11.7.4) *are true, for instance, if $F_n \xrightarrow{c} F = \mathcal{H}(a, H)$ and $\pm\varepsilon$ are continuity points of $H$.*

(ii) *If we assume infinite smallness and* (11.7.1), *then we have*

$$\lim_{\varepsilon\to+0} \overline{\lim_{n\to\infty}} \big(C_n(\varepsilon) - D_n(\varepsilon)\big) = 0\,.$$

For the proof of (i), (ii) we refer to the proof of Th. 8 in PETROV (1975), IV, § 2. We remark that Petrov's proof of (i) is slightly incorrect since the relation (11.7.4) is not assumed. However, in proving Prop. 11.2.5 from Prop. 11.2.4 it suffices to show the relation (ii).

**11.7.7. Problem.** Let us suppose infinite smallness. Will it be possible to find a representation of a defective limit function $F$ by means of inf.div.d.f. if the one-sided thightness of $(F_n)$ is not guaranteed; see Prop. 11.3.3 ?

The following result concerns r-convergence to an analytic d.f.

**11.7.8.** *Let $S \subset \mathbb{R}_1$ be a set with a finite limit point. Assume, moreover, that $\sum_k \mathsf{Var}\, X_{nk} \leq C$ for all $n \geq 1$ and that the c.f. $f_{nk}$ of $F_{nk}$ satisfy the condition*

$$|f_{nk}(t)| \leq \mathrm{e}^{-c_{nk}t}\,, \qquad |t| \geq 1\,,$$

*where $c_{nk} \geq 0$ and*

$$c = \underline{\lim}_{n\to\infty} \sum_k c_{nk} > 0\,.$$

*If $F_n \underset{S}{\rightarrow} \psi$ for a certain function $\psi$, then there exists an analytic d.f. $F$ such that $F_n \xrightarrow{c} F$* (JESIAK and ROSSBERG (1981)).

For the proof Vitali's convergence theorem is applied.

## Notes and bibliographical data

The single assertions of Th. 11.1.1 were given by various authors; the equivalence (i) ⇔ (ii) is well known in the case of symmetric d.f. $F_{nk}$, see KRUGLOV (1976); (i) ⇔ (iii) was formulated in a different form in BERGSTRÖM (1963), Th. 11.5; (i) ⇔ (iv) is due to SIEGEL (1978). — Cor. 11.1.5 and Prop. 11.1.7 can be found in SIEGEL (1978).

The powerful tool L.11.2.1 is partially included in the classic theory of summation of i.r.v.; namely, the formula (11.2.4) was shown under the condition (iii), see PETROV (1975), IV, § 1, Lemma 10. PETROV (1975) gave also the other results of Sect. 11.2 beginning with Th. 11.2.3. A first exposition of this theory was stated in GNEDENKO and KOLMOGOROV (1959).

The most of the material contained in Sect. 11.3 was formulated in SIEGEL (1978) with different proofs. The only-if part of Prop. 11.3.2 was also proved in TITOV (1981 a); further, in the same paper it is shown that Prop. 11.3.3 – 11.3.5 remain true even if the boundedness of the sequence $(a_n)$ is not assumed. Prop. 11.3.3 was correctly formulated by PETŠINKIN (1975) but his proof contains problems. The paper SIEGEL (1978) contains also a proof of the equivalence (i) ⇔ (ii) in Prop. 11.4.5.

The recent versions of limit theorems given in Prop. 11.5.3 and Cor. 11.5.4 concerning the convergence of $(F_n)$ to a Poisson d.f. were proved in SIEGEL (1981 c).

The results of Sect. 11.6 up to Rem. 11.6.4 are due to JESIAK (1978) and (1979), while Prop. 11.6.5 was shown in SIEGEL (1979).

# Chapter 12. The central limit theorem for triangular arrays

The central limit theorem (CLT) is of eminent practical and theoretical importance. It states that under very general assumptions the distributions of sums of r.v. tend to a normal distribution for an increasing number of summands. In the simplest situation these assumptions distinctly imply that the individual summands are "uniformly negligible" in a certain sense to be specified in each case. Further, research on more and more general and complicated formulations of the CLT has demanded new methods and the continuous sharpening of tools which are now widely used in probability theory. Zolotarev's conjecture (see Sect. 10.4) and its solution show that this process is still going on today.

In the present chapter we investigate important well-known limit relations for a triangular array of i.r.v. $(X_{nk})$. A considerable part of the text is also given over to material which has not previously appeared in monographs on probability theory. For instance, we treat the celebrated Lindeberg-Feller theorem together with recent versions of it; these results concern the case of bounded variances.

Here we use the same notation as in the preceding chapter; i.e., we write

$$X_{nk} \sim F_{nk}, \qquad S_n = \sum_k X_{nk} - A_n \sim F_n$$

for certain real numbers $A_n$, $n \geqq 1$, and put $a_{nk} = \int_{|x|<\tau} x \, \mathrm{d}F_{nk}(x)$. The symbol $F_n \xrightarrow{c} \Phi_{\mu,\sigma}$ means that there exists a sequence $(A_n)$ such that the complete convergence of $(F_n)$ takes place.

## 12.1. The general case

The normal d.f. is inf.div. and admits the representation $\Phi_{\mu,\sigma} = \mathcal{L}(\mu, \sigma^2, 0, 0)$. Thus we may derive a CLT by specifying results contained in Sect. 11.2.

**Theorem 12.1.1.** *We have infinite smallness and* $F_n \xrightarrow{c} \Phi_{\mu,\sigma}$ *for some* $\mu \in R_1$, $\sigma \geqq 0$, *iff the two conditions*

$$\lim_{n\to\infty} \sum_k \mathsf{P}(|X_{nk}| \geqq \varepsilon) = 0\,, \quad \textit{for all} \quad \varepsilon > 0\,, \tag{12.1.1}$$

*and*

$$\lim_{n\to\infty} \sum_k \Big[\int_{|x|<\varepsilon} x^2 \, \mathrm{d}F_{nk}(x) - \Big(\int_{|x|<\varepsilon} x \, \mathrm{d}F_{nk}(x)\Big)^2\Big] = \sigma^2\,, \quad \textit{for all} \quad \varepsilon > 0\,, \tag{12.1.2}$$

*hold true. Then the constants* $A_n$ *can be chosen as*

$$A_n = \sum_k \int_{|x|<\tau} x \, \mathrm{d}F_{nk}(x) - \mu\,, \qquad \textit{for some } \tau > 0\,. \tag{12.1.3}$$

Note that if $\mu = \sigma = 0$, then the conditions (12.1.1)–(12.1.3) are necessary and sufficient for the LLN (a similar result is given in Th. 11.4.3).

Before proving this theorem we state the following lemma the proof of which can be found in PETROV (1975), IV, § 4; it is purly technical.

**Lemma 12.1.2.** *If we have* (12.1.1) *and*

$$\lim_{\varepsilon \to +0} \overline{\lim_{n \to \infty}} \sum_k \Big[ \int_{|x|<\varepsilon} x^2 \, dF_{nk}(x) - \Big( \int_{|x|<\varepsilon} x \, dF_{nk}(x) \Big)^2 \Big]$$
$$= \lim_{\varepsilon \to +0} \underline{\lim_{n \to \infty}} \sum_k \Big[ \int_{|x|<\varepsilon} x^2 \, dF_{nk}(x) - \Big( \int_{|x|<\varepsilon} x \, dF_{nk}(x) \Big)^2 \Big] = \sigma^2 , \tag{12.1.4}$$

*then* (12.1.2) *holds.* □

Proof of Theorem 12.1.1. If we assume (12.1.1), then we obtain infinite smallness from

$$\max_k \mathsf{P}(|X_{nk}| \geqq \varepsilon) \leqq \sum_k \mathsf{P}(|X_{nk}| \geqq \varepsilon) \to 0 , \qquad n \to \infty . \tag{12.1.5}$$

On the other hand, (12.1.2) obviously implies (12.1.4). Thus we see that the suppositions of Prop. 11.2.5 are satisfied with $M = N = 0$. Hence, $F_n \xrightarrow{c} \Phi_{\mu,\sigma}$ for suitably chosen $A_n$.

Conversely, we now assume infinite smallness and $F_n \xrightarrow{c} \Phi_{\mu,\sigma}$. Applying again Prop. 11.2.5, we see that (11.2.8) implies (12.1.1); moreover, we also have (12.1.4). Now L. 12.1.2 tells us that (12.1.2) is true. Finally, in the relation (11.2.10) we have to put $a = \mu$; this yields the asserted representation (12.1.3). □

It is interesting to note that the assumption (12.1.1) represents the unavoidable condition for the CLT in the present model while (12.1.2) can be equivalently replaced by other conditions which are quite different. Such conditions contain, for instance, the r-convergence of $(F_n)$, see Prop. 12.1.5.

We give now a reformulation of condition (12.1.1), which describes a special uniform negligibility of the summands $X_{nk}$; the relation (12.1.5) shows that it is sharper than infinite smallness.

**Corollary 12.1.3.** *The relation* (12.1.1) *is equivalent to*

$$\lim_{n \to \infty} \mathsf{P}(\max_k |X_{nk}| \geqq \varepsilon) = 0 , \quad \text{for all} \quad \varepsilon > 0 . \tag{12.1.6}$$

Proof. Let us assume (12.1.1), then (12.1.6) follows from

$$\mathsf{P}(\max_k |X_{nk}| \geqq \varepsilon) = \mathsf{P}(\bigcup_k \{|X_{nk}| \geqq \varepsilon\}) \leqq \sum_k \mathsf{P}(|X_{nk}| \geqq \varepsilon) .$$

Conversely, we assume (12.1.6) and get (12.1.1) from

$$\mathsf{P}(\max_k |X_{nk}| \geqq \varepsilon) = 1 - \mathsf{P}(\bigcap_k \{|X_{nk}| < \varepsilon\})$$
$$= 1 - \prod_k \mathsf{P}(|X_{nk}| < \varepsilon) \geqq 1 - \exp\Big(- \sum_k \mathsf{P}(|X_{nk}| \geqq \varepsilon)\Big) \geqq 0 .$$ □

The point of the following investigation is that we make a better use of (12.1.1) than in the proof of Th. 12.1.1, namely we also derive the relative compactness of $(F_n)$.

**Theorem 12.1.4.** *We have infinite smallness and $F_n \xrightarrow{c} \Phi_{\mu,\sigma}$ for some $\mu \in R_1$, $\sigma \geqq 0$, iff the condition* (12.1.1) *is true and $F_n \xrightarrow{w} F$ for a certain non-constant function $F$.*

Proof. The necessity is obvious because of Th. 12.1.1. To prove the sufficiency we first assume (12.1.1). Then we get in view of the relation (11.6.5)

$$\lim_{x\to\infty} \sup_n \sum_k \mathsf{P}(|X_{nk} - m_{nk}| \geqq x) = 0 .$$

Since we also assume $F_n \xrightarrow{w} F \neq \text{const}$. Prop. 11.3.1 applies to conclude that $F$ is a non-defective d.f. Assumption (12.1.1) also implies infinite smallness so that $F$ is inf.div. by Hinčin's theorem, $F = \mathcal{L}(\mu, \sigma^2, M, N)$ say. Now we use Prop. 11.2.5 and compare (12.1.1) with (11.2.8); thus we obtain $M = N = 0$. This proves the desired assertion. □

**Proposition 12.1.5.** *Let $\mu \in \mathbb{R}_1$ and $\sigma > 0$. Then $F_n \xrightarrow{c} \Phi_{\mu,\sigma}$ and infinite smallness hold iff we have* (12.1.1) *and the* r-*convergence*

$$\lim_{n\to\infty} F_n(x_j) = \Phi_{\mu,\sigma}(x_j) , \qquad j = 1, 2 , \tag{12.1.7}$$

*for certain real numbers $x_1 \neq x_2$.*

Proof. Clearly, it is enough to prove the *if-part*. Let $(F_{n'})$ be a subsequence which satisfies $F_{n'} \xrightarrow{w} F$ for some monotone function $F$.

*Compactness.* From (12.1.7) we conclude that $F \neq$ const. Now we note that the subsequence $(n')$ defines a triangular array $(X_{n'k})$ to which Th. 12.1.4 can be applied. Thus $F_{n'} \xrightarrow{c} \Phi_{\mu',\sigma'}$ follows for certain numbers $\mu' \in \mathbb{R}_1$, $\sigma' \geqq 0$; i.e., the set of possible limit d.f. consists only of normal d.f.

*Continuation.* Using this result and the assumption (12.1.7) we have $\Phi_{\mu,\sigma}(x_j) = \Phi_{\mu',\sigma'}(x_j)$, $j = 1, 2$. Now we use Prop. 1.1.7 and get $\mu = \mu'$, $\sigma = \sigma'$. This result is independent of the special subsequence $(F_{n'})$ whence the assertion follows. □

## 12.2. The Lindeberg-Feller theorem

In this section we assume that the variances of $X_{nk}$ exist for all $n$, $k$. It will be seen that this assumption leads to remarkably simple modifications of the CLT; an example of such a result has already been given in Th. 3.7.2. More information in this direction can be found in Suppl. 12.5.1.

First we observe that there is no loss of generality in assuming that $\mathsf{E}X_{nk} = 0$ since the centering constants $A_n$ can be chosen appropriately. Then we will see that we may put $A_n = 0$. Hence we have only to consider the sums $S_n = \sum_k X_{nk}$ so that $\mathsf{E}S_n = 0$ follows. Next we write

$$\sigma_{nk}^2 := \mathsf{Var}\, X_{nk} , \qquad B_n^2 := \mathsf{Var}\, S_n = \sum_k \sigma_{nk}^2$$

and see that it is a mild restriction of generality to assume that $B_n^2 = 1$, $n \geqq 1$. In fact, we may introduce another triangular array of i.r.v. $(Y_{nk})$ by putting $Y_{nk} = B_n^{-1} X_{nk}$, $k = 1, 2, \dots, k_n$. Thus $\mathsf{Var}\,(\sum_k Y_{nk}) = 1$ is guaranteed for all $n$. This property of the model under consideration proves to be essential; among other things it implies also the relative compactness of $(F_n)$, see Prop. 2.4.8.

The above construction of norming and centering constants is particularly natural in the special case of sums of mutually i.r.v. $X_1, X_2, \dots$ with finite variances. Then the numbers $a_n$, $b_n$ occurring in (10.1.1) can be chosen as

$$a_n = \sum_{k=1}^n \mathsf{E}X_k , \qquad b_n^2 = \sum_{k=1}^n \mathsf{Var}\, X_k .$$

Next we state some preliminary results.

**Lemma 12.2.1.** *Suppose infinite smallness,* $\mathsf{E}X_{nk} = 0$ *for all* $n, k$, *and let there exist a number* $\eta \geqq 1$ *such that*

$$\sup_n \sum_k \mathsf{E}\,|X_{nk}|^\eta < \infty . \tag{12.2.1}$$

*Then the following assertions hold true:*

(i) $\sup\limits_n \sum\limits_k |a_{nk}| < \infty$ ;

(ii) $\sum\limits_k a_{nk}^2 \to 0 , \qquad n \to \infty .$

(iii) *Assume further* $\eta > 1$ *and let* (12.1.1) *be satisfied. Then*

$$\sum_k |a_{nk}| \to 0 , \qquad n \to \infty . \tag{12.2.2}$$

Proof. (i) is easily seen since $\mathsf{E}X_{nk} = 0$ implies

$$\sum_k |a_{nk}| = \sum_k \Big| \int_{|x| \geqq \tau} x \,\mathrm{d}F_{nk}(x) \Big| \leqq \sum_k \int_{|x| \geqq \tau} |x| \,\mathrm{d}F_{nk}(x)$$
$$\leqq \tau^{1-\eta} \sum_k \int_{|x| \geqq \tau} |x|^\eta \,\mathrm{d}F_{nk}(x) \leqq \tau^{1-\eta} \sum_k \mathsf{E}\,|X_{nk}|^\eta . \tag{12.2.3}$$

To prove (ii) we make use of L. 10.2.3 so that the required assertion follow from (i) and

$$\sum_k a_{nk}^2 \leqq \max_k |a_{nk}| \sum_k |a_{nk}| \to 0 .$$

(iii) Using $\mathsf{E}X_{nk} = 0$ and (12.2.3) we get for fixed $T$, $T > \tau$,

$$\sum_k |a_{nk}| \leqq \sum_k \int_{\tau \leqq |x| \leqq T} |x| \,\mathrm{d}F_{nk}(x) + \sum_k \int_{|x| \geqq T} |x| \,\mathrm{d}F_{nk}(x)$$
$$\leqq T \sum_k \mathsf{P}(|X_{nk}| \geqq \tau) + T^{1-\eta} \sum_k \int_{|x| \geqq T} |x|^\eta \,\mathrm{d}F_{nk}(x) .$$

In view of (12.2.1) and $\eta > 1$ we can make the second term on the right arbitrarily small for sufficiently large (fixed) $T$ and all $n$; the first tends to zero as $n \to \infty$ by (12.1.1). □

Now we introduce the notation

$$L_n(\varepsilon) = \sum_k \int_{|x| \geqq \nu} x^2 \,\mathrm{d}F_{nk}(x) , \qquad \varepsilon > 0 ,$$

and define the celebrated *Lindeberg condition* as

$$\lim_{n \to \infty} L_n(\varepsilon) = 0 \quad \text{for all} \quad \varepsilon > 0 . \tag{12.2.4}$$

Further, we will need the relation

$$\lim_{n \to \infty} \max_k \sigma_{nk}^2 = 0 . \tag{12.2.5}$$

These conditions have an intuitive meaning. Roughly speaking, (12.2.4) requires that the variances $\sigma_{nk}^2$ be due mainly to masses in an interval whose length is small in comparison with $B_n^2 = 1$. On the other hand, the condition (12.2.5) intuitively implies that the individual variances $\sigma_{nk}^2$ are uniformly small as compared to their sum $B_n^2 = 1$ in the sense that for given $\varepsilon > 0$ and all $n$ sufficiently large we have

$$\max_k \sigma_{nk}^2 \leqq \varepsilon B_n^2 .$$

Next we state several consequences of the Lindeberg condition.

**Lemma 12.2.2.** *For every n,k we assume* $\mathsf{E}X_{nk} = 0, \sigma_{nk}^2 < \infty, B_n^2 = 1$.

(i) *The Lindeberg condition* (12.2.4) *implies* (12.1.1), (12.2.2), (12.2.5).

(ii) *If we have* (12.1.1) *or* (12.2.5), *then infinite smallness is satisfied.*

Proof. By (12.2.4), (12.1.1) is an immediate consequence of

$$\sum_k \mathsf{P}(|X_{nk}| \geqq \varepsilon) = \sum_k \int_{|x|\geqq\varepsilon} \mathrm{d}F_{nk}(x) \leqq \varepsilon^{-2} L_n(\varepsilon)\,.$$

Next we use $\mathsf{E}X_{nk} = 0$ and (12.2.3). Then we get analogously

$$\sum_k |a_{nk}| \leqq \tau^{-1} L_n(\tau)$$

whence (12.2.2) follows. The asserted relation (12.2.5) is easily seen from

$$\max_k \sigma_{nk}^2 \leqq \varepsilon^2 + \max_k \int_{|x|\geqq\varepsilon} x^2\, \mathrm{d}F_{nk}(x) \leqq \varepsilon^2 + L_n(\varepsilon)\,.$$

(ii) By (12.1.1), infinite smallness follows from (12.1.5). On the other hand, if we assume (12.2.5), then infinite smallness follows from Čebyšev's inequality. □

Now we state the main result of this section.

**Theorem 12.2.3** (Lindeberg-Feller). *For every n, k we suppose* $\mathsf{E}X_{nk} = 0, \sigma_{nk}^2 < \infty$; $B_n^2 = 1$, *and put* $A_n = 0$. *Then the following assertions are equivalent.*

(i) *The Lindeberg condition* (12.2.4) *is satisfied.*

(ii) $F_n \xrightarrow{c} \Phi$ *and the summands* $X_{nk}$ *are infinitely small.*

(iii) $F_n \xrightarrow{c} \Phi$ *and* (12.2.5) *is satisfied.*

Proof. (iii) ⇒ (ii) follows from L. 12.2.2, (ii).

(ii) ⇒ (i) By assumption we can apply the only-if part of Th. 12.1.1 so that we have (12.1.2). On the other hand, since $B_n^2 = 1$ we verify (12.2.1) for $\eta = 2$. Now L. 12.2.1 (ii), implies

$$\sum_k a_{nk}^2 \to 0\,, \quad \text{as} \quad n \to \infty\,. \tag{12.2.6}$$

Thus, (12.1.2) can be reformulated as

$$1 = \sigma^2 = \lim_{n\to\infty} \sum_k \int_{|x|<\varepsilon} x^2\, \mathrm{d}F_{nk}(x)\,, \qquad \varepsilon > 0\,.$$

Using this and the trivial relation

$$1 = B_n^2 = L_n(\varepsilon) + \sum_k \int_{|x|<\varepsilon} x^2\, \mathrm{d}F_{nk}(x)\,,$$

we arrive at the required condition (12.2.4).

(i) ⇒ (iii) We apply the if-part of Th. 12.1.1. From L. 12.2.2 (i), it is clear that (12.1.1) is true. Accordingly, the $X_{nk}$ are infinitely small. On the other hand, we again obtain the relation (12.2.6) from L.12.2.1 (ii) and get the required condition (12.1.2) from

$$\sum_k \Big[\int_{|x|<\varepsilon} x^2\, \mathrm{d}F_{nk}(x) - a_{nk}^2\Big] = 1 - L_n(\varepsilon) - \sum_k a_{nk}^2 \to 1\,.$$

This proves $F_n \xrightarrow{c} \Phi$ where $A_n$ is given by (12.1.3). The asserted relation (12.2.5) follows from L. 12.2.2 (i).

It remains to show that we may put $A_n = 0$. Indeed, since (12.1.1) is guaranteed we conclude from L. 12.2.1 (iii), that $\sum_k a_{nk} \to 0$, as $n \to \infty$. Thus in view of $\mu = 0$ we get from (12.1.3) that $A_n = \sum_k a_{nk} \to 0$. This in connection with Prop. 2.3.1 shows that the special choice $A_n = 0$ is also possible. □

The reader will easily recognize that the Lindeberg-Feller theorem implies the Lévy theorem (Th. 3.7.2) since the Lindeberg condition (12.2.4) is then trivially satisfied. However, it is often hard to verify (12.2.4). Hence it is noteworthy that (12.2.4) is satisfied under the additional assumption

$$\lim_{n\to\infty} \sum_k \mathsf{E}\,|X_{nk}|^{2+\delta} = 0\,, \quad \text{for some} \quad \delta > 0\,, \tag{12.2.7}$$

provided that the absolute moments $\mathsf{E}\,|X_{nk}|^{2+\delta}$ exist for all $n$, $k$. The condition (12.2.7) is called the *Ljapunov condition*. It turns out that it is sharper than the Lindeberg condition; we state this assertion in a precise way.

**Theorem 12.2.4.** *For every $n, k$ we suppose $\mathsf{E}X_{nk} = 0$, $\mathsf{E}\,|X_{nk}|^{2+\delta} < \infty$, $B_n^2 = 1$, and put $A_n = 0$. Then the Ljapunov condition* (12.2.7) *is satisfied iff we have infinite smallness, $F_n \xrightarrow{c} \Phi$, and*

$$\lim_{T\to\infty} \sup_n \sum_k \int_{|x|\geqq T} |x|^{2+\delta}\, \mathrm{d}F_{nk}(x) = 0\,. \tag{12.2.8}$$

Proof. Let us first assume (12.2.7). Then the estimates

$$L_n(\varepsilon) \leqq \frac{1}{\varepsilon^\delta} \sum_k \int\limits_{|x|\geqq\varepsilon} |x|^{2+\delta}\, \mathrm{d}F_{nk}(x) \leqq \frac{1}{\varepsilon^\delta} \sum_k \mathsf{E}\,|X_{nk}|^{2+\delta}$$

yield the Lindeberg condition. Hence $F_n \xrightarrow{c} \Phi$ and infinite smallness is fulfilled. Moreover, we easily see that (12.2.8) follows from (12.2.7).

Conversely, we have the estimates

$$\begin{aligned}\sum_k \mathsf{E}\,|X_{nk}|^{2+\delta} &\leqq \varepsilon^\delta \sum_k \int_{|x|<\varepsilon} x^2\, \mathrm{d}F_{nk}(x)\\ &\quad + T^\delta \sum_k \int_{\varepsilon\leqq|x|<T} x^2\, \mathrm{d}F_{nk}(x) + \sum_k \int_{|x|\geqq T} |x|^{2+\delta}\, \mathrm{d}F_{nk}(x)\\ &\leqq \varepsilon^\delta B_n^2 + T^\delta L_n(\varepsilon) + \sum_k \int_{|x|\geqq T} |x|^{2+\delta}\, \mathrm{d}F_{nk}(x)\,.\end{aligned}$$

By $B_n^2 = 1$, the first term on the right is arbitrarily small for sufficiently small (but fixed) $\varepsilon$. The third term is arbitrarily small for large (but fixed) $T$ in view of (12.2.8). Finally, so is the second term since the assumptions $F_n \xrightarrow{c} \Phi$ and infinite smallness imply the Lindeberg condition. □

## 12.3. Recent modifications of the Lindeberg-Feller theorem

We continue using the notation of Sect. 12.2 and again assume that the variances of $X_{nk}$ exist. Our aim is to deduce CLT assuming the r-convergence of $(F_n)$. A basic idea of the present section is to make better use of the condition $B_n^2 = 1$ than in Sect. 12.2. In fact, we obtain from Prop. 2.4.8 the following result.

**Lemma 12.3.1.** *The conditions $\mathsf{E}S_n = 0$ and $B_n^2 = 1$ imply the relative compactness of $(F_n)$.*

The assumed r-convergence will permit us to replace the Lindeberg condition by significantly weaker conditions. First we impose the relation

$$\lim_{\varepsilon\to+0} \overline{\lim_{n\to\infty}}\, L_n(\varepsilon) =: \eta < 1\,. \tag{12.3.1}$$

It is clear that $L_n(\varepsilon) \leqq B_n^2 = 1$. Hence we generally have $\eta \leqq 1$. Note that the limit $\varepsilon \to +0$ exists by monotonicity.

**Proposition 12.3.2.** *For every n, k we suppose* $\mathsf{E}X_{nk} = 0$, $\sigma_{nk}^2 < \infty$, $B_n^2 = 1$, *and put* $A_n = 0$. *Further, we assume infinite smallness. Then* $F_n \xrightarrow{c} \Phi$ *holds iff* (12.3.1) *is satisfied and there exists a set S of real numbers having a finite limit point such that*

$$F_n \underset{S}{\rightarrow} \Phi\,, \qquad n \to \infty\,. \tag{12.3.2}$$

Proof. We proceed according to what was said in Sect. 10.4. By L. 12.3.1, the relative compactness of $(F_n)$ is granted. Thus there exists a subsequence $(F_{n'})$ satisfying $F_{n'} \xrightarrow{c} G$ for a certain d.f. $G$ which is inf.div. in view of the assumed infinite smallness; we write $G = \mathcal{L}(a, \gamma^2, M, N)$.

We have to solve the continuation problem. The first step is to determine $\gamma^2$ by means of Prop. 11.2.5. As in the proof of (12.2.6) we get

$$\sum_k a_{nk}^2 \to 0\,, \quad \text{as} \quad n \to \infty\,.$$

Now (11.2.9) yields

$$\begin{aligned}\gamma^2 &= \lim_{\varepsilon \to +0} \underline{\lim}_{n' \to \infty} \sum_k \int_{|x| < \varepsilon} x^2 \, \mathrm{d}F_{n'k}(x) \\ &= 1 - \lim_{\varepsilon \to +0} \overline{\lim}_{n' \to \infty} L_{n'}(\varepsilon) = 1 - \eta > 0\,;\end{aligned}$$

this means that $G$ has a normal component, i.e., $G$ is integral by Prop. 7.5.1. Accordingly, $F_{n'}(x) \to G(x)$ for all $x$. Now we verify from our assumption (12.3.2) that $G(x) = \Phi(x)$, $x \in S$. Hence, $G$ coincides with $\Phi$ by analytic continuation. This proves the required assertion. □

**Remark 12.3.3.** There are some modifications of Prop. 12.3.2 in the sense that other conditions are used instead of (12.3.1) while the assumption (12.3.2) is not changed. Here we mention two examples.

(i) Let us suppose that the set $S$ is unbounded. Then Prop. 12.3.2 remains true without assuming the relation (12.3.1), see Cor. 12.4.2.

(ii) Let $\mathsf{E}X_{nk} = 0$, $B_n^2 = 1$, $A_n = 0$ for all $n$, $k$. Assume infinite smallness and suppose that the r.v. $X_{nk}$ are symmetric. Further, let $a = 0$ be a limit point of $S$. Then (12.3.2) is equivalent to $F_n \xrightarrow{c} \Phi$. □

In Prop. 12.3.2 we have studied the effect of the condition (12.3.1). Next we remind the reader of the condition (12.1.1) which is also weaker than the Lindeberg condition.

**Proposition 12.3.4.** *For every n, k we suppose* $\mathsf{E}X_{nk} = 0$, $\sigma_{nk}^2 < \infty$, $B_n^2 = 1$, *and put* $A_n = 0$. *Then we have infinite smallness and* $F_n \xrightarrow{c} \Phi$ *iff* (12.1.1) *is satisfied and there exists a number* $x_0 \neq 0$ *such that*

$$F_n(x_0) \to \Phi(x_0)\,, \qquad n \to \infty\,. \tag{12.3.3}$$

Proof. In view of the Lindeberg-Feller theorem it is enough to prove the if-part. As in the proof of Prop. 12.3.2, L. 12.3.1 implies the relative compactness of $(F_n)$. Next we select a subsequence $(F_{n'})$ which satisfies $F_{n'} \xrightarrow{c} G$ for some d.f. $G$. Now Th. 12.1.4 yields $G = \Phi_{\mu,\sigma}$ for certain $\mu \in \mathbb{R}_1$, $\sigma \geqq 0$. We turn to the continuation problem. By L. 12.2.1 (iii), we have $\sum_k a_{nk} \to 0$, $n \to \infty$. Using Th. 12.1.1 and noting that $A_n = 0$ we easily see from (12.1.3) that $\mu = 0$. Thus (12.3.3) implies $\Phi_{0,\sigma}(x_0) = \Phi(x_0)$; hence $\sigma = 1$ follows since $x_0 \neq 0$. □

Finally, we direct the readers attention to Prop. 12.1.5 in which no moment conditions are assumed but convergence in two points is needed.

## 12.4. Asymptotic conditions

Here we use the notation of Sect. 12.1. The present section aims at deriving CLT from assumptions imposed only on the one- or two-sided tails of $F_n$, while no moment conditions are assumed. The corresponding proofs are based on characterizations of the normal d.f. given in Sect. 9.2 and 9.3.

First we use one-sided asymptotic conditions. To this end, we introduce the functions

$$\underline{F}(x) = \varliminf_{n\to\infty} F_n(x)\,, \qquad \overline{F}(x) = \varlimsup_{n\to\infty} F_n(x)\,.$$

The following result shows that it suffices to assume only that the one-sided asymptotic behaviour of $\underline{F}$ and $\overline{F}$ is closely related to that of the normal d.f.

**Theorem 12.4.1.** *Let $\mu \in R_1$, $\sigma > 0$. Let us further suppose infinite smallness and put $A_n = \sum_k a_{nk}$. Then $F_n \xrightarrow{c} \Phi_{\mu,\sigma}$ holds iff there exists a sequence of real numbers $(x_k)$, $x_k \to -\infty$, such that, as $k \to \infty$,*

$$\begin{aligned} &\underline{F}(x_k) \geqq \Phi_{\mu,\sigma}(x_k)\,\big(1 + o(1)\big)\,, \\ &\overline{F}(x_k) \leqq \Phi_{\mu,\sigma}(x_k)\, e^{o(x_k)}\,. \end{aligned} \tag{12.4.1}$$

Before proving this result we state some easy consequences where the conditions imposed on the left tail of $F_n$ are essentially simpler than in (12.4.1).

**Corollary 12.4.2.** *Let $\mu \in R_1$, $\sigma > 0$. Let us further suppose infinite smallness and put $A_n = \sum_k a_{nk}$. Then $F_n \xrightarrow{c} \Phi_{\mu,\sigma}$ holds iff there is an unbounded set $S$ of real numbers such that*

$$F_n \xrightarrow[S]{} \Phi_{\mu,\sigma}\,, \qquad n \to \infty\,. \tag{12.4.2}$$

Proof. If $S$ is left-sided unbounded, then the assertion follows from $\underline{F}(x) = \overline{F}(x) = \Phi_{\mu,\sigma}(x)$, $x \in S$. Otherwise, if $S$ is right-sided unbounded we use the right-sided counterpart of Th. 12.4.1. □

**Corollary 12.4.3.** *Let $\mu \in R_1$, $\sigma > 0$. Let us further suppose infinite smallness and put $A_n = \sum_k a_{nk}$. Then $F_n \xrightarrow{c} \Phi_{\mu,\sigma}$ holds iff there is a function $F$ such that*

$$F_n \xrightarrow{w} F\,, \qquad F(-\infty) = 0\,, \qquad F > 0\,,$$

*and*

$$1 \leqq \varliminf_{x\to-\infty} \frac{F(x)}{\Phi_{\mu,\sigma}(x)}\,, \qquad \varlimsup_{x\to-\infty} \frac{F(x)}{\Phi_{\mu,\sigma}(x)} < \infty\,. \tag{12.4.3}$$

Proof. In this case $\underline{F} = \overline{F} = F$ so that (12.4.3) implies (12.4.1). □

It is easily verified that Cor. 12.4.2 implies the following assertion.

**Corollary 12.4.4.** *Let $\mu \in R_1$, $\sigma > 0$. Let us further suppose infinite smallness and put $A_n = \sum_k a_{nk}$. Then $F_n \xrightarrow{c} \Phi_{\mu,\sigma}$ holds iff there is a number $\tau \in R_1$ such that*

$$F_n \xrightarrow[(-\infty,\tau)]{} \Phi_{\mu,\sigma}\,, \qquad n \to \infty\,.$$

Proof of Theorem 12.4.1. It is enough to prove the sufficiency of (12.4.1). Let $(F_{n'})$ be a subsequence which satisfies $F_{n'} \xrightarrow{w} F'$ for certain function $F'$.

In this case we are in a position to treat the compactness problem and the uniqueness problem in one step. We obviously have $\underline{F}(x) \leqq F'(x) \leqq \overline{F}(x)$. The first relation in (12.4.1) implies $F' > 0$ while the second implies $F'(-\infty) = 0$. So we get $\varrho := F'(\infty) > 0$ and, by Prop. 11.3.3, the d.f. $G = (1/\varrho)\, F'$ is inf. div. Then we see from (12.4.1) that, as $k \to \infty$,

$$G(x_k) \geqq \Phi_{\mu,\sigma}(x_k) \left(\frac{1}{\varrho} + o(1)\right) \geqq \Phi_{\mu,\sigma}(x_k)\, \left(1 + o(1)\right),$$

$$G(x_k) \leqq \frac{1}{\varrho}\, \Phi_{\mu,\sigma}(x_k)\, e^{o(x_k)} = \Phi_{\mu,\sigma}(x_k)\, e^{o(x_k)}\,.$$

Now Prop. 9.3.5 tells us that $G = \Phi_{\mu,\sigma}$. Hence we have $\varrho = 1$ so that even $F' = \Phi_{\mu,\sigma}$ is valid. □

To formulate a result with two-sided asymptotic conditions we introduce the function

$$\overline{T}(x) = \overline{\lim_{n\to\infty}}\, T_{F_n}(x) = \overline{\lim_{n\to\infty}}\, \mathsf{P}(|S_n| \geqq x)\,.$$

**Proposition 12.4.5.** *Let $\mu \in R_1$, $\sigma > 0$. Let us further suppose infinite smallness. Then $F_n \xrightarrow{c} \Phi_{\mu,\sigma}$ holds iff there is a sequence of real numbers $(x_k)$, $x_k \to \infty$, such that*

$$\lim_{k\to\infty} \frac{x_k \log x_k}{\log \overline{T}(x_k)} = 0\,, \tag{12.4.4}$$

*and there are $y_1, y_2 \in R_1$, $y_1 \neq y_2$, such that*

$$\lim_{n\to\infty} F_n(y_j) = \Phi_{\mu,\sigma}(y_j)\,, \qquad j = 1, 2\,. \tag{12.4.5}$$

Proof. We only have to prove the sufficiency of (12.4.4) and (12.4.5). Clearly, (12.4.4) implies $\overline{T}(x_k) \to 0$, as $k \to \infty$. Since $\overline{T}$ is monotone we also obtain $\overline{T}(x) \to 0$, i.e., $(F_n)$ is tight and therefore relatively compact. Next we select a subsequence $(F_{n'})$ satisfying $F_{n'} \xrightarrow{c} G$ where $G$ is of course inf.div. We write $G = \mathcal{L}(a, \gamma^2, M, N)$. Since $T_G(x) \leqq \overline{T}(x) \leqq 1$ we see that

$$0 \geqq \frac{x_k \log x_k}{\log T_G(x_k)} \geqq \frac{x_k \log x_k}{\log \overline{T}(x_k)} \to 0\,, \qquad k \to \infty\,.$$

Keeping Prop. 9.2.1 in mind we get $G = \Phi_{a,\gamma^2}$. Now we obtain $\mu = a$, $\sigma^2 = \gamma^2$ from (12.4.5) using the simple continuation theorem Prop. 1.1.7. □

## 12.5. Supplement

**12.5.1.** Convergence of $(F_n)$ under the assumption of bounded variances. *Let us suppose $\mathsf{E}X_{nk} = 0$, $\sigma_{nk}^2 = \mathsf{Var}\, X_{nk} < \infty$, $B_n^2 = \sum_k \sigma_k^2 \leqq C$ for every $n$, $k$, and put $A_n = 0$, $n \geqq 1$. Assume, moreover, infinite smallness and let $F = \mathcal{L}(a, \gamma^2, M, N)$ be inf.div. If $F$ has finite variance and the number $a$ is given by the relation*

$$a = -\int_{-\infty}^{-0} \frac{x^3}{1+x^2}\, dM(x) - \int_{+0}^{\infty} \frac{x^3}{1+x^2}\, dN(x)\,, \tag{12.5.1}$$

*then* $F_n \xrightarrow{c} F$ *holds iff we have*

$$\sum_k F_{nk}(x) \xrightarrow[(-\infty,0)]{} M(x)\,, \qquad \sum_k \left(F_{nk}(x) - 1\right) \xrightarrow[(0,\infty)]{} N(x)\,,$$

*and*

$$\gamma^2 = \lim_{\varepsilon\to+0} \overline{\lim_{n\to\infty}} \sum_k \int_{|x|<\varepsilon} x^2\,\mathrm{d}F_{nk}(x)$$
$$= \lim_{\varepsilon\to+0} \underline{\lim_{n\to\infty}} \sum_k \int_{|x|<\varepsilon} x^2\,\mathrm{d}F_{nk}(x)\,.$$

Note that the integrals in (12.5.1) exist by Cor. 7.7.9 (i).

This result is well known and was formulated, for instance, in GNEDENKO and KOLMOGOROV (1959) and LOÈVE (1960) in terms of Kolmogorov's representation of inf.div. d.f. We remark that a simple proof of this assertion can be obtained also from L. 12.2.1 and Prop. 11.2.5 by letting $\tau \to \infty$.

**12.5.2. The CLT without infinite smallness.** *Let us denote by L the Lévy metric and assume* $\mathsf{E}X_{nk} = 0$, $\sigma_{nk}^2 < \infty$, $B_n^2 = 1$ *for every n, k. Further we put* $\Phi_{nk} = \Phi_{0,\sigma_{nk}}$ *and* $A_n = 0$. *Then* $F_n \xrightarrow{c} \Phi$ *holds iff we have*

$$\lim_{n\to\infty} \sup_k L(F_{nk}, \Phi_{nk}) = 0$$

*and*

$$\lim_{n\to\infty} \sum_k \int_{|x|\geqq\varepsilon} x^2\,\mathrm{d}\left(F_{nk}(x) - \Phi_{nk}(x)\right) = 0\,, \quad \text{for all} \quad \varepsilon > 0\,,$$

(ZOLOTAREV (1967)).

**12.5.3. The CLT and the convergence of moments.** *For every n, k we assume* $\mathsf{E}X_{nk} = 0$, $\sigma_{nk}^2 < \infty$, $B_n^2 = 1$, *and put* $A_n = 0$. *Then the following assertions hold.*

(i) *The Lindeberg condition* (12.2.4) *implies*

$$\lim_{n\to\infty} \int_{-\infty}^{\infty} |x|^\eta\,\mathrm{d}F_n(x) = \int_{-\infty}^{\infty} |x|^\eta\,\mathrm{d}\Phi(x)\,, \qquad 0 < \eta \leqq 2\,. \tag{12.5.2}$$

(ii) *Let* $\delta > 0$. *Then the Ljapunov condition* (12.2.7) *implies the relation* (12.5.2) *for* $0 < \eta \leqq 2 + \delta$.

For more general results of this type see KRUGLOV (1976a).

**12.5.4.** An example of the CLT in the case of bounded variances where the Lindeberg condition is not guaranteed. We put

$$F_{n1} = \frac{1}{4n^2} U_{-n} + \left(1 - \frac{1}{2n^2}\right) U_0 + \frac{1}{4n^2} U_n\,,$$
$$F_{nk} = \Phi_{0,1/\sqrt{2n}}\,, \qquad k = 2, 3, \dots, n+1 =: k_n\,.$$

Then we easily verify that $\mathsf{E}X_{nk} = 0$, $B_n^2 = 1$, and by Th. 12.1.1,

$$F_n \xrightarrow{c} \Phi_{0,1/\sqrt{2}}\,.$$

Moreover, we have infinite smallness. However, the Lindeberg condition is not true since, for all $\varepsilon > 0$,

$$L_n(\varepsilon) = \sum_k \int_{|x|\geqq\varepsilon} x^2\,\mathrm{d}F_{nk}(x) \to \tfrac{1}{2}\,, \qquad n \to \infty\,.$$

Hence all three conditions of the Lindeberg-Feller theorem are violated.

**12.5.5.** Let us assume $\mathsf{E}X_{nk} = 0$, $\sigma_{nk}^2 < \infty$, $B_n^2 = 1$ for every $n$, $k$, and consider the following assertions:

(i) The Ljapunov condition (12.2.7) is valid for some $\delta > 0$.

(ii) We have (12.1.1) and

$$\sup_n \sum_k \mathsf{E}\,|X_{nk}|^{2+\delta} < \infty\,, \quad \text{for some} \quad \delta > 0\,.$$

Then it is easily seen that (i) implies (ii). However the converse is not true in general. But it can be shown that (ii) implies (i) for all $\delta'$, $0 < \delta' < \delta$. The proofs are left to the reader.

**12.5.6.** Special models in which the Ljapunov condition are satisfied. Let $X_1, X_2, \ldots$ be i.r.v. with finite variances and put

$$S_n = \frac{1}{b_n}\Big(\sum_{k=1}^{n} X_k - a_n\Big)$$

where

$$a_n = \sum_{k=1}^{n} \mathsf{E}X_k\,, \qquad b_n^2 = \sum_{k=1}^{n} \mathsf{Var}\, X_k\,.$$

We assume that $b_n \to \infty$, as $n \to \infty$. Then we have the following assertions for the r.v. $X_{nk} = 1/b_n\,(X_k - \mathsf{E}X_k)$, $k = 1, 2, \ldots, n$:

(i) If there are constants $\delta > 0$, $C > 0$, such that

$$\frac{\mathsf{E}\,|X_k - \mathsf{E}X_k|^{2+\delta}}{\mathsf{Var}\, X_k} \leqq C\,, \quad \text{for all} \quad k\,, \tag{12.5.3}$$

then (12.2.7) follows. For instance, (12.5.3) is satisfied if $X_1, X_2, \ldots$ are i.i.d.r.v. with $\mathsf{P}(X_1 \neq 0) > 0$ and $\mathsf{E}\,|X_1|^{2+\delta} < \infty$.

(ii) If there is a constant $C > 0$ such that

$$\mathsf{P}(|X_k| \leqq C) = 1\,, \quad \text{for all} \quad k\,,$$

then (12.2.7) follows for any $\delta > 0$ (Loève (1960)).

**12.5.7. Problem.** Carry CLT for sums $S_n \sim F_n$ in a triangular array under the assumption of r-convergence over to the case in which the summands $X_{nk}$ are weakly dependent in a certain sense.

For summands forming a Markov chain Heinrich (1983) gave sufficient conditions guaranteeing that the limit d.f. of $(F_n)$ is inf.div. Since we have a number of uniqueness theorems for the normal d.f. at hand (see Chap. 9) the above problem amounts to solving the corresponding compactness problem (cf. Sect. 10.4).

## Notes and bibliographical data

Th. 12.1.1 and Cor. 12.1.3 represent standard results in the theory of the summation of i.r.v. A similar result to that in Th. 12.1.4 was given in Gnedenko and Kolmogorov (1959) under the sharper assumption $F_n \xrightarrow{c} F$ instead of $F_n \xrightarrow{w} F \neq$ const. Prop. 12.1.5 was proved in Jesiak and Rossberg (1980) under the additional assumption of bounded variances.

The results of Sect. 12.2 are well known facts.

Prop. 12.3.2 is due to Jesiak and Rossberg (1981). The subject of Sect. 12.3 has also been treated in Jesiak, Rossberg, and Siegel (1979).

The general result given in Prop. 12.4.1 and its corollaries are contained in Rossberg and Siegel (1981); a special case of Cor. 12.4.2 is contained in Rossberg (1979) while, particular cases of Cor. 12.4.3 were already proved in Riedel (1977) and Siegel (1978).

An initial survey on various statements concerning the CLT under the assumption of r-convergence has been given in Rossberg, Jesiak and Siegel (1981 b).

The first multidimensional CLT involving r-convergence is proved in Siegel (1983).

# Chapter 13. Independent identically distributed random variables

In the present chapter we consider i.i.d.r.v. $X_k$ subject to the common non-degenerate d.f. $V$. For appropriately chosen norming and centering constants $b_n > 0$, $a_n$, we form the sums

$$S_n = \frac{1}{b_n} \left( \sum_{k=1}^{n} X_k - a_n \right) \sim F_n , \qquad n \geqq 1 ,$$

and see that they are subject to the d.f.

$$F_n(x) = V^{*n}(b_n x + a_n) , \qquad n \geqq 1 . \tag{13.0.1}$$

This model has already been discussed in Sect. 10.1; it is a special triangular array which is easily seen by setting

$$X_{nk} = X_k/b_n \sim V(b_n x) ; \qquad A_n = a_n/b_n , \qquad k = 1, 2, \dots , n . \tag{13.0.2}$$

So we can handle this model using some results given in Chap. 10—12.

In Sect. 13.1 we study the compactness of the sequence $(F_n)$. We state not only consequences of the general results proved in Sect. 11.1, but we also treat compactness criteria which are typical for the sequence (13.0.1).

Sect. 13.2 includes a collection of standard results for the present model. In particular, we show that the set of all stable d.f. coincides with the set of all possible limit d.f. of $(F_n)$. Moreover, we find necessary and sufficient conditions on the d.f. $V$ such that the sequence $(F_n)$ completely converges to a certain non-degenerate stable d.f. provided the $a_n$, $b_n$ are appropriately chosen.

A final solution of the problem of r-convergence on a half axis is presented in Sect. 13.3. It is shown that this r-convergence implies complete convergence whence it follows that the limit d.f. is stable. The formulations of these results are elegant since no additional compactness conditions are needed.

We encourage the reader to have another look at Sect. 2.3 since these results are repeatedly applied here.

## 13.1. Compactness of sums

We first make another use of the concentration function and consider the condition

$$\inf_n Q(S_n, h) > 0 , \qquad \text{for some } h > 0 , \tag{13.1.1}$$

which we have already used in Chap. 11. Intuitively it means that not the whole probability mass of $F_n$ dissipates to infinity.

**Lemma 13.1.1.** *Let* (13.1.1) *be satisfied. Then* $(\sqrt{n}/b_n)$ *is bounded; in particular, we have* $b_n \to \infty$ *as* $n \to \infty$.

Proof. Assume indirectly that there is a subsequence $(b_{n'})$ of $(b_n)$ satisfying $b_{n'} = o(\sqrt{n'})$ as $n' \to \infty$. Then from Cor. 5.4.5 we get

$$Q(S_{n'}, h) = Q(\sum_{k=1}^{n'} X_k, b_{n'}h) = o(1) .$$

This contradicts our assumption which completes the proof. □

In view of Prop. 5.2.1 it is clear that (13.1.1) is necessary for the shift compactness of $(F_n)$. Hence and by L. 13.1.1 we have the following fact.

**Lemma 13.1.2.** *The relation* $b_n \to \infty$ *is necessary for the shift compactneess of* $(F_n)$.

Next we state two sufficient compactness conditions assuming the r-convergence of $(F_n)$ on a half axis.

**Proposition 13.1.3.** *Let* $\tau \in R_1$ *and* $F > 0$ *be a d.f. Assume the r-convergence*

$$F_n \xrightarrow[(-\infty,\tau)]{} F , \qquad n \to \infty , \tag{13.1.2}$$

*and let the subsequence* $(F_{n'})$ *weakly trend to some monotone and left -continuous function* $H$, $F_{n'} \to H$. *Then, for every* $k \geqq 2$, *there exist constants* $A_k \in R_1$, $B_k > 0$ *such that*

$$H(x) = H^{*k}(B_k x + A_k) , \qquad x < \tau , \tag{13.1.3}$$

and $H(\infty) = 1$, i.e., *the sequence* $(F_n)$ *is relatively compact.*

For this result the assumption $F > 0$ is important, see Ex. 13.1.5 in which $F > 0$ is violated.

Proof. We introduce the sums

$$S_n^j = b_n^{-1}(X_{jn+1} + \dots + X_{jn+n} - a_n) , \qquad j = 0, 1, \dots , k-1 .$$

Since the $X_k$ are i.i. d.r.v. this goes also for the $S_n^j \sim F_n$. With the abbreviations $B_n^k := b_{kn}/b_n$, $A_n := (a_{kn} - ka_n)/b_n$ we now find

$$S_{kn} = b_n^{-1}(X_1 + \dots + X_{kn} - a_{kn}) = (B_n^k)^{-1} (\sum_{j=0}^{k-1} S_n^j - A_n^k) .$$

Hence we have for all $k, n \geqq 1$

$$F_{kn}(x) = P(S_{kn} < x) = P(\sum_{j=0}^{k-1} S_n^j < B_n^k x + A_n^k) = F_n^{*k}(B_n^k x + A_n^k) .$$

Now we conclude from our assumption $F_{n'} \xrightarrow{w} H$ that $H(x) = F(x)$, $x < \tau$, so that $H(-\infty) = 0$. We recall that by L. 7.4.3 this condition suffices to obtain $F_{n'}^{*2} \xrightarrow{w} H^{*2}$, $n' \to \infty$, even if $H(\infty) < 1$. By induction we obtain

$$F_{n'}^{*k} \xrightarrow{w} H^{*k} , \qquad k \geqq 2 .$$

On the other hand, we see from (13.1.2) that

$$F_{kn'}(x) = F_{n'}^{*k}(B_{n'}^k x + A_{n'}^k) \xrightarrow[(-\infty,\tau)]{} H(x) .$$

Now we are ready to apply Prop. 2.3.4 with the notation $x_2 = \tau$, $G = H^{*k}$, $G' = H > 0$; thus we obtain $A_{n'}^k \to A_k \in R_1$, $B_{n'}^k \to B_k > 0$. Now we apply Prop. 2.3.1 to the weakly convergent sequence $(F_{n'}^{*k})$ so that $I = \mathrm{R}_1$. Thus

$$F_{kn'}(x) = F_{n'}^{*k}(B_{n'}^k x + A_{n'}^k) \xrightarrow{w} H^{*k}(B_k x + A_k),$$

and (13.1.3) immediately follows.

To prove that $H(\infty) = 1$ we use the left-continuity of $H$ and write

$$0 < H(\tau) = H^{*k}(B_k\tau + A_k) \leqq H^{*k}(\infty) \leqq (H(\infty))^k .$$

Thus the assumption $H(\infty) < 1$ leads to an obvious contradiction as $k \to \infty$. Accordingly, $H(\infty) = 1$ so that all assertions have now been proved. □

Now we turn to an important counterpart.

**Proposition 13.1.4.** *Let $\tau > 0$ and $F$ be a d.f. with* lext $(F) = 0$, $F(+0) = 0$. *Assume the r-convergence* (13.1.2) *and let the subsequence $(F_{n'})$ weakly tend to some monotone and leftcontinuous function $H$, $F_{n'} \xrightarrow{w} H$. Then, for every $k \geqq 2$, there exists a constant $B_k > 0$ such that*

$$H(x) = H^{*k}(B_k x), \qquad x < \tau, \tag{13.1.4}$$

*and $H(\infty) = 1$, i.e., the sequence $(F_n)$ is relatively compact.* □

The assumption $F(+0) = 0$ is essential for the above result; this is shown by Ex. 13.1.5 in which $F(0) = 0$ and $F(+0) = \mathrm{e}^{-1/2} > 0$ hold.

The proof of this proposition can be carried out like the above proof where Prop. 2.3.5 applies instead of Prop. 2.3.4, for details cf. ROSSBERG (1983b). □

**Example 13.1.5.** Let the d.f. $V$ be defined by the probabilities

$$p_k = \mathsf{P}(X_1 = k!\sqrt{k}) = (2k(k+1))^{-1}, \qquad k = 1, 2, \ldots,$$

$$p_0 = \mathsf{P}(X_1 = 0) = 1/2$$

and put $a_n = 0$, $b_n = n!$. Then we will see that

$$S_n \sim F_n \xrightarrow{w} F = \varrho U_0 \quad \text{where} \quad \varrho = \mathrm{e}^{-1/2}.$$

To show this we fix $x > 0$ and choose $n$ large enough so that $\sqrt{n-1}/n < x < \sqrt{n}$. Now we obtain

$$\psi_n(x) := n\mathsf{P}(X_1 \geqq n!\,x) = n \sum_{k=n}^{\infty} p_k = n \sum_{k=n}^{\infty} \left(\frac{1}{2k} - \frac{1}{2(k+1)}\right) = \frac{1}{2}.$$

Accordingly, $\psi_n \xrightarrow[(0,\infty)]{} \frac{1}{2}$. In a similar way we can verify

$$\frac{n}{(n!)^2} \int\limits_{|x|<n!\varepsilon} x^2 \,\mathrm{d}V(x) \leqq n \sum_{k\leqq n} \left(\frac{k!}{n!}\right)^2 p_k \to 0,$$

provided that $\varepsilon < \sqrt{n}$, and for $\tau < \sqrt{n}$

$$\frac{n}{n!} \int\limits_{|x|<n!\tau} x \,\mathrm{d}V(x) \leqq n \sum_{k\leqq n} \frac{k!}{n!} p_k \to 0.$$

So we are ready to apply Prop. 11.3.5 to the r.v. $X_{nk} = X_k/n!$ whence the required convergence follows. □

Before formulating a necessary and sufficient compactness condition we prove the following lemma.

**Lemma 13.1.6.** *Let $V$ be a d.f. with $\sigma_V^2 = \infty$. Then we have*

$$\Big(\int_{|y|<x} y\,\mathrm{d}V(y)\Big)^2 = o\Big(\int_{|y|<x} y^2\,\mathrm{d}V(y)\Big)\,, \qquad x \to \infty\,.$$

Proof. It is always possible to find a symmetric function $h \geqq 1$ which is non-decreasing on $(0, \infty)$ such that

$$h(y) \to \infty \quad \text{as} \quad y \to \infty\,, \quad \text{and} \quad C := \int_{-\infty}^{\infty} h(y)\,\mathrm{d}V(y) < \infty;$$

e.g., $h(y) = T_V(|y|)^{-1/2}$. Now let $\varepsilon > 0$ be arbitrary. Then we choose first $y_0 > 0$ such that $h(y_0) \geqq C/\varepsilon$ and then (using $\sigma_V^2 = \infty$) $x > y_0$ such that

$$\int_{|y|<y_0} \frac{y^2}{h(y)}\,\mathrm{d}V(y) \leqq \frac{\varepsilon}{C} \int_{|y|<x} y^2\,\mathrm{d}V(y)\,.$$

So we obtain, by the Schwarz inequality,

$$\Big(\int_{|y|<x} y\,\mathrm{d}V(y)\Big)^2 \leqq \int_{|y|<x} h(y)\,\mathrm{d}V(y) \int_{|y|<x} \frac{y^2}{h(y)}\,\mathrm{d}V(y)$$

$$\leqq C\Big[\int_{|y|<y_0} \frac{y^2}{h(y)}\,\mathrm{d}V(y) + \frac{1}{h(y_0)} \int_{y_0 \leqq |y| < x} y^2\,\mathrm{d}V(y)\Big] \leqq 2\varepsilon \int_{|y|<x} y^2\,\mathrm{d}V(y)\,. \qquad \square$$

The rest of the present section is not need in the sequal.

**Proposition 13.1.7.** *The following assertions are equivalent:*

(i) *$(F_n)$ is shift compact.*

(ii) *We have the three relations*

$$\sup_n \frac{n}{b_n^2} \int_{|x|<b_n} x^2\,\mathrm{d}V(x) < \infty\,, \tag{13.1.5}$$

$$\sup_n nT_V(b_n) < \infty\,, \tag{13.1.6}$$

$$\lim_{x\to\infty} \sup_n nT_V(b_n x) = 0\,. \tag{13.1.7}$$

(iii) *The two relations* (13.1.1) *and* (13.1.7) *are satisfied.*

In the special case of Ex. 13.1.5 we see that the conditions (13.1.1), (13.1.5), and (13.1.6) are satisfied, while (13.1.7) is violated. Hence $(F_n)$ is not shift compact.

Proof. We show this result by specifying Th. 11.1.1 for the r.v. $X_{nk} = X_k/b_n$ with medians $m_{nk} = m/b_n$ where $m$ is a median of $X_1$.

(i) $\Rightarrow$ (ii) From the assertion (i) $\Rightarrow$ (iii) in Th. 11.1.1 we get (11.1.5) which can be written in the form

$$\lim_{x\to\infty} \sup_n n\big(1 - V(b_n x + m) + V(-b_n x + m)\big) = 0\,. \tag{13.1.8}$$

By L. 13.1.2 we have $b_n \to \infty$ so that $x \geqq |m|/b_n$ provided that $x \geqq x_0 > 0$ and $n \geqq n_0(x_0)$. Thus

$$nT_V(b_n 2x) \leqq nT_V(b_n x + |m|) \leqq n\big(1 - V(b_n x + m) + V(-b_n x + m)\big)\,.$$

This proves (13.1.7) in view of (13.1.8).

To prove (13.1.6) we use the function $w(x) = \min\{1, x^2\}$ and obtain in a similar way as above

$$n T_V(b_n) \leqq n\left(1 - V\left(\frac{b_n}{2} + m\right) + V\left(-\frac{b_n}{2} + m\right)\right) \leqq 4n \int_{-\infty}^{\infty} w(x) \, \mathrm{d}V(b_n x + m) \; ; \tag{13.1.9}$$

the last estimate follows from $1 \leqq 4w(x)$, $|x| \geqq \frac{1}{2}$. Since we also have the relation (11.1.4) the term on the right of (13.1.9) is bounded whence (13.1.6) follows.

Next we show (13.1.5). Let first $\sigma_V^2 < \infty$. Then we note that (i) implies (13.1.1). Hence $n/b_n^2$ is bounded by virtue of L. 13.1.1. This proves (13.1.5). If $\sigma_V^2 = \infty$, then we indirectly assume that there is a subsequence $(n')$ such that

$$\frac{n'}{b_{n'}^2} \int\limits_{|x|<b_{n'}} x^2 \, \mathrm{d}V(x) \to \infty\,, \qquad n' \to \infty\,.$$

Now L. 13.1.6 implies

$$\frac{n'}{b_{n'}^2} \int\limits_{|x|<b_{n'}} x^2 \, \mathrm{d}V(x) = \frac{n'}{b_{n'}^2}\left[\int\limits_{|x|<b_{n'}} x^2 \, \mathrm{d}V(x) - \left(\int\limits_{|x|<b_{n'}} x \, \mathrm{d}V(x)\right)^2\right](1 + o(1))\,. \tag{13.1.10}$$

By the assumption of shift compactness we can find a subsequence $(F_{n''})$ of $(F_{n'})$ such that $F_{n''} \xrightarrow{c} F$ for some d.f. $F$ where $a_n$ is suitably chosen. Since $b_n \to \infty$ the r.v. $X_{nk} = X_k/b_n$ are infinitely small. Hence $F$ is inf.div. So we are able to apply Prop. 11.2.5; the relation (11.2.9) then yields the boundedness of the right of (13.1.10) for the sequence $(n'')$ contrary to our assumption.

(ii) $\Rightarrow$ (i) Since $V$ is non-degenerate we easily see that (13.1.5) and (13.1.6) imply $b_n \to \infty$. Hence we have $x/2 \geqq |m|/b_n$ provided that $x \geqq x_0 > 0$ and $n \geqq n_0(x_0)$, and we may write

$$n(1 - V(b_n x + m) + V(-b_n x + m)) \leqq n T_V(b_n x - |m|) \leqq n T_V\left(b_n \frac{x}{2}\right).$$

Thus we conclude (13.1.8) from the assumption (13.1.7). On the other hand, from Cor. 1.4.8 we get the estimate

$$n \int_{-\infty}^{\infty} w(x) \, \mathrm{d}V(b_n x + m) \leqq 8n \int_{-\infty}^{\infty} w(x) \, \mathrm{d}V(b_n x)\,.$$

The right term is bounded because of (13.1.5) and (13.1.6). Now we observe that all suppositions of Th. 11.1.1 (iii), are satisfied whence the shift compactness of $(F_n)$ follows.

(iii) $\Rightarrow$ (i) By L. 13.1.1, the relation (13.1.1) implies $b_n \to \infty$. In the same way as in the proof of (ii) $\Rightarrow$ (i) we then verify the relation (13.1.8). Thus the assumptions (iv) of Th. 11.1.1 are satisfied whence the shift compactness of $(F_n)$ follows.

(i) $\Rightarrow$ (iii) By Prop. 5.2.1 we see that (i) implies (13.1.1). The relation (13.1.7) is obvious since (i) $\Rightarrow$ (ii) has already been shown. The proof is complete.

**Proposition 13.1.8.** *If* (13.1.1) *and*

$$\lim_{x\to\infty} \sup_{u>0} \frac{T_V(ux)}{T_V(u)} = 0 \tag{13.1.11}$$

*hold, then* $(F_n)$ *is shift compact.*

Proof. In view of Prop. 13.1.7 it is enough to show that (13.1.7) is satisfied. As in the above proof we introduce the r.v. $X_{nk} = X_k/b_n$ with medians $m_{nk} = m/b_n$. By L. 11.1.2 we then see that the sequence $\left(n \int\limits_{-\infty}^{\infty} w(x)\, dV(b_n x + m)\right)$ is bounded. Using the estimates in (13.1.9) we obtain for a certain constant $C$

$$nT_V(b_n) \leqq C\,, \quad \text{for all} \quad n\,. \tag{13.1.12}$$

Now the desired relation (13.1.7) immediately follows from (13.1.11) and (13.1.12) since

$$\sup_n nT_V(b_n x) \leqq C \sup_n \frac{T_V(b_n x)}{T_V(b_n)} \to 0\,, \qquad x \to \infty\,.$$ □

**Proposition 13.1.9.** *Assume* (13.1.1) *and let* $k > 1$ *be a fixed integer. Then*

$$\sup_n \frac{b_{kn}}{b_n} < \infty \tag{13.1.13}$$

*implies the shift compactness of* $(F_n)$.

Proof. We again wish to apply Prop. 13.1.7. As in the proof of Prop. 13.1.8 it remains to derive (13.1.7). We first conclude from (13.1.13) that

$$b_{k^j n} = b_n \prod_{l=1}^{j} \left(\frac{b_{k^l n}}{b_{k^{l-1} n}}\right) \leqq b_n D^j \tag{13.1.14}$$

for certain constant $D > 1$ and all integers $n$, $j > 1$. Since $T_V$ is non-increasing we have

$$\lim_{x\to\infty} \sup_n nT_V(b_n x) = \lim_{j\to\infty} \sup_n nT_V(b_n D^j) =: \eta \geqq 0\,.$$

As is shown in the proof of Prop. 13.1.8, (13.1.1) implies the relation (13.1.12). Using first (13.1.14) and then (13.1.12) we now obtain

$$\eta \leqq \lim_{j\to\infty} \frac{1}{k^j} \sup_n nk^j\, T_V(b_{k^j n}) \leqq C \lim_{j\to\infty} \frac{1}{k^j} = 0\,.$$

This implies (13.1.7). □

**Remark 13.1.10.** The assertions which have been stated in L.13.1.1, L.13.1.2, Prop. 13.1.7, Prop. 13.1.8, and Prop. 13.1.9 remain true if we replace the natural numbers $(n)$ by some subsequence $(n')$. But in contrast to this fact for the results given in Prop. 13.1.3, and Prop. 13.1.4 it is essential that the assumed r-convergence (13.1.2) is satisfied for the whole sequence $(n)$. □

## 13.2. General convergence theorems

In this section we will describe the class of limit functions for sums of i.i.d.r.v. $X_k$ subject to the d.f. $V$, and we formulate conditions on $V$ under which there exist constants $a_n$ and $b_n > 0$ such that the sequence $(F_n)$ defined in (13.0.1) converges to a d.f. of that class.

In the following three lemmas, $(F_{n'})$ denotes a certain subsequence of the sequence $(F_n)$ under consideration.

**Lemma 13.2.1.** *Assume $F_{n'} \xrightarrow{w} G$ for some monotone function $G \neq$ const. Then $b_{n'} \xrightarrow{w} \infty$.*
Proof. First of all, because $G \neq$ const. we find a real $x$ such that $x - 1$ and $x + 1$ are continuity points of $G$ which satisfy

$$p := G(x + 1) - G(x - 1) > 0.$$

Thus we obtain for the concentration function $Q$ of $S_{n'}$, by virtue of $F_{n'} \xrightarrow{w} G$,

$$Q(S_{n'}, 1) \geqq F_{n'}(x + 1) - F_{n'}(x - 1) \to p > 0\,, \qquad n' \to \infty\,.$$

Accordingly, we have $\inf_{n'} Q(S_{n'}, 1) > 0$. Now we can apply L. 13.1.1 to $(b_{n'})$ to obtain the assertion. □

**Lemma 13.2.2.** *Assume $F_{n'} \xrightarrow{c} G$ for some d.f. $G$. Then $G$ is inf.div.*

Proof. We consider the r.v. $X_{n'k} = X_k/b_{n'}$, $1 \leqq k \leqq n'$, which form a triangular array. Now L. 13.2.1 tells us that $b_{n'} \to \infty$; hence, we obtain for every $\varepsilon > 0$, as $n' \to \infty$,

$$\max_k \mathsf{P}(|X_{n'k}| \geqq \varepsilon) = \max_k \mathsf{P}(|X_k| \geqq b_{n'}\varepsilon) = \mathsf{P}(|X_1| \geqq b_{n'}\varepsilon) \to 0\,.$$

Accordingly, the r.v. $X_{n'k}$ are infinitely small so that the assertion follows from Hinčin's theorem 11.2.3. □

**Lemma 13.2.3.** *Assume $F_{n'} \xrightarrow{c} G_1$ and $F_{kn'} \xrightarrow{c} G_k$ for non-degenerate d.f. $G_1$ and $G_k$ where $k > 1$ is a fixed natural number. Then there exist constants $A_k$ and $B_k > 0$ such that*

$$H_k(x) = H_1^{*k}(B_k x + A_k)\,, \qquad x \in R_1$$

Proof. We start like in the proof of Prop. 13.1.3. But since in the present case complete convergence is assumed we need not apply L. 7.4.3. Namely, it is now the continuity theorem that permits us to conclude

$$F_{n'}^{*k} \xrightarrow{c} H_1^{*k}\,, \qquad k \geqq 2\,;$$

moreover, we obviously have

$$F_{kn'}(x) = F_{n'}^{*k}(B_{n'}^k x + A_{n'}^k) \xrightarrow{c} H_k(x)\,.$$

Now we are in a position to apply Prop. 2.3.5 (with notations $G = H_1^{*k}$, $G' = H_k$) where complete convergence is supposed. It leads us to

$$A_n^k \to A_k \in \mathrm{R}_1\,, \qquad B_n^k \to B_k > 0\,.$$

Finally, since the $F_{kn'}$ and the $F_{n'}^{*k}$ are of the same type, the limits $H_k$ and $H_1^{*k}$, which are non-degenerate, must also be of the same type by Prop. 2.3.3. □

Now we are able to prove our first main result.

**Theorem 13.2.4.** *A d.f. $F$ is the complete limit of a sequence $(F_n)$ formed by means of (13.0.1) iff $F$ is stable.*

Proof. Let $F$ be stable. Then we see from (8.1.1) that

$$F(x) = F^{*n}(b_n x + a_n) =: F_n(x)\,.$$

This means that $F$ is the limit of a sequence $(F_n)$ of the type (13.0.1) where we have to put $V = F$.

Conversely, assume now that $F_n \xrightarrow{c} F$ for such a sequence, then we have to show that $F$ is stable. This is obvious if $F$ is degenerate. Otherwise, we choose an arbitrary natural number $k > 1$ and have

$$F_n \xrightarrow{c} F, \qquad F_{kn} \xrightarrow{c} F.$$

Now L. 13.2.3 tells us that for every $k > 1$ there exist constants $A_k$ and $B_k > 0$ such that

$$F(x) = F^{*k}(B_k x + A_k) \qquad x \in \mathbb{R}_1 \tag{13.2.1}$$

Accordingly, $F$ is stable by Def. 8.1.1. □

**Corollary 13.2.5.** *If $F_n \xrightarrow{c} F$ for some non-degenerate d.f. $F$, then $F$ is stable and we have, for every $k > 1$,*

$$\frac{b_{kn}}{b_n} \to k^{1/\alpha}, \qquad n \to \infty,$$

*where $\alpha \leqq 2$ is the characteristic exponent of $F$.*

Proof. By Cor. 8.1.4 (i), the constants $B_k$ in (13.2.1) satisfy $B_k = k^{1/\alpha}$. Further, from the proof of L. 13.2.3 we see that $B_k = \lim B_n^k = \lim_{n\to\infty} b_{kn}/b_n$. □

Now we are going to seek conditions on $V$ under which the convergence $F_n \xrightarrow{c} F$ takes place. For this purpose we introduce the following notion.

**Definition 13.2.6.** A d.f. $V$ is said to belong to the *domain of attraction* of stable d.f. $F$ if there exist constants $a_n$ and $b_n > 0$ such that $F_n \xrightarrow{c} F$ for the d.f. $F_n$ defined in (13.0.1). □

The first part of the proof of Th. 13.2.4 shows that every stable d.f. belongs to its own domain of attraction.

With regard to Prop. 2.3.1 and Prop. 2.3.3 it is clear that if $V$ belongs to the domain of attraction of $F$, then we may say that the type of $V$ belongs to the domain of attraction of the type of $F$.

First of all we wish to formulate a criterion for $V$ to belong to the domain of attraction of the normal type. Of course, it represents the CLT for the model under consideration, of our remark about (10.4.5)

**Theorem 13.2.7.** *A d.f. $V$ belongs to the domain of attraction of the normal type iff*

$$x^2 T_V(x) = o\Big(\int_{|y|<x} y^2 \, dV(y)\Big), \qquad x \to \infty. \tag{13.2.2}$$

Proof. *Only-if part*: Assume that there exist constants $a_n$ and $b_n > 0$ such that

$$F_n(x) = V^{*n}(b_n x + a_n) \xrightarrow{c} \Phi_{0,1}(x);$$

then we have to show that (13.2.2) is valid. This is trivial if $\sigma_V^2 < \infty$ since in this case

$$x^2 T_V(x) = x^2 \int_{|y|\geqq x} dV(y) \leqq \int_{|y|\geqq x} y^2 \, dV(y) \to 0, \qquad x \to \infty.$$

Assume now that $\sigma_V^2 = \infty$. Then we apply Th. 12.1.1 to our model; this is justified by virtue of (13.0.2); moreover, L. 13.2.1 implies $b_n \to \infty$ so that the summands $X_k/b_n$ are infinitely small. We thus obtain

$$\lim_{n\to\infty} nT_V(b_n\varepsilon) = 0 \quad \text{for all} \quad \varepsilon > 0, \tag{13.2.3}$$

and

$$\lim_{n\to\infty} n\Big[\int_{|y|<\varepsilon} y^2 \, \mathrm{d}V(b_n y) - \Big(\int_{|y|<\varepsilon} \mathrm{d}V(b_n y)\Big)^2\Big] = 1, \quad \text{for all} \quad \varepsilon > 0.$$

In view of L. 13.1.6 the latter relation can be rewritten as

$$\lim_{n\to\infty} n \int_{|y|<\varepsilon} y^2 \, \mathrm{d}V(b_n y) = 1, \quad \text{for all} \quad \varepsilon > 0. \tag{13.2.4}$$

Since $b_n \to \infty$ it is possible for every $x > b_1$ to define the natural number $n = n(x)$ by means of

$$n(x) = \max\{j : b_j \leq x\}, \tag{13.2.5}$$

then we have $b_n \leq x < b_{2n}$. Of course, the conditions $x \to \infty$ and $n \to \infty$ are equivalent. Now we obtain

$$\frac{x^2 T_V(x)}{\int_{|y|<x} y^2 \, \mathrm{d}V(y)} \leq \frac{b_{2n}^2 T_V(b_n)}{\int_{|y|<b_n} y^2 \, \mathrm{d}V(y)} \leq \frac{b_{2n}^2}{b_n^2} \, \frac{nT_V(b_n)}{n \int_{|y|<1} y^2 \, \mathrm{d}V(b_n y)}.$$

Using (13.2.3), (13.2.4), and Cor. 13.2.5 we see that the last expression tends to 0 as $n \to \infty$. Hence, (13.2.2) is verified.

*If-part*: We may suppose that $\sigma_V^2 = \infty$ since otherwise Th. 3.7.2 can be applied to conclude that $V$ belongs to the domain of attraction of $\Phi$.

We will show that there exist constants $b_n > 0$ such that (13.2.3) and (13.2.4) are valid; then the assertion follows from Th. 12.1.1 since L. 13.1.6 again applies.

For fixed $\delta > 0$ we choose some number $c_n = c_n(\delta)$ which satisfies

$$nT_V(c_n + 0) \leq \delta \leq nT_V(c_n)$$

and put

$$b_n^2 = b_n^2(\delta) = n \int_{|y|\leq c_n} y^2 \, \mathrm{d}V(y).$$

Clearly, we have $c_n \to \infty$ and $b_n \to \infty$ as $n \to \infty$. Further, using (13.2.2) we obtain for every fixed $\delta > 0$

$$\frac{c_n^2}{b_n^2} = \frac{c_n^2 T_V(c_n)}{\int_{|y|\leq c_n} y^2 \, \mathrm{d}V(y)} \cdot \frac{1}{nT_V(c_n)} \to 0, \qquad n \to \infty.$$

Since $\delta > 0$ is arbitrary we can select a sequence $(\delta_n)$ such that both $\delta_n \to 0$ and $c_n(\delta_n)/b_n(\delta_n) \to 0$ as $n \to \infty$. Accordingly,

$$\frac{c_n(\delta_n)}{b_n(\delta_n)} + \frac{1}{nb_n(\delta_n)} \to 0,$$

so that, for fixed $\varepsilon > 0$ and sufficiently large $n$,

$$c_n(\delta_n) + \frac{1}{n} \leq \varepsilon b_n(\delta_n).$$

Thus we obtain, as $n \to \infty$,

$$nT_V(b_n\varepsilon) \leqq nT_V\left(c_n + \frac{1}{n}\right) \leqq \delta_n \to 0;$$

hence, (13.2.3) is verified. Finally, by the definition of $b_n$ we get, as $n \to \infty$,

$$\left|n \int\limits_{|y|<\varepsilon} y^2 \,\mathrm{d}V(b_n y) - 1\right| = \frac{n}{b_n^2}\left|\int\limits_{|u|<\varepsilon b_n} u^2 \,\mathrm{d}V(u) - \frac{b_n^2}{n}\right|$$

$$= \frac{n}{b_n^2} \int\limits_{c_n<|u|<\varepsilon b_n} u^2 \,\mathrm{d}V(u) \leqq \varepsilon^2 n T_V(c_n + 0) \leqq \varepsilon^2 \delta_n \to 0 .$$

This shows that (13.2.4) is valid. □

**Remark 13.2.8.** If (13.2.2) is not satisfied but

$$\lim_{x\to\infty} \frac{x^2 T_V(x)}{\int\limits_{|y|<x} y^2 \,\mathrm{d}V(y)} = 0 ,$$

then there exist constants $a_n$ and $b_n > 0$ and a subsequence $(F_{n'})$ of $(F_n)$ which converges to $\Phi$. This result is due to P. Lévy; a proof can be found in Feller (1971).

Now we are going to prove a condition for $V$ to belong to the domain of attraction of a non-normal stable d.f. $F$. By virtue of (8.1.4), such d.f. have the Lévy representation

$$F = \mathscr{L}(a, 0, d_1 |u|^{-\alpha}, -d_2 u^{-\alpha}) ,$$

where $a \in \mathbb{R}_1$, $0 < \alpha < 2$, $d_1 \geqq 0$, $d_2 \geqq 0$, and $d_1 + d_2 > 0$. — In the following we put $d_1/d_2 = \infty$ if $d_2 = 0$.

**Proposition 13.2.9.** *A d.f. $V$ belongs to the domain of attraction of a stable d.f. $F$ with characteristic exponent $\alpha < 2$ iff*

$$\lim_{x\to\infty} \frac{V(-x)}{1 - V(x)} = \frac{d_1}{d_2} \tag{13.2.6}$$

*and*

$$\lim_{x\to\infty} \frac{T_V(x)}{T_V(ux)} = u^\alpha \quad \textit{for all}, \quad u > 0 . \tag{13.2.7}$$

Proof. *Only-if part*: We assume $F_n(x) = V^{*n}(b_n x + a_n) \xrightarrow{c} F(x)$. Then we apply Th. 11.2.5 to the r.v. $X_{nk} = X_k/b_n$ and thus obtain, for every $u > 0$,

$$nV(-b_n u) \to d_1 u^{-\alpha} , \qquad n \to \infty , \tag{13.2.8}$$

$$n[1 - V(b_n u)] \to d_2 u^{-\alpha} , \qquad n \to \infty; \tag{13.2.9}$$

accordingly,

$$nT_V(b_n u) \to (d_1 + d_2)\, u^{-\alpha} , \quad n \to \infty . \tag{13.2.10}$$

Since $b_n \to \infty$ (see L. 13.2.1) and because of $d_1 + d_2 > 0$, we conclude from Cor. 2.5.5 that $T_V$ is of regular variation with exponent of regularity $r = -\alpha$ whence (13.2.7) follows. Further, for $x > b_1$ we define the natural number $n = n(x)$ by means of (13.2.5); then we have $b_n \leqq x < b_{n+1}$ and therefore

$$\frac{n}{n+1} \frac{(n+1)\, V(-b_{n+1})}{n[1 - V(b_n)]} \leqq \frac{V(-x)}{1 - V(x)} \leqq \frac{n+1}{n} \frac{nV(-b_n)}{(n+1)\,[1 - V(b_{n+1})]} .$$

Letting $x \to \infty$, we have $n \to \infty$ so that (13.2.6) immediately follows from (13.2.8) and (13.2.9).

*If-part*: We again apply Th. 11.2.5. Hence, we have to show that, for certain constants $b_n > 0$, the assumed conditions (13.2.6) and (13.2.7) imply (13.2.8), (13.2.9), and (11.2.9) for $\gamma^2 = 0$.

First we show that there exist constants $b_n > 0$ such that (13.2.8) and (13.2.9) are valid. — Since the assumed relations have only meaning if $T_V > 0$ for all $x > 0$ we can define $b_n$ by means of

$$nT_V(b_n + 0) \leqq d_1 + d_2 \leqq nT_V(b_n)$$

and see that $b_n \to \infty$ as $n \to \infty$. Thus we obtain by virtue of (13.2.7) for fixed $u > 0$

$$nT_V(b_n u) \geqq (d_1 + d_2)\,\frac{T_V(b_n u)}{T_V(b_n)} \to \frac{d_1 + d_2}{u^\alpha}, \qquad n \to \infty\,.$$

On the other hand, we easily see that $b_n u \geqq (b_n + \varepsilon)\,(u - \varepsilon)$ for any $\varepsilon > 0$ and sufficiently large $n$; therefore

$$nT_V(b_n u) \leqq (d_1 + d_2)\,\frac{T_V\big((b_n + \varepsilon)\,(u - \varepsilon)\big)}{T_V(b_n + \varepsilon)} \to \frac{d_1 + d_2}{u},$$

where the limit is to be taken on letting first $n \to \infty$ and then $\varepsilon \to +0$. So we get

$$nV(-b_n u) + n[1 - V(b_n u)] = \frac{d_1 + d_2 + o(1)}{u^\alpha}, \quad n \to \infty\,.$$

Further, from (13.2.6) it follows that

$$nV(-b_n u) = n[1 - V(b_n u)]\left[\frac{d_1}{d_2} + o(1)\right], \quad n \to \infty;$$

now the relations (13.2.8) and (13.2.9) can easily be verified.

It remains to show that (11.2.9) is valid for $\gamma^2 = 0$, where $F_{nk}(x) = V(b_n x)$. Clearly, because of the Schwarz inequality we have

$$\Big(\int\limits_{|y|<\varepsilon} y\, \mathrm{d}V(b_n y)\Big)^2 \leqq \int\limits_{|y|<\varepsilon} y^2\, \mathrm{d}V(b_n y);$$

thus it is enough to prove that

$$\lim_{\varepsilon \to +0}\ \overline{\lim_{n \to \infty}}\ \frac{n}{b_n^2} \int\limits_{|u|<\varepsilon b_n} u^2\, \mathrm{d}V(u) = 0\,. \tag{13.2.11}$$

To verify this let $\varepsilon > 0$ be fixed and satisfy $\varepsilon < \min\{\alpha, 2 - \alpha\}$. Then we fix an $x_0 = x_0(\varepsilon) > 0$ such that

$$2^{\alpha-\varepsilon} \leqq \frac{T(x)}{T(2x)} \leqq 2^{\alpha+\varepsilon}, \qquad x \geqq x_0; \tag{13.2.12}$$

this is possible by virtue of (13.2.7). As a consequence, by induction we find that, for all $j \geqq 0$ and $m \geqq 0$,

$$T(2^j x_0) \leqq 2^{(\alpha+\varepsilon)m}\, T(2^{j+m} x_0)\,. \tag{13.2.13}$$

Now we define, for fixed $n \geqq 1$, the natural number $N = N(n, \varepsilon)$ by means of

$$2^N x_0 < \varepsilon b_n \leqq 2^{N+1} x_0\,. \tag{13.2.14}$$

Thus we obtain

$$I_n(\varepsilon) := \int\limits_{x_0 \leq |u| < \varepsilon b_n} u^2 \, \mathrm{d}V(u) \geq \sum_{k=0}^{N-1} \int\limits_{2^k x_0 \leq |u| < 2^{k+1} x_0} u^2 \, \mathrm{d}V(u)$$

$$\geq x_0^2 \sum_{k=0}^{N-1} 2^{2k} [T(2^k x_0) - T(2^{k+1} x_0)] .$$

Now we apply first the left inequality in (13.2.12) and then (13.2.13) for $j = 0$ and $m = N$; hence, for $0 \leq k \leq N - 1$,

$$T(2^k x_0) - T(2^{k+1} x_0) = \left[ \frac{T(2^k x_0)}{T(2^{k+1} x_0)} - 1 \right] T(2^{k+1} x_0)$$

$$\geq (2^{\alpha - \varepsilon} - 1) \, T(2^N x_0) \geq (2^{\alpha - \varepsilon} - 1) \, T(x_0) \, 2^{-(\alpha + \varepsilon) N} .$$

This yields

$$I_n(\varepsilon) \geq x_0^2 (2^{\alpha - \varepsilon} - 1) \, T(x_0) \, 2^{-(\alpha + \varepsilon) N} \frac{2^{2N} - 1}{3} ,$$

and since $T(x_0) > 0$ we get, by virtue of $2 > \alpha + \varepsilon$ and $N(n, \varepsilon) \to \infty$

$$\lim_{n \to \infty} I_n(\varepsilon) = \infty . \tag{13.2.15}$$

Note, that this means that $\sigma_V^2 = \infty$. Next we give an upper bound for $I_n(\varepsilon)$. — From (13.2.14) we see that

$$I_n(\varepsilon) \leq \sum_{k=0}^{N} \int\limits_{2^k x_0 \leq |u| < 2^{k+1} x_0} u^2 \, \mathrm{d}V(u) \leq x_0^2 \sum_{k=0}^{N} 2^{2(k+1)} T(2^k x_0) .$$

Now we apply (13.2.13) for $j = k$ and $m = N - k + 1$. Hence,

$$I_n(\varepsilon) \leq x_0^2 T(2^{N+1} x_0) \sum_{k=0}^{N} 2^{2(k+1)} 2^{(\alpha + \varepsilon)(N - k + 1)}$$

$$= x_0^2 T(2^{N+1} x_0) \, 2^{2(N+1) + \alpha + \varepsilon} \sum_{k=0}^{N} 2^{(\alpha + \varepsilon - 2)(N - k)} .$$

Next we make use of (13.2.14) which tells us that $x_0 \leq \varepsilon b_n 2^{-N}$ and $T(2^{N+1} x_0) \leq T(\varepsilon b_n)$. Accordingly,

$$I_n(\varepsilon) \leq \varepsilon^2 b_n^2 2^{-2N} T(\varepsilon b_n) \, 2^{2(N+1) + \alpha + \varepsilon} \sum_{k=0}^{N} 2^{(\alpha + \varepsilon - 2)(N - k)} \leq \varepsilon^2 b_n^2 T(\varepsilon b_n) \frac{2^{2 + \alpha + \varepsilon}}{1 - 2^{\alpha + \varepsilon - 2}}$$

where the last estimate is due to the fact that $\alpha + \varepsilon - 2 < 0$ so that the series which dominates the above sum converges. Now the relation (13.2.10), which has already been shown, can be applied to conclude that

$$\lim_{\varepsilon \to +0} \overline{\lim_{n \to \infty}} \frac{n}{b_n^2} I_n(\varepsilon) \leq \lim_{\varepsilon \to +0} \varepsilon^2 \frac{d_1 + d_2}{\varepsilon^\alpha} \frac{2^{2 + \alpha + \varepsilon}}{1 - 2^{\alpha + \varepsilon - 2}} = 0 . \tag{13.2.16}$$

Finally, from (13.2.15) we see that, for fixed $\varepsilon$ and for the corresponding $x_0$, there exists an $n_0 = n_0(\varepsilon)$ such that

$$\int\limits_{|u| < x_0} u^2 \, \mathrm{d}V(u) \leq I_n(\varepsilon) , \qquad n \geq n_0 .$$

Accordingly, for $n \geq n_0$ we have

$$\frac{n}{b_n^2} \int\limits_{|u| < \varepsilon b_n} u^2 \, \mathrm{d}V(u) = \frac{n}{b_n^2} \left( \int\limits_{|u| < x_0} u^2 \, \mathrm{d}V(u) + I_n(\varepsilon) \right) \leq \frac{2n}{b_n^2} I_n(\varepsilon) .$$

Now (13.2.16) leads us to the desired relation (13.2.11). □

**Remark 13.2.10.** Comparing the conditions of Th. 13.2.7 and Prop. 13.2.9 we see the following. The domain of attraction of the normal d.f. contains all d.f. $V$ such that $T_V(x) \to 0$, as $x \to \infty$, sufficiently rapidly; this is a very large set of d.f. Contrary to this fact, all the other stable d.f. $F$ have domains of attraction the d.f. of which have an asymptotic behaviour similar to that of $F$; this is obvious from the above remark saying that $F$ belongs to its own domain of attraction. Thus we recognize the preponderating importance of the normal d.f. in the set of stable d.f. □

## 13.3. Restricted convergence

We adhere to the notation $S_n = 1/b_n\,(X_1 + X_2 + \dots + X_n - a_n) \sim F_n$ where $X_1, X_2, ..$ are i.i.d.r.v. subject to the d.f. $V$. In the present section we wish to prove complete convergence of $(F_n)$, assuming the r-convergence on a half axis, i.e., we assume that there is a number $\tau \in \mathbf{R}_1$ and a certain monotone function $F$ on $(-\infty, \tau)$ such that

$$F_n \xrightarrow[(-\infty,\tau)]{} F\,, \qquad n \to \infty\,. \tag{13.3.1}$$

The general subject of r-convergence of sums has already been described in Sect. 10.4 We recall that the first result in which the assumption (13.3.1) is used is connected with Zolotarev's conjecture; it has been formulated in the CLT Th. 10.4.1. So the r-convergence $F_n \xrightarrow[(-\infty,\tau)]{} \Phi$ implies $F_n \xrightarrow{c} \Phi$. Without changing the proof sketched in Sect. 10.4 we are now in a position to state an even more general result. For this purpose it is useful to introduce the set $\mathbf{F}_\tau := \mathbf{F}\big((-\infty, \tau)\big)$ of all inf.div.d.f. $F$ which are uniquely defined by their values $F(x)$, $x \in S = (-\infty, \tau)$, see Suppl. 7.9.12.

**Proposition 13.3.1.** *Let* $\tau \in \mathbf{R}_1$ *and* $F \in \mathbf{F}_\tau$ *where* $F > 0$. *Then the r-convergence* (13.3.1) *implies the complete convergence* $F_n \xrightarrow{c} F$; *moreover,* $F$ *is stable.*

Proof. We select some subsequence $(F_{n'})$ such that $F_{n'} \xrightarrow{w} H$, where $H$ stands for a monotone function. Then we obviously have $F(x) = H(x)$, $x < \tau$. The compactness problem has been solved in Prop. 13.1.3, so that $H$ is a non-defective d.f. Now L. 13.2.2 tells us that $H$ is inf.div.

In the present case the continuity problem is settled by the definition of $\mathbf{F}_\tau$ from which it follows that $H = F$ holds. This means that all weakly converging subsequences $(F_{n'})$ tend to the same limit d.f., that is, $F_n \xrightarrow{c} F$ holds. Finally, it follows from Th. 13.2.4 that $F$ is stable. □

It is clear from Th. 7.6.1 that $\Phi \in \mathbf{F}_\tau$. Thus Th. 10.4.1 is contained in the above result. But it is also easily verified that all stable d.f. $F = \mathcal{S}(\alpha, \beta; a, b)$ with $1 \leq \alpha < 2$, $\beta = -1$, $a \in \mathbf{R}_1$, $b > 0$, belong to $\mathbf{F}_\tau$. Indeed, we then have $F > 0$, see Prop. 8.1.12 (iv); further, $\beta = \dfrac{d_1 - d_2}{d_1 + d_2} = -1$ ensures that $d_1 = 0$, i.e., the left spectral function $M$ of $F$ vanishes; see Prop. 8.1.5. Now we get from Prop. 7.3.2 (applied to $N$) that the c.f. $f$ of $F$ is continuable to the upper half plane $S(0, \infty)$. So the desired uniqueness of $F$ follows from Prop. 7.6.2.

Unfortunately, we know that $\mathbf{F}_\tau$ does not coincide with the class of all inf.div.d.f. (an inf.div.d.f. $F > 0$ with $F \notin \mathbf{F}_\tau$ can be found in IBRAGIMOV (1977)). Further, there are even stable d.f. $F$ which are not contained in $\mathbf{F}_\tau$, see Cor. 8.5.2. This is the reason why we have to refine the above proof by adding a "reduction lemma" saying that — under (13.3.1) — all limit d.f. for subsequences $(F_{n'})$ belong to a certain subclass of inf.div.d.f. for which the continuation problem may be solved.

An essential tool will be the set (13.3.2) of functional equations which are defined for all real numbers $s > 0$. This replaces the set (13.1.3) of equations which hold only for natural numbers $s = k \geqq 2$. The crucial idea of the following investigation will then be *how to apply these equations using the fact that $G$ is inf.div.* (cf. the reduction lemma).

Before formulating a first lemma we recall that (13.3.1) implies the relative compactness of the sequence $(F_n)$ by Prop. 13.1.3. Thus Helly's selection theorem guarantees the existence of a completely converging subsequence $(F_{n'})$.

**Lemma 13.3.2.** *Let* (13.3.1) *be satisfied for some d.f.* $F > 0$ *and* $\tau \in \mathbb{R}_1$. *If* $F_{n'} \xrightarrow{c} G$ *for a certain subsequence* $(F_{n'})$, *then $G$ is inf.div. and admits the representation*

$$G(B_s x + A_s) = G_s(x)\,, \qquad s > 0\,, \qquad x < (\tau - A_s)/B_s\,, \tag{13.3.2}$$

*where $G_s$ denotes the d.f. with c.f. $g^s$ and $B_s > 0$, $A_s \in \mathbb{R}_1$ are certain numbers.*

Proof. For $sn \geqq 1$ we set $a_{n,s} = ([sn]/n)\, a_n$ (entier). Further, it is clear that $s_n := [sn]/n \to s$, as $n \to \infty$. Now we introduce the d.f.

$$H_{n,s}(x) = \mathsf{P}\left(\frac{X_1 + X_2 + \dots + X_{[sn]} - a_{n,s}}{b_n} < x\right)$$

and obtain for the corresponding c.f.

$$h_{n,s}(t) = \exp\left(-it\frac{a_{n,s}}{b_n}\right) v^{[sn]}\left(\frac{t}{b_n}\right) = \left[\exp\left(-it\frac{a_n}{b_n}\right) v^n\left(\frac{t}{b_n}\right)\right]^{s_n} = (f_n(t))^{s_n}.$$

Assume now $F_{n'} \xrightarrow{c} G$. Then L. 13.2.2 tells us that $G$ is inf.div. Hence, $f_{n'}(t) \to g(t) \neq 0$. Since the convergence is uniform in any bounded interval we have $f_{n'}(t) \neq 0$ $|t| \leqq t_0$, $n' \geqq n_0(t_0)$. Thus we get, as $n' \to \infty$,

$$h_{n',s}(t) = e^{s_{n'} \log f_{n'}(t)} \to e^{s \log g(t)} = g^s(t)\,.$$

Next we put $B_{n,s} = b_n/b_{[ns]}$, $A_{n,s} = a_{n,s} - a_{[sn]}$. Then we obtain by the definition of $H_{n,s}$ and in view of the above convergence

$$F_{[sn']}(B_{n',s}x + A_{n',s}) = H_{n',s}(x) \xrightarrow{c} G_s(x)\,. \tag{13.3.3}$$

On the other hand, (13.3.1) yields $F(x) = G(x)$, $x < \tau$, and

$$F_{[sn']} \xrightarrow[(-\infty,\tau)]{} G \quad \text{for all} \quad s > 0\,. \tag{13.3.4}$$

By virtue of (13.3.3) and (13.3.4) we are able to apply Prop. 2.3.4 according to which $B_{n',s} \to B_s > 0$, $A_{n',s} \to A_s \in \mathbb{R}_1$ hold. Hence (13.3.4) can be rewritten in the form

$$F_{[sn']}(B_{n',s}x + A_{n',s}) \xrightarrow[I']{} G_s(B_s x + A_s)\,, \qquad I' := \left(-\infty, (\tau - A_s)/B_s\right),$$

see Prop. 2.3.1. Combining this with (13.3.3) we obtain (13.3.2) as asserted. □

Now we are going to analyse the equations (13.3.2).

**Lemma 13.3.3.** *Let $G$ be inf.div. satisfying $G > 0$, $G(\tau) < 1$ and assume* (13.3.2). *Then for every $x < 0$ there exists $s_0 > 0$ such that*

$$B_s x + A_s < \tau\,, \qquad 0 < s < s_0\,.$$

*Moreover, we have $B_s \to \infty$ and $A_s/B_s \to 0$, as $s \to +0$.*

Proof. Let $\eta < \tau$ and define $x_s = x_s(\eta)$ by $B_s x_s + A_s = \eta$. Then (13.3.2) implies the relation

$$0 < G(\eta) = G_s(x_s) \leqq G(\tau) < 1\,. \tag{13.3.5}$$

Since $g^s \to 1$, $s \to +0$, we easily see that for all $y > 0$

$$G_s(-y) \to 0\,, \qquad G_s(y) \to 1\,. \tag{13.3.6}$$

Now we will show indirectly that

$$x_s \to 0 \quad \text{as} \quad s \to +0\,. \tag{13.3.7}$$

Assume that there is a sequence $s_n \to +0$ such that $|x_{s_n}| \geqq \delta > 0$ for all $n$. Then we get a contradiction between (13.3.5) and (13.3.6). Thus (13.3.7) follows. So we have for every $x < 0$ and sufficiently small $s > 0$

$$B_s x + A_s \leqq B_s x_s + A_s = \eta < \tau\,.$$

To prove the other assertions we choose two different numbers $\eta^j$, $\eta^j < \tau$, and define $x_s^j$ by $B_s x_s^j + A_s = \eta^j$, $j = 1, 2$. From $B_s(x_s^1 - x_s^2) = \eta^1 - \eta^2 \neq 0$ and $x_s^j \to 0$ it is seen that $B_s \to \infty$. Now the last assertion follows from $A_s/B_s = \eta/B_s - x_s \to 0$. □

**Lemma 13.3.4** (Reduction lemma). *Let $G = \mathcal{L}(a, \gamma^2, M, N)$ be an inf.div.d.f. which satisfies* (13.3.2); *then there exist numbers $c \geqq 0$, $0 < \alpha < 2$, such that*

$$M(x) = c\,|x|^{-\alpha}\,, \qquad x < 0\,.$$

Proof. We recall that $G = ({}^nG)^{*n}$. Thus we see from the definition of $G_s$ that $G_{1/n} = {}^nG$. Hence Cor. 7.2.7 tells us that

$$M_n := nG_{1/n} = n\,{}^nG \xrightarrow[(-\infty,\,0)]{} M\,, \qquad n \to \infty\,. \tag{13.3.8}$$

Now we use (13.3.2) for $s = 1/n$ and obtain

$$M_n\left(x - \frac{A_{1/n}}{B_{1/n}}\right) = nG_{1/n}\left(x - \frac{A_{1/n}}{B_{1/n}}\right) = nG(B_{1/n}x)\,, \qquad x < 0\,. \tag{13.3.9}$$

From L. 13.3.3 we get $A_{1/n}/B_{1/n} \to 0$, as $n \to \infty$. Now Prop. 2.3.1 tells us that the expressions (13.3.8) and (13.3.9) have the same limit function, i.e.,

$$M_n\left(x - \frac{A_{1/n}}{B_{1/n}}\right) = nG(B_{1/n}x) \xrightarrow[(-\infty,\,0)]{} M(x)\,.$$

If $M = 0$, then the assertion is obvious with $c = 0$. Otherwise $M$ is positive on some interval. Thus it follows from $G > 0$ that $B_{1/n} \to \infty$. Now we are ready to apply the Cor. 2.5.5 on functions of regular variation; it tells us that $M(x) = c\,|x|^{-\alpha}$ for certain real numbers $c, \alpha$. From $M \neq 0$ we get $c > 0$. Finally, we recall that the spectral function $M$ satisfies $M(-\infty) = 0$ and $\int_{-1}^{-0} x^2\,\mathrm{d}M(x) < \infty$ so that $0 < \alpha < 2$ follows. □

We now state the first main result of this section.

**Proposition 13.3.5.** *Suppose that the sequence $(F_n)$ r-converges on $(-\infty, \tau)$ for a certain $\tau \in R_1$ to some d.f. $F > 0$. Then there exists a stable d.f. $G$ such that the complete convergence $F_n \xrightarrow{c} G$ holds.*

Proof. First we conclude from Prop. 13.1.3 that $(F_n)$ is relatively compact. Thus there exists a subsequence $(F_{n'})$ and a d.f. $G$ such that $F_{n'} \xrightarrow{c} G$. By L. 13.2.2, $G$ is inf.div and by L. 13.3.2 in satisfies (13.3.2).

We turn now to the continuation problem. By the reduction lemma L. 13.3.4 we have to solve it within the class $\mathcal{L}(a, \gamma^2, c\,|x|^{-\alpha}, N)$. For this class it is Prop. 7.6.4 which implies the complete solution, i.e., $G$ is uniquely defined.

Thus every converging subsequence $(F_{n'})$ tends to the same limit d.f. $G$ so that $F_n \xrightarrow{c} G$ follows. $G$ proves to be stable by virtue of Th. 13.2.5. □

So far we have deduced consequences of the r-convergence (13.3.1) under the hypothesis $F > 0$, i.e., if lext $(F) = -\infty$. Our next aim is to derive similar results for lext $(F) > -\infty$. Clearly, it is no restriction of generality to assume that lext $(F) = 0$ and $\tau > 0$.

Now we formulate the counterpart to Prop. 13.3.5.

**Proposition 13.3.6.** *Suppose that the sequence $(F_n)$ r-converge on $(-\infty, \tau)$ for a certain $\tau > 0$ to some d.f. $F$ with $F(+0) = 0$, $F(x) > 0$, $x > 0$. Then there exists a stable d.f. $G$ such that the complete convergence $F_n \xrightarrow{c} G$ holds.*

Proof. The relative compactness of $(F_n)$ is guaranteed in view of Prop. 13.1.4. Thus there exists a subsequence $(F_{n'})$ and a d.f. $G$ such that

$$F_{n'} \xrightarrow{c} G \text{ and } G(x) = F(x), \qquad x < \tau.$$

Moreover, Prop. 13.1.4 tells us that

$$G(B_k^{-1}y) = G^{*k}(y), \qquad y < \tau B_k, \qquad k \geqq 2. \tag{13.3.10}$$

L. 13.2.2 implies also in this case that $G$ is inf.div.

We turn now to the continuation problem.

First we recall from Prop. 7.3.9 that $G = \mathscr{L}(a, 0, 0, N)$. Accordingly, it is seen from Lévy's canonical representation formula and Cor. 7.2.11 that

$$G^{*k} = \mathscr{L}(ka, 0, 0, kN) \quad \text{and} \quad G(B_k^{-1}y) = \mathscr{L}\big(r_k, 0, 0, N(B_k^{-1}u)\big)$$

for a certain $r_k \in R_1$. Next we apply Prop. 7.6.10 and see from (13.3.10) that

$$kN(u) = N(B_k^{-1}u), \qquad 0 < u < \tau B_k, \qquad k \geqq 2.$$

Since $N \neq 0$ is follows that $B_k \to \infty$ and we have

$$N(v) = \lim_{k \to \infty} k^{-1}N(B_k^{-1}v), \qquad 0 < u < \infty.$$

Thus, putting $H = -N$ we are now in a position to apply Rem. 2.5.6 on regular variation; it tells us that the

$$H(v) = -N(v) = cv^{-\alpha}, \qquad 0 < v < \infty,$$

here $c > 0$ since $G$ is non degenerate by assumption and $0 < \alpha < 1$ since

$$\int_0^1 u \, \mathrm{d}N(u) < \infty$$

(by Prop. 7.3.9) and $N(\infty) = 0$. By virtue of $F(x) = G(x)$, $x < \tau$, the constants $c, \alpha$ are uniquely defined, see Th. 8.4.1 or Prop. 7.6.10. □

**Remark 13.3.7.** If – contrary to the assumption of Prop. 13.3.6 – only lext $(F) = 0$ is assumed where $F(+0) \geqq 0$, then the limit theorem is not true in general. This can be seen by Ex. 13.1.5 in which $(F_n)$ is not relatively compact but (13.3.1) is satisfied. □

## 13.4. Supplement

**13.4.1.** Relative compactness of $(F_n)$ under the assumption of left-tightness. We introduce the functions

$$\underline{F}(x) = \underline{\lim}_{n\to\infty} F_n(x)\,, \qquad \overline{F}(x) = \overline{\lim}_{n\to\infty} F_n(x)\,.$$

(i) *If* $\underline{F} > 0$ *and* $\overline{F}(-\infty) = 0$, *then* $(F_n)$ *is relatively compact* (Rossberg (1979)).

Note that $\overline{F}(-\infty) = 0$ is equivalent to the left tightness of $(F_n)$.

(ii) *If* $\overline{F}(+0) = 0$ *and* $\underline{F}(x) > 0$, $x > 0$, *then* $(F_n)$ *is relatively compact* (Rossberg (1983b)).

Special cases of these two assertions are formulated as Prop. 13.1.3 and 13.1.4, resp.

**13.4.2.** Compactness of sums and non-degenerate limit d.f. *The following two assertions are equivalent:*

(i) *There exist sequences* $a_n$, $b_n > 0$, *such that the sequence of d.f.* $(F_n)$ *is relatively compact and any limit d.f.* $F$ *of subsequences* $(F_{n'})$ *of* $(F_n)$ *is non-degenerate.*

(ii) *We have the relation*

$$\overline{\lim_{x\to\infty}} \frac{x^2 T_V(x)}{\int\limits_{|y|\leq x} y^2\,\mathrm{d}V(y)} < \infty\,, \tag{13.4.1}$$

(Feller (1967)); compare Rem. 13.2.8.

**13.4.3.** The class of limit d.f. for sequences $(F_n)$ which are relatively compact. Let $F = \mathscr{L}(a, \gamma^2, M, N)$ be an inf.div.d.f. and put

$$G(x) = x^2\Big(\int\limits_{y<-x} \mathrm{d}M(y) + \int\limits_{y>x} \mathrm{d}N(y)\Big)\,,$$

$$K(x) = \gamma^2 + \int\limits_{-x\leq y<0} y^2\,\mathrm{d}M(y) + \Big(\int\limits_{0<y\leq x} y^2\,\mathrm{d}N(y)\Big)\,.$$

Then we say that $F$ belongs to the set $\boldsymbol{F}$ if

$$\overline{\lim_{x\to\infty}} \frac{G(x)}{K(x)} < \infty\,, \qquad \overline{\lim_{x\to+0}} \frac{G(x)}{K(x)} < \infty\,.$$

*We have* $F \in \boldsymbol{F}$ *iff there exists a d.f.* $V$ *which satisfies* (13.4.1) *and moreover, there are numbers* $b_n > 0$, $a_n \in \boldsymbol{R}_1$, *and a subsequence* $(F_{n'})$ *such that* $F_{n'} \xrightarrow{c} F$ *holds* (Pruitt (1983)).

**13.4.4.** *Let* $(F_n)$ *be left-tight, that is,*

$$\lim_{x\to-\infty} \sup_n F_n(x) = 0\,.$$

*If* $F_n \xrightarrow{w} F \neq$ *const., then either* $F$ *is a non-defective limit d.f. and the convergence is complete, or* $F$ *is a step function* $F = \varrho U_{x_0}$ *for some numbers* $\varrho$, $0 < \varrho = F(\infty) < 1$, *and* $x_0 \in \boldsymbol{R}_1$ (Siegel (1978)); compare Ex. 13.1.5.

**13.4.5.** The r-convergence of $(F_n)$ on a bounded set. *Let* $S$ *be a set with a finite limit point. Further, let* $G$ *be stable with characteristic exponent* $\alpha$, $1 \leq \alpha \leq 2$. *Then the r-convergence*

$$F_n \xrightarrow[S]{} G\,, \qquad n \to \infty\,,$$

*and*

$$0 < c_1 \leqq x^\alpha T_F(x) \leqq c_2 < \infty\,, \quad \textit{for all} \quad x > 0\,,$$

*imply* $F_n \xrightarrow{c} G$ (Jesiak (1979)).

**13.4.6. Problem.** Let $\tau \in R_1$ and $G > 0$ be inf.div. If there exist real numbers $B_n > 0$, $A_n$, such that

$$G(x) = G^{*n}(B_n x + A_n)\,, \qquad n \geqq 1\,, \qquad x < \tau\,,$$

then it would be of interest to know whether $G$ is stable.

This problem concerns the proof of Prop. 13.3.5, cf. L. 13.3.2—13.3.4.

**13.4.7. Problem.** Will it be possible to generalize Prop. 13.3.5 such that the half axis in assumption (13.3.1) is replaced by some left-sided unbounded subset? By the compactness theorem quoted in 13.4.1 (i), this amounts to solving the corresponding continuation problem.

13.4.8. Consider a sequence $(X_k)$ of i.r.v. where the r.v. $X_k$ are not necessarily identically distributed and form the sums

$$S_n = \frac{1}{b_n}\Big(\sum_{k=1}^{n} X_k - a_n\Big) \sim F_n\,, \qquad n \geqq 1\,,$$

where the summands are assumed to be infinitely small. The class of limit d.f. for this model is well known and usually denoted by $L$ (see Sect. 10.1).

**Problem.** Treat the model from the viewpoint of r-convergence.

In a first approach Bischoff (1984) considered l-sequences $(X_k)$ of i.r.v. which are subject to one of the d.f. $V_1, V_2, \ldots, V_l$, $l \geqq 2$. His result generalizes Prop. 13.1.3 and 13.1.4.

*Let* $X_1, X_2, \ldots$ *be an* l-*sequence of i.r.v. Assume the r-convergence* (13.3.1) *where* $F$ *is a monotone function on* $(-\infty, \tau)$ *such that either* lext $(F) = -\infty$ *or* $-\infty <$ lext $(F) = x_0 < \tau$, $F(x_0 + 0) = 0$. *Then the sequence* $(F_n)$ *is relatively compact and all possible limit d.f. are inf.div.* (Bischoff (1985)).

13.4.9. Convergence of extreme values. *We set*

$$Y_n = \max\{X_1, X_2, \ldots, X_n\}\,, \qquad X_k \sim V\,, \qquad n \geqq 1\,. \tag{13.4.2}$$

*For appropriately norming and centering constants* $b_n > 0$, $a_n$, *we put* $G_n(x) = \mathsf{P}\left(1/b_n\,(Y_n - a_n) < x\right)$ *and assume* $G_n \xrightarrow{c} G$, *where* $G$ *is non-degenerate. Then* $G$ *belongs to exactly one of the following types of d.f.*:

(i) $\tilde{F}_1(x) = \exp\left(-\exp(-x)\right)\,, \quad x \in R_1;$

(ii) $\tilde{F}_2(x) = \begin{cases} 0\,, & x \leqq 0\,, \\ \exp(-x^{-\alpha})\,, & x > 0,\ \alpha > 0\,; \end{cases}$

(iii) $\tilde{F}_3(x) = \begin{cases} \exp\left(-(-x)^\alpha\right)\,, & x \leqq 0\,, \\ 1\,, & x > 0\,, \end{cases} \quad \alpha > 0\,,$

(Gnedenko (1943)).

13.4.10. Convergence of extreme values and r-convergence. Th. 10.4.1 stimulated B. V. Gnedenko to apply Zolotarev's idea to extreme values. *Assume the r-convergence* $G_n \xrightarrow[I]{} \psi$, $n \to \infty$, *on some finite interval* $I = (x_1, x_2)$ *and a certain non-decreasing function* $\psi$ *satisfying* $\psi(x_1) < \psi(x_2)$. *Then* $G_n \xrightarrow{c} G$ *holds for a certain d.f.* $G$ (Gnedenko.

and SENUSI-BEREKSI (1982a)). This theorem was carried over to more general order statistics in GNEDENKO and ŠERIF (1983).

**13.4.11.** Convergence of extreme values with random indices. We define $Y_n$ as in (13.4.2) and consider integer-valued r.v. $N_n$ subject to

$$\mathsf{P}(N_n = k) = \frac{p}{n}\left(1 - \frac{p}{n}\right)^{k-1}, \quad k \geqq 1,$$

*for some* $p > 0$, *and* $n \geqq p$. *Let* $N_n, X_1, X_2, \ldots$ *be i.r.v. for each* $n$ *and put* $H_n(x) = \mathsf{P}\,(1/b\,(Y_{N_n} - a_n) < x)$.

(i) *If* $H_n \xrightarrow{c} \hat{F}$ *then* $\hat{F}$ *belongs to one of the following types of d.f.*

$$\hat{F}_1(x) = \frac{1}{1 + \mathrm{e}^{-x}}, \qquad x \in \mathbf{R}_1;$$

$$\hat{F}_2(x) = \begin{cases} 0, & x \leqq 0, \\ \dfrac{x^\alpha}{1 + x^\alpha}, & x > 0, \end{cases} \qquad \alpha > 0,$$

$$\hat{F}_3(x) = \begin{cases} \dfrac{1}{1 + (-x)^\alpha}, & x \leqq 0, \\ 1, & x > 0, \end{cases} \qquad \alpha > 0,$$

(GNEDENKO (1983)).

(ii) *If* $H_n \xrightarrow[I]{} \hat{F}$ *where* $\hat{F}$ *is one of the d.f. stated above and* $I = (x_1, x_2)$, $\hat{F}(x_2) > \hat{F}(x_1)$, *then the complete convergence* $\hat{H}_n \xrightarrow{c} \hat{F}$ *holds true* (GNEDENKO and SENUSI-BEREKSI (1982b)).

## Notes and bibliographical data

Prop. 13.1.3 and 13.1.4 are due to ROSSBERG (1979) and (1983), resp. A compactness theorem analogous to Prop. 13.1.7 can be found in FELLER (1967), see PRUITT (1983).

Prop. 13.1.7 has been proved in SIEGEL (1978).

Sect. 13.2 contains only well-known results (see e.g. GNEDENKO and KOLMOGOROV (1959)); Sect. 13.3 gives results contained in ROSSBERG (1979) and (1983).

# Appendices

## A. Appendix on real functions

In Prop. A.1—4 we quote some important results of integration theory. All the integrands are supposed to be Borel measurable functions on either $R_1$ or $R_2$; the integrations are carried out with respect to distribution functions (d.f.) so that they are Lebesgue-Stieltjes integrals.

In the first two theorems we consider a sequence $(s_n)_1^\infty$ of real functions on $R_1$.

**Proposition A.1** (Fatou). *If $s_n \geqq 0$ and $s_n$ is integrable with respect to the d.f. $F$, then*

$$\varliminf_{n\to\infty} \int_{-\infty}^{\infty} s_n(u)\, F(du) \geqq \int_{-\infty}^{\infty} \varliminf_{n\to\infty} s_n(u)\, F(du)\,. \qquad \square$$

In the next theorem we combine two well-known sufficient conditions permitting the passage to the limit under the integral sign.

**Proposition A.2.** *Let $s_n$ be integrable with respect to the d.f. $F$ and $s_n \to s$ pointwise almost everywhere with respect to the measure defined by $F$. Then either of the following two conditions is sufficient for $s$ to be integrable and the formula*

$$\lim_{n\to\infty} \int_{-\infty}^{\infty} s_n(u)\, F(du) = \int_{-\infty}^{\infty} s(u)\, F(du)\,.$$

(i) Dominated convergence principle: *There exists an integrable function $S$ such that $|s_n| \leqq S$ for all $n$.*
(ii) Monotone convergence principle: *$s_1 \leqq s_2 \leqq \ldots$ and the sequence of the integrals*

$$\int_{-\infty}^{\infty} s_n(u)\, F(du)$$

*is bounded.*

**Proposition A.3** (Fubini). *If $h \geqq 0$ then*

$$\int_{-\infty}^{\infty} F(dx) \int_{-\infty}^{\infty} h(x, y)\, G(dy) = \int_{-\infty}^{\infty} G(dy) \int_{-\infty}^{\infty} h(x, y)\, F(dx)$$

*in the sense that if one side converges, then so does the other. If $h_1 \geqq 0$ and $h_2 \geqq 0$, then the assertion holds also for $h = h_1 + ih_2$.* $\square$

Let $G$ be a d.f. and $H(x, y)$ be a family of functions on $R_2$ with the following properties:
(i) for almost all values of $y$ the function $H(x, y)$ is a d.f. in $x$;
(ii) $H(x, y)$ is a bounded and Borel measurable function of $y$.

We form the expression

$$G_0(x) := \int_{-\infty}^{\infty} H(x, y)\, G(\mathrm{d}y)$$

and see by dominated convergence that $G_0$ is a d.f.

**Proposition A.4.** *If $h$ is a bounded Borel measurable function, then*

$$\int_{-\infty}^{\infty} h(x)\, G_0(\mathrm{d}x) = \int_{-\infty}^{\infty} \Big[ \int_{-\infty}^{\infty} h(x)\, H(\mathrm{d}x, y)\Big]\, G(\mathrm{d}y)\,. \qquad \square$$

**Proposition A.5** (RIEMANN-LEBESGUE). *If $s$ is L-integrable over $\mathbb{R}_1$ and*

$$\hat{s}(t) := \int_{-\infty}^{\infty} \mathrm{e}^{\mathrm{i}tx}\, s(x)\, \mathrm{d}x$$

*then $\hat{s}(t) \to 0$ as $t \to \pm\infty$.* $\square$

**Proposition A.6.** *The expression* $(2/\pi) \int_0^y \sin x\, (\mathrm{d}x/x)$ *is bounded for* $y > 0$ *and approaches* 1 *as* $y \to \infty$. $\square$

**Proposition A.7.** *Let $h > 0$ be a function on $\mathbb{R}_1$ which is bounded on some finite interval $I$. If $h$ satisfies*

$$h(t_1 + t_2) = h(t_1)\, h(t_2)$$

*for all real $t_1$, $t_2$, then $h(t) = \mathrm{e}^{ct}$ for a certain $c \in \mathbb{R}_1$.* $\square$

For the proofs of Prop. A.1—A.3 and Prop. A.5 see FELLER (1971); the proofs of Prop. A.4 and Prop. A.7 are contained in KAWATA (1972), Sect. 3.10, and FELLER (1968), resp.

# B. Appendix on analytic functions

## B.1. Schwarz' reflection principle

**Proposition B.1.1.** *Let $h$ be a function analytic in a simply connected $z$-domain $D$ intersected by $\mathbb{R}_1$ which is real on $\mathbb{R}_1 \cap D$. Then $h(z) = \overline{h(\bar{z})}$ for $z, \bar{z} \in D$.* □

The proof amounts to showing that the coefficients of the expansion

$$h(z) = \sum_{n=0}^{\infty} a_n (z - z_0)^n$$

are real provided that $z_0 \in \mathbb{R}_1 \cap D$.

The next proposition is often called *Schwarz' reflection principle.* It is something of a converse to the previous theorem and provides a method of constructing the analytic continuation of certain analytic functions. For a brief formulation we put

$$S(0, \infty) := \{z = t + is : s > 0\}\,, \qquad S(-\infty, 0) := \{z = t + is : s < 0\}$$

$$D_+ := D \cap S(0, \infty)\,, \qquad D_- := D \cup S(-\infty, 0)\,, \qquad C := D \cap \mathbb{R}_1\,.$$

**Proposition B.1.2.** *Let $h_+$ be a function analytic in $D_+$ and continuous in $D_+ \cup C$ which is real on $C$. Define $h_-(z) := \overline{h_+(\bar{z})}$ for $z \in D_- \cap C$ and $\bar{z} \in D_+ \cup C$. Then $h_-$ is an analytic continuation of $h_+$.* □

In this way $h_+$ and $h_-$ define a function $h$ analytic in $D_- \cup C \cup D_+$. It is seen that its singularities are located symmetrically with respect to $C$ so that its entire domain of regularity is also symmetric with respect to $C$.

In our context the following definition is useful.

**Definition B.1.3.** Let us consider a complex function $R$ defined on $C = D \cap \mathbb{R}_1 \neq \emptyset$. Assume that there exists a function $A$ analytic in $D_+$ which is continuous in $D_+ \cup C$ and satisfies $A(t) = R(t)$, $t \in C$. In this case we will say that $R$ is *continuable to $D_+$*. □

The continuation is then also denoted by $R$; this is permissible by the

**Corollary B.1.4.** *If a function $R$ defined on $C$ is continuable to $D_+$ then the continuation is unique.*

Proof. Assume that there are two continuations $R^1$ and $R^2$. Then the function $d := R^1 - R^2$ vanishes on $C$. By Prop. B.1.2 it can be continued to $D_-$ whence $d = 0$ follows from the uniqueness theorem for analytic functions. □

We also need another criterion.

**Proposition B.1.5.** *Let $h_\pm$ be two functions analytic in $D_\pm$ and continuous in $D_\pm \cup C$, resp. If $h_+(t) = h_-(t)$, $t \in C$, then $h_\pm$ are analytic continuations of each other.* □

For details see Rühs (1971) and Titchmarsh (1932).

## B.2. Fourier-Stieltjes and Laplace-Stieltjes transforms of functions of bounded variation

Let $r$ be a real function on $R_1$. We decompose the interval $[a, b]$ by the division points $x_0 = a < x_1 < \dots < x_N = b$ and denote this decomposition by $Z$. In the expression

$$T(a, b) := \sup_Z \sum_{k=1}^{N} |r(x_k) - r(x_{k-1})|$$

the supremum is taken over all $Z$ of $[a, b]$; if it is finite then $r$ is said to have *bounded variation* on $[a, b]$ and $T(a, b)$ is called the *total variation* of $r$ on $[a, b]$.

Since

$$T(a, b) = T(a, c) + T(c, b), \qquad a < c < b,$$

the function $T$ is monotone with respect to both variables. If

$$T(-\infty, b) := \lim_{a \to -\infty} T(a, b); \qquad T(-\infty, \infty) := \lim_{b \to \infty} T(-\infty, b) \tag{B.2.1}$$

exist finitely then $r$ is said to have *bounded variation on* $(-\infty, b]$ and on $R_1$ with *total variations* $T(-\infty, b)$ and $T(-\infty, \infty)$, resp.

In this case $r(-\infty)$ trivially exists. It can easily be shown that then

$$r(x) = T(-\infty, x) - \big(T(-\infty, x) - r(x)\big) \tag{B.2.2}$$

is a decomposition into monotone and bounded functions so that $r$ has at most a countable set of discontinuities.

In the present book we suppose functions of bounded variation to be defined on $R_1$ and left-continuous on $R_1$, satisfying $r(-\infty) = 0$. Then from (B.2.2) we immediately obtain

**Proposition B.2.1.** *A real function $r$ defined on $R_1$ has bounded variation iff it can be represented as*

$$r = c_+F_+ - c_-F_- \tag{B.2.3}$$

*where $c_\pm \geqq 0$ and $F_\pm$ are certain distribution functions.* □

For all functions of bounded variation we can introduce the function

$$r^*(t) := \int_{-\infty}^{\infty} e^{itx} \, dr(x) . \tag{B.2.4}$$

**Definition B.2.2.** The function $r^*$ is called the *Fourier-Stieltjes transform* of $r$. □

From (B.2.3) and (B.2.4) we directly get the following simple fact.

**Corollary B.2.3.** *We have the representation*

$$r^* = c_+f_+ - c_-f_- \tag{B.2.5}$$

*in terms of the characteristic functions $f_\pm$ of $F_\pm$.* □

In view of our considerations in Sect. 1.4 it is now clear what we mean by the *convolution* $r_1 * r_2$ of two functions of bounded variation; namely, we put

$$r_1 * r_2(x) := \int_{-\infty}^{\infty} r_1(x - u) \, r_2(du) . \tag{B.2.6}$$

Now we find by Cor. 1.4.3 and the representation (B.2.3)

$$r_1 * r_2 = r_2 * r_1 .$$

The following lemma can easily be established.

**Lemma B.2.4.** *The convolution $r_1 * r_2$ of two functions $r_1$ and $r_2$ of bounded variation is a function of bounded variation.* □

**Proposition B.2.5.** *Let $r_1$ and $r_2$ be two functions of bounded variation and set $r = r_1 * r_2$. Then we have for the corresponding Fourier-Stieltjes transforms*

$$r^* = r_1^* r_2^* .$$

Proof. In view of L. B.2.4 the transform $r^*$ exists. Now the assertion can be directly verified by (B.2.5) and the convolution theorem for characteristic functions. □

In Sect. 1.1 we defined the left extremity lext $(F)$ of a distribution function $F$. Now we carry this definition over to functions $r$ of bounded variation putting

$$\text{lext}\,(r) := \sup \{x: r(y) = 0, y < x\} ;$$

we write lext $(r) = -\infty$ if the set $\{y: r(y) = 0, y < x\}$ is empty for all $x \in R_1$.

**Definition B.2.6.** Let $r$ be a function of bounded variation such that $l :=$ lext $(r) > -\infty$, and put $\sigma = u + iv$. Then the function

$$\tilde{r}(\sigma) := \int_l^\infty e^{-\sigma x}\, dr(x) , \qquad u \geqq 0 , \tag{B.2.7}$$

is called the *Laplace-Stieltjes transform* of $r$. □

Obviously $\tilde{r}(-it) = r^*(t)$. In view of (B.2.3) we can now apply the integral representation (Prop. 4.2.3) and the ridge property of characteristic functions analytic in a strip to obtain the following result.

**Proposition B.2.7.** *The Laplace-Stieltjes transform $\tilde{r}$ of $r = c_+F_+ - c_-F_-$ is analytic in $\{s: u > 0\}$ and continuous in $\{s: u \geq 0\}$ satisfying*

$$|\tilde{r}(u + iv)| \leqq (c_+ + c_-) \exp(-ul) , \qquad u \geqq 0 . \quad \square \tag{B.2.8}$$

Laplace-Stieltjes transforms are very applicable by virtue of the uniqueness and the convolution theorems. We formulate them as follows.

**Theorem B.2.8.** *Distinct functions of bounded variation with finite left extremum have distinct Laplace-Stieltjes transforms.* □

**Theorem B.2.9.** *Let $r_1$ and $r_2$ be two functions of bounded variation with finite left extremities and put $r = r_1 * r_2$. Then for the corresponding Laplace-Stieltjes transforms $\tilde{r}$, $\tilde{r}_1$, $\tilde{r}_2$ we have*

$$\tilde{r}(\sigma) = \tilde{r}_1(\sigma)\, \tilde{r}_2(\sigma) , \qquad u \geqq 0 .$$

Proof. The function $d := \tilde{r} - \tilde{r}_1\tilde{r}_2$ is analytic in $\{\sigma = u + iv: u > 0\}$ and continuous in $\{\sigma = u + iv: u \geqq 0\}$. By Prop. B.2.5 it vanishes on the imaginary axis so that $d = 0$ follows like in the proof of Cor. B.1.4. □

For details on Laplace-Stieltjes transforms see Widder (1946).

## B.3. The growth of an analytic function

Let us consider a domain $D$ the boundary $\partial D$ of which we take to be a simply closed contour.

**Proposition B.3.1** (Maximum modulus principle). *If $w$ is a non-constant function analytic in $D$ continuous in $D \cup \partial D$, then there exists $z_0 \in \partial D$ such that $\sup_{z \in D} |w(z)| = |w(z_0)|$. Moreover, $|w(z)| < |w(z_0)|$ for all $z \in D$.* □

Now we put $D := \{z\colon |z| < R\}$ for some $R > 0$. Then the expression

$$M(w, r) := \max_{|z| \leq r} |w(z)|$$

is defined for $0 \leqq r < R$; it is called the *maximum modulus of $w$ in the circle* $\{z\colon |z| \leqq r\}$. From the maximum modulus principle it follows that $M$ is strictly increasing provided that $w$ does not reduce to a constant.

If $w \neq$ const is integral, then $M$ is defined for all $r \geqq 0$ and by Liouville's theorem it increases indefinitely as $r \to \infty$. The growth of $M$ gives us some information about the behaviour of $w$. It is natural to describe it by comparing $M$ with a simple function like $\exp(\tau r^\varrho)$, $\tau > 0$, $\varrho > 0$. This motivates the

**Definition B.3.2.** The order $\varrho$ $(0 \leqq \varrho \leqq \infty)$ of an integral function $w$ is given by

$$\varrho := \overline{\lim_{r\to\infty}} \frac{\log\log M(w, r)}{\log r}. \tag{B.3.1}$$

In the case that $0 < \varrho < \infty$ the function $w$ is said to be of type $\tau$ $(0 \leqq \tau \leqq \infty)$ if

$$\tau := \overline{\lim_{r\to\infty}} \frac{\log M(w, r)}{r^\varrho}. \quad \square \tag{B.3.2}$$

For $0 < \varrho < \infty$ this means that to every $\eta > 0$ there corresponds $r_0(\eta)$ such that

$$M(w, r) < \exp(r^{\varrho+\eta}), \qquad r \geqq r_0, \tag{B.3.3}$$

while there exists a sequence $r_i(\eta) \to \infty$ such that

$$M(w, r_i) > \exp(r_i^{\varrho-\eta}) \qquad i = 1, 2, \ldots$$

For $0 < \tau < \infty$ we have obviously an analogous interpretation; $\tau = 0$ means that for every $\varepsilon > 0$

$$M(w, r) < \exp(\varepsilon r^\varrho)$$

for all sufficiently large $r$.

As an example we mention the function $\exp(\tau z^k)$ with $\tau > 0$ and $k$ a natural number; it has order $k$ and type $\tau$. Polynomials obviously have order 0.

We need the following results on order and type of an integral function $w$.

**Proposition B.3.3.** *A necessary and sufficient condition that*

$$w(z) = \sum_{k=0}^{\infty} a_n z^n$$

*should be an integral function of finite order $\varrho \geqq 0$ is that*

$$\underline{\lim_{n\to\infty}} \frac{\log(|a_n|^{-1})}{n \log n} = \frac{1}{\varrho}.$$

*If $w$ has order $\varrho > 0$ the type $\tau$ is given by*

$$(\tau \mathrm{e} \varrho)^{1/\varrho} = \overline{\lim_{n\to\infty}} n^{1/\varrho} \sqrt[n]{|a_n|}. \qquad \square$$

We state next an easy consequence of these formulas.

**Corollary B.3.4.** *An integral function and its derivative have the same order $\varrho \geqq 0$ and, if $0 < \varrho < \infty$, the same type $\tau$.* $\square$

We turn now to the Carathéodory inequality. This is a result which directly follows from the maximum modulus principle, see TITCHMARSH (1932). It enables us to deduce an upper bound for the modulus of a function $w$ on a circle $\{z: |z| = r\}$ from bounds for its real part valid on a larger concentric circle $\{z: |z| = R\}$, $R > r$.

**Proposition B.3.5** (CARATHÉODORY). *Let $w$ be a function analytic in $\{z: |z| < R\}$ and continuous in $\{z: |z| \leqq R\}$ and set*

$$A(w, r) := \max_{|z| \leqq r} \operatorname{Re} w(z) .$$

*Then we have, for $0 < r < R$,*

$$M(w, r) \leqq \frac{R + r}{R - r} \left(A(w, R) + |w(0)|\right) . \qquad \square$$

As an application we derive a special case of Hadamard's factorization theorem for integral functions of finite order; we need it in the proof of Cramér's theorem.

**Proposition B.3.6.** *An integral function $w$ without zeros and with a finite order permits the representation $w = \exp(P_m)$ where $P_m$ stands for a polynomial of a certain degree $m$.*

Proof. Since $w$ has a finite order there exists a number $C > 0$ and a natural number $n$ such that

$$|w(z)| \leqq C \exp(|z|^n) \quad \text{for all } z .$$

Thus we have for the integral function $v := \log w$ (principal determination vanishing at 0 if $w(0) = 1$)

$$\operatorname{Re} v(z) = \log |w(z)| \leqq |z|^n + \log C .$$

Hence Prop. B.3.5 tells us for $R = 2r$ that $M(v, r) = O(|r|^n)$, as $r \to \infty$. Now $v = P_m$ follows from a well-known characteristic property of polynomials where $m$, $0 \leqq m \leqq n$, is an integer. $\square$

The growth of functions $w$ analytic in an angle $A_{ab} := \{z = r\,e^{i\varphi}: a < \varphi < b\}$ can also be described by means of order and type using the maximum modulus

$$M_{ab}(w, r) := \max_{|z| \leqq r,\, z \in A_{ab}} |w(z)| .$$

We have only to take a precaution due to the fact that a non-constant function $w$ analytic in $A_{ab}$ can be bounded.

**Definition B.3.7.** Let the function $w$ be analytic in the angle $A_{ab}$. If it is unbounded then its order and type are defined by the formulas (B.3.1) and (B.3.2), resp., on replacing $M$ by $M_{ab}$. Bounded functions $w$ have order 0. $\square$

## B.4. Estimates of the remainder term of the exponential function

**Proposition B.4.1.** *For the functions*

$$r_n(z) := e^{iz} - 1 - \dots - \frac{(iz)^{n-1}}{(n-1)!}, \qquad n = 1, 2, \dots, \qquad z = t + is$$

*we have*

$$|r_n(z)| \leqq \begin{cases} |z^n|/n!, & z \in S[0, \infty), \\ \exp(|z|)\,|z|^n/n!, & z \in S(\infty, 0]. \end{cases}$$

**Proof.** Obviously we have

$$r_1(z) = \mathrm{i}z \int_0^1 \mathrm{e}^{\mathrm{i}wz}\,\mathrm{d}w = \mathrm{e}^{\mathrm{i}z} - 1\,.$$

Hence we get

$$|r_1(z)| = |z| \int_0^1 \exp(\varrho - ws)\,\mathrm{d}w\,.$$

Thus

$$|r_1(z)| \leqq \begin{cases} |z|\,, & 0 \leqq s \\ |z|\,\mathrm{e}^{|s|}\,, & s \leqq 0 \end{cases}$$

Generally we see by inspection that

$$r_n(z) = \mathrm{i}z \int_0^1 r_{n-1}(wz)\,\mathrm{d}w$$

whence the assertion follows by induction. □

We list a number of further estimates.

(i) *Function* $k_1(z, u) := \mathrm{e}^{\mathrm{i}zu} - 1 - \mathrm{i}zu = r_2(zu)$:

$$|k_1(z, u)| \leqq \mathrm{e}^{|su|} \frac{|z|^2}{2} u^2 \quad \text{for all complex } z \text{ and real } u \tag{B.4.1}$$

(follows from Prop. B.4.1 and $1 \leqq \mathrm{e}^{|su|}$).

$$|k_1(z, u)| \leqq \frac{|z|^2}{2} u^2\,, \qquad z \in S(-\infty, 0]\,, \qquad u < 0\,, \tag{B.4.2}$$

(follows from Prop. B.4.1 since $zu \in S[0, \infty)$).

$$k_1(\mathrm{i}s, u) \geqq 0 \quad \text{for all real } s, u\,. \tag{B.4.3}$$

**Proof.** The function $h(x) := \mathrm{e}^{-x} - 1 + x$ decreases for $x \leqq 0$ and increases for $x \geqq 0$. Accordingly, $k_1(\mathrm{i}s, u) = h(su) \geqq h(0) = 0$. □

$$\frac{k_1(\mathrm{i}s, u)}{su} \geqq \frac{1}{\mathrm{e}}\,, \qquad s > 0\,, \qquad u \geqq \frac{1}{s}\,. \tag{B.4.4}$$

**Proof.** The function $h(x)/x$ increases for $x > 0$. Hence

$$\frac{k_1(\mathrm{i}s, u)}{su} = \frac{h(su)}{su} \geqq h(1) = \frac{1}{\mathrm{e}}\,, \qquad su \geqq 1\,.$$ □

(ii) *Function* $k(z, u) := \mathrm{e}^{\mathrm{i}zu} - 1 - \dfrac{\mathrm{i}zu}{1 + u^2} = k_1(z, u) + \dfrac{\mathrm{i}zu^3}{\mathrm{i} + u^2}$:

$$k(\mathrm{i}s, u) \geqq 0\,, \qquad s > 0\,, \qquad u < 0\,. \tag{B.4.5}$$

**Proof.** By virtue of (B.4.3) we have

$$k(\mathrm{i}s, u) = k_1(\mathrm{i}s, u) - \frac{su^3}{1 + u^2} \geqq 0\,, \qquad su < 0\,.$$

## B.5. Stirling formula

We consider the *gamma-function*

$$\Gamma(x) = \int_0^\infty t^{x-1} e^{-t} dt\,, \qquad x > 0\,.$$

It satisfies the equation $\Gamma(x+1) = x\Gamma(x)$, so that $\Gamma(n+1) = n!$, $n = 1, 2, \ldots$ As is well known it possesses an analytic continuation to the complex plane and proves to be meromorphic. Its asymptotic behaviour is described in the following statement.

**Proposition B.5.1.** *We have*

$$\log \Gamma(z) = (z - \tfrac{1}{2}) \log z - z + \tfrac{1}{2} \log 2\pi + o(1)$$

*as* $|z| \to \infty$ *uniformly in every angle* $|\arg z| < d < \pi$. □

## B.6. Weierstrass convergence theorem

**Proposition B.6.1.** *Suppose that each member of the sequence of functions* $(h_1)_1^\infty$ *is analytic inside a domain* $D$ *and let the series* $\sum_{i=1}^\infty h_i$ *be uniformly convergent throughout every domain* $D' \subset D$. *Then the function*

$$h(z) := \sum_{i=1}^\infty h_i(z)\,, \qquad z \in D\,,$$

*is analytic inside* $D$, *and all its derivatives may be calculated by term-by-term-differentiation.* □

For details we refer the reader to Levin (1956), Rühs (1971), Titchmarsh (1932).

# C. Appendix on the theory of Phragmén and Lindelöf

## C.1. Purely analytic theory

The maximum modulus principle as stated in B.3 is violated in the example of the function

$$w(z) = \exp\left(\frac{1+z}{1-z}\right).$$

Namely, it is analytic in the open unit circle $C$ and continuous in the domain $\{z: |z| \leqq 1, z \neq 1\}$; moreover, we have $|w(e^{i\varphi})| = 1$, $\varphi \neq 0$. Thus all assumptions of the principle are satisfied with the only exception that $w$ is not defined at the point $z_1 = 1$, and the assertion is not true either: It is easily seen that $w(z)$ quickly tends to infinity when $z$ approaches this point from the interior of $C$.

In spite of this the maximum modulus principle can be generalized. Namely, it was the discovery of Phragmén and Lindelöf that an exceptional point $z_1 \in \partial D$ of some domain $D$ may be admitted where the function $w$ need not be defined. Even a certain slow growth of $|w|$ in $D$ is permissible when the variable tends to $z_1$; it depends on the configuration of $\partial D$ in the neighbourhood of $z_1$. This is illustrated by Cor. C.1.5, since the angle $\gamma$ determines the configuration of $\partial D$ at infinity on Riemann's sphere of complex numbers. As a conclusion it then turns out that $w$ is bounded in $D$. In the present appendix theorems of this type will be called PL-theorems. We give an example where $D_+$ stands for the semi-disk

$$D_+ := \{z = t + is: s < 0, |z| < 1\}$$

and $z_1 = 0$.

**Theorem C.1.1.** *Assume that $h$, $h \neq 0$, and $w$ are functions analytic in $D_+$ and continuous in $D_+ \cup \partial D_+$ except for the point 0 where they need not exist. Suppose further that*

$$\begin{aligned} |h| \leqq 1\,, \qquad |w| \leqq 1 \quad &\text{on} \quad \partial D_+ \setminus \{0\} \\ |h| \leqq 1\,, \qquad |hw| \leqq 1 \quad &\text{in} \quad D_+\,. \end{aligned} \tag{C.1.1}$$

*If*

$$\overline{\lim_{s\to 0}}\, s \log |h(is)| = 0 \tag{C.1.2}$$

*then $|w| \leqq 1$ also in $D_+$.* □

An elementary proof for this result has not yet been found, and hence we cannot give a proof here. Th. C.1.1 is equivalent to the last statement of Th. C.1.14 in view of the conformal mapping

$$z' = (1 - z)(1 - z^{-1})/4$$

which maps the half plane $S(0, \infty)$ to $D_+$. Th. C.1.1 is rather trivial if $|h|$ is bounded away from 0, cf. Cor. C.1.4. But if $|h(z)| \to 0$ as $z \to 0$, $z \in D_+$, then condition (C.1.1) admits an indefinite growth of $|w(z)|$. It is the point of the condition (C.1.2) that $|h(iy)|$ does not decrease too quickly; namely, in view of (C.1.1), this implies a restriction on the growth of $|w(iy)|$. The assertion means that unbounded functions $w$ with such a slow growth do not exist.

Th. C.1.1 is the only PL-theorem valid for a bounded region $D_+$ which we need in the present book. Namely, it is a salient feature of the maximum modulus principle modified in such a way that it can be extended to unbounded domains $D$. Then the point $\infty$ takes the role of the exceptional point.

In particular we need results of this kind which are valid in the upper half plane $S(0, \infty) = \{z = t + is\colon s \in (0, \infty)\}$. Let us therefore introduce the sets of functions

$$M_+ := \left\{\begin{matrix} w\colon \text{analytic in } S(0, \infty), \text{ continuous in } S[0, \infty), \\ |w(t)| \leqq 1 \quad \text{for} \quad t \in R_1 \end{matrix}\right\}$$

$$N_+ := \{w \in M_+\colon |w(z)| \leqq 1 \text{ for } z \in S(0, \infty)\} .$$

The sets $M_-$ and $N_-$ are defined analogously with $S(0, \infty)$ and $S[0, \infty)$ replaced by $S(-\infty, 0)$ and $S(-\infty, 0]$, resp.

**Example C.1.2.** $w(z) := \exp(-i\tau z) \in M_+$ for any $\tau \in R_1$. Here we have $w \in N_+$ for $\tau \leqq 0$, but $w \notin N_+$ for $\tau > 0$. □

We will see that unbounded functions with a growth slightly less than that of the exponential function do not exist in $M_+$. In general we have to seek conditions restricting the growth of $w \in M_+$ such that $w \in N_+$ follows. According to the intention of the present book we will prove a number of very elementary results of this type.

We start with a result which can be considered a key theorem of the elementary PL-theory. We will see how the problem can be reduced to the maximum modulus principle in an elegant way.

**Theorem C.1.3.** *If $w \in M_+$ is of order at most $\varrho < 1$, then $w \in N_+$.*

Proof. We choose a number $g$ such that $\varrho < g < 1$. Then by the argument which led us to (B.3.3) we see that there is a number $r_0 = r_0(g, \varrho)$ such that for $\eta := g - \varrho$

$$\log |w(r\,e^{i\varphi})| \leqq r^{\varrho+\eta/2}, \qquad 0 \leqq \varphi \leqq \pi, \qquad r \geqq r_0 .$$

For an arbitrary $\varepsilon > 0$ we have trivially $(\varepsilon/2)\, r^{\eta/2} \cos(\pi g/2) \geqq 1$ for all sufficiently large $r$. Thus it follows that

$$\log |w(r\,e^{i\varphi})| \leqq \frac{\varepsilon}{2} r^g \cos\left(\frac{\pi g}{2}\right), \qquad 0 \leqq \varphi \leqq \pi, \qquad r \geqq r_1(\varepsilon, g, \varrho) .$$

Now we introduce the auxiliary function

$$W_\varepsilon(z) := w(z) \exp\left(-\varepsilon(-iz)^g\right) \tag{C.1.3}$$

and note that

$$\operatorname{Re}(-ir\,e^{i\varphi})^g = \operatorname{Re}\left(r^g\,e^{i(\varphi-(\pi/2))g}\right) = r^g \cos\left(\left(\varphi - \frac{\pi}{2}\right)g\right) > 0, \qquad 0 \leqq \varphi \leqq \pi,$$

because $g < 1$. Accordingly, $|W_\varepsilon(t)| \leqq |w(t)| \leqq 1$, i.e. $W_\varepsilon \in \boldsymbol{M}_+$. For $0 \leqq \varphi \leqq \pi$ we have by (C.1.3)

$$\log |W_\varepsilon(r\,e^{i\varphi})| = \log |w(r\,e^{i\varphi})| - \varepsilon r^g \cos\left(\left(\varphi - \frac{\pi}{2}\right) g\right)$$
$$\leqq \frac{\varepsilon}{2} r^g \cos\left(\frac{\pi g}{2}\right) - \varepsilon r^g \cos\left(\frac{\pi g}{2}\right) < 0\,, \qquad r \geqq r_1\,.$$

Hence

$$|W_\varepsilon(r\,e^{i\varphi})| \leqq 1\,, \qquad 0 \leqq \varphi \leqq \pi\,, \qquad r \geqq r_1\,.$$

By the maximum modulus principle it follows that

$$|W_\varepsilon(z)| \leqq 1 \quad \text{for} \quad z = r\,e^{i\varphi}\,, \qquad 0 \leqq r \leqq r_1\,, \qquad 0 \leqq \varphi \leqq \pi\,,$$

so that $W_\varepsilon \in \boldsymbol{N}_+$ for arbitrary $\varepsilon > 0$. Thus we obtain $|w(z)| \leqq \exp\left(\varepsilon r^g \cos\left(\left(\varphi - \frac{\pi}{2}\right) g\right)\right)$.
Now the desired result follows as $\varepsilon \to 0$. □

As a very simple, but important, consequence we state

**Corollary C.1.4.** *If $w \in \boldsymbol{M}_+$ is bounded, then $w \in \boldsymbol{N}_+$.* □

The idea of the proof also yields an analogous statement for arbitrary angles where the order is given in Def. B.3.7. It contains Th. C.1.3 as a particular case.

**Corollary C.1.5.** *Let $w$ be a function which is analytic in the angle*

$$A = \left\{z = r\,e^{i\varphi} : 0 < |z| < \infty,\ \left|\left(\varphi - \frac{\pi}{2}\right)\right| \leqq \frac{\gamma}{2}\right\}, \tag{C.1.4}$$

*and continuous in $A \cup \partial A$. If $w$ possesses at most the order $\varrho < \pi/\gamma$ and is bounded by 1 on the boundary $\partial A$, then $|w| \leqq 1$ also in $A$.*

1. **Proof.** We consider the function (C.1.3) in $A$ and put $\varrho < g < \pi/\gamma$. Then $W$ can be seen to be bounded by 1 in the same way as above since in $A$ we have $\cos\left(\left(\varphi - \frac{\pi}{2}\right) g\right)$ $\geqq \cos(\gamma g/2) > 0$.

2. **Proof.** For the following it is convenient to carry out the proof by the mapping of the half plane $S(0, \infty)$ to $A$ which is performed by the transform

$$-iz' = (iz)^{\gamma/\pi}\,, \qquad z' = r'\,e^{i\varphi'}\,.$$

Thus $w(z) = w\left((-iz')^{\pi/\gamma}\,e^{-i(\pi/2)}\right) = w'(z')$, say, and we obtain

$$M_{0\pi}(w, r) = \max_{0 \leqq \varphi \leqq \pi} |w(r\,e^{i\varphi})| = \max_{|\varphi' - (\pi/2)| < \gamma/2} |w'(r'\,e^{i\varphi'})| = M_{ab}(w', r')$$

where $a, b = \pi/2 \pm \gamma/2$. Now it follows that

$$\varrho = \overline{\lim_{r\to\infty}} \frac{\log\log M_{0\pi}(w, r)}{\log r} = \frac{\pi}{\gamma} \overline{\lim_{r'\to\infty}} \frac{\log\log M_{ab}(w', r')}{\log r'}\,.$$

Hence we see that the function $w$ defined in $S(0, \infty)$ has order $\varrho < 1$ iff the function $w'$ defined in $A$ has order $\varrho' < \gamma/\pi$. Thus Cor. C.1.5 is equivalent to Th. C.1.3. □

**Example C.1.6.** The function $w(z) = \exp\left((-iz)^\varrho\right)$, $\varrho = \pi/\gamma$ is unbounded in the angle (C.1.4), but bounded on $\partial A$. This means that in the restriction $\varrho < \pi/\gamma$ made in Cor. C.1.5 and Th. C.1.3 (here $\gamma = \pi$) we cannot replace $<$ by $=$. For $\gamma = \pi$ we have the Example C.1.2. □

In spite of this we can generalize Th. C.1.3, and for this purpose we introduce the symbol

$$k_w := \overline{\lim_{s\to\infty}} \frac{\log |w(\mathrm{i}s)|}{s}. \tag{C.1.5}$$

Cor. C.1.12 will tell us that all conditions implying $w \in \boldsymbol{N}_+$ also guarantee that $k_w$ is finite. At the moment we admit the possibility $k_w = -\infty$. It is characteristic for the following proof that we need a decomposition of $S(0, \infty)$ into its two quadrants $Q_1$ and $Q_2$.

**Proposition C.1.7.** *Suppose that $w \in \boldsymbol{M}_+$, $w \neq 0$, has at most order $\varrho < 2$. Then*
(i) *$w \in N_+$ if $k_w \leqq 0$*
(ii) *$w$ has order 1 and type $k_w$ if $k_w > 0$.*

Proof. (i) We consider the function

$$w_A(z) := w(z)\, \mathrm{e}^{\mathrm{i}Az} \tag{C.1.6}$$

for an arbitrary $A > 0$ and note that in view of $k_w \leqq 0$ we have

$$|w_A(\mathrm{i}s)| = \exp\left(s\left(\frac{\log|w(\mathrm{i}s)|}{s} - A\right)\right) \to 0\,, \quad \text{as} \quad s \to \infty\,. \tag{C.1.7}$$

Hence $|w_A(\mathrm{i}s)|$, $s \geqq 0$, is bounded, and using an appropriate number $m_A \geqq 1$ we find that $\tilde{w}_A := w_A/m_A$ satisfies the assumptions of Cor. C.1.5 (with $\gamma = \pi/2$) both in $Q_1$ and in $Q_2$. Accordingly, we have $\tilde{w}_A \in \boldsymbol{N}_+$; now $w_A \in \boldsymbol{N}_+$ follows from Cor. C.1.4, so that

$$|w(z)| \leqq \exp\,(As)\,. \tag{C.1.8}$$

Thus the assertion becomes obvious as $A \to 0$.

(ii) Now let $k_w > 0$. Putting $A = k_w$ we see that the function (C.1.6) satisfies the assumptions of part (i). Accordingly, $w_A \in \boldsymbol{N}_+$, that is $|w(z)| \leqq \mathrm{e}^{k_w s}$, $s \geqq 0$. Thus we conclude that

$$M_+(w, r) = \max_{0 \leqq \varphi \leqq \pi} |w(r\,\mathrm{e}^{\mathrm{i}\varphi})| \leqq \mathrm{e}^{k_w r}\,, \qquad r \geqq 0\,.$$

On the other hand, by the definition of $k_w$ there exists a sequence $r_\nu \to \infty$ such that

$$M_+(w, r_\nu) \geqq w(\mathrm{i}r_\nu) \geqq \mathrm{e}^{(k_w - \varepsilon) r_\nu}\,.$$

Combining the last two estimates we see that $w$ has order 1 and, moreover,

$$\overline{\lim_{r\to\infty}} \frac{\log M_+(w, r)}{r} = k_w > 0\,. \qquad \square$$

The above result implies a well-known fact.

**Corollary C.1.8.** *Functions $w \in \boldsymbol{M}_+$ with an order $\varrho$, $0 < \varrho < 1$, or of order 1 and type 0, do not exist.* $\square$

The reason is that such functions would possess the property $k_w = 0$. The method permits a further generalization for which we have to map Prop. C.1.7 to the quadrant $Q_1$. This yields

**Lemma C.1.9.** *Let the function $w$, $w \neq 0$, be analytic in the open domain $Q_1$ and continuous in $Q_1 \cup \partial Q_1$, $|w| \leqq 1$ on $\partial Q_1$. If it has at most order $\varrho < 4$ and satisfies*

$$\overline{\lim_{r\to\infty}} \frac{\log w(r \exp(i\pi/4))}{r^2} \leqq 0, \tag{C.1.9}$$

*then $w$ is bounded in $Q_1$.* □

Now we turn to the previously mentioned generalization.

**Proposition C.1.10.** *If $w \in \boldsymbol{M}_+$, $w \neq 0$, has at most order $\varrho < 4$ and satisfies $k_w \leqq 0$,* (C.1.9), *and*

$$\overline{\lim_{r\to\infty}} \frac{\log w(r \exp(3i\pi/4))}{r^2} \leqq 0,$$

*then $w \in \boldsymbol{N}_+$.*

Proof. As in the proof of Prop. C.1.7 we note that $w_A$ is bounded by 1 on the sides of $Q_1$ and $Q_2$. Hence L. C.1.9 applies in $Q_1$ and $Q_2$ and yields $w_A \in \boldsymbol{N}_+$, whence $w \in \boldsymbol{N}_+$ follows as above. □

The next corollary is by far weaker than Prop. C.1.10, but it is of interest for its simplicity.

**Corollary C.1.11.** *If $w \in \boldsymbol{M}_+$, $w \neq 0$, has either order $\varrho = 2$ and type $\tau = 0$ or an order $\varrho < 2$ and if, moreover, $k_w \leqq 0$, then $w \in \boldsymbol{N}_+$.* □

Our method obviously permits us to prove an infinite sequence of theorems, each of which is a generalization of the preceding one. For the consequences of this fact see App. C.3. We direct the reader's attention to the fact that Cor. C.1.11 immediately applies to integral d.f. We refer the reader to L. 6.4.1.

Next we turn to a remark on the symbol $k_w$ introduced in (C. 1.5).

**Corollary C.1.12.** *For functions $w$, $w \neq 0$, defined on $\mathbb{R}_1$ and continuable to the half plane $S(0, \infty)$ satisfying*

$$|w(z)| \leqq c\,|z|^n, \qquad z \in S[0, \infty),$$

*for some $c > 0$ and a certain integer $n \geqq 0$, that is, in particular, for $w \in \boldsymbol{N}_+$, we have $k_w > -\infty$.*

Proof. For $w \neq 0$ we introduce the function

$$w'(z) := w(z)\, M^{-1}(z + i)^{-n}.$$

By assumption we can choose $M > 0$ such that $w' \in \boldsymbol{N}_+$. Assuming $k_{w'} = -\infty$ we reconsider the proof of Prop. C. 1.7 (i) (with $w'$ instead of $w$). Then (C. 1.7) is true even for an arbitrary $A < 0$, and as in that proof we obtain $w_A \in \boldsymbol{N}_+$, i.e.,

$$w'(z)\, e^{-As} \leqq 1, \qquad s \geqq 0.$$

On letting $A \to -\infty$ we obtain $w' = 0$, a contradiction. □

Next we apply Th. C. 1.3 to obtain an elementary result of a different type. For probabilistic applications see also Linnik and Ostrovskij (1972) and Lukacs (1970).

**Proposition C.1.13** (Ostrovskij). *Suppose that $w \in \boldsymbol{M}_+$, $w \neq 0$, satisfies the conditions $k_w \leqq 0$ and*

$$|w(z)| \leqq M \exp(ds^2), \qquad s \geqq 0, \tag{C. 1.10}$$

*where $M$ and $d$ are certain positive numbers. Then $w \in \boldsymbol{N}_+$.*

Proof. From $k_w \leqq 0$ we see in the same way as above that there exists a number $m_A \geqq 1$ such that the function $w_A$ defined in (C. 1.6) is bounded by $m_A$ on the ray $l_+ := \{z = t + is : s \geqq 0, t = 0\}$. All that we have to do is prove that $w_A \in N_+$ for all $A > 0$.

With this aim in mind we select a point $z_0 = t_0 + is_0$ such that $t_0 > 0$, $s_0 > 0$. Let $A_c$ be the angle with the sides $l_+$ and $l_c := \{z = t + is : t \geqq 0, s = t/2c\}$; we choose $c > 0$ such that $z_0 \in A_c$. Now we introduce the function

$$w_{Ac}(z) := w_A(z) \exp(idz^2/4c),$$

so that $|w_{Ac}(is)| = |w_A(is)| \leqq m_A$, $s \geqq 0$. By (C. 1.10) we also have

$$|w(r e^{i\varphi})| \leqq M \exp(dr^2 \sin^2 \varphi), \qquad z = r e^{i\varphi} \in A_c, \tag{C. 1.11}$$

so that the function $w_{Ac}$ has at most order 2 in $A_c$. Now we obtain

$$|w_{Ac}(z)| \leqq |w(z)| \, |\exp(idz^2/4c)| \leqq M \exp(ds^2 - tsd/2c) = M, \qquad z \in l_c.$$

Hence an easy application of Cor. C. 1.5 (with $\gamma < \pi/2$) tells us that

$$|w_{Ac}(z)| \leqq \max\{m_A, M\}, \qquad z \in A_c.$$

Since this bound does not depend on $c$ it is true also in the quadrant $Q_1$. An analogous argument yields the same estimate in $Q_2$. Now it is Cor. C. 1.4 that leads us to the desired statement $w_A \in N_+$. □

Finally, we state without proof two non-elementary theorems which permit noteworthy probabilistic applications.

**Theorem C.1.14.** *Assume that $h \in N_+$, $h \neq 0$, and let there exist a function $w \in M_+$, such that*

$$hw \in N_+ .$$

*Then either $w \in N_+$ or $w$ belongs to a positive type $\tau$ of order 1, $\tau \leqq |k_h|$. If $k_h = 0$, then $w \in N_+$.* □

**Example C.1.15.** Let $h(z) = e^{iz} \in N_+$, $w(z) = e^{-iz} \notin N_+$, so that $hw = 1 \in N_+$; this shows that the condition $k_h = 0$ is essential for the conclusion $w \in N_+$. □

**Remark C.1.16.** Th. C.1.14 is particularly useful for probability theory since $h$ may be a c.f. whose d.f. $H$ satisfies lext $(H) = 0$. Namely, it is seen from Cor. 4.4.2 that this property is equivalent to $k_h = 0$. A number of applications is given in C. 2. □

**Proposition C.1.17.** *Assume that $f_\pm \in N_\pm$, $f_\pm(0) > 0$, $k_{f_\pm} = 0$. Consider an integral function $w$ without zeros satisfying*

$$\frac{wf_+}{d} \in N_+, \qquad \frac{wf_-}{d} \in N_-$$

*for some $d > 0$. Then $w =$* const. □

Note that this proposition follows directly from theorem C. 1.14 if we add the assumption $|w(t)| \leqq 1$; so the point of this result is that we do not in fact need this additional assumption. For an application see Prop. C. 2.5.3.

## C.2. Applications

We are going to make use of the above results and prove a number of analytic theorems which we need in the main part of the present book. In this way we introduce the reader to the reasoning typical for this field of analysis; moreover, we avoid disrupting the probabilistic discussions in the text. Finally we review some probabilistic applications of the PL-theory which do not fit into the framework of this book. We recall that the symbols $M_\pm$ and $N_\pm$ are explained in C. 1.

### C.2.1. A continuation problem

**Proposition C.2.1.1.** *Assume that* $P_+, Q_+, R_+ \in N_+$ *and* $R_- \in N_-$. *Let, moreover,* $Q_+/P_+$ *be a function analytic in* $S(0, \infty)$ *and continuous in* $S[0, \infty)$. *If the equation*

$$P_+(t)\left(R_+(t) + R_-(t)\right) = Q_+(t), \qquad t \in R_1, \tag{C. 2.1.1}$$

*is satisfied then* $R_-$ *is continuable to* $S(0, \infty)$ *so that it represents an integral function which is either a constant or has order* 1 *and type* $\tau$, $0 < \tau \leqq |k_{P_+}|$ *where*

$$k_{P_+} = \overline{\lim_{s\to\infty}} \frac{\log |P_+(is)|}{s}.$$

Proof. We define the function

$$A(z) := Q_+(z)/P_+(z) - R_+(z), \qquad z \in S[0, \infty),$$

which obviously satisfies

$$A(t) = R_-(t), \qquad t \in R_1.$$

In view of Prop. B.1.5 the function $A$ is the continuation of $R_-$ in $S(0, \infty)$. Accordingly, $R_-$ is integral, and we may write

$$P_+(z)\, R_-(z) = Q_+(z) - P_+(z)\, R_+(z), \qquad z \in S[0, \infty). \tag{C.2.1.2}$$

This means that equation (C.2.1.1) can be continued to $S[0, \infty)$, and we see that $P_+R_- \in N_+$. Putting $h = P_+$ and $w = R_-$ we can apply Th. C.1.14; from this the assertions follow. □

**Corollary C.2.1.2.** *If we know that* $R_- \in N_-$ *is integral then the assumption on the ratio* $Q_+/P_+$ *can be dropped.*

Proof. Instead of $A$ we define the function

$$B(z) := P_+(z)\left(R_+(z) + R_-(z)\right) - Q_+(z), \qquad z \in S[0, \infty),$$

which vanishes for $z \in R_1$ by (C.2.1.1). Reasoning as in the proof of Cor. B.1.4 we get $B = 0$, i.e., equation (C.2.1.2), whence the assertions follow as before. □

### C.2.2. Transforms of functions of bounded variation

In this section the notation $r$ (with or without affixes) stands for a function of bounded variation, $T(-\infty, \infty) \leqq 1$, and we consider its Fourier-Stieltjes transform $r^*$. If $l = \text{lext}\,(r) > -\infty$ then the Laplace-Stieltjes transform $\tilde{r}$ also exists, by Prop. B.2.7;

this means that $r^*$ is continuable to $S(0, \infty)$ in the form

$$r^*(z) = \int_l^\infty e^{izx} \, dr(x) , \qquad z \in S(0, \infty) , \tag{C.2.2.1}$$

so that $r^* \in \boldsymbol{M}_+$. We adhere to the notation (C.1.5).

**Proposition C.2.2.1.** *The transform* $r^* \neq 0$ *is continuable to* $S(0, \infty)$ *as a function which is bounded or has order* 1 *and type,* $0 < \tau < \infty$, *iff* $r$ *is bounded on the left. In either case we have*

$$k_{r*} = -\operatorname{lext}(r); \tag{C.2.2.2}$$

*finally, if* $\operatorname{lext}(r) < 0$, *then the order equals* 1 *and* $\tau = k_{r*}$.

Proof. If $l = \operatorname{lext}(r) > -\infty$ we have remarked just now that $r^*$ is continuable in the form (C.2.2.1). In view of (B.2.8) we see that (for a certain $c > 0$)

$$r^*(z) \leqq c \, e^{-sl} , \qquad z \in S(0, \infty) , \tag{C.2.2.3}$$

so that $r^*$ has an order $\leq 1$. From Cor. C.1.8 we see that $r^*$ is then either bounded or has a positive type of order 1 as asserted.

For the converse we first conclude from the assumption (in view of Cor. C.1.12) that the limit

$$k_{r*} = \overline{\lim_{s \to \infty}} \frac{\log |r^*(is)|}{s}$$

exists finitely. Now we introduce the function

$$w(t) := r^*(t) \, e^{ik_{r*}t} = \int_{-\infty}^{-0} e^{itu} \, dr(u - k_{r*}) + \int_{-0}^{\infty} e^{itu} \, dr\,(u - k_{r*}) = I_1(t) + I_2(t)$$

say. $I_1$ is trivially continuable to $S(-\infty, 0)$ and represents a bounded function here, while $I_2$ has the same properties in $S(0, \infty)$. From the assumption it is clear that $w$ is also continuable to $S(0, \infty)$ and increases not faster than a function of order 1 and a finite type. Obviously, we have $k_w = 0$ so that Prop. C.1.7 tells us that $w \in \boldsymbol{N}_+$. Now we see from Prop. B.1.5 that $I_1$ can be continued to $S(0, \infty)$ in the form $I_1 = w - I_2$ so that $I_1$ is an integral function which is bounded everywhere. Thus $I_1 = \text{const.}$, i.e., $r(u - k_{r*}) = 0$, $u < 0$, by an obvious counterpart of Th. B.2.8. Accordingly, $r$ is bounded to the left, as asserted.

Further, it follows that $k_{r*} \geqq -\operatorname{lext}(r)$. On the other hand, we see from (C.2.2.3) that $k_{r*} \leqq -\operatorname{lext}(r)$ whence (C.2.2.2) is obvious.

The very last assertion is an immediate consequence of Prop. C.1.7. □

From this we immediately obtain a well-known result where the right extremum of $r$ is defined by

$$\operatorname{rext}(r) = \inf \{x : r(y) = \text{const}, y > x\} .$$

**Proposition C.2.2.2** (Pólya). *Assume that* $r^*$ *is continuable to the complex plane, belonging to the type* $\tau > 0$ *of order* 1 *in* $S(0, \infty)$ *and to the type* $\sigma > 0$ *of the same order in* $S(-\infty, 0)$. *Then* $\operatorname{lext}(r) = -\tau$, $\operatorname{rext}(r) = \sigma$. □

Next we turn to convolutions.

**Proposition C.2.2.3.** *Consider two functions $r_1$ and $r_2$ of bounded variation as before and assume that* lext $(r_1) = 0$,

$$r_1^*(z) := \int_{-0}^{\infty} e^{izu}\, dr_1(u) \neq 0\,, \qquad z \in S[0, \infty)\,, \tag{C.2.2.4}$$

$$r(x) := \int_{-\infty}^{\infty} r_1(x-u)\, r_2(du) = 0\,, \qquad x < \tau\,, \tag{C.2.2.5}$$

*for some $\tau \in R_1$. Then $r_2(u) = 0$, $u < \tau$.*
Proof. Let be $\tau = 0$. We introduce the transforms

$$r_{2-}^*(z) := \int_{-\infty}^{-0} e^{izu}\, dr_2(u)\,, \qquad z \in S(-\infty, 0]\,,$$

$$r_{2+}^*(z) := \int_{-0}^{\infty} e^{izu}\, dr_2(u)\,, \qquad z \in S[0, \infty)\,,$$

$$r^*(z) := \int_{-0}^{\infty} e^{izu}\, dr(u)\,, \qquad z \in S[0, \infty)\,,$$

so that

$$r_2^*(t) := \int_{-\infty}^{\infty} e^{itu}\, dr_2(u) = r_{2-}^*(t) + r_{2+}^*(t)\,,$$

$$r_{2\pm}^* \in N_\pm\,, \qquad r_1^* \in N_+\,, \qquad r^* \in N_+\,.$$

By the convolution theorem (Prop. B.2.5) we have

$$r^*(t) = r_1^*(t)\,\big(r_{2-}^*(t) + r_{2+}^*(t)\big)\,.$$

Applying Prop. C.2.1.1 we see that $r_{2-}^*$ is integral. Now Prop. C.2.2.1 tells us that $k_{r_1^*} = 0$ so that $r_{2-}^* = \text{const}$ by Prop. C.2.1.1. Now the assertion follows from the uniqueness theorem (Prop. B.2.8). If $\tau \neq 0$ we put $y := x - \tau$ and obtain from (C.2.2.5)

$$r(y+\tau) = \int_{-\infty}^{\infty} r_1(y+\tau-u)\, r_2(du)$$

$$= \int_{-\infty}^{\infty} r_1(y-v)\, r_2\big(d(v+\tau)\big) = 0\,, \qquad y < 0\,.$$

Hence the general case be reduced to the result of the first part of the proof. □

Finally we quote a result which can be derived by similar reasoning.

**Corollary C.2.2.4.** *Let $r$ be a function of bounded variation on $R_1$, $r(x) = 0$, $x < 0$. Denote by $F$ a d.f. such that* lext $(F) = 0$ *whose c.f. $f$ has no zeros in $S[0, \infty)$. If*

$$\int_{-0}^{\infty} F(u-x)\, r(du) = 0\,, \qquad x \geqq \tau \geqq 0$$

*then* $r(x) = \text{const}$ *for $x > \tau$.* □

## C.2.3. A corollary to the uniqueness theorem for characteristic functions

Since the continuation of d.f. is a main subject of this book, it is also of interest to decide under which conditions — in terms of c.f. — two d.f. $F_1$ and $F_2$ coincide on a half axis such that

$$F_1(u) = F_2(u)\,, \qquad u < \tau\,. \tag{C.2.3.1}$$

**Proposition C.2.3.1.** *The d.f.* $F_1$ *and* $F_2$ *with c.f.* $f_1$ *and* $f_2$, *resp., satisfy* (C.2.3.1) *for a certain number* $\tau$ *iff the function*

$$D(t) := f_1(t) - f_2(t)$$

*is continuable as an analytic function* $D$ *defined in* $S(0, \infty)$ *which is bounded or of order* 1 *and satisfies (in either case)*

$$D(\mathrm{i}s) = O(\mathrm{e}^{-\tau s}) \quad as \quad s \to \infty . \tag{C.2.3.2}$$

Proof. If (C.2.3.1) is true then the continuation

$$D(z) = \int_\tau^\infty \mathrm{e}^{\mathrm{i}zu}\, \mathrm{d}\big(F_1(u) - F_2(u)\big) , \qquad z \in S[0, \infty) ,$$

of $D$ is obvious, and we have trivially (by the ridge property of c.f.)

$$|D(t + \mathrm{i}s)| \leqq \int_\tau^\infty \mathrm{e}^{-su}\, \mathrm{d}\big(F_1(u) + F_2(u)\big)$$

so that (C.2.3.2) is true.

For the converse we confine ourselves to the case $D \neq 0$; this is possible in view of the uniqueness theorem for c.f. We first note that

$$k_D = \overline{\lim_{s\to\infty}} \frac{\log |D(\mathrm{i}s)|}{s} \leqq -\tau$$

by (C.2.3.2). Thus we can apply Prop. C.2.2.1 to obtain

$$k_D = -\operatorname{lext}(F_1 - F_2) \leqq -\tau$$

whence (C.2.3.1) follows which was to be proved. □

## C.2.4. A theorem on integral functions of order one and positive type

The main result of the present section serves as an essential tool for the proof of Prop. 9.4.2 (theorem of Hardy).

We introduce the integral function

$$d(z) = \exp\big((a - \mathrm{i}b)\, z\big) ,$$

where $a, b \in R_1$, so that in polar coordinates $z = r\,\mathrm{e}^{\mathrm{i}\varphi}$, $0 \leqq \varphi < 2\pi$,

$$|d(z)| = \exp\big((a \cos\varphi + b \sin\varphi)\, r\big) .$$

**Definition C.2.4.1.** Let $k$ be an analytic function whose domain of regularity contains the angle

$$A = \{z : 0 < |z| < \infty, \varphi_1 \leqq \varphi \leqq \varphi_2\} , \qquad 0 \leqq \varphi_1 < \varphi_2 \leqq 2\pi .$$

We assume $k$ is of order 1. Then the function

$$h_k(\varphi) := \overline{\lim_{r\to\infty}} \frac{\log |k(r\,\mathrm{e}^{\mathrm{i}\varphi})|}{r}, \qquad \varphi_1 \leqq \varphi \leqq \varphi_2 \tag{C.2.4.1}$$

is called the *indicator of* $k$ *in* $A$. □

**Example C.2.4.2.** The indicator of the above function $d$ is obviously

$$h_d(\varphi) = a \cos\varphi + b \sin\varphi , \qquad 0 \leqq \varphi < 2\pi . \qquad \square$$

An indicator of this form is said to be *trigonometric.*

If two real numbers $h^1$ and $h^2$ are given, then it is easy to find a trigonometric indicator $H$ such that $H(\varphi_1) = h^1$ and $H(\varphi_2) = h^2$, $0 \leqq \varphi_1 < \varphi_2 < \pi$; namely we put

$$H(\varphi) = \frac{h^1 \sin(\varphi_2 - \varphi) + h^2 \sin(\varphi - \varphi_1)}{\sin(\varphi_2 - \varphi_1)}, \quad \varphi_1 \leqq \varphi \leqq \varphi_2 \qquad \text{(C.2.4.2)}$$

This special indicator possesses the following important property.

**Lemma C.2.4.3.** *Assume that the indicator $h_k$ introduced in Def.* C.2.4.1 *satisfies*

$$h_k(\varphi_1) \leqq h^1, \qquad h_k(\varphi_2) \leqq h^2,$$

*where $\varphi_2 - \varphi_1 = \gamma < \pi$. Then it follows that*

$$h_k(\varphi) \leqq H(\varphi), \qquad \varphi_1 \leqq \varphi \leqq \varphi_2.$$

Proof. We consider the trigonometric indicator

$$H_u(\varphi) := \frac{(h^1 + u)\sin(\varphi_2 - \varphi) + (h^2 + u)\sin(\varphi - \varphi_1)}{\sin(\varphi_2 - \varphi_1)}, \quad \varphi_1 \leqq \varphi \leqq \varphi_2,$$

$u > 0$, which satisfies $H_u(\varphi_1) = h^1 + u$, $H_u(\varphi_2) = h^2 + u$. Then we have, with certain coefficients $a_u$ and $b_u$,

$$H_u(\varphi) = a_u \cos\varphi + b_u \sin\varphi, \qquad \varphi_1 \leqq \varphi \leqq \varphi_2.$$

Now we introduce the function

$$g(z) := k(z) \exp\left(-(a_u - ib_u) z\right)$$

and remind the reader of a simple general fact: If we we have two functions $f_1$, $f_2$ on $(0, \infty)$ and $\lim_{r\to\infty} f_2(r)$ exists, then

$$\overline{\lim_{r\to\infty}} \left(f_1(r) \pm f_2(r)\right) = \overline{\lim_{r\to\infty}} f_1(r) \pm \lim_{r\to\infty} f_2(r).$$

Thus we obtain the indicator of $g$ in the form

$$h_g(\varphi) = h_k(\varphi) - H_u(\varphi)$$

so that $h_g(\varphi_1) \leqq -u$ and $h_g(\varphi_2) \leqq -u$. Accordingly, it can be seen from the definition of the indicator that $\lim_{r\to\infty} g(r\,e^{i\varphi}) = 0$ for $\varphi = \varphi_1$ and $\varphi = \varphi_2$. Now Cor. C.1.5 applies so that $|g(z)|$ is bounded in the angle $A$ and, as a consequence, we get $h_k(\varphi) \leqq H_u(\varphi)$. The assertion now follows as $u \to 0$ since $h_k$ does not depend on $u$. □

We consider now an integral function $A$ growing at most like a function of order 1 and type $\tau$, $0 < \tau < \infty$, so that

$$\overline{\lim_{r\to\infty}} \frac{\log M(A, r)}{r} \leqq \tau, \qquad 0 < \tau < \infty. \qquad \text{(C.2.4.3)}$$

**Lemma C.2.4.4.** *Let $A$ be integral satisfying* (C.2.4.3) *and assume, moreover,*

$$\overline{\lim_{x\to\infty}} \frac{\log|A(x)|}{x} \leqq -b$$

*for some $b \geqq \tau$. Then the corresponding indicator satisfies*

$$h_A(\varphi) \leqq -\tau\cos\varphi, \qquad 0 \leqq \varphi < 2\pi.$$

Proof. We clearly have $h_A \leqq \tau$, $h_A(0) \leqq -b$. Let $d$ be some number, $0 < d < \pi$, and introduce the trigonometric indicator

$$H_d(\varphi) := -\tau \cos\varphi + \tau \tan(d/2) \sin\varphi = -\tau \frac{\cos(\varphi + d/2)}{\cos d/2}.$$

Then we see that $H_d(0) = -\tau$, $H_d(\pi - d) = \tau$, i.e., we have by assumption $h_A(0) \leqq H_d(0)$ and $h_A(\pi - d) \leqq H_d(\pi - d)$. Now L. C.2.4.3 indicates that

$$h_A(\varphi) \leqq H_d(\varphi), \qquad 0 \leqq \varphi \leqq \pi - d.$$

Hence we obtain the assertion for $0 \leqq \varphi \leqq \pi$ as $d \to 0$ since $h_A$ does not depend on $d$.

For $\pi < \varphi < 2\pi$ we can apply an analogous argument. □

**Proposition C.2.4.5.** *Let $A$ be an integral function satisfying* (C.2.4.3). *Further, assume, for some $C > 0$ and certain numbers $a_1, a_2 \geqq 0$;*

$$\begin{aligned} |A(x)| &\leqq C x^{a_1} e^{-bx}, \qquad x > 0, \qquad b \geqq \tau, \\ |A(x)| &\leqq C |x|^{a_2} e^{\tau|x|}, \qquad x < 0. \end{aligned} \tag{C.2.4.4}$$

*Then $b = \tau$ and*

$$A(z) = e^{-\tau z} P_q(z)$$

*where $P_q$ stands for a polynomial of degree not higher than $q := [\min\{a_1, a_2\}]$ (entier).*

Proof. We put $Q = \max\{a_1, a_2\}$ and consider the function

$$B(z) := c \frac{A(z) \exp(\tau z)}{(i + z)^Q}$$

which belongs to $M_+$ for an appropriate $c > 0$. Thus by L. C.2.4.4 we have $h_B(\pi/2) \leqq 0$; accordingly, Prop. C.1.7 (i) can be applied to obtain $B \in N_+$. Hence

$$|A(z) e^{\tau z}| = O(|z|^Q) \tag{C.2.4.5}$$

for $z \in S[0, \infty)$. Since an analogous argument applies in $S(-\infty, 0]$ we see that (C.2.4.5) is true in the whole plane. A well known characterization theorem for polynomials (see, for example, Saks and Zygmund (1952)) yields now

$$A(z) e^{\tau z} = P_{[Q]}(z),$$

where $P_{[Q]}$ denotes a polynomial of degree not greater than $[Q]$. Now $b > \tau$ is trivially ruled out by (C.2.4.4); hence $b = \tau$. Now we note that $|A(z) e^{\tau z}| = r^l(1 + o(1))$, as $|z| = r \to \infty$, where $l$ stands for an integer, $0 \leqq l \leqq Q$. Thus it is clear by assumption that the true degree $l$ of $P_{[Q]}$ is not greater than $q$. □

## C.2.5. Survey on the role played by the Phragmén-Lindelöf theory in the statistical characterization theory

Unfortunately the applications of the PL-theory in the literature on probability theory are sporadic. In the present book reference to appendix $C$ is made in connection with the following subjects

a) Existence of moments (Sect. 4.2)
b) Integral c.f. and d.f. (Sect. 4.5 and 6.4, resp.)
c) Counterpart to the theorem of Cramér and some modifications (Sect. 4.7)

d) Hinčin's phenomenon (Sect. 4.8)
e) Continuation problem for $P_\lambda$ (Sect. 7.6)
f) Theorem of Hardy (Sect. 9.4)

Concerning applications in queueing theory we refer the reader to the survey article ROSSBERG (1983a).

But this enumeration does not give a sufficiently convincing impression of the power of the tools provided by the PL-theory. To acquaint interested reader with a number of further applications, we wish to enter into a short discussion of the possibilities for using the PL-theory which we meet in another field of stochastics, namely in the statistical characterization theory of d.f., which is devoted to the properties of r.v. which are subject to important d.f. To begin with we outline certain types of problems which are treated in this field of research.

We consider the i.i.d.r.v. $X_1, \dots, X_n$ with d.f. $F$. In mathematical statistics there are many reasons for studying the properties of certain functions $S(X_1, \dots, X_n)$. The branch known as distribution theory is particularly interested in the d.f. $G$ to which these r.v. are subject when $F$ is already explicitely known. Sometimes different functions $S_i(X_1, \dots, X_n)$, $i = 1, 2$, are needed, and it is important to know under which conditions — i.e., for which special choice of $F$ — they prove to be either identically distributed or independent. Such results are useful for statistical tests.

In contrast to this theory, characterization theory starts from the converse assumptions, e.g., from $S \sim G$ where $S$ and $G$ are given explicitly. Moreover, if we consider two functions $S_1$ and $S_2$ like those above, it is sometimes of interest to draw conclusions from $S_1 \sim S_2$ or from the independence of $S_1$ and $S_2$. Of course, these two assumptions are not the only ones possible, but in this short survey we confine ourselves to them. It is not trivially seen that each of them imposes strong restrictions on $F$. But it is the point of characterization theorems that they tell us under which conditions such assumptions define the d.f. $F$ uniquely — except for a few parameters.

In the same way as the CLT explains why the normal d.f. possesses an outstanding importance, this theory tells us with quite different tools which d.f. are worthy of the particular attention of the statistician. Besides the normal d.f., the exponential d.f. proves to be of great interest. In this way characterization theory is an indispensible tool for understanding the missing links between the mathematical structure of certain d.f. and the actual behaviour of real world random phenomena.

The following two simple examples are due to Bernstein and Pólya, resp. Note that it is the only-if parts of these theorems that represent the characterization theorems in the strict sense. Here we have $n = 2$.

**Proposition C.2.5.1.** *Assume* $\operatorname{Var} X_1 < \infty$. *Then* $S_1 := X_1 + X_2$ *and* $S_2 := X_1 - X_2$ *are independent iff* $F = \Phi_{a,\sigma^2}$.

**Proposition C.2.5.2.** *Assume* $\mathsf{E} X_1 = 0$ *and* $\operatorname{Var} X_1 < \infty$. *Then* $S_1 := X_1$ *and* $S_2 := 2^{-1/2}(X_1 + X_2)$ *are identically distributed iff* $F = \Phi_{0,\sigma^2}$. □

For an example where the characterization of $F$ is made by means of the property $S \sim G$ we refer the reader to Prop. C.2.5.3; here $n = 2$ and $S = X_1 - X_2$. Also, both Cramér's theorem 4.6.4 and its elementary counterpart Cor. 4.7.6 can be interpreted in this way; here $S$ is the sum of i.r.v.

We return now to the PL-theory, where we have an analogous situation. Namely, the assumption $g \in \boldsymbol{M}_+$ is made and some restrictions on the growth of $g$ are imposed. They do not trivially imply that $g \in \boldsymbol{N}_+$, i.e., that the class of functions $g$ under consideration is comparatively small and well explored. But actually the PL-theorems do permit the conclusion $g \in N_+$. Clearly, the best results can be obtained if $g$ is integral and if also $g \in N_-$, since then we have $g = \text{const}$. At first sight this analogy seems to be far-fetched, but it can really be used.

As a first example we treat a very simple but deep result whose proof amounts to reformulating the given statistical assumptions to those of Prop. C.1.17. It is remarkable that Prop. C.2.5.3 can be reformulated in such a way that it gives the complete answer to a boundary value problem of Borovkov concerning a Wiener-Hopf factorization.

**Proposition C.2.5.3.** *Let $X_1$ and $X_2$ be i.r.v. subject to the d.f. $F_1$ and $F_2$ with c.f. $f_1$ and $f_2$, resp. Assume further* $\operatorname{lext}(F_i) = 0$ *and $f_i(z) \neq 0$ in $S[0, \infty)$, $i = 1, 2$, and define the d.f. $H$ by*

$$H(x) := \mathsf{P}(X_1 - X_2 < x) . \tag{C.2.5.1}$$

*Let $X_1'$ and $X_2'$ be two more i.r.v. such that their d.f. $G_1$ and $G_2$ and their c.f. $g_1$ and $g_2$ satisfy the same assumptions as above. Then from*

$$H(x) = \mathsf{P}(X_1' - X_2' < x) . \tag{C.2.5.1'}$$

*it follows that $F_1 = G_1$ and $F_2 = G_2$.*

**Remark C.2.5.4.** It can be shown by examples that the assumption „$f_i(z) \neq 0$ in $S[0, \infty)$" is essential. □

**Remark C.2.5.5.** If $F_1$ and $F_2$ are given and $S = X_1 - X_2$ then it is a distribution problem to calculate $H$. Now let $F_1$, $F_2$ and $H$ be such that (C.2.5.1) is satisfied. Then the theorem states under which assumptions we may conclude that there is no other pair of d.f. $G_1$ and $G_2$ such that (C.2.5.1′) holds with the same d.f. $H$; in other words, the theorem solves a characterization problem. □

Proof. Let $h$ stand for the c.f. of $H$. Then we have by assumption

$$h(t) = f_1(t)\, f_2(-t) = g_1(t)\, g_2(-t) . \tag{C.2.5.2}$$

Now we introduce the analytic functions.

$$K_+(z) := \frac{f_2(-z)}{g_2(-z)} , \quad z \in S[0, \infty) ,$$

$$K_-(z) := \frac{g_1(z)}{f_1(z)} , \quad z \in S(-\infty, 0] .$$

By virtue of (C.2.5.2) it follows that $K_+(t) = K_-(t)$, $t \in R_1$. Thus our assumptions imply by Prop. B.1.5 that $K_+$ and $K_-$ define together an integral function $w$ coinciding with $K_+$ in $S[0, \infty)$ and with $K_-$ in $S(-\infty, 0]$. It obviously satisfies the assumptions of Prop. C.1.17. Since $w(0) = 1$ it follows $w = 1$, as asserted. □

Next we present a result whose proof is based on Cor. C.2.2.4, which in its turn heavily rests on Th. C.1.14. It characterizes the exponential d.f. As before we consider the i.i.d.r.v. $X_1, \dots, X_n \sim F$ and introduce the corresponding order statistics $Y_1 \leqq Y_2 \leqq \dots \leqq Y_n$. Putting $S(X_1, \dots, X_n) := Y_{h+1} - Y_h$ (for $1 \leqq h \leqq n-1$, fixed) we can verify the formula

$$\mathsf{P}(S \geqq x) = \binom{n}{h} \int\limits_{-\infty}^{\infty} (1 - F(u + x))^{n-h} \mathrm{d}F^h(u) \tag{C.2.5.3}$$

by combinatorial methods.

**Proposition C.2.5.6.** *Assume* lext $(F) = 0$, $F(+0) = 0$, *and suppose that the c.f.*

$$\int_0^\infty e^{izu}\, dF^h(u)$$

*has no zeros in* $S[0, \infty)$; *let* $\tau \geqq 0$ *be fixed. Then the equation*

$$\mathsf{P}(S \geqq s) = \exp(-(n-h)\, s)\,, \qquad s \geqq \tau\,, \tag{C.2.5.4}$$

*holds true iff* $F(y) = 1 - e, y^{-y} \geqq \tau$.

Proof. In view of (C.2.5.3) it is enough to prove the only-if part of the theorem.

Let us assume (C.2.5.4). We put $s = x + y$, $x > \tau$, $y > 0$ and introduce the function $r$ of bounded variation by

$$r(u) := \left(1 - F(u + y)\right)^{n-h} - e^{-(n-h)y} \left(1 - F(u)\right)^{n-h}, \qquad u > 0\,,$$

$r(u) := 0$, $u \leqq 0$. Then we obtain from (C.2.5.3) and (C.2.5.4) the condition

$$\int_0^\infty F^h(u - x)\, dr(u) = 0\,, \qquad x \geqq \tau\,,$$

as the equivalent analytical expression of our statistical assumption. This means that we can apply Cor. C.2.2.4 — with $F$ replaced by $F^h$ — to get the conclusion $r(u) = \text{const}$ for $u > \tau$. Since $r(\infty) = 0$ it follows that $r(u) = 0$ for $u > \tau$, i.e.,

$$1 - F(u + y) = \exp(-y)\left(1 - F(u)\right), \qquad u \geqq \tau\,, \qquad y > 0\,.$$

Now we see that also

$$1 - F(u + y) = \exp(-u)\left(1 - F(y)\right), \qquad u \geqq \tau\,, \qquad y > 0\,.$$

On letting $y \to 0$ we obtain the assertion, since $F(+0) = 0$. □

A slight change of this argument shows the following. If we admit that $\delta := F(+0) > 0$, then (C.2.5.4) implies $F(y) = \delta U_0(y) + (1 - \delta)(1 - e^{-y})$, $y \geqq \tau$.

In Rossberg (1975) there is another characterization theorem connected with order statistics. There a problem is treated by reformulating the statistical assumptions to those of Cor. C.2.2.4.

But there are also other cases where the statistical assumption cannot immediately be reformulated in terms of the PL-theory or its direct consequences. Then it is only in the course of the proof that there arises the occasion of applying results of that theory. We are going to explain three very different cases.

The first two concern characterizations of the normal d.f. $\Phi$. As before $X_1, \ldots, X_n$ stands for $n$ i.i.d.r.v. $\sim F$. We start with a result of Linnik (1953), who participated in the study of a problem which arose from Prop. C.2.5.1: Under which additional assumptions can $\Phi$ be characterized by the independence of two linear forms

$$S_1 := a_1X_1 + \ldots + a_nX_n \quad \text{and} \quad S_2 := b_1X_1 + \ldots + b_nX_n\,?$$

For a brief formulation of Linnik's result it is convenient to introduce the notation

$$c := \max\{|a_j|, |b_j| : 1 \leqq j \leqq n\}$$

and

$$K(x) := \sum_{j=1}^n \left(|a_j|^x - |b_j|^x\right), \qquad -\infty < x < \infty\,.$$

Further we need the following 4 conditions:

(i) card $\{j: |a_j| = c\} \neq$ card $\{j: |b_j| = c\}$, i.e., the two equations $|a_j| = c$ and $|b_j| = c$ have a different number of solutions.

(ii) All negative roots of $K$ being even integers have a multiplicity not exceeding 2. There exists at most one double root $x_0$ of this kind, and if it exists, then $x_0$ is the smallest root of $K$. Further the number $[|x_0|/2]$ (entier) is odd.

(iii) $K$ has at most one root $x_1 < 0$ which is not an even integer. If such a root exists, then it is simple and it is the smallest root; further $[|x_1|/2]$ is odd.

(iv) $K(2) = 0$.

**Proposition C.2.5.7.** *Suppose* (i). *Then the two assertions*

(A) *$F$ is a normal d.f.*

(B) *$S_1$ and $S_2$ are identically distributed*

*are equivalent iff the conditions* (ii)—(iv) *are true.* □

Unfortunately Linnik's proof of this theorem is extremely difficult. So it is to be welcomed that Riedel (1979) succeeded in simplifying that part of the proof which derives (A) from (B) and (i)—(iv). This author essentially changed the plan of this proof. Namely, using Th. C.1.14 he showed that all moments of $F$ exist, so that finally a theorem of Marcinkiewicz[1]) yields the desired result. It is to be emphasized that this method also permits us to treat the case $n = \infty$.

These results can be extended to stochastic integrals which depend on a stochastic process $X(t)$, $t \geqq 0$. We first remind the reader of certain concepts. The r.v. $X(t_2) - X(t_1)$ is called the *increment* of the process over the interval $[t_1, t_2]$. A process $X(t)$ is called *homogeneous with independent increments* if the d.f. of the increment $X(t + s) - X(t)$ depends only on $s$ and if the increments over non-overlapping integrals are independent. A process $X(t)$ is said to be *continuous* if for every $t > 0$ we have $X(t + s) \to X(t)$ in probability as $s \to 0$. A continuous homogeneous process with independent increments is called a *Wiener process* if the d.f. of the increments is normal and $X(0) = 0$. For a homogeneous continuous process with independent increments $X(t)$, $t > 0$, and for a continuous function $a(t)$ defined on $[A, B]$, the *stochastic integral* $\int_A^B a(t)\, dX(t)$ is defined as the limit of Riemann sums in the sense of convergence in probability.

We quote now a characterization of Wiener processes by identically distributed stochastic integrals. The analogy with Prop. C.2.5.7 is obvious, but the formulation of the following result is by far simpler. This is due to the fact that the stochastic integral can only be defined for continuous homogeneous processes with independent increments; this implies that only inf.div.d.f. have to be considered.

Let us first introduce the function

$$K(z) := \int_A^B (|a_1(t)|^z - |a_2(t)|^z)\, dt\,, \qquad \operatorname{Re} z > 0\,,$$

where $a_1$ and $a_2$ are functions continuous on $[A, B]$.

**Proposition C.2.5.8.** *Suppose that $K(z)$ is analytic in a neighbourhood of the origin. Then the two assertions*

(A) *$X(t)$ is a Wiener process,*

(B) *for the continuous homogeneous process with independent increments $X(t)$, the integrals $\int_A^B a_i(t)\, dX(t)$, $i = 1, 2$, are identically distributed*

*are equivalent iff $K(x)$ $(0 < x \leqq 2)$ has a zero only at $x = 2$.*

[1]) see e.g. Rossberg-Jesiak (1983), Theorem 10.3.

The proof contains another occasion to apply Th. C.1.14.

In Prop. C.2.5.8 one process $X(t)$ and two i.d. stochastic integrals with different integrands were considered. Now the following question can be asked. Do there exist different processes such that the corresponding stochastic integrals with the same integrand are i.d.? Prop. C.2.5.9 gives an answer in the negative under appropriate conditions.

We are going to describe the characterization problem precisely. Let $b$ be a function continuous on $[A, B]$ and write $w$ for a non-decreasing non-negative and left-continuous function which is also defined on $[A, B]$.

Let us consider two continuous homogeneous processes $X(t)$ and $Y(t)$, $t \geqq 0$, with independent increments. Then the stochastic integral

$$S(X) := \int_A^B b(t) \, \mathrm{d}X(w(t))$$

exists. Now, the important point: We call $X(t)$ *completely defined by the integral* $S(X)$ if from $S(X) \sim S(Y)$ it follows that $X(t) = Y(t)$ for all $t \geqq 0$.

Now we are in a position to formulate

**Proposition C.2.5.9.** *Suppose that* $\mathsf{E}X(1) < \infty$ *for a certain and* $X(0) = 0$. *Then the stochastic integral* $S(X)$ *determines complctely the process* $X(t)$ *iff the following conditions are satisfied:*

(i) $$\int_A^B |b(t)|^z \, \mathrm{d}w(t) \neq 0 \qquad \text{for} \quad 0 < \lambda \leqq \operatorname{Re} z < 2;$$

(ii) $$\int_A^B |b(t)|^{z-1} \, b(t) \, \mathrm{d}w(t) \neq 0 \quad \text{for} \quad 0 < \lambda \leqq \operatorname{Re} z < 2;$$

(iii) $$\int_A^B b(t) \, \mathrm{d}w(t) \neq 0 .$$ □

The proof is based on a certain Wiener-Hopf factorization and the PL-theorems C.1.14 and C.1.7.

## C.3. Final remarks

The problem of under which additional suppositions $w \in N_+$ follows from the assumption $w \in M_+$ has proved to be inexhaustible and an astonishing number of different conditions has been found. As more detailed introductions we recommend the books Cartwright (1962) and Evgrafov (1979); these authors have also made remarkable attempts to develop the general theory. But so far there is no complete survey of all the material that has been accumulated since the classic paper of Phragmén and Lindelöf (1908) appeared.

In our view the importance of the PL-theory has been underestimated in probability theory so far. This is shown by the sporadic use of PL-theorems even in monographs which are devoted to analytic aspects of probability theory. Contrary to the usual habit, we collect together quite a number of applications in the present book.

In other boundary fields of classic analysis there is possibly a similar underestimation. In order to facilitate more applications it would be worth-while surveying all the results of the PL-theory. For a comparison it would be particularly important to list all versions of known PL-theorems which are valid in the half plane $S(0, \infty)$

or in an arbitrary angle. At the same time it could be discovered whether a great many of the results known so far follow from a few basic theorems, provided that all possibilities of generalizing are exhausted. One of these basic theorems will presumably be the result of Fuchs (1981).

As to the possibilities of generalizing a known result, we direct the reader's attention to what we call the *method of successive mapping*, which has been developed in Rossberg (1974a), Lehmann and Rossberg (1976). To make it clear what we mean by this, we note that the procedure which led from Th. C.1.3 to Prop. C.1.10. can be continued. It yields a sequence of PL-theorems such that the $m$-th, $m \geqq 1$, theorem generalizes the foregoing one. The method was brought to an end only in Rossberg and Jesiak (1983). The final result obtained there concerns a function $w \in \boldsymbol{M}_+$, $w \neq 0$; it states that either $w$ has infinite order or $w \in \boldsymbol{N}_+$ provided that certain restrictions on the growth of $w$ are satisfied; these restrictions are imposed on a countable set of rays starting from 0 and contained in the half plane $S(0, \infty)$. It is remarkable that the method of successive mapping applied to this theorem does not give a more general result, but yields the same statement.

## Notes and bibliographical data

The pioneer work of Phragmén and Lindelöf is contained in their common paper (1908). Introducing the auxiliary function $W_\varepsilon$ in the proof of Th. C.1.3 was one of their decisive ideas in this field; it can also be applied to finite domains, see e.g. Rühs (1971). The results C.1.2—C.1.9 can be found in many textbooks. The "method of successive mapping" was elaborated in Lehmann and Rossberg (1976). It was brought to an end only in Rossberg and Jesiak (1983) ,where a theorem was proved with restrictions only on a countable set of rays; it cannot be generalized by another mapping.

Theorem C.1.14 was elaborated, step by step, in Rossberg (1968), (1978), and Riedel (1977). Prop. C.1.17 is due to Seiffert and Riedel (1979). Prop. C.2.1.1 and Cor. C.2.1.2 were proved in Rossberg and Chu Duc (1980). Prop. C.2.2.3 and its Cor. C.2.2.4 is contained in Rossberg (1975).

A detailed representation of the statistical characterization theory is contained in Lukacs and Laha (1964); Kagan, Linnik, and Rao (1972). For simple proofs of Prop. C.2.5.1 and C.2.5.2 see Rényi (1962). Special cases of Prop. C.2.5.3 were obtained in Rossberg (1968) and (1975), the general case (together with a boundary value problem of Borovkov) was treated in Seiffert and Riedel (1979). For Prop. C.2.5.6 see Rossberg (1972) and (1975). Prop. C.2.5.8 and C.2.5.9 are due to Riedel (1980) and (1985), resp.

# Table of distribution functions

In the following table, we collect together a number of well-known d.f., which are often met in practice, with their c.f., expectations, and variances.

Lattice d.f. $F$ will be described by their atoms $x_k$ and the corresponding jumps $p_k = F(x_k + 0) - F(x_k)$, and absolutely continuous d.f. $F$ by their density $p(x)$. Unless otherwise stated, densities are concentrated on $\mathbb{R}_1$.

If there occur location and/or scale parameters, they are mentioned explicitely; otherwise, the transformation rules for c.f. (Prop. 3.1.2 (ii)) and moments (Prop. 1.3.4) can be applied.

If, for a c.f. or for a moment, no formula is given, then no "elementary" expression is available; the star $*$ means that the corresponding moment does not exist. Further, $\varrho$ and $\tau$ denote the order and the type, resp., of an integral d.f.

| name, notation parameters | atoms $x_k$ jumps $p_k$ | characteristic function | $\mu_F$ | $\sigma_F^2$ | remarks special properties |
|---|---|---|---|---|---|
| degenerate $U_a$ <br> $a \in \mathbb{R}_1$ | $a$ <br> $1$ | $e^{ita}$ | $a$ | $0$ | $a$ — location parameter <br> inf.div. (Ex. 7.1.12 (i)) <br> stable |
| uniform <br> $n \geqq 1$ | $1 \leqq k \leqq n$ <br> $\frac{1}{n}$ | $\frac{1}{n} \sum_{k=1}^{n} e^{itk}$ | $\frac{n+1}{2}$ | $\frac{n^2-1}{12}$ | |
| binomial $B_{n,p}$ <br> $n \geqq 1, 0 < p < 1$ | $0 \leqq k \leqq n$ <br> $\binom{n}{k} p^k (1-p)^{n-k}$ | $(1 - p + p\, e^{it})^n$ | $np$ | $np(1-p)$ | $B_{n_1,p} * B_{n_2,p} = B_{n_1+n_2,p}$ |
| Poisson $P_\lambda$ <br> $\lambda > 0$ | $k \geqq 0$ <br> $e^{-\lambda} \frac{\lambda^k}{k!}$ | $e^{\lambda(e^{it}-1)}$ | $\lambda$ | $\lambda$ | inf.div. (Ex. 7.2.12 (iii)) <br> $P_{\lambda_1} * P_{\lambda_2} = P_{\lambda_1+\lambda_2}$ |
| geometric <br> $0 < p < 1$ | $k \geqq 0$ <br> $(1-p)\, p^k$ | $\frac{1-p}{1-p\, e^{it}}$ | $\frac{p}{1-p}$ | $\frac{p}{(1-p)^2}$ | inf.div. |
| hypergeometric <br> $N \geqq l, 0 \leqq n,$ <br> $M \leqq N$ | $\max\{0, n+M-N\}$ <br> $\leqq k \leqq \min\{n, M\}$ <br> $\frac{\binom{M}{k}\binom{N-M}{n-k}}{\binom{N}{n}}$ | ${}_2F_1(-M, -n, -N, 1 - e^{it})$ | $\frac{Mn}{N}$ | $\frac{Mn(N-M)(N-n)}{N^2(N-1)}$ <br> (if $N \geqq 2$) | ${}_2F_1$ — Gaussian hypergeometric function |

| name, notation parameters | density $p(x)$ | characteristic function | $\mu_F$ | $\sigma_F^2$ | remarks special properties |
|---|---|---|---|---|---|
| normal $\Phi_{\mu,\sigma}$ <br> $\mu \in R_1$, <br> $\sigma^2 > 0$ | $\frac{1}{\sqrt{2\pi}\,\sigma} \exp\left\{-\frac{1}{2} \times \right.$ <br> $\left. \times \left(\frac{x-\mu}{\sigma}\right)^2\right\}$ | $e^{it\mu - \sigma^2 t^2/2}$ | $\mu$ | $\sigma^2$ | $\mu$ — location parameter <br> $\sigma^2$ — scale parameter <br> inf.div. (Ex. 7.2.12 (i)) <br> stable, $\alpha = 2$ <br> integral d.f., $\varrho = 2$, $\tau = 1/(2\sigma^2)$ |
| lognormal <br> $\mu \in R_1$, <br> $\sigma^2 > 0$ | $\frac{1}{x} \varphi_{\mu,\sigma}(\log x)$ <br> $(x > 0)$ | | $e^{\mu + \sigma^2/2}$ | $(e^{\sigma^2} - 1)\,\mu_F^2$ | inf.div. |
| double exponential | $e^{-x}\, e^{-e^{-x}}$ | $\Gamma(1 - it)$ | $-\Gamma'(1)$ <br> $\approx 0.577$ | $\frac{\pi^2}{6}$ | d.f. $F(x) = e^{-e^{-x}}$ <br> integral d.f., $\varrho = \infty$ |
| Student $T_n$ <br> ($t$-distr.) <br> $n \geqq 1$ | $c\left(1 + \frac{x^2}{n}\right)^{-(n+1)/2}$ | $\frac{1}{\pi\Gamma\left(\frac{n}{2}\right)} \left(\frac{\lvert t\rvert \sqrt{n}}{2}\right)^{n/2} \times$ <br> $\times K_{n/2}(\lvert t\rvert \sqrt{n})$ | $0$ <br> $(n \geqq 2)$ | $\frac{n}{n-2}$ <br> $(n \geqq 3)$ | $c = \Gamma\left(\frac{n+1}{2}\right) \Big/ \Gamma\left(\frac{n}{2}\right) \sqrt{\pi n}$ <br> $K_{n/2}$ — modified Bessel function <br> inf.div. <br> analytic in $S(\sqrt{n})$ |
| Cauchy <br> $a \in R_1$, $b > 0$ | $\frac{1}{\pi} \frac{b}{b^2 + (x-a)^2}$ | $e^{ita - b\lvert t\rvert}$ | * | * | $a$ — location parameter <br> $b$ — scale parameter <br> d.f. $F(x) = \frac{1}{2} + \frac{1}{\pi} \arctan \frac{x-a}{b} =$ <br> $= T_1\left(\frac{x-a}{b}\right)$ <br> inf.div. (Ex. 7.2.12 (ii)) <br> stable (Ex. 8.1.11 (i)) <br> analytic in $S(b)$ |
| beta$_1$ $B_{p,q}^{(1)}$ <br> $p, q > 0$ | $\frac{x^{p-1}}{B(p,q)} (1-x)^{q-1}$ <br> $(0 < x < 1)$ | ${}_1F_1(p, p+q, -it)$ | $\frac{p}{p+q}$ | $\frac{pq}{(p+q)^3 + p + q}$ | $B(p,q) = \Gamma(p)\,\Gamma(q)/\Gamma(p+q)$ <br> ${}_1F_1$ — degenerate Gaussian hypergeometric function |

| name, notation parameters | density $p(x)$ | characteristic function | $\mu_F$ | $\sigma_F^2$ | remarks special properties |
|---|---|---|---|---|---|
| uniform $a, b \in R_1$, $a < b$ | $\frac{1}{b-a}$ $(a < x < b)$ | $\frac{e^{itb} - e^{ita}}{it(b-a)}$ | $\frac{b+a}{2}$ | $\frac{(b-a)^2}{12}$ | $a$ — location parameter<br>$b - a$ — scale parameter<br>d.f. $F(x) = B_{1,1}^{(1)}\left(\frac{x-a}{b-a}\right)$ |
| beta$_2$ $B_{p,q}^{(2)}$ $p, q > 0$ | $\frac{1}{B(p,q)} \frac{x^{p-1}}{(1+x)^{p+q}}$ $(x > 0)$ | ${}_1F_1(p, -q, -it)$ | $\frac{p}{q-1}$ $(q > 1)$ | $\frac{p(p+q-1)}{(q-1)^2(q-2)}$ $(q > 2)$ | $B(p, q)$, ${}_1F_1$ — see beta$_1$<br>inf.div. |
| Fisher $F_{m,n}$ $m, n \geqq 1$ | $\frac{c\, e^{mx}}{(n + m\, e^{2x})^{(m+n)/2}}$ | $\frac{\Gamma\left(\frac{n-it}{2}\right)\Gamma\left(\frac{m+it}{2}\right)}{\Gamma\left(\frac{m}{2}\right)\Gamma\left(\frac{n}{2}\right)}\left(\frac{n}{m}\right)^{it}$ | | | $c = 2m^{m/2}n^{n/2}/B(m/2, n/2)$<br>$F_{m,n}(x) = B_{m/2,n/2}^{(2)}((m/n)\, e^{2x})$<br>inf.div.<br>analytic in $S(\pi/2)$ |
| general gamma $\Gamma_{\lambda,\alpha}$ $\lambda, \alpha > 0$ | $\frac{\alpha}{\Gamma\left(\frac{\lambda}{\alpha}\right)} x^{\lambda-1} e^{-x^\alpha}$ $(x > 0)$ | | $\frac{\Gamma\left(\frac{\lambda+1}{\alpha}\right)}{\Gamma\left(\frac{\lambda}{\alpha}\right)}$ | $\frac{\Gamma\left(\frac{\lambda+2}{\alpha}\right)}{\Gamma\left(\frac{\lambda}{\alpha}\right)} - \mu_F^2$ | inf.div., if $\alpha \leqq 1, \lambda \leqq 1$<br>not inf.div., if $\alpha > 1$ |
| special gamma $\Gamma_\lambda$ $\lambda < 0$ | $\frac{1}{\Gamma(\lambda)} x^{\lambda-1} e^{-x}$ $(x > 0)$ | $\frac{1}{(1-it)^\lambda}$ | $\lambda$ | $\lambda$ | $\Gamma_\lambda = \Gamma_{\lambda,1}$<br>inf.div. (Ex. 7.1.2 (iii)) |
| Erlang $n \geqq 1$ | $\frac{x^{n-1}}{(n-1)!} e^{-x}$ $(x > 0)$ | $\frac{1}{(1-it)^n}$ | $n$ | $n$ | d.f. $F(x) = \Gamma_n(x)$<br>inf.div. |
| exponential $b > 0$ | $b\, e^{-bx}$ $(x > 0)$ | $\frac{b}{b-it}$ | $\frac{1}{b}$ | $\frac{1}{b^2}$ | $b$ — scale parameter<br>d.f. $F(x) = \Gamma_1(bx)$<br>inf.div. |

| name, notation parameters | density $p(x)$ | characteristic function | $\mu_F$ | $\sigma_F^2$ | remarks special properties |
|---|---|---|---|---|---|
| chi square $\chi_n^2$<br>$n \geqq 1$ | $\frac{x^{(n/2)-1}}{2^{n/2}\Gamma(n/2)} e^{-x/2}$<br>$(x > 0)$ | $\frac{1}{(1-2it)^{n/2}}$ | $n$ | $2n$ | $\chi_n^2(x) = \Gamma_{n/2}\left(\frac{x}{2}\right)$<br>inf.div. |
| Weibull<br>$\lambda > 0$ | $\lambda x^{\lambda-1} e^{-x^\lambda}$<br>$(x > 0)$ | | $\Gamma\left(1+\frac{1}{\lambda}\right)$ | $\Gamma\left(1+\frac{2}{\lambda}\right) - \mu_F^2$ | d.f. $F(x) = \Gamma_{\lambda,\lambda}(x) = 1 - e^{-x^\lambda}$<br>inf.div. iff $\lambda \leqq 1$ |
| Maxwell | $\frac{4}{\sqrt{\pi}} x^2 e^{-x^2}$ $(x > 0)$ | | $\frac{2}{\sqrt{\pi}}$ | $\frac{3}{2} - \frac{4}{\pi}$ | d.f. $F(x) = \Gamma_{3,2}(x)$ |
| Laplace<br>$a \in \mathbb{R}_1$, $b > 0$ | $\frac{1}{2b} e^{-\frac{\lvert x-a\rvert}{b}}$ | $\frac{e^{ita}}{1+b^2t^2}$ | $a$ | $2b^2$ | $a$ — location parameter<br>$b$ — scale parameter<br>inf.div. |
| triangular<br>$b > 0$ | $\frac{1}{b}\left(1 - \frac{\lvert x\rvert}{b}\right)$<br>$(-b < x < b)$ | $\left(\frac{\sin\frac{bt}{2}}{\frac{bt}{2}}\right)^2$ | $0$ | $\frac{b^2}{6}$ | $b$ — scale parameter |
| $b > 0$ | $\frac{b}{\pi}\left(\frac{\sin bx}{bx}\right)^2$ | $\max\left(1 - \frac{\lvert t\rvert}{2b}, 0\right)$ | * | * | $b$ — scale parameter<br>integral d.f., $\varrho = 1$, $\tau = 2b$ |
| | $\frac{x^{-3/2}}{\sqrt{2\pi}} e^{-1/2x}$ $(x > 0)$ | $e^{-\sqrt{\lvert t\rvert}(1 - i \operatorname{sgn} t)}$ | * | * | stable (Ex. 8.1.11 (ii))<br>Pearson density (type IV) |
| Pareto<br>$b, \lambda > 0$ | $\frac{\lambda}{b}\left(\frac{b}{x}\right)^{\lambda+1}$<br>$(x > b)$ | | $\frac{\lambda b}{\lambda - 1}$<br>$(\lambda < 1)$ | $\frac{\lambda b^2}{(\lambda-1)^2(\lambda-2)}$<br>$(\lambda > 2)$ | $b$ — scale parameter<br>inf.div. |

# References


Amosova, N. N. (1979) (Н. Н. Амосова), О вероятностях умеренных уклонений сумм независимых случайных величин. Теор. вероятн. примен. **24**, 858—865.

Amosova, N. N. (1981) (Н. Н. Амосова), Узкие интегральные зоны нормального притяжения. Lietuvos Matematikos Rinkings **XXII**, 1—15.

Antonov, S. N. (1980a) (С. Н. Антонов), Аналитические свойства безгранично делимых законов. Кандидатская диссертация, МГУ.

Antonov, S. N. (1980b) (С. Н. Антонов), К асимптотическому поведению безгранично делимых законов. Мат. заметки **28**, 939—946.

Antonov, S. N. (1981) (С. Н. Антонов), Об аналитических свойствах квазиустойчивых распределений. Мат. заметки **30**, 595—602.

Bergström, H. (1963), Limit Theorems for Convolutions. Wiley, New York.

Bergström, H. (1973), The limit problem for sums of independent random variables which are not uniformly asymptotically negligible. Transactions of the Sixth Prague Conference Prague 1971, 125—135.

Bethmann, J. (1985), Characterization of power-mixtures by means of their one-sided asymptotic behaviour. Теор. вероятн. примен **30**, 136—140.

Billingsley, P. (1968), Convergence of Probability Measures. Wiley, New York.

Bischoff, D. (1982), On the continuation of convolutions with stable and left-stable factors, Preprint, Karl-Marx-Universität Leipzig 1982. (See also: Stabilis és balstabilis kompensü konvolúciók falyathatóságához. Matematikai Lapok Budapest, 30. évfolyam. 1978—1982, 1—3, 181—189.)

Bischoff, D. (1985) (Д. Бишоф), Проблема компактности для сумм независимых случайных величин с функциями распределения при условии ограниченной сходимости на полуоси. Теор. вероятн. и мат. статистика **33**, in print

Blank, N. M. (1981) (Н. М. Бланк), Об однозначной определенности функций ограниченной вариации и функции распределения асимптотическим поведением при $x \to -\infty$. In: Проблемы устойчивости стохастических моделей, Производственно-издательский комбинат ВИНИТИ, Москва, 10—15.

Blank, N. M. (1982) (Н. М. Бланк), Об условиях однозначной определенности функций ограниченной вариации и функций распределения значениями на полупрямой. Теор. вероятн. примен. **27**, 784—787.

Blank, N. M. (1983), (Н. М. Бланк), Распределения, свертки которых совпадают на полуоси. ДАН УССР, Сер. А, № 6, 3—4.

Blank, N. M., and I. V. Ostrovskij (1981) (Н. М. Бланк и И. В. Островский), О функциях ограниченной вариации, близких к нормальной функции распребеления. Теория операторов в функциональных пространствах и ее приложения, Киев, 3—10.

Boas, R. P., and F. Smithies (1938), On the characterization of a distribution function by its Fourier transform, Amer. J. Math. **60**, 523—531.

Brown, B. M. (1970), Characteristic functions, moments, and the central limit theorem. Ann. Math. Stat. **41**, 658—664.

Callaert, H., and J. W. Cohen (1972), A lemma on regular variation of a transient renewal function. Z. Wahrscheinlichkeitstheorie verw. Geb. **24**, 275—278.

Cuppens, R., and E. Lukacs (1970), On the domains of definition of analytic characteristic functions. Ann. Math. Statist. **41**, 1096—1101.

DAVIES, L. (1976), Tail probabilities for positive random variables with entire characteristic functions of very regular growth. Z. angew. Math. Mech. 56, T 334–T 336.

DEWESS, M. (1977a), A note on the asymptotic behaviour of an entire characteristic function of a multivariate probability law. J. multivariate analysis 7, 608–613.

DEWESS, M. (1977b), Asymptotisches Verhalten von Verteilungsfunktionen und Wachstum der zugehörigen charakteristischen Funktion. Ph. D. Thesis, Leipzig.

DEWESS, M. (1978a), The tail behaviour of a distribution function and its connection to the growth of its entire characteristic function. Math. Nachr. 81, 217–231.

DEWESS, M. (1978b), On applications of dominated variation to queuing theory. Math. Operationsforsch. Statist., Ser. Optimization, 9, 601–609.

DEWESS, M. (1980), On a transient renewal function of dominated variation. Math. Operationsforsch. Statist., Ser. Optimization, 11, 499–505.

DEWESS, M. (1982), Über ganze charakteristische Funktionen mit vorgegebenem Wachstum. Serdica 8, 307–312.

DEWESS, M., and M. RIEDEL (1977), The connection between the proximate order of an entire characteristic function and the corresponding distribution function. Czechoslovak. Math. J. 27(102), 173–185.

ESSEEN, C. G. (1968), On the concentration function of a sum of independent random variables. Z. Wahrscheinlichkeitstheorie verw. Geb. 9, 290–308.

ESSEEN, C.-G. (1981), On the tails of a class of infinitely divisible distributions. *In*: GANI, J., and V. K. ROHATGI (eds.), Contributions to Probability. A Collection of Papers Dedicated to Eugene Lukacs, Academic Press, New York, 115–122.

EVGRAFOV, M. A. (1979) (М. А. Евграфов), Асимптотические оценки и целые функции. Наука, Москва.

FELLER, W. (1967), On regular variation and local limit theorems. Proc. Fifth Berkeley Symp. Math. Statist. Probab., Vol. II, Part 1. University of California Press, Berkeley, 373–388.

FELLER, W. (1968), An Introduction to Probability Theory and its Applications, Vol. 1. 3rd ed., Wiley, New York.

FELLER, W. (1971), An Introduction to Probability Theory and its Applications, Vol. 2. 2nd ed., Wiley, New York.

FUCHS, W. H. G. (1981), A Phragmén-Lindelöf theorem conjectured by D. J. Newman. Trans. Amer. Math. Soc. 267, 285–293.

GALAMBOS, J., and S. KOTZ (1978), Characterization of Probability Distributions. Lecture notes in mathematics, vol. 675, Springer, Berlin–Heidelberg–New York.

GIRLICH, H.-J., and H.-J. ROSSBERG (1971), Eine neue Methode zum Studium des Zusammenhangs zwischen charakteristischen Funktionen und Momenten sowie Semi-Invarianten. Math. Nachr. 51, 123–147.

GNEDENKO, B. V. (1943), Sur la distribution du term maximum d'une série aléatoire. Ann. Math. 44, 423–453.

GNEDENKO, B. V. (1980), Lehrbuch der Wahrscheinlichkeitsrechnung, 8. Aufl., Akademie-Verlag, Berlin.

GNEDENKO, B. V. (1983), On some stability theorems. In KALAŠNIKOV, V. V., and V. M. ZOLOTAREV (eds.), Stability problems for stochastic models. Lecture Notes in Mathematics, vol. 982. Springer, Berlin–New York, 24–31.

GNEDENKO, B. V., and A. N. KOLOMOGOROV (1959), Grenzverteilungen von Summen unabhängiger Zufallsgrößen. Akademie-Verlag, Berlin.

GNEDENKO, B. V., and L. SENUSI-BEREKSI (1982a) (Б. В. Гнеденко и Л. Сенуси-Берекси), Об одном свойстве предельных распределений для максимального и минимального членов вариационного ряда. ДАН СССР 267, 1039—1040.

GNEDENKO, B. V., and L. SENUSI-BEREKSI (1982b) (Б. В. Гнеденко и Л. Сенуси-Берекси), Об одном свойстве логистического распределения. ДАН СССР 267, 1293—1295.

GNEDENKO, B. V., and A. ŠERIF (1983) (Б. В. Гнеденко и А. Шериф), Предельные теоремы для крайних членов вариационного ряда. ДАН СССР 270, 523—525.

GOL'DBERG, A. A., and I. V. OSTROVSKIĬ (1974) (А. А. Гольдберг и И. В. Островский), О росте целых хребтовых бункций с действительными нулями. Мат. физика и функциональный анализ „Вищая школа", Харьков.

HAYMAN, W. K. (1956), Questions of regularity connected with the Phragmén-Lindelöf principle. J. Math. Pures Appl. 35, 115–126.
HEINRICH, L. (1982), Infinitely divisible functions as limit laws for sums of random variables connected in a Markov chain. Math. Nachr. 107, 103–121.
HENGARTNER, W., and R. THEODORESCU (1973), Concentration Functions. Academic Press, New York–London.
HILLE, E., and R. S. PHILLIPS (1957), Functional Analysis and Semi-Groups. Amer. Math. Soc., Providence, Rhode Island.

IBRAGIMOV, I. A. (1977) (И. А. Ибрагимов), Об определении безгранично делимой функции распределения по ее значениям на полупрямой. Теор. вероятн. примен. 22, 393—399.
IBRAGIMOV, I. A., and JU. V. LINNIK (1965) (И. А. Ибрагимов и Ю. В. Линник), Независимые и стационарно связанные величины. Наука, Москва.

JAKOVLEVA, N. I. (1972) (Н. И. Яковлева), О росте целых характеристических функций вероятностных законов. Теор. функ. анализ прилож. 15, 43—49.
JAKOVLEVA, N. I. (1976) (Н. И. Яковлева), Вопросы математической физики и функционального анализа: Материалы науч. семинаров. АН УССР, физ.-Техн. Инст. низких Температуров. — Киев: Наук. Думка.
JESIAK, B. (1978), Fortsetzung von Verteilungsfunktionen und Grenzwertsätze für Summen unabhängiger Zufallsgrößen unter der Voraussetzung eingeschränkter Konvergenz. Ph. D. Thesis A, Leipzig.
JESIAK, B. (1979a), On analytic distribution functions and analytic properties of infinitely divisible distribution functions. Теор. вероятн. примен. 24, 825–831.
JESIAK, B. (1979b), Extension of JESIAK (1979a). Preprint, Karl-Marx-Universität Leipzig.
JESIAK, B. (1979c), A uniqueness theorem for stable laws. Math. Nachr. 92, 243–246.
JESIAK, B. (1980), Limit theorems under the assumptions of restricted convergence on a set with a finite limit point. Math. Nachr. 95, 187–194.
JESIAK, B. (1981), On the unique determination of Lévy's spectral functions and parameters of infinitely divisible distributions. Теор. вероятн. примен. 26, 160–165.
JESIAK, B. (1982), A characterization theorem for a class of distribution functions. Preprint, Karl-Marx-Universität Leipzig.
JESIAK, B. (1985), Further uniqueness theorems on stable distribution functions. Preprint, Karl-Marx-Universität Leipzig.
JESIAK, B., H.-J. ROSSBERG, and G. SIEGEL (1979), On the central limit theorem for the triangular array and the Lindeberg-Feller theorem. Preprint, AdW der DDR.
JESIAK, B., and H.-J. ROSSBERG (1981), New versions of the Lindeberg-Feller theorem. Теор. вероятн. примен. 26, 858–862.

KAGAN, A. M., JU. V. LINNIK, and S. R. RAO (1972) (А. М. Каган, Ю. В. Линник и С. Р. Рао), Характеризационные задачи математической статистики. Наука, Москва.
KAWATA, T. (1972), Fourier Analysis in Probability Theory. Academic Press, New York–London.
KESTEN, H. (1969), A sharper form of the Doeblin-Lévy-Kolmogorov-Rogozin inequality for concentration functions. Math. Scand. 25, 133–144.
KRUGLOV, V. M. (1970) (В. М. Круглов), Замечания к теории безгранично-делимых законов. Теория вероятн. примен. 15, 330—336.
KRUGLOV, V. M. (1974a) (В. М. Круглов), О безгранично делимых распределениях в Гильбертовом пространстве. Мат. заметки 16, 585—594.
KRUGLOV, V. M. (1974b) (В. М. Круглов), Характеризация одного класса безгранично делимых распределений в гильбертовом пространстве. Мат. заметки 16, 777—783.
KRUGLOV, V. M. (1974c), Weak convergence of distributions for sums of independent Hilbert space valued random variables. Studia Sci. Math. Hung. 9, 33–44.

KRUGLOV, V. M. (1976a), The method of accompanying infinitely divisible distributions. *In*: MARUYAMA, G., and J. V. PROKHOROV (eds.), Proceedings of the Third Japan-USSR Symposium on Probability Theory. Lecture Notes in Mathematics, vol. 550, Springer, Berlin—Heidelberg—New York, 316—323.

KRUGLOV, V. M. (1976b) (В. М. Круглов), О сходимости распределений сумм случайного числа независимых случайных величин к нормальному распределению. Вестник Моск. ун-та 5, 5—12.

KRUGLOV, V. M. (1976)b) (В. М. Круглов), Новая характеризация пуассоновского распределения. Мат. заметки **20**, 876—882.

KRUGLOV, V. M., and S. N. ANTONOV (1982) (В. М. Круглов и С. Н. Антонов), Об асимптотическом поведении безгранично делимых распределений в Банаховом пространстве. Теория вероятн. примен. **27**, 625—642.

LAHA, R. G., and V. K. ROHATGI (1979), Probability theory. Wiley, New York.

LAUE, G. (1973), Zur Theorie der charakteristischen Funktionen nichtnegativer Zufallsgrößen. Transactions of the Sixth Prague Conference, Prague 1971, 1—18.

LAUE, G. (1974), Zur Theorie der charakteristischen Funktionen nichtnegativer Zufallsgrößen. Ph. D. Thesis A, Leipzig.

LAUE, G. (1977), Results on the theory of characteristic functions of non-negative random variables and their application to the factorization $\varepsilon_0 - F = (1 - \xi)(\varepsilon_0 - H^+) * (\varepsilon_0 - H^-)$. Math. Nachr. **80**, 279—288.

LAUE, G. (1979), The importance of the zeros of characteristic functions of positive random variables. Preprint, Karl-Marx-Universität Leipzig.

LAUE, G. (1980), Remarks on the relation between fractional moments and derivatives of characteristic functions. J. Appl. Prob. **17**, 456—466.

LAUE. G. (1983a), Remarks on the theory of characteristic functions of non-negative random variables. Sankhya, Ser. A, **45**, 44—55.

LAUE, G. (1983b), Characteristic functions of non-negative random variables: The connection between Re $f$ and Im $f$. Transactions of the Ninth Prague Conference, Prague 1982, 49—56.

LAUE, G. (1983), Existence and representation of density functions. Math. Nachr. **114**. 7—21

LAUE, G. (1985a), Duality of distribution functions, Math. Nachr. in print.

LAUE, G. (1985b), Results on moments of non-negative random variables, Sankhya in print

LAUE, G. (1986), New Tauberian and Abelian theorems for distribution functions, Теория вероятн. и. мат. статистика **34**, in print.

LEHMANN, I., and H.-J. ROSSBERG (1976), Elementary generalizations of a theorem of Phragmén-Lindelöf with applications. Math. Nachr. **70**, 87—93.

LÈVIN, B. J. (1962), Nullstellenverteilung ganzer Funktionen. Akademie-Verlag, Berlin.

LINNIK, JU. V. (1953), (Ю. В. Линник), Линеарные формы и статистические критерии, I, II. Украин. мат. ж. **5**, 207—243, 247—2890 (*Engl. transl. in*: Selected Transl. Math. Statistics and Prob., Amer. Math. Soc., Providence, R.I., **3**, (1963), 1—90).

LINNIK, JU, V., and I. V. OSTROVSKIĬ (1972) (Ю. В. Линник и И. В. Островский), Разложения случайных величин и векторов. Наука, Москва.

LOEVE, M. (1960), Probability Theory. D. van Nostrand, Princeton.

LOTOV, V. I. (1974) (В. И. Лотов), Предельные свойства функции концентрации. Теория вероятн. и мат. статистика. Вищая школа, Киев **10**, 109—115.

LUKACS, E. (1970), Characteristic Functions. 2nd ed., Griffin, London.

LUKACS, E. (1983), Developments in Characteristic Function Theory. Griffin, London.

LUKACS, E., and R. C. LAHA (1964), Applications of Characteristic Functions. Griffin, London.

MIROŠNIKOV, A. L., and B. A. ROGOZIN (1980) (А. Л. Мирошников и Б. А. Рогозин), Неравенства для функции концентрации. Теория вероятн. примен. **25**, 178—183.

MÜLLER, J. (1977), Conservation theorems for the factorization identity $\varepsilon_0 - K = (\varepsilon_0 - H) * (\varepsilon_0 - U)$. Math. Nachr. 78, 263—266.

OSTROVSKIĬ, I. V. (1977), The arithmetic of probability distributions. J. multivariate analysis **7**, 475—490.

OSTROVSKIĬ, I. V. (1979) (И. В. Островский), Об одном классе функций ограничен ной вариации на прямой, определяемых своими значениями на полупрямой. *In*: Записки научных сем. ЛОМИ, **92**: Исследования по линейным операторам и теории функций IX, 220—229.
OSTROVSKIĬ, I. V. (1981a), On infinitely divisible distribution functions coinciding on a half axis (ukrain.) Доп. АНУРСР Сер. А, 15—18.
OSTROVSKIĬ, I. V. (1981b) (И. В. Островский), О безгранично делимых функциях распределения, совпадающих на полуоси. ДАН УССР, Сер. А, 15—18.
OSTROVSKIĬ, I. V. (1982a). (И. В. Островский), Меры, имеющие бесконечно много безгранично делимых продолжений. ДАН УССР, Сер. А, 18—20.
OSTROVSKIĬ, I. V. (1982b) (И. В. Островский), О росте целых и аналитических в полуплоскости хребтовых функций. Мат. сб. **119**, 150—159.
OSTROVSKIĬ, I. V. (1982c) (И. В. Островский), О безгранично делимых распределениях, совпадающих с устойчивыми на полупрямой. Проблемы устойчивости стохастических моделей, Москва, 83—87.

PEČINKIN, A. V. (1975) (А. В. Печинкин), Об одном свойстве нормального закона. Труды МИЭМ **44**, 3—12.
PETROV, V. V. (1975), Sums of Independent Random Variables. Akademie-Verlag, Berlin.
PHRAGMÉN, E., and E. LINDELÖF (1908), Sur l'extension d'un principe classique de l'analyse. Acta Math. **31**, 381—406.
PROHOROV, J. V. (1956) (Ю. В. Прохоров), Сходимость случайных процессов и предельные теоремы теории вероятностей. Теор. вероятн. примен. **1**, 177—238.
PRUITT, W. E. (1983), The class of limit laws for stochastically compact normed sums. Ann. Probab. **11**, No. 4, 962—969.

RAĬKOV, D. A. (1939) (Д. А. Райков), О компонировании аналитических функций распределения. ДАН СССР **23**, 511—514.
RAĬKOV, D. A. (1940), Sur les fonctions positive définis. ДАН СССР **26**, 860—865.
RAMACHANDRAN, B. (1966), On one-sided distribution functions. Sankhya, Ser. A, **28**, 315—318.
RAMACHANDRAN, B. (1967), Advanced Theory of Characteristic Functions. Statistical publishing society, Calcutta.
RÉNYI, A. (1962), Wahrscheinlichkeitsrechnung. VEB Deutscher Verlag der Wissenschaften, Berlin.
RIEDEL, M. (1976), On the one-sided tails of infinitely divisible distributions. Math. Nachr. **70**, 155—163.
RIEDEL, M. (1977a), Some theorems of the Phragmén-Lindelöf theory for subharmonic functions. Math. Nachr. **78**, 13—20.
RIEDEL, M. (1977b), A new version of the central limit theorem. Теор. вероятн. примен. **22**, 187.
RIEDEL, M. (1979), On a characterization of the normal distribution by identically distributed linear statistics, Preprint Karl-Marx-Universität, Leipzig. J. multivariate Analysis **16**, 241—252 (1985).
RIEDEL, M. (1980a), Determination of a stochastic process by means of stochastic integrals. Теор. вероятн. примен. **25**, 339—349.
RIEDEL, M. (1980b), Representation of the characteristic function of a stochastic integral. J. Appl. Prob. **17**, 448—455.
RIEDEL, M. (1980c), Characterization of stable processes by identically distributed stochastic integrals. Adv. Appl. Prob. **12**, 689—709.
ROSSBERG, H.-J. (1966), Wachstum und Nullstellenverteilung ganzer charakteristischer Funktionen. Monatsber. DAW Berlin **8**, 275—286.
ROSSBERG, H.-J. (1967a), Ganze charakteristische Funktionen mit vollkommen regulärem Wachstum. Czechoslovak. Math. J. **17** (**92**), 335—346.
ROSSBERG, H.-J. (1967b), Der Zusammenhang zwischen einer ganzen charakteristischen Funktion einer verfeinerten Ordnung und ihrer Verteilungsfunktion. Czechoslovak. Math. J. **17** (**92**), 317—334.
ROSSBERG, H.-J. (1968), Eigenschaften der charakteristischen Funktionen von einseitig beschränkten Zufallsgrößen und ihre Anwendung auf ein Charakterisierungsproblem der mathematischen Statistik. Math. Nachr. **37**, 37—57.

Rossberg, H.-J. (1974a), Funktionentheoretische Möglichkeiten und Resultate, angeregt durch Probleme der Wahrscheinlichkeitsrechnung. Mitt. Math. Ges. DDR, H. 4, 62–72.

Rossberg, H.-J. (1974b), On a problem of Kolmogorov concerning the normal distribution. Теор. вероятн. примен. **19**, 824–828..

Rossberg, H.-J. (1975), An extension of the Phragmén-Lindelöf theory which is relevant for characterization theory. *In*: Patil, G. P., S. Kotz, and J. K. Ord (eds.), Statistical Distributions in Scientific Work, Vol. 3. Reidel, Dordrecht-Boston.

Rossberg, H.-J. (1976), New viewpoints, results and problems in the theory of Phragmén-Lindelöf. Ann. Polon. Math. **33**, 91–100.

Rossberg, H.-J. (1977), Phragmén-Lindelöf theorems in probability theory. Adv. Appl. Prob. **10**, 270–272.

Rossberg, H.-J. (1978), Non-elementary generalization of a theorem of Pólya-Szegö from the theory of Phragmén-Lindelöf. Math. Nachr. **82**, 269–276.

Rossberg, H.-J. (1979), Limit theorems for identically distributed summands assuming the convergence of the distribution functions on a half axis. Теор. вероятн. примен. **24**, 692–709.

Rossberg, H.-J. (1982), Generalization of Cramér's and Linnik's factorization theorems in the framework of the continuation theory of distribution functions. Preprint, Karl-Marx-Universität Leipzig. Z. für Analysis und ihre Anwendungen, **4**, 193–200 (1985).

Rossberg, H.-J. (1983a), Kombinatorische und funktionentheoretische Methoden in der Bedienungstheorie. *In*: Gnedenko, B. V., and D. König (eds.), Handbuch der Bedienungstheorie, Vol. I, Akademie-Verlag, Berlin, 307–336.

Rossberg, H.-J. (1983b), Limit theorems for sums of identically distributed random variables assuming restricted convergence of their distribution functions. Math. Nachr. **113**, 137–143.

Rossberg, H.-J., and Chu Duc (1980), Uniqueness theorems for components of a convolution given on a half-line. Math. Operationsforsch. Statist., Ser. Statistics, **11**, 361–371.

Rossberg, H.-J., and B. Jesiak (1978), On the unique determination of stable distribution functions (with discussion). Math. Nachr. **82**, 297–308.

Rossberg, H.-J., and B. Jesiak (1983), Complex analysis in probability theory. *In*: Lanckau, E., and W. Tutschke (eds.), Complex Analysis. Methods, Trends, and Applications. Akademie-Verlag, Berlin, 309–338.

Rossberg, H.-J., B. Jesiak, and G. Siegel (1981a), Continuation of distribution functions. *In*: Gani, J., and V. K. Rohatgi (eds.), Contributions to Probability: A Collection of Papers Dedicated to Eugene Lukacs. Academic Press, New York–London, 29–48.

Rossberg, H.-J., B. Jesiak, and G. Siegel (1981b), A new branch in the theory of sums of independent random variables. Serdica **7**, 93–106.

Rossberg, H.-J., and I. Lehmann (1974), Über Wachstumsbeschränkungen für analytische Funktionen, die die Beschränktheit dieser Funktionen zur Folge haben. Math. Nachr. **61**, 259–270.

Rossberg, H.-J., and G. Siegel (1975), Continuation of convergence in the central limit theorem. Теор. вероятн. примен. **20**, 885–887.

Rossberg, H.-J., and G. Siegel (1981), One-sided characterizations of the normal distribution in the set of infinite divisible distributions. Теор. вероятн. примен. **26**, 400–407.

Rühs, F. (1971), Funktionentheorie. VEB Deutscher Verlag der Wissenschaften, Berlin.

Ryžik, I. M., and I. M. Gradštejn (1957), Summen-, Produkt- und Integraltafeln. VEB Deutscher Verlag der Wissenschaften, Berlin.

Saks, S., and A. Zygmund (1952), Analytic Functions. Polskie Towarzystwo Matematiczno, Warszawa.

Scharm, R. (1983), Verallgemeinerungen elementarer Sätze der Theorie von Phragmén und Lindelöf und einige Anwendungen bei der Faktorisierung ganzer charakteristischer Funktionen. Diplom paper, Karl-Marx-Universität, Section of Mathematics, Leipzig.

Seiffert, I., and M. Riedel (1979), Uniqueness of two factorization problems. Теор. вероятн. примен. **24**, 207–211.

Seneta, E. (1976), Regularly Varying Functions. Lecture Notes in Mathematics, vol. 508. Springer, Berlin–Heidelberg–New York.

Siegel, G. (1978), Zero-one laws and weak convergence for sums of independent random variables. Math. Nachr. **86**, 333–346.

SIEGEL, G. (1979), Uniqueness of symmetric distribution functions defined on bounded sets. Теор. вероятн. примен. **24, 831–834.**

SIEGEL, G. (1980), Characterization of a Gaussian distribution in the set of infinitly divisible distributions on a Hilbert space. Preprint, Karl-Marx-Universität Leipzig. Math. Nachr. **110**, (1983), 37–42.

SIEGEL, G. (1981a) (Г. Зигель), Компактность сумм независимых случайных величин со значениями в Гильбертовом пространстве. Литов. мат. сб. 21, 123–136.

SIEGEL, G. (1981b) (Г. Зигель), Верхние оценки для функции концентрации в Гильбертовом пространстве. Теор. вероятн. примен. 26, 335—349.

SIEGEL, G. (1981c), Limit theorems for the Poisson distribution. Preprint, Karl-Marx-Universität Leipzig.

SIEGEL, G. (1982), Grenzwertsätze bei zufälligen Indizes. Ph. D. Thesis B, Leipzig.

SKOROHOD, A. V. (1954) (А. В. Скороход), Асимптотические формулы для устойчивых законов распределения. ДАН СССР 98, 731—734.

ŠOPF, G. (1979) (Г. Шопф), О сопряжённых порядках и сопряжённых типах убывания многомерного вероятностного закона. Теор. вероятн. примен. 24, 204—207.

STEUTEL, F. W. (1973), Some recent results in infinite divisibility. Stochastic Proc. Appl. 1, 125–143.

STEUTEL, F. W. (1974), On the tails of infinitely divisible distributions. Z. Wahrscheinlichkeitstheorie verw. Geb. 28, 273–276.

STEUTEL, F. W. (1979), Infinite divisibility in theory and praxis. Skand. J. Statist. 57–64.

TITCHMARSH, E. C. (1932), The Theory of Functions. Clarendon Press, Oxford.

TITOV, A. N. (1981a) (А. Н. Титов), Односторонние предельные теоремы. Кандидатская диссертация, МГИ Москва.

TITOV, A. N. (1981b) (А. Н. Титов), Об определении свертки одинаковых функций распределения по ее значениям на полупрямой. Теор. вероятн. примен. 26, 610—611.

TUTSCHKE, W. (1970), Über eine spezielle Form des Maximum-Prinzips für holomorphe Funktionen im Zusammenhang mit einem Satz von Rossberg. Ann. Polon. Math. 22, 341–343.

VINNICKIĬ, B. V. (1975) (Б. В. Винницкий), Об одном свойстве целых характеристических функций вероятностных законов. Изв. высш. учебн. завед. Мат. 155, 95—97.

WIDDER, D. V. (1946), The Laplace Transform. Princeton University Press, Princeton.

WOLFE, S. J. (1973), On the local behaviour of characteristic functions. Ann. Probab. 1, 862–866.

WOLFE, S. J. (1980), Mixtures of infinitely divisible distribution functions. Colloq. Math. Soc. János Bolyai 21, Analytic functions Methods in Probability Theory. Debrecen 1977, 359–367.

ZOLOTAREV, V. M. (1957) (В. М. Золотарев), Преобразования Меллина-Стильтьеса в теории вероятностей. Теор. вероятн. иримен. 2, 444—469.

ZOLOTAREV, V. M. (1963) (В. М. Золотарев), Аналитическое строение безгранично делимых законов класса $L$. Лит. мат. сб. 3, 123—140.

ZOLOTAREV, V. M. (1965) (В. М. Золотарев), Асимптотическое поведение распределений процессов с независимыми приращениями. Теор. вероятн. примен. 10, 30—50.

ZOLOTAREV, V. M. (1967) (В. М. Золотарев), Обобщение теоремы Линдеберга-Феллера. Теория вероятн. примен. 12, 666—677.

ZOLOTAREV, V. M. (1970), Théorémes limites géneraux pour les sommes de variables aléatoires indépendantes. C.R. Acad. Sci. Paris, Ser. A, 270, 14, 889–902.

ZOLOTAREV, V. M. (1978) (В. М. Золотарев), Несколько замечаний к статс Россберга-Есияка (1978). Math. Nachr. 82, 301–304.

ZOLOTAREV, V. M. (1983) (В. М. Золотарев), Одномерные устойчивые распределения. Наука, Москва.

# Appendix in proof reading

Since the manuscript of the present book was completed research on the subjects under consideration has been continued. We have also come to know more literature referring to this book and we feel obliged to inform the reader about this.

## Chapter 3

1. The notion of "weak convergence up to an additive constant" is given in Def. 3.4.6. Now for two sequences of d. f. $(F_n)$, $(G_n)$ we assume that $F_n \xrightarrow{w^*} F$, $G_n \xrightarrow{w^*} G$. Unfortunately, we then cannot generally infer the relation $F_n * G_n \xrightarrow{w^*} F * G$ without assuming additional conditions (examples can be found in SIEGEL (1986)). L.7.4.3 tells us that this conclusion is true provided that $(F_n)$ and $(G_n)$ are both left-tight. A purely analytic (two-sided) assumption is the following. *Let $F, G$ be non-decreasing, bounded, and left continuous, and denote by $f$ the Fourier-Stieltjes transform of $F$. Then from*

$$\lim_{n\to\infty} \int_0^t |f_n(u) - f(u))| \, du = 0 , \qquad t > 0 , \tag{1}$$

*and $G_n \xrightarrow{w^*} G$ it follows that $F_n * G_n \xrightarrow{w^*} F * G$.*

It is clear that (1) implies $f_n^* \to f^*$ and, accordingly, $F_n \xrightarrow{w^*} F$. The converse assertion is not true. However, we have the following statement. *The relation* (1) *holds iff $F_n \xrightarrow{w^*} F$ and, in addition,*

$$\lim_{T\to\infty} \sup_{|x|\leq T} \sup_n (F_n(x+h) - F_n(x-h)) = 0$$

*or some $h > 0$* (SIEGEL (1986)).

2. More precise results than those given in Suppl. 3.8.5 are contained in PITMAN (1968).

## Chapter 4

1. *Assume that for two d. f. $F$ and $G$, $G > 0$, and a natural number $n \geqq 3$ we have*

$$F^{*n}(x) = G^{*n}(x) , \qquad x \leqq x_0 .$$

*If the c. f. $g$ of $G$ is continuable to the half plane $S(0, \infty)$ then $F = G$. For $n = 2$ the same assertion is true provided that $x \log x = o(\log G^{-1}(-x))$, $x \to \infty$* (BLANK (1984)). This is a deep generalization of Prop. 4.7.5.

2. In Sect. 4.7 and Suppl. 4.10.9 and 4.10.12 we have treated Cramér's theorem from a new standpoint. But Prop. 4.7.4 is not quite satisfactory since its assumption

can hardly be expressed by a property of $F_1, F_2$. *Let* $\sigma_1, \sigma_2 > 0$ *such that* $\sigma_1^2 + \sigma_2^2 = 1$ *and let* $2k > \max(\sigma_1^{-2}, \sigma_2^{-2})$. *If*

$$|F_1 * F_2(x) - \Phi(x)| = O(e^{-kx^2}), \; x \to -\infty ,$$

*and*

$$F_k(x) = \Phi(x/\sigma_k), x \leqq x_0 , \qquad x_0 \in \mathbb{R}_1 , \qquad k = 1, 2 ,$$

*then* $F_k(x) = \Phi(x/\sigma_k)$, $x \in \mathbb{R}_1$, *provided that* $\sigma_1 \neq \sigma_2$; *otherwise, if* $\sigma_1 = \sigma_2 = \sigma$ *the d.f.* $F_k$ *are only connected by the condition*

$$(F_1(x) + F_2(x))/2 = \Phi(x/\sigma) , \qquad x \in \mathbb{R}_1 ,$$

*so that the solution is not unique* (ROSSBERG (1985)). This solves the problem stated in Suppl. 4.10.9.
The complete solution to the problem stated in Suppl. 4.10.9 runs as follows. *Let F tbe an arbitrary d.f. and* $F_k(x) = F(b_k x)$, $x \in \mathbb{R}_1$, $0 < b_1 < b_2$. *Then the condition*

$$F_1 * F_2(x) = \Phi(x) , \qquad x \leqq x_0 ,$$

*implies that F is normal* (Oral communication of A. M. ULJANOVSKIĬ).

A modification of this result is the following where $G$ stands for a d.f. and $G_k(x) := G(b_k x)$, $x \in \mathbb{R}_1$. *If*

$$F_1 * F_2(x) = G_1 * G_2(x) , \qquad x \leqq x_0 ,$$

*and*

$$\frac{F(-x) - G(-x)}{F(-x) + G(-x)} \to 0 , \qquad x \to \infty ,$$

*ten* $F = G$ (Oral communication of A. M. ULJANOVSKIĬ).

3. *Let F be an arbitrary d.f. satisfying*

$$\sum_{k=1}^{n} p_k F^{*(k+1)}(x) = \Phi(x) , \qquad x \leqq x_0 , \qquad x_0 \in \mathbb{R}_1 ,$$

*where* $n \geqq 1$, $p_k \geqq 0$, $\sum_{k=1}^{n} p_k = 1$. *Then there exists an integer* $m$, $1 \leqq m \leqq n$, *such that* $p_m = 1$ *and F is normal* (Oral communication of KRUGLOV and TITOV). This generalizes Titov's result Cor. 4.7.6; see also Suppl. 4.10.11.

4. It was D. DUGUÉ who stated the following problem. Determine all pairs $(f_1, f_2)$ of c.f. satisfying $(f_1 + f_2)/2 = f_1 f_2$. According to LUKACS (1983) (p. 97) it has not been solved yet. But a certain contribution to this subject is contained in the following result which was obtained by a Wiener-Hopf method. *Let* $a_1, a_2 > 0$, $a_1 + a_2 = 1$; *and* $F_1, F_2$ *be two d.f. such that* $F_1(+0) = 0$, $F_2(+0) = 1$. *If* $F_1$ *is no lattice d.f. and*

$$a_1 F_1 + a_2 F_2 = F_1 * F_2$$

*then*

$$F_1(x) = 1 - e^{-x/b_1} , \qquad F_2(-x) = e^{-x/b_2} , \qquad x > 0 ,$$

*where* $b_1 = b_2 a_2 / a_1$ *and* $b_2 > 0$ *remains undefined* (ROSSBERG (1972)).

5. Let $X$, $Y$ be arbitrary i.r.v. with the same d.f. $F$, $k \geqq 0$ an integer and $0 < \lambda < 1$. Estimates of $\mathsf{E}\,|X - Y|^{k+\lambda}$ give some information about $|f|^2$ where $f$ stands for the c.f. of $F$. There are several methods to estimate $\mathsf{E}\,|X - Y|^{k+\lambda}$ if $k = 1, 2$, or 3. But, in general, there is no simple method to obtain good estimates for higher moments of $|X - Y|$.

In the special case that $X$, $Y$ are non-negative r.v. estimates for the first moments of $|X - Y|$ can be used to obtain estimates for $\mathsf{E}\,|X - Y|^r$ with arbitrary $r > 0$. *Assuming that* $\mu_1 = \mathsf{E}X < \infty$ *and introducing r.v.* $X_1$, $Y_1$ *by their d.f.*

$$F_1(x) := \frac{1}{\mu_1} \int_0^x (1 - F(u))\, du$$

*we have the relations*

$$\mathsf{E}\,|X - Y|^r = 2\mu_r - r(r-1)\,\mu_1^2 \mathsf{E}\,|X_1 - Y_1|^{r-2}$$

*and*

$$\mu_r = \mathsf{E}X^r = r_1 \mu_1 \mathsf{E}X_1^{r-1}\,.$$

Obviously, these relations and the estimates mentioned above lead to the desired estimates for $\mathsf{E}|X - Y|^r$, $r > 0$ (Laue (1985)).

## Chapter 7

1. Kruglov (1984) has made a successful attempt to present the theory of inf.div. d.f. on Banach and Hilbert spaces in a unified way. Sect. 3.6 of this book contains results on the asymptotic behaviour of inf.div. d.f. which generalize some results of Sect.7.7 of the present book to the case of a separable vector space with arbitrary dimensions. Further, recent results of this type are given by Antonov (1983).

2. In Sect. 7.2 of Lukacs (1983) it is explained why integral c.f. of order 2 with a finite number of zeros are of interest; they occur also in the assertion of Prop. 9.4.2. We wish to point out that their d.f. possess the continuability property stated in Prop. 7.9.1.

3. In Sect. 7.4. we have treated the weak convergence of inf. div. d.f. $F_n$. For this purpose we have essentially used the left-tightness of $(F_n)$, see Prop. 7.4.4. If this condition is violated, then it may happen that a weak limit $F$ of $(F_n)$ cannot be represented in terms of inf. div. d.f. Examples for this fact can be found in Titov (1981) and Siegel (1986). Next we state another criterion which excludes this occurrence; it is also proved in Siegel (1986):

*The inf. div. d.f.* $F_n = \mathcal{H}(a_n, H_n)$, $n \geqq 1$, with *c.f.* $f_n$ *satisfy the relation* (1) *for some c.f.* $f$ *iff there are real numbers* $a$, $c$ *and a spectral function* $H$ *such that, as* $n \to \infty$,

$$a_n \to a, \qquad H_n \overset{w^*}{\to} H\,, \qquad H_n(\infty) - H(\infty) \to c\,,$$

*and*

$$\lim_{T\to\infty} \sup_{|x| \geqq T} \sup_n \,(H_n(x + h) - H_n(x - h)) = 0 \quad \textit{for some} \quad h > 0.$$

*In either case, the function* $F$ *corresponding to* $f$ *can be represented by*

$$F = \mathrm{e}^{-c}\, \mathcal{H}(a, H)\,.$$

This result is a first answer to the problem stated in Suppl. 11.7.5.

4. Bethmann (1984) gives a generalization of Prop. 7.6.14 concerning the characterization of the Poisson d.f. in the set of inf. div. lattice d.f.; he shows that this result remains true also in the more general set of lattice power mixtures. Further, statements on the asymptotic behaviour of lattice power mixtures are given.

To give some details let $G$ be a non-degenerate inf. div. lattice d.f. with jumps at $0, 1, 2, \dots$ and let $V$ be a d.f. with lext $(V) = 0$. Further, let $F$ be the power mixture of $G$ with respect to $V$ (see formula (7.8.2)) and let $p_k$ denote the jump of $F$ at $k$ $(k = 0, 1, \dots)$.

The following three conditions are equivalent

(i) *$F$ is a Poisson distribution,*

(ii) $\sum\limits_{k=0}^{n} p_k \geqq p_0 \sum\limits_{k=0}^{n} \frac{(-\log p_0)^k}{k!}$ *for some* $n \geqq 1$,

(iii) $\overline{\lim\limits_{k\to\infty}}\, (k!\, p_k)^{1/k} \leqq -\log p_0$

*In any case the equality signs occurring in* (ii) *and* (iii) *hold and* $\lambda := -\log p_0$ *represents the parameter of the Poisson d.f. $F$.*

The equivalence (i) $\Leftrightarrow$ (iii) is new even in the case of inf. div. lattice d.f.

5. *Let $f$ be a non-degenerate inf. div. c.f. and $V$ a d.f. with* lext $(V) = 0$. *If*

$$\exp\{-\sigma^2 t^2/2 - \lambda(e^{it} - 1)\} = \int\limits_0^\infty f^u(t)\, dV(u)\,, \qquad t \in R_1\,,$$

$\sigma \geqq 0$, $\lambda \geqq 0$, $\sigma + \lambda > 0$ *then $V$ is degenerate and $f$ has the structure of the left side* (Oral communication of Kruglov and Titov).

6. The following result is an interesting modification of Cor. 7.6.7. *Let $F_1$, $F_2$ be inf. div. d.f. satisfying*

$$\left|1 - \frac{F_2(-x)}{F_1(-x)}\right| \leqq x^{-1-\delta}\,, \qquad x \geqq x_0 > 0\,,$$

*for some* $\delta > 0$, *and*

$$\lim_{x\to\infty} \frac{\log\left(1/E_1(-x)\right)}{x\sqrt{\log x}} = 0\,.$$

*Then* $F_1 = F_2$ (Oral communication of N. M. Blank).

## Chapter 8

1. In Sect. 8.3 it was pointed out that stable c.f. can be analytically continued from the half axes $\{t > 0\}$ and $\{t < 0\}$. The class of c.f. sharing this property with the stable c.f. has not been sufficiently investigated so far, but Lukacs (1983) quotes an interesting result (Th. 5.1.2) in this direction.

## Chapter 9

1. In Sect. 3.6 of Kruglov (1984) some of the results of Sect. 9.2 are carried over to vector spaces of arbitrary dimensions.

2. Let $(X_{nk})$, $n, k = 1, 2, \dots$, be a triangular array (with infinite rows) of i.r.v. and assume, in addition, that the r.v. $X_{nk}$ $(k = 1, 2, \dots)$ are i.d. for fixed $n$. Further, let $(N_n)$ be a sequence of random indices such that $N_n \xrightarrow{P} \infty$ and assume that $N_n, X_{n1}, X_{n2}, \dots$ are independent for each $n$.

In BETHMANN (1985a) the limiting behaviour of random sums

$$\hat{S}_n := \sum_{k=1}^{N_n} X_{nk} \sim \hat{F}_n \tag{2}$$

is considered. *All weak limit functions $\hat{F}$ of $(\hat{F}_n)$ satisfying $\hat{F}(-\infty) = 0$ are (possibly defective) power mixtures.* This result is an extension of Prop. 11.3.3. Moreover, conditions for the convergence to a certain defective power mixture which are either necessary or sufficient are given.

As an application of these results limit theorems for the complete convergence of $\hat{F}_n$ to the normal and the Poisson d.f. are obtained. In the first case a pecular CLT (in analogy to Th. 12.4.1) is given; in the second case we have the following extension of Prop. 11.5.3. *Suppose* $\mathsf{P}(X_{nk} \in \{0, 1, \dots\}) = 1$, *$n, k = 1, 2, \dots$, and let $\lambda > 0$.*

*Then $F_n \xrightarrow{c} P_\lambda$ holds iff*

$$\lim_{n\to\infty} F_n(+0) = P_\lambda(+0) \quad \textit{and} \quad \varliminf_{n\to\infty} \hat{F}_n(x_0) \geqq P_\lambda(x_0)$$

*for some $x_0 > 1$.*

## Chapter 12

1. Contrary to the above mentioned paper, in BETHMANN (1985b) it is not assumed that the r.v. $X_{n1}, X_{n2}, \dots$ are i.d. Put $X_{nk} \sim V_{nk}$ and suppose $\mathsf{E}X_{nk} = 0$, $\mathsf{E}X_{nk}^2 < \infty$ $(n, k = 1, 2, \dots)$. The following result carries the Lindeberg-Feller theorem (Th. 12.2.3) over to the case of random sums $\hat{S}_n$. *Suppose $\mathsf{E}\hat{S}_n^2 = 1$ for all $n$. Then we have $\hat{F}_n \xrightarrow{c} \Phi$ and*

$$\max_{1\leqq k\leqq N_n} \mathsf{P}(|X_{nk}| > \varepsilon) \xrightarrow{P} 0 \quad \textit{for all} \quad \varepsilon > 0$$

*iff the following "random Lindeberg condition" is satisfied*

$$\sum_{k=1}^{N_n} \int_{|x|<\varepsilon} x^2 \, \mathrm{d}V_{nk}(x) \xrightarrow{P} 1 \quad \textit{for all} \quad \varepsilon > 0 .$$

An essential means for the proof of this theorem is the following statement. *For $q \in (0, 1)$ let $l_n(q)$ denote the $q$-th quantile of the r.v. $N_n$ and let $\sum_{k=1}^{l_n} X_{nk} \sim G_n^{(q)}$. Suppose $\mathsf{E}\hat{S}_n^2 = 1$ for all $n$. Then we have $\hat{F}_n \xrightarrow{c} \Phi$ iff $G_n^{(q)} \xrightarrow{c} \Phi$ for all $q \in (0, 1)$.*

2. Consider a sequence $(X_k)$ of i.r.v. $X_k \sim V_k$ with $\mathsf{E}X_k = 0$, $\operatorname{Var} X_k = \sigma_k^2$, $k = 1, 2, \dots$ Putting $B_n^2 = \sum_{k=1}^{n} \sigma_k^2$, $X_{nk} = B_n^{-1} X_k$ the sums (2) take the form

$$\hat{S}_n = \frac{1}{B_n} \sum_{k=1}^{N_n} X_k \sim \hat{F}_n .$$

The above modification of the Lindeberg-Feller theorem makes it possible to derive a CLT for this model (BETHMANN (1985b)).

3. A considerably different model is obtained if we use *random* norming constants $B_{N_n}$ instead of $B_n$. A random Lindeberg condition for this case was found by RYCHLIK (1976).

4. Zolotarev's CLT quoted in Suppl. 12.5.2 does not assume infinite smallness; the original proof is rather complicated. Thus it is to be welcomed that D. BISCHOFF and E. VOIGTLÄNDER (1984) found a streamlined method to prove that theorem; they

made use of the well-known fact that d.f. define certain operators thus showing that the power of this method has not completely been explored yet.

5. In Prop. 12.3.2 we have reformulated the Lindeberg-Feller theorem using the assumption of r-convergence. In BISCHOFF (1986) results of this type were carried over to Zolotarev's model mentioned above.

## Chapter 13

1. In Suppl. 13.4.9 we introduced the so-called l-sequences of i.r.v. with the sums $S_n \sim F_n$. We now consider the more general model of sequences $(X_k)$ of i.r.v., $X_k \sim V_k$, $k = 1, 2, \ldots$, with the well-known class $L$ of limit d.f. provided that the summands $X_k/b_n$ are infinitely small. BISCHOFF (1985) treated this model assuming r-convergence of $F_n$ to some monotone limit function $G$ on a half axis (both for lext $(G) = -\infty$ and $> -\infty$). He obtained an analogue to the self-decomposability of the (possibly defective) limit d.f. $G$; moreover, it follows that $b_n \to \infty$, $b_n/b_{n+1} \to 1$.

But the compactness problem is still unsolved in this general case.

## Appendix C1

1. The variety of the methods and results of the theory of Phragmen and Lindelöf suggest an effort to systematize this theory. In SCHARM (1985) the well-known majorization formula for log $|w|$ given by F. and R. NEVANLINNA is fundamental. SCHARM applies the principle of harmonic majorization in a generalized form and shows that it is more suitable than other approaches to derive a considerable number of known theorems of the Phragmén-Lindelöf type. The proofs become more straightforward and superfluous assumptions needed before can be omitted. Among others, a stremlined proof of Prop. C1.14 is given.

## References


ANTONOV, S. N. (1983), On the asymptotic behaviour of the tails of infinitely divisible distribution laws. Lecture Notes Math. **1021**, 41–44.

BETHMANN, J. (1984), Characterization of Poisson distributions in the set of power mixtures. Preprint, Karl-Marx-Universität Leipzig. *Submitted to* Math. Operationsforsch. Statist., Ser. Statistics.

BETHMANN, J. (1985a), Limiting behaviour of sums of a random number of independent and identically distributed random variables. Preprint, Karl-Marx-Universität Leipzig. *Submitted to* Теор. вероятн. примен.

BETHMANN, J. (1985b), The Lindeberg-Feller theorem for sums of a random number of, random variables in a triangular array. Preprint, Karl-Marx-Universität Leipzig. *Submitted to* Теор. вероятн. примен.

BISCHOFF, D. (1985), Zur eingeschränkten Konvergenz von Verteilungsfunktionsfolgen für Summen unabhängiger Zufallsgrößen. Ph. Thesis A, Karl-Marx-Universität Leipzig.

BISCHOFF, D. (1986), New versions of the Lindeberg-Feller theorem in the non-classic form of Zolotarev. Теор. вероятн. примен. (*in print*).

BISCHOFF, D., and E. VOIGTLÄNDER (1984), An operator proof for the Lindeberg theorem in the non-classic form of Zolotarev. Preprint, Karl-Marx-Universität Leipzig. *Submitted to* Math. Operationsforsch. Statist., Ser. Statistics. (*in print*)

BLANK, N. M. (1984) (Н. М. Бланк), О распределениях, свертки которых совпадают на полуоси. Теор. функций, функц. анализ и их приложения. Харьков, Издат. при ХГУ, **41**, 17—25.

Kruglov, V. M. (1984) (В. М. Круглов), Дополнительные главы теорий вероятностей. Издат. Высшая Школа, Москва.
Laue, G. (1985), Bounds for the $r$-th absolute moment of a difference of two random variables. Preprint, Karl-Marx-Universität Leipzig.
Lukacs, E. (1983), Developments in characteristic function theory. Griffin, London.
Pitman, E. J. G. (1968), On the behaviour of the characteristic function of a probability distribution in the neighbourhood of the origin. J. Austral. Math. Soc. **8**, 423—443.
Rossberg, H.-J. (1972), Characterization of the exponential and the Pareto distributions by means of some properties of the distributions which the differences and quotients of order statistics are subject to. Math. Operationsforsch. Statist. **3**, 207—216.
Rossberg, H.-J. (1985), Cramér's theorem within the framework of the continuation theory of distribution functions. Preprint, Karl-Marx-Universität Leipzig.
Rychlik, Z. (1976), A central limit theorem for sums of a random number of independent random variables. Colloquium Mathematicum **35**, 147—158.
Scharm, R. (1985), On the importance of the method of harmonic majorization in the theory of Phragmen and Lindelöf. Preprint, Karl-Marx-Universität Leipzig. *Submitted to* Math. Nachrichten.
Scharm, R. (1986), Remark on a paper of N. M. Blank concerning the unique continuability of functions of bounded variation. Теория вероятн. и мат. стат. **34** (*in print*).
Siegel, G. (1986), Vage convergence for convolutions and for infinitely divisible functions. Теор. вероятн. примен. (*in print*).
Titov, A. N. (1981) (А. Н. Титов), Односторонние предельные теоремы. Кандидатская Дисс., Москва.

# Subject index

# Name index

Bei Fragen zur Produktsicherheit wenden Sie sich bitte an:
If you have any questions regarding product safety,
please contact:

Walter de Gruyter GmbH
Genthiner Straße 13
10785 Berlin
productsafety@degruyterbrill.com